COASTAL EROSION AND PROTECTION IN EUROPE

Europe has a long history of managing coastal erosion through a variety of protection strategies, from the defences of the Venice lagoons to coastal land reclamation in the Netherlands.

This book provides a comprehensive review of the entire coastline of Europe and a comparative analysis of erosion problems and solutions in each country. Each chapter discusses the natural and anthropogenic factors in the erosion process and in defence projects design and maintenance, including coastal morphology and wave climate, land use changes and use of coastal areas, the evolution of coastal protection, climate change and political and administrative assessments.

Particular attention is paid to demographic and economic factors influencing coastal erosion in each country and to technical and administrative criteria influencing defence projects design. Lavishly illustrated in full colour throughout, the book represents a definitive reference work on its subject.

Enzo Pranzini is Professor of Physical Geography at the University of Florence, Italy. He is President of the Italian National Group of Environmental Coastal Research, and editor of the Italian *Studi Costieri* series.

Allan Williams is Emeritus Professor at Swansea Metropolitan University, University of Wales, UK. He is author or co-author of over 300 publications, including *Beach Management* (Earthscan, 2009).

'This is the first book that summarizes the critical issues of coastal erosion and shore defense measures that have been deployed in Europe. Leading scholars in each country have authored 22 chapters in this all-encompassing book that serves as an up-to-date resource for coastal professionals and others interested in European coastal changes and adaptive responses. I highly recommend this book.'

Stephen P. Leatherman, Director and Professor, Laboratory for Coastal Research, Florida National University, USA

'This book takes you on a tour of each of Europe's 25 coastal countries. It's a fascinating insight into both the physical characteristics (guess which is the longest) and major coastal processes of each coast, together with an overview of contemporary protective works and major coastal structures. A must for all students of the coast, academics and professionals.'

Andy Short, Professor, Coastal Studies Unit, University of Sydney, Australia

'Countries fronting the sea have been suffering from serious problems of beach and coastal erosion in response to increasing urban, industrial, recreational or tourism development. This book describes site-specific problems in each country of Europe, along with the physiographic background, and illustrates various types of remedial measures employed. The book will be of highest value to coastal scientists, engineers and planners not only in European countries but in other countries of the world.'

Tsuguo Sunamura, Professor Emeritus, Osaka University, Japan

COASTAL EROSION AND PROTECTION IN EUROPE

Edited by Enzo Pranzini and Allan Williams

LONDON AND NEW YORK

This first edition published 2013
by Routledge
2 Park Square, Milton Park, Abingdon, Oxon, OX14 4RN

Simultaneously published in the USA and Canada
by Routledge
711 Third Avenue, New York, NY 10017

Routledge is an imprint of the Taylor & Francis Group, an informa business

This book represents a contribution to the e-CEO Research Centre for
Geography and Regional Planning, FCSH Universidada Nova de Lisboa,
Lisboa, Portugal.

All royalties from sales of this book have been donated by the editors to the
Prostate Cymru Charity Trust, Wales.

British Library Cataloguing in Publication Data
A catalogue record for this book is available from the British Library

Library of Congress Cataloging-in-Publication Data
Coastal erosion and protection in Europe / edited by Enzo Pranzini and
Allan T. Williams. — 1st ed.
     p. cm.
Includes bibliographical references and index.
1. Coast changes—Europe. 2. Shore protection—Europe. I. Pranzini, Enzo, 1947–
II. Williams, A. T. (Allan Thomas), 1937–
TC330.C63 2013
627'.58094—dc23   2012023614

ISBN: 978-1-84971-339-9 (hbk)
ISBN: 978-0-203-12855-8 (ebk)

Typeset in Bembo by
Keystroke, Station Road, Codsall, Wolverhampton

Unless otherwise accredited, all illustrations have been provided by the chapter authors or the editors.

Dedicated to the memory of

Nick Kraus, a veritable Colossus, not only as a coastal engineer, but as a man and friend.

'Goodnight, sweet prince,
and flights of angels sing thee to thy rest!'
William Shakespeare, *Hamlet*, Act V, Scene 2

Contents

Notes on contributors

Enzo Pranzini (Italy) is a Full Professor of Physical Geography at Firenze University and has over 35 years' experience on coastal evolution and its connections to the environmental, economic, political and social issues arising in this context. He has coordinated 80 scientific and 55 research projects on assignments from public parties and worked as a consultant in Italy and abroad on 58 projects involving soft and hard coastal defences and nautical structures. He has published approximately 200 peer reviewed papers. He is President of the National Group for the Research on Coastal Environment and Editor of the national scientific journal *Studi Costieri*. enzo.pranzini@unifi.it

Allan Williams (Wales) is Emeritus Professor at Swansea Metropolitan University, Swansea, Wales. He has published more than 300 papers and reports, specializing in coastal processes and coastal management. He has worked mainly in Europe, but also Africa, China, India, the USA and Latin America. He was an Erskine Fellow at the University of Canterbury, Christchurch, New Zealand; Scientific researcher at the International Hurricane Research Centre, Miami, USA; Visiting Professor at the University of Virginia; Charlottesville, USA, Ocean University, Qingdao, China and

Southampton Solent University, UK. He is the holder of a Winston Churchill Fellowship award. He is an ex-editor of the *Journal of Coastal Conservation* and currently is a trustee and member of the editorial board of the *Journal of Coastal Research*. Consultancies have included work carried out for: Wardell-Armstrong Eng; Hyder Engineering; Robert West and Partners; Asia Development Bank; International Coral Reef Organisation Ltd.; UNDP; EUCC; PAPRAC, Split, Croatia. allan.williams@smu.ac.uk

George Alexandrakis (Greece) holds a B.Sc. in Geology (2003), an M.Sc. in Oceanography (2006) and a Ph.D. from the School of Sciences, National and Kapodistrian University of Athens (2011), with a thesis entitled: 'The development of a beach vulnerability index (BVI) for the assessment of beach erosion in Greece'. He is attached to the Department of Geography and Climatology, Faculty of Geology and Geoenvironment, National and Kapodistrian University of Athens, and is also a research fellow at the Foundation of Research and Technology, Hellas (Crete). His research focus is on coastal geological processes and sediment dynamics. He has participated in four research projects, and has been involved in the publication of 14 scientific papers in national and international journals and conference proceedings. g_alex@geol.uoa.gr

Giorgio Anfuso (Spain) graduated in Geological Sciences and holds a Ph.D. in Marine Sciences (2002) from the University of Cádiz, Spain. He is Professor of Coastal Geomorphology and Coastal Environmental Geology at the Faculty of Marine

Sciences, University of Cádiz. In past years, he has carried out investigations on coastal processes and evolution in Spain, Italy, Morocco, Colombia and other countries, and is the author of many publications in national and international journals. giorgio.anfuso@uca.es

Edward J. Anthony (France) is a Professor of Physical Geography at Aix-Marseille University CEREGE, Aix en Provence. Author of over 70 peer-reviewed papers in international earth-science journals, he has a 25-year research engagement in coastal geomorphology, morphodynamics and management at various spatial and timescales. Study themes have included mangroves, estuaries and sand barriers in West Africa; mud banks, mangroves, cheniers and beaches in South America; beaches, dunes and estuaries in the eastern English Channel and southern North Sea; and beaches and barriers in Mediterranean France. anthony@cerege.fr

Bas Arens (The Netherlands) is an Associate Scientist at the Delft University of Technology, Faculty of Civil Engineering and Geosciences Delft, as well as the Arens Bureau for Beach and Dune Research. He obtained a Ph.D. (University of Amsterdam, 1994) on Dutch foredune aeolian processes. After this he worked on foredune and dune dynamics in the Netherlands and other European countries (Portugal, Belgium, France, Wales), resulting in a number of peer reviewed publications. In 1997, he started his private research company, focusing on dune dynamics and monitoring of dune reactivation projects. He is a member of the Dutch expertise group 'Coast and Coastal Dune Landscape', formerly of the Ministry of Agriculture and Nature Conservation, and is frequently consulted about dune destabilisation, as well as by managers and policy-makers. arens@duinonderzoek.nl

Simon J. Blott (England) is a Principal Consultant at Kenneth Pye Associates Ltd, Crowthorne, Berkshire, UK. He obtained a First Class degree in Environmental Science of the Earth and Atmosphere at Reading University in 1996; a Masters of Research in Earth and Atmospheric Science at Reading in 1997; and a Ph.D. in coastal geomorphology, sedimentology and management from Royal Holloway University of London in 2002. His specialist interests include analysis of remotely sensed data and sediment analysis in relation to coastal and estuarine problems. s.blott@kpal.co.uk

Glicherie Caraivan (Romania) was Director of the Constanta Branch of the National Research-Development Institute for Marine Geology and Geoecology – Geoecomar and Associate Professor of Ovidius University, Constanta. His Ph.D., at the University of Bucharest in 1982, was concerned with the Romanian Black Sea beach and shelf sedimentology during the Holocene and recent times. His research focus is on beach dynamics, coastal environmental changes during the Upper Quaternary, submerged geo-archaeology and hydrogeology. gcaraivan@geoecomar.ro

Roger H. Charlier (Belgium) has an M.S. (Brussels, ULB), M.A. (Brussels, ULB), Ph.D. (Erlangen), D.Sc. (Paris), Lit.D. (Paris-Sorbonne), Ed. D. (Parsons, Iowa, USA), and an Emerging Management Nation Economy Diploma (Industrial College of the Armed Forces, Washington). He has dual US and Belgium citizenship and is a Knight Order of Leopold (Belgium), Knight Order of Academic Palms (France); Commander of the Order of Leopold II, Commander of the Order of the Crown; Member: New Jersey, Academy Science; Academy of Natural Sciences, Belles-Lettres (Fr.); National Academy of Romanian Scientists. Currently he is Professor Emeritus at the Vrije Universiteit, Brussels, and Research Professor at Florida Atlantic (Boca Raton) West Palm Beach, USA. Academic career: Professor, Universities of: Brussels (VUB), Bordeaux, Minnesota, North-eastern Illinois, DePaul (Chicago). He is the author of 12 books and more than 500 publications. Professional career: Scientific Advisor to CEO at HAECON Inc. rocharli @vub. ac.be

Boris Chubarenko (Russia) is Deputy Director for Science and Head of the Laboratory for Coastal Systems Study, the Atlantic Branch of the P.P. Shirshov Institute of Oceanology, Kaliningrad. He was employed as Head of the Ecological Laboratory and lecturer at the Immanuel Kant Baltic Federal University (Kaliningrad). As a guest-scientist, he worked at Darmstadt Technical University (Germany) at the Department of Environmental Flows. His research focuses on integrative studies and modeling of lagoons, estuaries and sandy coasts. chuboris@mail.ru

Andrew Cooper (Ireland) is Professor of Coastal Studies at the University of Ulster, Coleraine, Northern Ireland and Co-director of the Centre for Coastal and Marine Research and the Centre for Maritime Archaeology. His research centres on coastal geomorphology and coastal zone management and he is Course Director of the University of Ulster's M.Sc. in Coastal Zone Management. He is also chairman of the Northern Ireland Coastal and Marine Forum. JAG. Cooper@ulster.ac.uk

Merita Dollma (Albania) is a lecturer at the Faculty of History-Philology, Department of Geography, University of Tirana, Tirana, teaching World Economy, Human Ecology, and Economic Evaluation of Natural Resources. She studied geography at the University of Tirana and in 2004 received a Doctor of Science degree for urban environmental management. She has also studied at the Technical University Dresden, the University of Oxford and the University of Malta. From 1995 to 2007, she was a scientific researcher at the Geographic Studies Centre of the Albanian Academy of Sciences. In 2008/2009, she obtained a Fulbright postdoctorate scholarship at Eastern Michigan University, USA, carrying out research on geo-information for natural and cultural heritages. She is the author of two textbooks for high schools, *Geography 3 and 12*, author of the book Albanian regions, and co-author of a tourism guide-book, *Gate of the Orient*. meritadollma @hotmail.com

Vojislav Dragnić (Montenegro) is a Senior Associate for spatial planning and GIS systems in the Sustainable Development Department, Public Enterprise for Coastal Zone Management, Budva, Montenegro. His educational background is in geography and historical sciences, and he has been working in the Public Enterprise Department since 2002, focusing on coastal geomorphology and coastal processes, beaches, water management and preparation of data-bases, thematic maps, plans and programmes. vojislav.dragnic@morskodobro.com

Robert W. Duck (Scotland) is Professor of Environmental Geoscience and Dean of the School of the Environment in the University of Dundee, Dundee, Scotland. He has over 30 years of research experience in the physical sedimentology of lacustrine, estuarine and coastal environments worldwide and has written more than 100 refereed papers and chapters. He is the co-author of the best-selling text, *Practical Skills in Environmental Science*, and author of the recent (2011) popular science book, *This Shrinking Land: Climate Change and Britain's Coasts*. r.w. duck@dundee.ac.uk

Aysen Ergin (Turkey) is a Professor of Coastal Engineering in the Civil Engineering Department of the Middle East Technical University (METU), Turkey. Responsibilities include teaching courses in coastal hydraulics, marine and port engineering, physical and mathematical modeling of coastal processes and integrated coastal zone management. As a senior researcher of the 'Ocean Engineering Research Center, METU', she has participated in many national and international coastal engineering projects. She also lectures on national and international training courses on European Joint Master Programmes in Water and Coastal Management (Socrates-Erasmus). ergin@ metu.edu.tr

Óscar Ferreira (Portugal) is an Associate Professor at CIMA, Universidade do Algarve, Faro, lecturing on Coastal Dynamics, Coastal Management, Marine Geology, Wave and Tidal Energy and Marine Environmental Restoration. He was the coordinator of the Marine Sciences degree and Head of the Department on Environmental and Earth Sciences. He has more than 24 years of research experience, a Geology degree (1988), an Environmental and Applied Geology M.Sc. (1993), and a Marine Geology Ph.D. (1999). He has participated in more than 25 national and European research projects and published more than 100 articles on coastal and marine studies (*circa* 90 refereed at the SCI). He has been co-ordinator of several national projects and supervised 10 Ph.D. and several M.Sc. students. oferreir@ualg.pt

Kazimierz Furmańczyk (Poland) is a Professor at the University of Szczecin, Szczecin, lecturing on Remote Sensing methods and its application for Coastal Geomorphology and Coastal Management. He has over 40 years' experience on remote sensing application for coastal evolution and morphodynamic research and coastal management and has published over 100 reviewed papers and chapters. He obtained a Fulbright Grant at the University of Florida, is co-organiser of the Institute of Marine and Coastal Sciences at the University of Szczecin and is Head of Remote Sensing and Marine Cartography Unit. He is a Member of the Polish National Council on Marine Research and a Coordinator or PI of several European and national projects connected with coastal zone investigations. kaz@univ.szczecin.pl

George Ghionis (Greece) is a Research Associate of the Department of Geology of the University of Patras, and the Faculty of Geology and Geo-environment, of the National and Kapodistrian University of the University of Athens, and of the Foundation of Research and Technology, Hellas (Crete); he is also a coastal geomorphology consultant. He holds a B.Sc. in Earth Sciences (1983, Univ. of Athens), a M.Sc. in Coastal Geomorphology (1986, Univ. of Toronto) and a Ph.D. in Coastal Morphodynamics (2001, Univ. of Patras). Current research interests include coastal processes and sedimentation, photogrammetry and remote sensing, the impact of climate change on the coastal zone and the development of instruments and methods for coastal environmental measurements. He has participated in 32 national and EU research projects and has more than 35 publications in peer-reviewed journals and conference proceedings. gghionis@otenet.gr

Yuri N. Goryachkin (Ukraine) has a Ph.D. in Oceanology, and is Deputy Head of the Shelf Hydrophysics Department (Marine Hydrophysical Institute, National Academy of Sciences of Ukraine, Sevastopol). He has authored more than 100 research papers (including two monographs). Areas of expertise include: coastal process dynamics, variability of currents and sea level change. He lectures at the Moscow State University (Sevastopol branch) and currently is Head of the State Scientific Program, 'Sea Coasts of the Crimea'. yngor@yandex.ru

Vicente Gracia (Spain), geologist, obtained his Ph.D. degree in Marine Sciences at the Catalonia University of Technology (UPC). He works as a collaborating lecturer at the Department of Hydraulic, Maritime and Environmental Engineering (DEHMA), the Catalonia University of Technology (Universitat Politècnica de Catalunya UPC), Barcelona, and as a senior researcher at the Centre Internacional d'Investigació dels Recursos Costaners, Barcelona in the field of coastal engineering and morphodynamics. He has participated in 20 research projects and 12 advanced engineering contracts. He has published one book chapter, 12 research papers in national and international journals and more than 45 technical reports. He has participated in 12 international conferences and nine national conferences, which have produced 19 articles. vicente.gracia@upc.edu

Gjovalin Gruda (Albania) has worked as a geomorphology lecturer at the University of Tirana (1972–1977), and scientific researcher of the Geographic Studies Centre of the Albanian Academy of Sciences (1977–1982). He has been Rector of the University of Shkodra (1997–2000), Head of the Department of Physical Geography at the University of Tirana (2000–2002), and Dean of the Faculty of History-Philology, University of Tirana (2002–2005). Currently he is a lecturer in the Faculty of History-Philology, Department of Geography, University of Tirana, Tirana. He is the author of monographic studies such as: *Albanian Alps* (1982), *The Highland of Has and Gjakova* (1985), and *Mountain Range of Korab* (1987), as well as a university textbook, *General Geomorphology* (2002). tamalbania@hotmail.com

Saulius Gulbinskas (Lithuania) is a senior researcher at the Coastal Research and Planning Institute of Klaipėda University, Klaipėda. His educational background is geology, and his scientific interests cover marine sedimentology, integrated coastal zone management and environmental impact assessment. He has participated in many projects regarding geological mapping of the coastal zone, port development and implementation of coastal protections measures. Since 2008, he has been Director of the Baltic Valley Association, which coordinates research, academic and business activities in the Lithuanian maritime sector. saulius@corpi.ku.lt

Hans Hanson (Sweden) is a Full Professor and Head of the Department of Water Resources Engineering at Lund University, Lund, with research focuses on coastal processes and their mathematical descriptions, as well as Integrated Coastal Zone Management. He has been working with coastal model development over the last 30 years, often in close cooperation with researchers from the US Army Corps of Engineers. The shoreline evolution model GENESIS and the upcoming model GenCade are results of this cooperation. He has been a visiting researcher for extended time periods at the Technical University of Braunschweig, US Army Engineers Waterways Experiment Station, James Cook University, Texas A&M University, and the University of Granada. Author of more than 230 technical reports, conference papers, and journal articles, he is on the International Steering Board of the international conference series *Coastal Sediments*, as well as *Coastal Dynamics* and on the Editorial Boards of both *Coastal Engineering* and the *Journal of Coastal Research*. Hans.Hanson@tvrl.lth.se

Jürgen Jensen (Germany) is (since 1992) a Full Professor in Hydraulic Structures and Hydraulic Modelling at the University of Siegen, Siegen, and is Head of the Research Institute for Water and Environment. He has over 30 years of experience in coastal engineering and coastal zone management, especially in the field of Sea Level Rise and Storm Surges and has coordinated 150 scientific national and international research projects with a total funding of more than 10 million Euros. He is a member of several national and international boards. His main research fields are probabilistic and statistical methods and physical and numerical models for the optimization of hydraulic and coastal structures. He has been a Visiting Researcher, e.g. at the Centre for Interdisciplinary Research (ZIF) and at the Universities of Perth and Florida and has published approximately 180 papers in national and international journals. juergen.jensen@uni-siegen.de

Nikolaos A. Kampanis (Greece) is Senior Researcher at FORTH-IACM, and Head of the Laboratory for Coastal Research and Applications. He received his Ph.D. in Numerical Analysis (1992), from the University of Crete. His research area is computational wave propagation and fluid dynamics (sound propagation in fluids, environmental and physiological flows and climatic models). He is a member of the ERCIM Environmental Modeling Group. He has participated in several EU and national research projects and has several publications in refereed journals and conference proceedings. kampanis@ iacm.forth.gr

Ruben Kosyan (Russia) is Head of Department of the Coastal zone of the Southern Branch of the P.P. Shirshov Institute of Oceanology, Russian Academy of Science, Gelendzhik. He holds a Doctorate of Science and is an Academician of the Russian Academy of Ecology. Professor Kosyan has more than 40 years of scientific and engineering experience and has published more than 320 scientific papers, including ten books. He has led many international near-shore scientific experiments and projects and is engaged in problems of physical oceanography, marine geophysics, ocean engineering and methods of studying dynamical processes in coastal regions. rkosyan@hotmail.com

Marina Krylenko (Russia) is a leading researcher of the Southern Branch of the P.P. Shirshov Institute of Oceanology, Russian Academy of Science, Gelendzhik, and has published more than 50

scientific papers. She has more than 15 years experience in coastal fieldwork, specializing in hydro- and lithodynamical processes in the coastal zone, sediment transport, coastal erosion and protection, and ecology of the coastal zone, together with ICZM. krylenko@mail.ru

Jānis Lapinskis (Latvia) is a researcher and lecturer at the Faculty of Geography and Earth Sciences of the University of Latvia, Riga. He holds a B.Sc. and M.Sc. in Environmental Sciences (1997 and 1999, Univ. of Latvia). His Ph.D. thesis (2010) at the University of Latvia was on 'Dynamics of the Kurzeme coast of the Baltic'. From 1996 until 2009, he worked in coastal processes, monitoring programmes as a field researcher and analyst. Current research interests include coastal processes and dune development, the possible impact of climate change on the coast and 'human pressure'-related risk assessment. janisl@lu.lv

Magnus Larson (Sweden) is a Professor in Water Resources Engineering at Lund University, with a research focus on coastal processes and their mathematical description. He has been working with coastal model development over the last 25 years, often in close cooperation with researchers from the US Army Corps of Engineers. The profile evolution model SBEACH and the longshore current model NMLONG are results of this cooperation, as well as important components of the topographic analysis system RMAP and the Coastal Modeling System (CMS). He has been a visiting researcher for extended time periods at the University of Tokyo, the University of Sydney, the University of Tsukuba and the Technical University of Braunschweig. magnus.larson@tvrl.lth.se

Nemanja Malovrazić (Montenegro) is an Associate for Environmental Protection for the Sustainable Development Department, Public Enterprise for Coastal Zone Management, Budva, Montenegro. His education background is in Biology with a specialization in marine ecology. Since 2007, he has been responsible for monitoring coastal bathing water quality together with biological processes in the coastal area, preparation of environmental programmes and the management of protected areas. nemanja.malovrazic@morskodobro. com

Ana Matias (Portugal) is a post-doctoral Research Fellow at CIMA, Universidade do Algarve, Faro. She has a degree in Geology, an M.Sc. in Marine and Coastal Studies and a Ph.D. in Marine Geology. She devoted her earlier work to dune vulnerability, coastal evolution and beach nourishment. Since 2000, she has become a specialist in barrier island dynamics, especially overwash processes. She has publications of overwash measurements, both in the field and laboratory, over time-frames ranging from minutes to decades. ammatias@ualg.pt

Frank van der Meulen (The Netherlands) is Associate Professor for Integrated Coastal Zone Management at UNESCO-IHE Institute for Water Education and is also a senior advisor at Deltares/Delft Hydraulics Institute for Delta Research and Management, both in Delft, the Netherlands. He has more than 30 years of experience in multidisciplinary research and management of coastal lowlands and associated dunes in the Netherlands and various other parts of the world, both in temperate and (sub)tropical areas. He has been a University researcher for more than 20 years and worked for the Dutch Government policy-making Institute for Coastal and Marine Management for some ten years. frank.vandermeulen@hetnet.nl

Kaarel Orviku (Estonia) is an 'extraordinary researcher' at the Institute of Ecology at Tallinn University, Tallinn and is also a guest lecturer at Tartu University, Tallinn University of Technology, Tallinn University and at training courses organized by the Ministry of Education and Ministry of the Environment of the Estonian Republic. In 1993, he was awarded a Doct. Sci. Geol., from Tartu University and is the author of over 100 publications. From 1993 to 2009, he worked as an environmental expert for an engineering company (Merin Ltd). His main specialities are: marine and coastal geology; shore processes and coastal environment; erosion and accumulation control and monitoring; environmental impact assessment; and geo-radar investigations. kaarelorviku@hot.ee

Nicolae Panin (Romania) is a corresponding member of the Romanian Academy. He took his Ph.D. in Geology at the University Paris VI in 1974 and has been awarded the Officer of Palmes Académiques, a cavalier of the Romanian order 'Honest Service', and the prize of the Romanian Academy; he specializes in Marine Geology and Sedimentology, His research work encompasses the Danube Delta and the Black Sea, especially coast line changes during the Upper Pleistocene and the Holocene. He is the Founder and former General Director of the Romanian National Institute for Marine Geology and Geoecology – GeoEcoMar; a European Expert in Marine Sciences; a Member of the National IGBP Committee in Romania; a national focal point for the LOICZ Core Project; and a Member and President of the Romanian Committee of IAPSO (Intern. Ass. of Physical Sc. of the Oceans). panin@ geoecomar.ro

Michael R. Phillips (Wales) has a B.Sc. in Civil Engineering, a M.Sc. in Environmental Conservation Management and a Ph.D. in Coastal Processes and Geomorphology. He is Head of the School of the Built and Natural Environment Department at Swansea Metropolitan University, Swansea and has published more than 60 research papers. Research interests include coastal processes and morphological responses to climate change and sea level rise. He has been an invited speaker and presenter at many major international conferences and is a member of the Climate Change Working Group of the Global Forum on Oceans, Coasts, and Islands and amongst others, is vice-chairman of the Royal Geographical Society's Coastal and Marine Working Group. mike.phillips @smu.ac.uk

Kristina Pikelj (Croatia) is senior assistant in the Geology Department, Faculty of Science, University of Zagreb, where she received her degree in geology and geography. In 2010, she received a Ph.D. in Oceanology/Marine Geology from the Faculty of Science, University of Zagreb. Her main research topics are sediments, karst geomorphology and coastal processes, with a special focus on Adriatic Sea seabed surface sediments. She worked as a teacher in geology workshops for elementary school pupils and as a geography teacher in high school. kpikelj@ geol.pmf.hr

Nigel I. Pontee (Wales) currently heads the Coastal Science and Planning Team at Halcrow, Swindon. He studied Geological Oceanography at Bangor University, Wales, and, at the University of Reading, took a M.Sc. in Sedimentology and a Ph.D. on the morphodynamics of mixed beaches. Following this, he completed a post-doctoral research fellowship on Holocene evolution of saltmarsh and dune systems in the Medoc Region, France. For the last 14 years he has been working in coastal consultancy and has been involved in coastal process assessment, local and regional conceptual sediment transport models, habitat creation scheme design, feasibility studies, dredging assessments and strategies, impact assessment, coastal and estuarine shoreline management plans and strategies. He has authored over 190 consultancy reports and has extensive research experience, regularly supervises post-graduate research projects and contributes to the EA/Defra FCERM research programme and EU-funded projects. He is a peer reviewer for a number of ICE and Elsevier Journals and books and has written over 75 publications, including several book chapters, journal and conference papers, and best practice guides for CIRIA and PIANC. PonteeNI@halcrow. com

Serafim E. Poulos (Greece) is an Associate Professor in the Department of Geography and Climatology (Faculty of Geology and Geoenvironment), National and Kapodistrian University of Athens (NKUA), Athens. He holds a B.Sc. in Geology (NKUA, 1985) and a Ph.D. in Oceanography (University of Wales, 1989). His research interests focus on coastal geomorphology, oceanography and sedimentology, fluvial/marine interactions and environmental management/protection of the coastal zone. He is currently the President of the Hellenic Oceanographers' Association (HOA). He has participated as a principal scientist/investigator in 15 national (e.g. POSEIDON) and European research projects (e.g. CINCS, MATER, PDTD, EUMARSAND, etc.) and has authored/co-authored more than 100 publications in peer-reviewed international journals and conference proceedings. poulos@geol.uoa.gr

Kenneth Pye (England) is currently the Director of Kenneth Pye Associates Limited, Crowthorne, Berkshire, UK. He obtained a First Class Honours degree in Geography from Oxford University in 1977 and completed his Ph.D. in coastal geomorphology at Cambridge University in 1980. He subsequently held NERC Royal Society Research Fellowships in the Cambridge Department of Earth Sciences and held appointments as Lecturer, Reader and Personal Professor at Reading University between 1989 and 1999, when he took up the Chair of Environmental Geology at Royal Holloway University of London, a position he held until 2004. He was awarded a Sc.D. by Cambridge University in 1994 and became a Chartered Geologist in 1995. His specialist interests include coastal processes, sediments and morphological evolution in relation to coastal engineering and nature conservation. k.pye@kpal.co.uk

Daria Ryabchuk (Russia) is a leading scientist (with more than 50 publications) of the A. P. Karpinsky, Russian Geological Research Institute, St. Petersburg and graduated from Leningrad University (1990, M.Sc. Geology, geochemistry, lithology) followed by a post-graduate Ph.D. at VSEGEI, in 2002. Her main fields of study are sedimentology, environmental geology, coastal geology and marine coastal hazards. Academic posts have included: Chair of the Baltic Marine Geologists, 2009–2011 and since 2010, a Member of the Marine Geology Expert Group of EuroGeoSurveys. daria_ryabchuk@mail.ru

François Sabatier (France) is Assistant Professor of Coastal Geomorphology at Aix Marseille Université, CEREGE, Aix-en-Provence. Following his Ph.D. in 2001, he held successive post-doctoral positions at the French Ministry of Public Works and at Delft University. Since 2004, his publications have concerned large-scale coastal behaviour, delta morphodynamics, the performance of coastal protection works and beach-dune systems, mainly in the Mediterranean. He is involved in several national and European research programmes and is an expert of the French Ministry of the Environment. sabatier@cerege.fr

Agustín Sánchez-Arcilla (Spain) has a Ph.D. in Civil Engineering and is a Full Professor in the Laboratory of Maritime Engineering, Barcelona Tech, Barcelona. He has been the Director of the Laboratory of Maritime Engineering (Laboratori d'Enginyeria Marítima, LIM/UPC) since 1990; it is here that he has developed his research activity in the field of maritime (coastal and harbour) engineering. He was one of the promoters for the creation of the International Centre for Coastal Resources Research (CIIRC), of which he is the current Vice President. In the academic field, he lectures in several topics at the Barcelona Civil Engineering School. He has published over 15 books and over 100 research papers in national and international journals, supervised 21 Ph.D. theses, and taken part in more than 30 European research projects. director.lim@upc.edu

Klaus Schwarzer (Germany) is a Senior Scientist at the Institute of Geosciences, Sedimentology, Coastal and Continental Shelf Research of Kiel University where he is head of the Hydroacoustics group. He holds a Ph.D. degree in geology and has worked for more than 25 years on coastal evolution, coastal processes and coastal protection in Europe, South America and South East Asia. He has headed EU interdisciplinary coastal research in the Baltic Sea and a Tsunami research project in the Andaman Sea. He is a member of the Editorial Board of the *Journal of Continental Shelf Research* and is a member of the committee for offshore subsoil exploration for wind energy plants, as well as for hydromorphology. Appointed by IOC–UNESCO, he helped establish the Institute of Oceanographic (INOS) Sciences in Kuala Terengganu/Malaysia. kls@gpi.uni-kiel. de

Per Sørensen (Denmark) holds a master's degree in Civil Engineering from Aalborg University specializing in hydraulic engineering. His professional career spans more than 20 years in coastal engineering and he is now Head of Coastal Research at the Danish Coastal Authority. He has participated in many national and international projects on coastal engineering and has written more than 20 reviewed papers on coastal protection, in addition to teaching on several international courses in coastal engineering. pso@kyst.dk

Margarita Stancheva (Bulgaria) is Senior Research Assistant with 15 years experience in the fields of coastal processes, sand beaches/dunes, erosion and shoreline changes at the Department of Marine Geology and Archaeology, Institute of Oceanology, Bulgarian Academy of Sciences, Varna. In 2009, she received a Ph.D. with a thesis on 'Beach dynamics and modifications under the impact of maritime hydraulic constructions'. She currently coordinates the EU-funded SYMNET project under the Joint Operational Programme 'Black Sea 2007– 2013' (http://www.projectsymnet.eu/). She is a National Contact Point for Bulgaria in the European Dune Network (http://www.hope.ac.uk/coast) and convener of the Coastal Session at the EGU General Assembly ('Coastal zone geomorphologic interactions: Natural versus human-induced driving factors') and is the author of more than 35 scientific papers. stancheva@io-bas.bg

Adrian Stănică (Romania) is a Scientific Director with the Romanian National Institute for Marine Geology and Geoecology – GeoEcoMar, Bucharest. He graduated in Geology in 1993 at the University of Bucharest, was awarded a Ph.D. in Geology at the same university in 2003. He also holds a Master in Environmental Economics, from the Scuola Superiore 'Enrico Mattei', Milan, Italy (1993–1994); and from 2002 to 2005 he was a Marie Curie Fellow, NATO-CNR Senior Fellow, in Venice, Italy. He was a regional coordinator – Causal Analysis for the Black Sea Trans-boundary Diagnostic Analysis Report in 2007 and has worked as a researcher with GeoEcoMar since 1994. He has extensive experience of the Black Sea coast, Danube Delta and connected lagoon systems, and the Venice Lagoon area. astanica@geoecomar.ro

Hannes Tõnisson (Estonia) is a researcher at the Institute of Ecology at Tallinn University, Tallinn and guest lecturer at the same university. His Ph.D. thesis (2008) at Tallinn University was on 'Development of Spits on Gravel Beach Type in Changing Storminess and Sea Level Conditions'. He is the author of nearly 50 scientific publications and currently leads an Estonian Science Foundation financed project: 'Analysis of relationships between near-shore wave parameters and sediment movement on gravel-pebble shores'. His interests are in the effects on Estonian shorelines of global change and human pressure, together with coastal wetland ecosystems; he specializes mainly in gravel beaches. He is also involved in geo-radar investigations in various environments. hannest@gmail.com

Bert van der Valk (The Netherlands) studied Earth Sciences (Quaternary Geology), at the Vrije Universiteit, Amsterdam. In 1992, he received his Ph.D. degree on the geology of the Holocene beach barriers of the western Netherlands. He worked at the Geological Survey of the Netherlands, for CCOP in Bangkok, Thailand (1995–1999) and since 1999 for Deltares/Delft Hydraulics, Delft, both abroad and in the Netherlands. He is member of the dune management advisory board to Dunea (which produces drinking water for *circa* 1.5 million people). Bert.vanderValk@deltares.nl

Lilian Wetzel (Brazil) holds a B.Sc. in Oceanography, a B.Law and a M.Sc. in Environmental Engineering/ Watershed Planning. She has specialised in coastal management and planning, having focused on subjects such as marine debris pollution, coastal protected areas, ICZM and beach erosion. She has participated in many European research groups on coastal erosion management issues during EU-funded projects; she currently works as a technical consultant in a Brazilian NGO, from where she keeps in collaboration with European research institutions. lbwetzel@yahoo.com.br

Rimas Žaromskis (Lithuania) is a Professor and senior researcher of the Coastal Research and Planning Institut, Klaipėda University, Klaipėda. He has investigated Baltic Sea coasts for more than 30 years and participated in many scientific expeditions exploring the Baltic, the Black, Caspian, White and Barents seas. Fields of scientific interest include coastal processes, hydrotechnical constructions and integrated coastal zone management. rimas.zaromskis@cablenet.lt

Rien van Zetten (The Netherlands) studied Civil Engineering, Environmental Management and Philosophy and works for Rijkswaterstaat as a senior Project Manager. For the last seven years he has been the Project Manager for PMR (Project Mainport development, Rotterdam) and since 2010, is also Deputy Director, Delta Programme Rivers. He also works in southern Africa and in Asia for Rijkswaterstaat. He is working on a Ph.D. thesis concerning the role of culture in handling floods. rien.van.zetten@rws.nl

Foreword

Ayşen Ergin

'Although nature begins with the cause and ends with experience, we must follow the opposite course, namely begin with experience and by means of it investigate the cause.'

Leonarda Da Vinci, *Notebooks*, 1512

In the world, coastal areas can be scenes of peaceful resolution or volatile confrontation. The fearless sea is powerful, dynamic and meets the land at the coast where the interaction of natural physical processes with underlying geological elements are clearly seen as waves, currents and tides; all act to mould coastal features. Any single coast is the result of processes at different time scales from seconds to millions of years in duration. They are continually shaped and reshaped by *long-term forces* (tectonic plate movement) and sea-level changes (falling and then rising in response to global climate change), creating different coastal formations over extensive areas such as deltas and barrier islands by deposition, or coastlines where erosion dominates, e.g. soft glacial tills. More rapid changes, *day-to-day, and year-to-year*, which are associated with short-term physical processes (the action of winds, waves, currents and tides), produce changes that might extend only a few metres to many kilometres. One should also recognize the uneasy alliance between man and sea, *i.e.* the human influence, which plays a role in most coastal systems and needs to be considered as anthropogenic impacts when studying coastal processes.

Coastal studies are probably as old as any study, as, throughout history, people have always been fascinated by the coast. The first true scientific interest in the coast began after the Renaissance, in the context of navigation beyond the Mediterranean.

The nineteenth century was a time of increased scientific awareness of shoreline diversity. At the close of the nineteenth century, W. M. Davis (1899, 1909) played an important role in recognizing the importance of coastal features and processes in spatial and temporal scales. However, it was in the early twentieth century that the concept of coastal equilibrium was viewed within a context of the evolution of coastal geomorphology. This concept was extended and applied to coastal areas by series of his successors, most notably Douglas Johnson (1919), who undertook the systematic evolution from initial to sequential coastal forms.

During the mid-twentieth century, coastal development was considered to be based on geographical cycles of erosion. World War II was a milestone in the course of scientific studies on coastal processes and the dynamic nature of coastal beach environments, when investigations were undertaken for military beach landings. The Beach Erosion Board of the US Army Corps of Engineers (established 1944 and later the Coastal Engineering Research Center), Wallingford Hydraulic Research Laboratory in Britain, and Delft Hydraulics in the Netherlands where coastal engineers and geomorphologists focused on coastal processes based on laboratory and field experiments. In these studies, collected data were based on theoretical premises with the objective of testing between competing hypotheses. Late-twentieth-century studies on coastal evolution were expanded in the context of climate and sea level changes. Based on field and laboratory research work, in all these conceptual developments, coastal evolution was basically regarded as the result of impacts of wind, waves, and currents with varying boundary conditions, including climate and sea-level changes. In the final decades of the twentieth century, coastal studies were heavily focused on the USA and Northwest Europe, broadened by major overseas coastal research projects within other countries, in particular, Japan and Australia. Over the last two decades, extensive field experiments have been conducted and comprehensive and careful data of those experiments has provided insights and ideas for further coastal evolution studies with special emphasis on the human impact together with the natural dynamics of coasts.

It is well known that coastal zones are important social and profitable regions with high population densities, and that the utilization of coastal areas is very critical and complex, and should be managed by interdisciplinary approaches. Throughout history, people have made their best efforts to work with nature when building coastal protection structures, by using judgement based on experience gained by examining available historical records of the area.

In coastal regions, settlements, industry, tourism, agriculture and transportation sectors thrive. Human impact, by utilization of the sea and coastal areas through land reclamation, extraction of natural resources, building of coastal structures, can be manipulative of coastal processes by controlling the magnitude and direction of these processes. Changing land use patterns, so that the catchment yield is altered, often leads to a significant direct or indirect decrease or increase in sediment supplied to the coast, resulting in erosion or accretion. It is clear that successful coastal zone management for sustainable development, where a balance is attempted between minimized impacts of human usage and protected human interests, depends on a clear understanding of both natural and anthropogenic processes and the resultant coastal problems.

Today, many coastal areas in the world are facing problems of land loss due to severe erosion. There are several causes of erosion: the adverse effects of coastal structures, the decrease of sediment supply from rivers as a result of dam construction, sand mining from coastal areas, mean sea-level change, etc. all extremely adequately dealt with in the various chapters in this magnificent book. Therefore, at many places there exists an acute need for shore protection; however, the technical information necessary to make appropriate coastal defence works, based on reliable wave and current data, is often inadequate.

It is appropriate to ask: 'Where do we stand with the problem of coastal processes?' Coastal research has definitely moved a significant step forward, but studies are far from being finalised. Three things are required for a clear understanding: theories correctly based on basic hydrodynamic principles, a scientific data base constructed on reliable field and model measurements, and a rational proof showing that observation and measurements are in agreement with theory. Research and monitoring conducted over the last two decades has substantially advanced knowledge of coastal processes and the design capabilities for coastal defence projects. However, most coastal areas lack continuous data collection or monitoring systems, which hinders implementation of many of the available coastal morphological models. Most of the time, the decision-making process generally depends on the local experience of practitioners, as well as those historical events making up the concept of expert opinion. The state of the art in coastal engineering practice today demands detailed, accurate information on various design and environmental conditions before making planning decisions so that the optimum and most economic projects can be formulated.

The coast is strong, persistent yet fragile so not indestructible, but its destruction will not come via the waves and currents only, but from human intervention in natural systems. The major future challenges of coastal areas are twofold. First, coastal processes are site specific and coastal defence measures might vary significantly from site to site. The second is ensuring the physical and ecological sustainability of the region, which will require a more co-operative and effective joint work between engineers and environmentalists than currently exists.

It is not an easy endeavour to develop reliable sources to shed light on coastal processes, but this book is a scientific journey along European coasts. It is an illuminating scientific look at selected critical coastal erosion issues and protective works that have been implemented on 25 European coasts. The book's goal is to assemble a state of the art work on coastal erosion problems and performance together with readily applied remedial measures for coastal defence, which can be used as an up-to-date resource for coastal professionals and academics. It provides a lucid, original contribution attuned to the needs of those usually engaged in coastal engineering through the many examples containing descriptive, site-specific studies of the processes and sufficient qualitative and quantitative methodology for coastal defence structures. It fulfils the general goal of providing a suitable reference to phenomena only partially amenable to theoretical approaches.

The book is organized in an easy format – individual country-based chapters – intended to be of maximum benefit to the clear understanding of coastal erosion problems by giving specific attention to the development of insight both from a physical and analytical viewpoint, through the selected critical issues of coastal erosion along European coasts. It is my privilege to acknowledge the outstanding work of the editors and cooperation received from all authors for an extremely difficult undertaking. I hope this exceptional book will bring the reader's understanding to a point where the given cases will be helpful in their engineering practice and encourage them to start their own research, because 'In the world of time and change there is no last chapter. The solution of one problem creates other problems which have not been foreseen and certainly have not been produced intentionally' (Toynbee, 1969, 331).

In the end, as Neruda put it in his immortal poem, 'The Sea', 'Its essence–fire and cold; movement, movement'. Coastal problems related to the complex and dynamic nature of coastal processes are what drive many researchers from many disciplines to continue searching to 'understand a grain of sand so that they would understand everything . . .' (Muggeridge, 1966).

Ayşen Ergin
Professor, Ocean Engineering Research Center, Civil Engineering Department, Middle East Technical University, Ankara, Turkey, November 14, 2012

References

Toynbee, A. J., 1969, *Experiences*, OUP, New York, 417 pp.
Muggeridge, Malcolm, 1966, What I believe: Postscript, *The Observer*, 26 June, (http://www.thewords.com/articles/mugger9.htm).

Preface

Enzo Pranzini and Allan Williams

'I may confidently ask, where can we find nobler or more elevated pursuits than our own; whether it be to interpose a barrier against the raging ocean, and provide an asylum for our fleets.'

(Rennie, 1845, 24)

'If wave motion is arrested by any imposing barrier, a part at least of the energy of the wave will be exerted against the barrier itself, and unless the latter is strong enough to resist the successive attacks of the waves, its destruction will ensue.'

(Bascom, 1964, 243, quoting the words of a Capt. Gaillard)

'Powerful ocean waves pulverize rocks, dismantle cliffs, swallow islands. The anatomy of a wave attack mounted against some solid object – a cliff say, or a breakwater – is an awesome thing.'

(Lenček and Bosker, 1998, 13)

The Nobel Peace Prize 2007 quoted an excerpt from an IPCC (2001) report, which stated: 'Assessments of adaptation strategies for coastal zones have shifted emphasis away from hard protection structures of shorelines (e.g. seawalls, groins) towards soft protection measures (e.g. beach nourishment), managed retreat, and enhanced resilience of biophysical and socio-economic systems in coastal regions. Adaptation options for coastal and marine management are most effective when incorporated with policies in other areas, such as disaster mitigation plans and land use plans.'

These four quotations encapsulate the theme of this book, as waves, encompassing an energy spectrum that has an enormous range, wash up in a continuum against cliffs, beaches, and so on, and any anthropogenic barriers. The book is the result of a long trip carried out along the European coast, accompanied by old and new friends, exchanging the baton at country borders; these friends guided and drove us via hell, purgatory and paradise, through this fascinating environment, like Dante's Virgil (Alighieri, 1304–1321). The many authors – the Virgils – of individual chapters are all at the forefront of coastal research in their respective countries and present the most up-to-date information available on coastal erosion processes and protection policies in these countries, resulting in many examples of coastal strategies and policies. Differences in morphology, climate, wave energy and tide on one side, and on the other, history, urbanization, social and economic development, create conditions for a variety of natural and anthropogenic coastal landscapes, which provides a unique and complete repertory of European coastal case studies.

Coastal erosion, as is often remarked, is a natural

phenomena but it is usually with respect to accelerated erosion that barriers of one form or another are introduced, with an aim of slowing down the landward retreat rate or eliminating it altogether. The aim of this book was to bring some clarity to these myriads of coastal protection schemes that have been emplaced in order to offset erosion trends in European countries. As such, it delves deeply into the erosional history and strategy employed in these countries in order to counteract erosion and therefore represents an eclecsis of information for these countries.

However, over this non-homogenous area arises a shared technical knowledge, some could say a technical evolutionary convergence, in finding the same solutions to similar problems in areas which lie distant from each other. However, some regional tendencies do exist, with certain countries loving specific solutions so much, that this impinges upon and influences their coastal landscapes. Currently, with opposing globalization trends, differences are increasing due to a centralized strategy, which drives many countries towards soft shore protection, others continuing with old hard engineering defences. The pros and cons of differing ideologies of some countries are contrasted with the laissez-faire attitude adopted by others. Far from being a handbook of shore protection projects, this virtual trip along the European coast shows the reader the many different solutions found for this problem.

A quote attributed to the comedian George Burns runs: 'The secret of a good sermon is to have a good beginning and a good ending; and to have the two as close together as possible'. Each chapter can 'stand alone' and this book is not in any way a sermon, but we hope that it is not only the beginning (Chapter 1) and end (Chapter 22) that are good but that what lies between is even better. We leave the last word again to Bascom (1964), as his classic book inspired a generation of coastal scientists, as well as providing the general public with a textbook that read like an exciting novel: 'Is there anyone who can watch without fascination the struggle for supremacy between land and sea?' (Bascom, 1964, Prologue, 1).

References

Alighieri, Dante, 1304–1321. *Commedia*. First printed edition, 1472, by Johann Numeister, Foligno, Italy.

Bascom, Willard, 1964. *Waves and Beaches*, Doubleday Anchor, New York. 267 pp.

IPCC, 2001. Inter-Governmental Panel on Climate Change, *Working Group II:* 3.4, Coastal Zones and Marine Ecosystems, *'Summary for Policymakers, Climate Change 2000. Impacts, Adaptation and Vulnerability.'* Available at: http://www.grida.no/publications/other/ipcc_tar/ (accessed 7 August, 2012).

Lenček, Lena and Gideon Bosker, 1998. *The Beach*, Secker and Warburg, London. 310 pp.

Rennie, Sir John, 1845. *Presidential address to the Ordinary General Meeting*, Institute of Civil Engineers, Vol. 4, Feb. 4, 23–25.

1 Introduction

Enzo Pranzini and Allan Williams

The sea is all around us and brings to mind 'Newton with his prism and silent face . . . voyaging through strange seas of thoughts, alone' (Wordsworth, 1896, 61). The seas written about in this book may be strange or familiar, but the voyage across them to the European coastline has not been made by us alone, but together with a band of erudite and thoughtful companions. Coastal erosion and consequent deposition has been a natural process for aeons but sometimes human intelligence leaves a footprint on this landscape of transience and recurring cycles; frequently it is greed, when not ignorance, that leaves such landmarks. Beach erosion resulting from human intervention – the inevitability of human folly – is an example of the above.

Erosion is a generic term used to describe landform recession or lowering brought about by natural processes, e.g. water, ice, wind, all involving movement (Figure 1.1a) that is sometimes trigged or intensified by human action. Since prehistory, coastal areas favoured human settlements and, later, infrastructures protected these assets, as well as people, from erosion and flooding. For coasts to erode, more material has to be removed than supplied; consequently shorelines most susceptible to erosion are typified by soft or loose sediments, under high wave energy and sea level rise. However, in the past few centuries, anthropogenic influences have played a large part in this process. It is the *increased* erosion rate, which affects the coastal infrastructure (roads, buildings, etc.; Figure 1.1b), that usually necessitates an engineering solution, or management decision to do nothing and 'let nature rule'. Coastal protection features associated with the above are all given greater depth in analyses within this book.

The causes of erosion are many.

■ **Figure 1.1a** *Shore platform at Southerndown, Wales*

■ **Figure 1.1b** *Coastal erosion, Menfi, Sicily, Italy*

Natural changes occur on many scales (long vs. short term) and are functions of, for example, sea level rise producing increased wave attack; climate change possibly resulting in more severe storms; water table rise as a result of higher rainfall; coastal geology (hard vs. soft rocks with the latter more easily removable); high energy sea states (wind/wave) and/or change in angle of dominant wave attack, which in turn influences the efficiency of longshore currents to remove/deposit material; subsidence; etc.

Anthropogenic factors are particularly effective when they reduce sediment input to a coastline. Some examples include: river damming thereby cutting off sediment that should reach the sea (Figure 1.2); increased changes in land practice, e.g. farmers leaving the land to work in a burgeoning tourist industry, thereby ensuring that previously farmed land reverts to its natural state (grass, shrubs, trees), which in turn cuts down sediment input to rivers and ultimately beaches. The above affect beach sediment budgets, which can be further changed by, for example, groin construction, piers, jetties or breakwaters, which interrupt alongshore sediment transport;

construction of coastal defences – 'coastal concretisation' – i.e. seawalls, which prevent coastal retreat and reduce sediment input to the downcoast segments; seaboard deepening by dredging; wrong engineering practices; dune destabilisation; drainage modification.

It must be emphasised that erosion – and its counterpart, deposition – is a natural phenomenon and it is usually the accelerated rate of change imposed by man that causes anxiety. An introduction of a barrier between land and sea has been seen in many areas, not just in Europe but across the globe. Coastal areas have been exploited since time immemorial, in that early harbours utilised natural geographical features such as sheltered bays, lagoons, areas away from strong winds/seas, etc. Mediterranean harbour engineering (jetties and breakwaters) followed patterns set down by Egyptians, Phoenicians, Greeks, Etruscans and Romans, but with the exception of Vitruvius (30 BC), few sources exist in writing. Pre-Roman Mediterranean proto-harbours have been written about by Lehmann-Hartleben (1923) and little changed until Napoleonic times (Franco and

■ **Figure 1.2** *Monte Cotugno dam (Basilicata, Italy) is the largest earth dam in Europe, enclosing a large part of the Sinni river catchment area and triggering erosion of the Ionian coast (Upper photograph by A. Trivisani)*

Verdesi, 1993), so modern coastal protection could date from this era.

So what is the length of the European coast, considering that each part is potentially subject to the erosion process? It is extremely difficult to measure an irregular feature, *i.e.* coastal length, as it is scale-dependent (Mandelbrot, 1967). The lengths given in Table 1.1 were obtained from the World Vector Shoreline GIS database at 1:250,000 km, using a single source, which used a constant scale; they are approximations, which must be used with caution (http://www.mrj. com). For the above reason, these figures give a value that can vary from the others given in this book, e.g. the 2,600 km for the Ukraine coastline (chapter 21), as well as the fact that coastal features are constantly changing through time because of erosion/ deposition.

In this book, Bosnia and Herzegovina, Croatia, Montenegro and Slovenia have been considered as one entity – the Eastern Adriatic (with a coastal length of 6,021 km), Similarly the countries of Latvia, Estonia and Lithuania have been addressed as the Baltic States (with a coastal length of 3,378 km).

An EC (2004) project, which studied European coastal erosion levels, formulated four main recommendations:

1. Restore the sediment balance and provide space for coastal processes.
2. Internalise coastal erosion cost and risk in planning and investment decisions.
3. Make responses to coastal erosion accountable.
4. Strengthen the knowledge base of coastal erosion management and planning to coastal erosion.

With regard to the latter recommendation, findings suggested 'a rating of European coastal regions according to their exposure to coastal erosion together with dissemination of best practice,' *i.e.* what works/does not work (EC 2004, 36). The present book looks in detail at erosion issues associated with the many countries that constitute coastal Europe, together with the resulting engineering strategies introduced as possible solutions to the problem. Both 'good' and 'bad' practices are described in the various chapters. The authors are all key coastal research experts in their respective countries and present coastal erosion statistics and protection strategies (old and recent) that have been utilized in their countries. Work on coastal erosion and engineering structures has significantly developed over the past few decades and the 20 chapters covering

■ **Table 1.1** Coastline lengths of the considered European countries (km)

Albania	649	Lithuania	257
Belgium	76	Montenegro*	293
Bosnia and Herzegovina	23	Netherlands	1,913
Bulgaria	457	Poland	1,032
Croatia	5,664	Portugal	2,830
Denmark	5,316	Romania	695
Estonia	2,956	Russia*	38,000
France	7,330	Slovenia	41
Germany	3,623	Spain	7,268
Greece	15,147	Sweden	26,383
Ireland	6,437	Ukraine	4,953
Italy	9,226	United Kingdom	19,716
Latvia	565		

Source: Data are based on the World Vector Shoreline, United States Defense Mapping Agency, 1989. Figures were calculated by L. Pruett and J. Cimino, unpublished data, Global Maritime Boundaries Database (GMBD), Veridian–MRJ Technology Solutions (Fairfax, Virginia, January 2000)

Note: * not included in this database but obtained from authors.

25 countries span the spectrum of coastal protection measures, from 'hard' to 'soft' engineering practices.

Increased sea surface temperatures together with a sea level rise will probably result in more frequent and severe extreme weather events. Coastal storms will develop bigger waves than normal, which, together with surges, will inevitably result in increased coastal erosion rates and flooding of low-lying lands. After some 6,000 years of almost stable eustatic level, although locally glacio-isostasy and tectonics have played an important role, at the start of the nineteenth century this global phenomenon accelerated to 1.7–1.8 mm/yr (Gornitz, 2007). Throughout the nineteenth and twentieth centuries, global average sea levels have continued this acceleration to 3.1 mm/yr from 1993–2003 and it is further postulated that globally, sea level will rise a further 18–59 cm by the end of the twenty-first century (IPCC, 2007).

The scenarios described can on a global basis cause untold misery and death to many thousands of people, e.g. the catastrophic floods (Watersnoodramp), 'flood disaster', of 31 January to 1 February 1953, which killed 836 persons in the Netherlands, 307 in the UK and 22 in Belgium. This led to the Dutch Government introducing the Delta Plan (full closure of three northern inlets) designed to prevent a repetition of this disaster (Gerritsen, 2005). The aim was to connect the coastal high dune system in an almost continuous line, so that low-lying areas would be safe. After any major disaster

> the traditional model of post-disaster financing, relying on slow and unreliable assistance from the international community, the diversion of budget allocations from development to recovery, or raising new debt in expensive post-disaster capital markets, is increasingly inefficient as disaster occurrence and the magnitude of loss increase.
>
> (UNISDR, 2009)

The key is to undertake measures to increase resilience before any disaster strikes.

On the non-European front, Ashdown (2011) has commented that in 2006, Mozambique requested Sterling £2 million from the International Community (IC) in order to carry out engineering coastal protection countermeasures for potential flooding. This was turned down in 2007, and since then the IC has spent £60 million because of extreme flooding in the region, mainly for risk reduction, in this case by engineering or utilising natural structures, such as mangroves. This would have been a very cost-effective (£2 million vs. £60 million) approach rather than introducing humanitarian aid *after* storms and floods. The UNDP (2007) has shown that just US$1 invested in management activities in developing countries before any disaster prevents some US$7 in loss. In Grenada, Hurricane Ivan caused damages equivalent to 200 per cent of the country's GDP (World Bank, 2009). Points 3 and 4 of the Hyogo Framework for Action (UNISDR, 2005) relate to the utilisation of knowledge, innovation and education in order to build a culture of safety and resilience at all levels, together with a reduction of the underlying risk factors, and it is here that coastal protection techniques can play a massive role. Economics does come into this equation, as to protect a coastal area against a surge, tsunami, or hurricane with a return period of 400 years, might prove too costly, However, with hindsight, the recent (2011) tsunami in Japan has indicated the cost of not having a high enough tsunami wall (10 m) to shield the Fukushima nuclear power plant.

Engineering solutions come in a variety of forms (Figure 1.3). Essentially, they can be divided into 'hard', e.g. seawalls, and 'soft', where measures refer primarily to beach and/or dune replenishment with new material; both are described in detail in the various book chapters. Emplacement of permanent/hard structures, which tend to 'preserve' upland property and structures, is but one solution. Others include retreat and the relocation of threatened structures away from the coastline, or even to protect at all cost, e.g. Montauk Light, on Long Island, USA, a historical structure that is being protected with successive coastal engineering structures, all being built by the U.S. Army Corps of Engineers.

It is axiomatic that, any structure placed in the sea alters the dynamics of the system (Figure 1.4).

■ **Figure 1.3** *Hard shore protection in Denmark (left) and beach nourishment in Italy (right)*

■ **Figure 1.4** *Seawall at Porthcawl, Wales, circa 1980 with toe exposure. Compare with Figure 10.8*

One hundred years ago, the emphasis for protection was seawalls (especially in towns) and groins. In the twenty-first century, rising sea level, storm waves and coastal development pressures have led to the construction of many other types of coastal protection structures, in many different environments. Increasingly, alternative soft engineering techniques are being used, especially sand nourishment, as a response to the burgeoning growth of global coastal tourism (Figures 1.5; 1.6). We still construct seawalls and recently, a series of papers appeared in *Shore and Beach*, giving examples of the structural response to erosion. A salutary thought is that

many of us have recognised for many decades that both natural and engineered structures play a crucial part in maintaining beach width where there would be less or no beach without them, . . . to say that all structures on the coast are 'bad' is, well, structurally deficient, and downright wrong.

(Flick, 2010, 2)

Occasionally some structures, notably the many groin fields that abound on the European coast, become part of the area's heritage, especially in the UK, where people tend to find them aesthetically pleasing as,

■ **Figure 1.5a** *Holiday business is always looking for new coastal strips to develop as soon as political and economical conditions open new frontiers; here is an example from the Montenegrin coast six months after the country became independent*

■ **Figure 1.5b** *Dubai: one of the most explicit examples of how much the narrow coastal strip can be worth for global coastal tourism. Here, plans to increase the number of tourist arrivals from 2.8 million (in 2000) to 400 million (in 2015) have been made (Mocke and Smit, 2008)*

for example, it reminds them of happy childhood days spent at the beach (Williams *et al.*, 2005).

Harbours and marinas invariably have breakwaters and jetties associated with them and this will undoubtedly continue into any foreseeable future and interfere with longshore sediment transport. An old problem was faced at Durban (South Africa) from 1880 to 1937, where 55,000,000 m^3 of sediments closed the harbour entrance, which was dredged, and the sediments lost to the system. A bypass project was then started which, from 1938 to 1946, filled the downcoast segment with 8,000,000 m^3 of aggregates. Even earlier is the project carried out in Atlantic City (New Jersey), where, from 1935 to 1943, 3,200,000 m^3 of dredged sediments were dumped in depths of 4.5–6 m, and had no effect on the beach (Gallareto, 1960).

A large project was carried out in Harrison County (Mississippi) in 1951, where 5,500,000 m^3 of sediments were deposited in front of a seawall creating a 43 km long beach. However it was the expansion of Copacabana beach, Rio de Janeiro, Brazil, designed by the Portuguese Abecasis during the late 1960s, with gulf dredged sediments, that made beach nourishment a worldwide known technology. This now appears to be the first thought of practitioners as an erosion antidote. An adequate and cheap sand supply is an absolute necessity here, and the Netherlands is blessed with copious quantities and utilises it for dune building, but in other areas, this issue can be problematic. For example, nourishment of the Italian coastline (from northern Liguria to southern Sicily) utilises crushed quarry material in order to nourish beaches with coarse sediments (e.g. Caucana in Sicily; Cala Gonone in Sardinia; Marina di Pisa in Tuscany; Bergeggi in Liguria), although the use of shelf aggregates is continuously growing.

Norway, Finland and Iceland are not considered in this book. Even though Finland has a coastal length of some 14,000 km, an archipelago comprising thousands of islands, rock shores that make up 50% of the coastline, plus isostatic uplift of a maximum of 9 mm a year and an absence of tides, the end result is that 'coastal erosion is not an issue on the Finnish shoreline' (http://ec.europa.eu/maritime affairs/documentation/studies/documents/finland_climate_change_en.pdf). Additionally, in a Eurosion case study project of Finland, Sistermans and Nieuwenhuis (2005, 7) commented on the fact that 'coastal defence is not an issue'. Also in an advantageous position, 'it appears that Norway – as a whole – will not be seriously affected by accelerated sea level rise' (Aunan and Romstad, 2008, 403), due to steep slopes, high elevation and rocks resistant to erosion. Eastern Norway (around Oslo fiord) has a maximum isostatic uplift of 0.3 mm/yr, and the low-lying land to the south west is the only area that could be susceptible to a sea level rise and possible erosion. Similar conditions exist in Iceland, where coastal erosion should not be an issue, but glacial front retreat is starting to create problems. At Breiðamerkursandur, outwash rivers once fed the coast but now deposit their bedload in a newly formed lake. From 1904 to 2003 beach retreat was 770 m, *i.e.* 8 m/yr, a process that necessitated a US$13 million project to protect the coastal road (Jòhannesson and Sigurðarson, 2005). New ice-marginal lakes at Vatnajökull ice cap can trap sediments once delivered to the coast in the same way (Schomacker, 2010). However, in a land continuously accreting thanks to volcanic activity associated with the mid Atlantic ridge, coastal erosion seems to be a minor point!

References

Ashdown, P. (Chair), 2011. *Humanitarian Emergency Response Review.* Independent Review commissioned by the UK Government, March. (online) http://www.dfid.gov.uk/Documents/publications1/HERR.pdf (accessed 30 August 2012).

Aunan, K. and B. Romstad, 2008. Strong coasts and vulnerable communities: potential implications of accelerated sea-level rise for Norway. *Journal of Coastal Research*, 24(2), 403–409.

European Commission (EC), 2004. *Living with Coastal Erosion in Europe: Sediment and Space for Sustainability*, Luxembourg: Office for Official Publications of the European Communities, 40 pp.

Flick, R. F., 2010. On structures and the coast. *Shore and Beach*, 78(4), 2.

Franco, L. and Verdesi, G., 1993. Ancient Mediterranean harbours: A heritage to protect. In E. Ozhan (ed.), *Proceedings of the First International Conference on the Mediterranean Environment, MEDCOAST '93*, 255–272, Ankara, Turkey: Middle East Technical University.

Gallareto, E., 1960. *La difesa della spiagge e delle coste basse*. Milano: Hoepli Editore, 303 pp.

Gerritsen, H., 2005. What happened in 1953? The Big Flood in the Netherlands in retrospect. *Philosophical Transactions of the Royal Society A*, 363, 1271–1291.

Gornitz, V., 2007. Sea level rise after the ice melted and today. (online) www.giss.nasa.gov/research/briefs/gornitz_09 (accessed 31 July, 2011).

Intergovernmental Panel on Climate Change (IPCC), 2007. *Observations of Climate Change: Climate Change*. Synthesis report. A contribution of working group I, II and III to the Fourth Assessment report.

Jòhannesson, H. and Sigurðarson, S., 2005. Coastal erosion and coastal protection near the bridge across Jökusla river, *Breiðamerkursandur*, Iceland. Höfn í Hornafirði, Iceland: ICS, 15 pp.

Lehmann-Hartleben, K., 1923. *Die Antiken Hafenlagen des Mittelmeera*, Leipzig: Klio, Beiheft XIV.

Mandelbrot, B., 1967. How long is the coast of Britain? Statistical self-similarity and fractional dimensions, *Science*, 155, 636–638.

Mocke, G. P. and Smit, F., 2008. Challenges and impact associated with coastal megareclamation developments. *Shore and Beach*, 76(4), 5–16.

Schomacker, A., 2010. Expansion of ice-marginal lakes at the Vatnajökull ice cap, Iceland, from 1999 to 2009. *Geomorphology*, 119, 232–236.

Sistermans, P. and Nieuwenhuis, O., 2005. The western coast of Finland. www.eucc-d.de/.../000113 Eurosion.

United Nations Development Programme (UNDP), 2007. *Fighting Climate Change: Human Solidarity in a Divided World*. Human Development Report 2007/08. (online) http://hdr.undp.org/en/reports/global/hdr2007-2008 (accessed 8 August 2012).

United Nations International Strategy for Disaster Reduction (UNISDR), 2005. *Hyogo to Disasters*. Final Report of the World Conference on Disaster Risk Reduction, 'Framework for Action 2005–2015: Building the Resilience of Nations and Communities', 18–22 January, Kobe, Japan.

United Nations International Strategy on Disaster Reduction (UNISDR), 2009. *Global Assessment Report on Disaster Risk Reduction*, Geneva:United Nations.

Vitruvio, M. P., 30 BC. *de Architectura*, 11, ch V, ch ch XII. (in Latin)

Williams, A. T., Ergin, A., Micallef, A. and Phillips, M. R., 2005. The perception of coastal structures: Groyned beaches. *Zeitschrift für Geomorphologie*, 141, 111–122.

Wordsworth, W., 1896, The Prelude, Third book, Residence at Cambridge, (ed), W. Knight, www. bartleby.com/145/ww289.html (accessed on August 8, 2012).

World Bank, 2009. *Development and Climate Change*. World Development Report. Int Bank fr reconstruction and development, Washington DC, 417pp. (online) http://wdronline.worldbank.org/worldbank/a/c.html/world_development_report_2010/abstract/WB.978-0-8213-7987-5.abstract (accessed August 2011).

2 Russia

Ruben Kosyan, Marina Krylenko, Daria Ryabchuk and Boris Chubarenko

Introduction: Russian coastal research

In the Russian Federation, 25,000 km of the 61,000 km total length of the marine coast suffers from severe erosion. Coastal research has been carried out for several centuries, but intensive post-war development resulted in the creation of the Russian Coastal Research National Centre, which coordinated domestic coastal studies. In 1952, it became the Coastal Section, and later the Sea Coasts working group, created by the Russian Academy of Sciences, under the direction of Professor V. P. Zenkovich. During its semi-centennial activity, this public organization promoted domestic research and increased its standing on the international scene (Kosyan *et al.*, 2009).

Currently, basic and theoretical research is supported by many different funding schemes and programmes, with more than 100 organizations (often commercial), engaged to some extent in coastal studies. However, with the disintegration of the USSR, the number of coastal research groups brought up in the traditions of the Russian school of coastal scientists was reduced noticeably. Nevertheless, academic studies are actively developed in the P. P. Shirshov Institute of Oceanology of the Russian Academy of Sciences (both Southern and Atlantic branches), the Geographical Faculty of Moscow State University, the A. P. Karpinsky Russian Research Geological Institute and the Russian State Hydrometeorological University.

The Coastal Zone Department of the Southern Branch of the P. P. Shirshov Institute of Oceanology, Russian Academy of Science (SBSIORAS) is concerned with regular monitoring observations on the coasts of the Black Sea and the Sea of Azov in the Krasnodar region. Specialists from the Sea Coasts Research Center, Rostov State University and others, carry out different aspects of coastal studies in the Black Sea. In St. Petersburg, regular state monitoring of the eastern Gulf of Finland geological environment is fulfilled by the A. P. Karpinsky Russian Research Geological Institute (VSEGEI). Monitoring in the Kaliningrad Oblast is carried out by the Atlantic Branch of the P. P. Shirshov Institute of Oceanology (ABSIORAS) monitoring network through yearly measurements of the Kaliningrad marine coast (150 km) and partly the lagoon shores of the Vistula and Curonian spits. Nine segments of the Kaliningrad marine coast have been assigned as 'segments in danger' and are targets of specific monitoring activities (length, rate of erosion, etc.).

The Black Sea coast

The Russian Black Sea coast length (Krasnodar region) from Kerch Strait to the Psou River is *circa* 500 km. This coast's anthropogenic development both past and present is virtually defined by physiographic factors, especially the Caucasian mountain ridge that stretches along the east coast of the Black Sea. These mountains are the cause of the poor

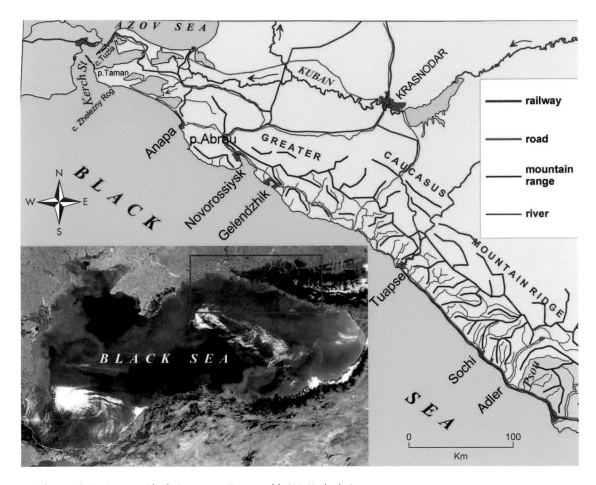

■ **Figure 2.1** *Russian Black Sea coast (Prepared by V. Krylenko)*

transport networks and the complexity associated with any settlement development. Today, a significant part of the population lives in the coastal strip, which houses important communications (Federal and International), as well as large industrial and civil construction complexes. Large ports (Novorossiysk, Tuapse and Taman) are located here and the region is the main recreational area for Russia (Figure 2.1).

Hydro-meteorological characteristics

The climate is a result of the influence of macro-circulating processes occurring in the Mediterranean region. Basic seasonal features are defined by interaction of the Siberian and Azores maxima, the Asian minimum and the Mediterranean winter cyclone. Local physical features and coastline irregularities create appreciable climatic differences in some areas.

The climate is similar to the Mediterranean (warm damp winters, hot and dry summers), but winters are sometimes subject to the influence of northern winds. Against the background of a relatively 'soft' winter with positive monthly average temperatures, practically every winter produces temperatures of −10 to −15°C, and in severe winters as low as −25°C; whilst average January temperatures range from +1 to +4°C; in July it is about +23°C. The mean yearly rainfall is 400–600 mm with a winter maximum and summer dryness and the annual number of sunny days is >300.

During the fall–winter period, winds from the N, NE, E and SE directions prevail in the Taman peninsula area, where there are no mountains; in summer, there are more winds from the NW to SE. South of Anapa, the mountain heights that protect the coast from northern winds are insufficient, so cold air mass penetration occurs as far as Tuapse, with the coast from Novorossiysk to Gelendzhik being most susceptible. The gusty northeast wind, which flows down hillsides, brings a significant cold snap and is called the *Bora of Novorossiysk* (Italian *bora*, from Latin *boreas*, Greek *boréas* meaning the northern wind). In Novorossiysk the annual average number of days with a *Bora* is 46 and wind speed reaches 40–65 m/s. The wind regime from Anapa to Tuapse is characterized by the prevalence of winds from the NE, S, SW and W.

The southeast (south of Tuapse) is protected by mountains; it has a prevalence of S, SW and W winds and is characterized by a damp subtropical climate (an abundance of precipitation, warm winters and hot summers). Average January temperatures are between +5°C to +7°C (average daytime temperature for January is +10°C); average July temperature is *circa* +23°C. The mountain chain's barrier role is even more evident with the increase in height of the Greater Caucasus Mountain Ridge. The coast is closed to northern cold air masses, and moisture-laden air masses originating in the west are stopped by the mountains; the Western Caucasus hillside is the dampest place in Europe. The Achishkho ridge slopes receives more than 3,000 mm of precipitation annually; the coast some 2,000 mm, with a maximum in winter which leads to high river water levels.

The Black Sea is non-tidal and sea level variation is defined by changes in water balance components. On the eastern coast, these are due to river drainage fluctuations and storm activity, with maximum levels in June, lowest in October/November. The average annual variation along the coast does not exceed 1 m.

The Basic Black Sea Current (BBSC) is the main structural element of the Black Sea water general circulation. In quasi-stationary condition, the BBSC occupies a strip about 50–55 km wide which moves in a generally northwest direction. In the upper continental slope, anticyclonic vortex structures (AVS) are formed with sizes from 3 5 to 25–30 km. Due to the AVS, the Russian shelf current regime is characterized by changes from a northwest (west) direction to southeast (east) direction and *vice versa*, producing a bimodal regime of currents. The maximum current speed is less than 100–120 cm/s (Krivosheya *et al.*, 1998). In depths greater than 50 m under the influence of the coastal anticyclone, the coastal alongshore current has speeds up to 40–50 cm/s directed to the southeast.

Generally, for all Black Sea Russian coastal sectors, the prevalence of hurricane waves having W, SW or S directions (Figure 2.2) is characteristic. Although the strongest winds with a speed up to 40 m/s and above are observed originating from the northeast, these winds have a weak influence on wave climate, since the small acceleration fetch distance limits wave

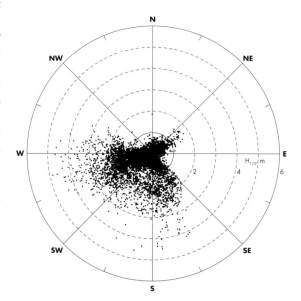

■ **Figure 2.2** *Wave direction and height repeatability distribution (Source: Divinskiy et al., 2003)*

growth (Bukhanovskiy *et al.*, 2000). Winds from the S, SW and W have the greatest influence on hurricane wave formation, with wave heights exceeding 6 m; the maximum wave height registered during field studies being 8.43 m, with a wave length of 120–200 m (Divinskiy *et al.*, 2003). The northeast part of the Black Sea has three seasons in terms of dynamics (Figure 2.2).

Evolution of coastal settlements

Two thousand years ago, the flat Taman territory, supplied with fresh water from the Kuban River, was attractive to *Homo sapiens*. The crossing of trading routes from the Don River basin to the Black Sea and Mediterranean regions together with overland routeways from Asia through Crimea to Europe was the setting for the Bospor Empire, one of the major foodstuff suppliers to Ancient Greece. Land uplift together with climate changes led to an agricultural downswing and during subsequent centuries, the Taman peninsula became practically uninhabited. Coastal development began only in the twentieth century, where the presence of unused territory attracted the attention of investors. Between Capes Panagiya and Zhelezny Rog, modern terminals for transshipment of various cargoes have been constructed, together with railways, road and other communications (Figure 2.3).

The region has a mountainous relief and the coastline is erosional. This territory south of Anapa was always difficult for economic development, as high relief and a complex climate (severe winters with frost and hot dry summers) limited agricultural development, but ancient settlements existed at two unique natural bays – Novorossiysk (Tsemes Bay) and Gelendzhik bays. Prior to the beginning of the nineteenth century, slaves were the basic 'goods' taken from the region but after inclusion of this coast into the structure of the Russian empire, military and economic development commenced. Almost all large settlements originally represented military coastal fortifications, as expansion of territory was inland from the coast and the road network developed from this period connecting coastal settlements. At the end of the nineteenth century the railway was constructed, cutting through the Caucasian ridge and exiting at Novorossiysk bay. At the beginning of the twentieth century, the railway connecting central Russia with Georgia was constructed, facilitating development of Tuapse and Novorossiysk ports. Pipelines connecting these ports with Caspian Sea oil deposits also pro-

■ **Figure 2.3** *Port Taman (Photograph by V. Krylenko)*

■ **Figure 2.4** *Landslide coasts near Cape Zhelezny Rog (Photograph by V. Krylenko)*

moted development of sea transport in the region. Creation of transport networks connecting the coast with the central part of the country also had a huge impact on recreational development.

The coastal landscape changes southwards. Coastal ridges become flatter, the climate softer and more humid, creating conditions ripe for agricultural development. The large rivers fed from mountain glaciers bring large amounts of sediments to form extensive deltas, which sometimes occupy the entire narrow shelf. The forefront of these deltas is located very close to the edge of the abrupt continental slope pierced by underwater canyons. The constant inflow of sediment together with the sea's influence creates a very complex picture of coastal evolution. This site does not have natural sheltered bays, but during ancient times had transport connections with Greece. In the twentieth century, together with agriculture, recreational activity began to develop. The uniqueness of location, presence of the transport system, the favorable climate, and mountains with permanent snow cover located nearby, have given

this region the site for the 2014 Winter Olympics, the main events being located on the Imeretinskaya lowlands. In the construction period, it will be necessary to solve extremely intricate problems regarding coastal protection, especially against washouts and flooding.

Coastal morphology

The most northerly part of the Black Sea Russian coast is Cape Tuzla (Figure 2.1). A bay-bar having the same name and consisting of quartz sand, ground shells and pebbles adjoins it to the north but further south occurs the 65 m high Cape Zhelezny Rog. The coast, except for the Cape areas, is composed of easily erodible rocks – loams and clays, with intercalation of sand, argillites and iron-rich sandstones. Wave impact in combination with coastal slope rock properties, have defined erosion processes (average coastal recession is 0.7 m/yr), the primary one being landslides, which with displaced ledge material sometimes blocks the entire 5–20 m wide beaches

■ **Figure 2.5** *Anapa beaches (Photograph by V. Krylenko)*

(Figure 2.4). From a lithodynamic viewpoint, the coast represents a closed system, as the stream of deposits through the Capes is practically absent (Kosyan and Krylenko, 2009).

Southward of Cape Zhelezny Rog, the coast becomes lower and between Solenoe Lake and Anapa becomes a sand bay-bar system with a length of 50 km. Few cliffs occur and sand beach widths are 30–250 m, a prominent feature being dunes with heights of 3 to 12 m (Figure 2.5).

The flysch zone, south of Anapa, with rocks whose layers are intensely dislocated, is distinguished by intercalation of limestone, chalk, clays, sandstones and slates, which have low durability and are susceptible to abrasion. In the surf zone, fragmentary material disintegrates quickly and is easily carried by currents in suspension prior to being deposited far from the coast. At Anapa, close to the Sukko River, the Paleogene flysch is replaced by Upper Cretaceous flysch, which in turn, is replaced by the Paleogene flysch to the south of Gelendzhik. At Tuapse the Paleogene flysch wedges out to Cretaceous flysch which spreads towards the Psezuapse River. Novorossiysk and Gelendzhik bays coinciding with transition zones between various structures.

The northern extremity of the Caucasian range projects into the sea between Novorossiysk and Anapa, in the form of the Abrau peninsula (Figure 2.6) which has a dismembered relief, maximum heights reaching 500 m. Grandiose collapse and landslips have occurred in the geological past and beaches with widths from 1 to 50 m coincide with small bays and outflows of temporary water currents (Krylenko, 2009).

An area of subdued mountain relief occurs from Gelendzhik bay to Tuapse city. The watershed extends inland from the sea to a distance of 20 to 25 km. On a number of sites, spurs of the Greater Caucasus Mountain Range recede away from the sea from a narrow strip of ancient sea terraces with cliffs reaching heights up to 60 m. The average width of gravel/pebble beaches on the open part of the coast

■ **Figure 2.6** *Abrau peninsula coast*

does not exceed 3 to 5 m (Figure 2.7). Sediment material from eroding cliffs is taken away by small rivers, which dry up in summer, so are not enough for longshore sediment flux saturation.

At Tuapse city, the coast line becomes almost linear, representing an area of subdued mountains and a terrace zone commences at the Shakhe River. Alluvium taken away by the Tuapse, Ashe, Psezuapse, Shakhe, Sochi and Khosta rivers, causes some aggradation at estuary sites. The Mzymta–Psou (8 km) interstream area (lowland Imeretinskaya) is the classic example of an accumulation coast, generated on a narrow shelf near the abrupt con-

tinental slope. Natural sand and shingle beaches have an average width of 50–60 m. The underwater slope is cut almost everywhere by submarine canyons, with the largest located at the mouth of the Mzymta River, with the Novy and Konstantinovsky canyons occurring some 2 km to the southeast. The delta's geological history began in the Neogene and almost 200 m thickness of alluvial/marine deposits has accumulated since this time. As the shelf width in this area does not exceed several kilometres, the delta sea border has reached the shelf edge where it has become stabilized. The beach barrier has a height of 4 to 4.5 m which protects it from periodic flooding.

■ **Figure 2.7** *Abrasion cliffs near Gelendzhik (Photograph by V. Krylenko)*

Coastal protection

The first experience of coastal protection at the Black Sea Caucasian coast can be attributed to sea wall construction in 1914–1916. Due to the planned economy conditions that existed during this period, the coastal protection problem was solved by development of *uniform* coastal protection schemes for *all* coasts and in 1961 the first coastal protection master plan was developed. Realization of this and subsequent master plans have appeared to be ineffective. The rather narrow set of prescribed coastal protection methods (groins and sea walls) frequently did not take into account site-specific lithodynamic conditions.

Tuzla spit is a large body composed of quartz sand, shell and pebble. At present the spit is degrading. At the close of 2003, a boulder dam (580,000 m³) with stone rip-rap was constructed (Peshkov, 2003). In order to prevent outflow of deposits, two 25 m long rock-fill spur dikes were constructed in 2004. In mid 2005, the width of some Tuzla beaches reached 30 m and restoration of the Tuzla has substantially reduced penetration by Black Sea storm waves, thereby reducing erosion.

Intense erosion processes are specific for the coast from Tuzla Cape to the Anapa bay-bar coastal sector. No coastal protection structures are present and there is only one protected part near Taman port (Figure 2.3), all construction being post 2001. Port and yacht harbour structures in Anapa city are protected by gravitational type and stone fill vertical walls. To the south of the city boundaries, artificial shingle beaches have been constructed for recreation (Figure 2.8) and to prevent pebble removal, some stone fill spurs have been built.

On the Abrau peninsula coast, recreational areas exist only on small river estuaries where natural pebble beaches occur in coastal concavities. On a number of open coast sites, artificial pebble beaches under the protection of stone backfill spurs are being created for recreational purposes, as at Anapa. The port of Novorossiysk is protected from most hazardous wave directions. Therefore, hydraulic engineering constructions are mainly represented by vertical walls, reinforced concrete groins and piers. Southward, within the limits of the same bay, is located the resort settlement of Kabardinka. Earlier beaches in the central part of this settlement were protected from washouts by the method traditional for the region – reinforced concrete groin pilings with a backfill of local weak rock and pebbles inserted between groin spaces.

A 1 km sand beach (10 m to 70 m in width; erosion at 5 cm/yr) was created in the 1970s from 250,000 m³ of material taken from the sea bed in the central part of Gelendzhik bay. In 1991, at Inal, a shallow bay near Tuapse, was backfilled with about 250,000 m³ of pebble material to an average width of 25–30 m, and a stable beach generated within a few years.

However, usage of artificial beaches for coastal protection and recreational complexes is limited due to coastline uniformity and the negative balance of longshore sediment flux. The combination of artificial beaches with various types of wave-suppressing and sediment-restraining constructions, allows coastal protection of any lithodynamical conditions without reducing its economic-geographical potential. An example of the traditional Black Sea artificial beach is presented in Figure 2.9, the beach created being under the protection of two or more groins.

The railway line runs along practically the whole length of the coastline south of Tuapse. Since construction in 1914 to the present day, special attention was given to the protection of these railway

■ **Figure 2.8** *Anapa shingle artificial beach (Photograph by V. Krylenko)*

■ **Figure 2.9** *Traditional pattern of an artificial beach on the Russian Black Sea coast (Photograph by V. Krylenko)*

tracks against the destructive influence of the sea and the preservation of pebble beaches was planned and carried out by means of beach supporting structures – groins and breakwaters (Kosyan *et al.*, 2005). Intensive construction of these proceeded until the mid-1980s. Currently the beach from Tuapse to the Mzymta River mouth is practically built up by various coastal protection structures. There are about 1,000 groins, 80 km of sea walls, and 60 underwater and surface breakwaters, having a length of 13 km between the rivers Tuapse and Mzymta. For example, 84 per cent of the coast between Magri and the mouth of the Ashe River consists of seawalls and also contains 33 groins of different design. For the coastal section between the Shakhe and Sochi Rivers, 111 groins, 11 submerged and four inter-mittent breakwaters were built. The coastal section between the rivers Sochi and Bzugu (3.6 km) has large numbers of coastal protection and beach-holding constructions. Sea walls have been built along the whole of this section and the port of Sochi is situated here. Municipal and sanatorium

embankments are protected from erosion with 52 groins and breakwaters (2.2 km long).

A general introduction of reinforced concrete has not led to coastal stabilization, and sites where shore-protecting constructions have functioned for the longest time are in bad condition. A longshore transport stream of deposits has been interrupted by a system of groins and breakwaters, which intercept practically all pebble and gravel material migration along the coast, so beach restoration by natural ways is impossible.

On a significant area (90 per cent) of the Imeretinskaya lowland the beach sediment supply was limited by artificial dam construction during 1980–1990 which protect coastal settlements from flood surges. Berms, tetrapod concrete blocks and sea walls are used for strengthening and protecting dams. In the last ten years dam backfilling was practised, as was replenishment by building waste and/or stone piles. Nevertheless, a significant number of dams are in an unsatisfactory condition and flooding of coastal apartments and houses occurs during storms.

In 2008, to the south of the Mzymta River mouth, construction commenced on protective structures of the cargo port designed for the 2014 Olympics. The port location was selected by local administrative decisions, without taking into account physical and geographical conditions. On December 14, 2009, partial destruction of the protective structures occurred, as a result of a Beaufort 6 storm which damaged coastal structures. New protection structures against extreme storm impacts are being created for the Imeretinskaya lowland coast and Olympic projects.

In essence, protection of a greater part of the Russian Black Sea coast is resolved by means of reinforced concrete rigid structures, but without taking into account existing beach widths and the influence of such constructions. Insufficient attention has been paid to preservation or improvement of coastal landscapes and this coast is being intensively developed with the recreational branch growing exponentially, mainly because of beaches which promote maximum usage for any recreational potential. A shortage of beaches, especially in recreational areas with the highest demand, essentially limits opportunities to increase the capacity of a resort. One way of solving the problem of beach shortage is by creating artificial ones but there is a deficiency of beach-forming material, and beach creation/ broadening is possible only due to replenishment with borrow pit material. Constant replenishment of material washed from a beach is economically unprofitable, so creation of stable artificial beaches is possible only in bays under the protection of capes. Artificial pebble beaches with a protection of stone backfill retaining structures can protect the coast from wave impact, by suppressing wave energy, thereby keeping the recreational potential. Artificial beaches under protection of beach-retaining structures are the optimal coastal protection method for the Russian Black Sea coast.

The Baltic Sea coast

Two segments of the Baltic Sea coastal zone belong to the Russian Federation: the Kaliningrad Region (South-East Baltic) and the Eastern Gulf of Finland (Figure 2.10). The length of the Gulf of Finland Russian shoreline (not accounting for islands), is *circa* 520 km. Its north-western section (to the west of Berjozovy Archipelago, around Vyborg) is made up of *skerries* (small rocky islands). The most common type of coast in the south-western section (in particular, Luga and Koporye bays) consists of embayments straightened by erosion and consequent deposition. The easternmost parts are largely open to the sea and affected by storm surges. The eastern coast of Neva Bay has been completely transformed by technogenic processes, because of 300 years of development of St. Petersburg, one of the largest cities of Russia. Traditionally, the coastal zone of the easternmost part of the Gulf of Finland has not been considered as an area of active litho- and morphodynamics, but recent studies have shown that within some coastal segments, shoreline recession rates reach 2–2.5 m/yr (Ryabchuk *et al.*, 2011). Natural erosion problems are added to by anthropogenic impacts, an old ineffective system of coast protection and intense recreational infrastructure development.

The South-Eastern Baltic Sea shore is a typical example of an open sand coast with eroded cliffs and coastal lagoons separated by spit sand barriers from the sea. The length of the Baltic Sea coastline within the Kaliningrad Oblast region is some 150 km, including 48 km of the Curonian Spit and 35 km of the Vistula Spit. Russia shares the Curonian Lagoon, with Lithuania to the north and the Vistula lagoon with Poland to the south. The mainland shore (Sambian Peninsula) is under intensive permanent erosion, while the shores of both sand spits are characterized by alternation of eroded/accreted segments (Boldyrev and Bobykina, 2008).

Hydro-meteorological characteristics

Gulf of Finland

The climate is moderate with high humidity, considerable cloudiness and frequent precipitation, and is strongly affected by the Atlantic Ocean. Average temperature in the coldest month (February) is

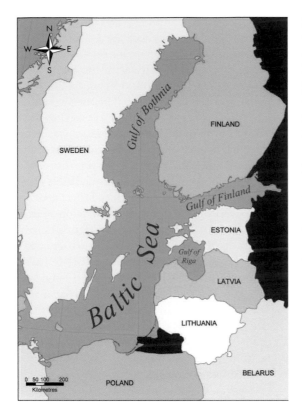

September to February. The most significant sea level variations apparently occur because of the combined effect of wind-induced storm surges and progressive long waves. Neva Bay area geometry is such that propagation of long waves into the eastern part of the Gulf is accompanied by a rapid water level increase. During the history of St Petersburg (1703–2008), 307 floods higher than 160 cm occurred (Floods Catalogue, www.nevariver.ru/flood_list.php). Three of them, one in each century, were catastrophic. On 21 September 1777 the water level reached 3.21 m. The highest surge ever (4.21 m) occurred on 19 November 1824, and the highest water level of the twentieth century (3.80 m) was 23 September 1924. The number of floods in a single year varies from zero to ten. Tidal water level changes are very small, just 1–5 cm; typically, the largest impact on the Baltic Sea coasts is observed in the case of storms that simultaneously cause high surges and rough seas.

South-east Baltic

Kaliningrad Oblast belongs to a moderate humid climate zone . The mean annual air temperature is about 7.5°C, the average temperature of the coldest month (January) ranges from –3°C to –5°C, and of the warmest one (July) is between 17°C and 20°C. Winters are usually without hard frost and ice cover occurs only in the Curonian and Vistula lagoons. The annual precipitation is *circa* 620–780 mm with a maximum in summer and autumn. The Curonian Lagoon is an estuary lagoon formed by the Neman River (runoff of *circa* 30 km³ a year). A westerly wind direction dominates the area and average wind speed varies between 3.6 to 7.3 ms⁻¹. Low winds (<5 ms⁻¹) occur for 56–60 per cent of the time, moderate winds of 6–9 ms⁻¹ for 27–32 per cent, and storm winds are observed for 5–7 per cent.

On average, a predominance of westerly winds forms, in general, two longshore sand drift currents: eastward and southward from the extreme north-western corner of the Sambian Peninsula. Maximum waves may reach 10 m height in deep water, while at a depth of 10 m, wave height is less than 4–5 m.

■ **Figure 2.10** *Russian parts of the Baltic Sea are indicated in red and represent the eastern part of the Gulf of Finland (the St Petersburg (Leningrad) area) and the south-eastern part of the Baltic Sea (Kaliningrad area, map prepared by A. Sergeev)*

–7.8 °C to 8.5 °C (minimum values –35 °C to –40 °C). The warmest month is July, with the average temperatures +17.4 to + 18.0 °C and maximum +35 °C; the mean annual rainfall is 517–557 mm.

Ice cover is annual but recently, due to warmer winters, average winter temperatures from 1870 to 2009 show an increasing trend. The seasonal length of solid ice cover has decreased and ice absence during months when severe storms are frequent substantially modifies both the course and intensity of coastal processes. The annual number of storms varies from 40 to 50 with wind speeds of 15–25 m/s; the average wave height in the Russian part of the Gulf is 1–2 m from March to August, 2–3 m from

Tidal amplitude does not exceed 5–10 cm and maximum storm surges (1.3–1.8 m) occur at the western coast of Kaliningrad Oblast (Gurova et al., 2008).

Evolution of coastal settlements

Gulf of Finland

The earliest archaeological sites could be considered as Early Mesolithic, *i.e.* the Ancylus transgression period. During the Littorina transgression, which corresponded to climate warming (8,300–4,500 years BP), a significant amount of late Mesolithic and Neolithic sites were located close to the shorelines of the Littorina Sea. In 2009, the Petersburg Archaeological Expedition found a Neolithic settlement in the city (3,000–4,000 years BC) with more than 40 ancient constructions, burial places, large amounts of pottery, stone implements and timber stakes and other equipment for fishery and hunting (Sorokin *et al.*, 2009).

During early medieval times (eighth to eleventh centuries), the Neva River and eastern Gulf of Finland were part of the trade route between Northern and Eastern Europe, but most settlements were located away from the coast. In the twelfth to sixteenth centuries, a number of small settlements (Russian, Finnish, Izhorian and Karelian) plus several Russian fortresses were situated around the Gulf of Finland, for example, within the territory of St. Petersburg (www.spbae.ru) and in the seventeenth century, several Swedish fortresses were built here.

The first stage of active technogenic coastal transformation commenced when Peter the Great founded St. Petersburg as a new capital of Russia. Building of parks and palaces along the Neva Bay coast (in Pertodvorets, Strelna, Oranienbaum, etc.) was accompanied by the building of coastal protection structures and embankments. Seventeen artificial islands with fortresses were constructed in 1704 at Neva Bay around Kotlin Island. In 1721, the Sister River was dammed for construction of the Sestroretsk Armoury factory and Lake Sestroretsky Razliv was formed, resulting in a change in the sediment balance for the northern Gulf.

The east coast of Neva Bay changed in the second half of the nineteenth century, when St. Petersburg's harbour was built. In the 1880s–1890s, the northern Gulf of Finland coast become a popular resort and a recreation infrastructure was begun, e.g. a railway line from St. Petersburg to Helsinki, as was coastal protection construction, especially in small villages such as Terijoki (now Zelenogorsk), Kellomjaki (now Komarovo), Kuokkala (now Repino) and Olila (now Solnecjnoye).

In the twentieth century, St. Petersburg (Leningrad) became one of the largest cities of Russia with a population of about 5 million people, having a high development of industry and transportation. In 1967–1973, a nuclear power station was built in Sosnovy Bor on the southern gulf coast. In 1979, the Flood Protection Facility (FPF) was inaugurated and in the 1960–1980s, because of hydro-engineering work, hundreds of hectares of new territories were created in the eastern part of Neva Bay.

Over the last decade, several big oil and coal port terminals have been constructed. The newly built Ust-Luga port complex is planned to be one of the world's ten biggest ports with a carrying capacity of general cargo reaching 120 million tonnes per year (www.ust-luga.ru). The Primorsk oil-terminal (part of the Baltic Pipeline System, of which the first segment commenced operations in 2001) is the biggest oil exporting port of north-west Russia and since 2006, 65 million tonnes of oil is exported annually (www.mtp-primorsk.ru). In 2006, construction commenced of a new Passenger Harbour on 476.7 ha of new territory in Neva Bay (www.mfspb.ru).

South-east Baltic

The Kaliningrad Oblast shoreline has been predefined by tectonic development: the Vistula and Curonian lagoons are confined to inherited depressions, whereas the Sambian Peninsula represents an upstanding block of the earth's crust (Zhamoida, pers. comm.). However, the main coastal zone features were formed during degradation in the last glacial ice cover and essentially transformed during the post-glacial period by the existence of different water bodies. The earliest archaeological sites in the Kaliningrad region are confined to the Yoldia Sea stage or Younger Dryas time, 8,800–8,100 years BC situated along river valleys (www.archaeology.ru). The recent coastal line was mainly formed at the initial Littorina Sea transgression stage, about 6,000–7,000 years BP (Zhindarev and Kulakov, 1996). Temporary settlements were found within the Curonian Spit (3–4,000 thousand years BC) and burial places within the Sambian Peninsula (1–2 thousand years BC). It is possible to surmise that settlements during the early stages of colonization were partially controlled by sea level oscillations during Littorina and post-Littorina times. During the Viking epoch, one of the factors controlling Curonian Spit settlements was the existence of the natural straits connecting the Curonian Lagoon and the Baltic Sea.

The first coastal settlement, where Baltiysk (was named Pillau before 1945) is located, was formed at the end of the thirteenth and beginning of the fourteenth century. In 1510, after a huge storm, the new Vistula Lagoon inlet moved approximately 10 km south. Due to human activities, the new inlet location was fixed and concrete moles, which exist today and border the Baltiysk (Pillau) Strait, were built from the end of the eighteenth century until the end of the nineteenth century. From 1901 to the present, the Baltiysk Strait is the entrance to a navigable canal connecting the Port of Kaliningrad (was named Königsberg before 1945), located 40 km inland, with the Baltic Sea.

Coastal geomorphology

Gulf of Finland

The coastal zone is located at the boundary between the Baltic Shield and Russian Platform. Geological processes caused by the Weichselian glaciation and following alternations of continental, lacustrine and marine environments during the last 14,500 years, played an important role in relief formation and sediment distribution. Coastal zone relief on both land and nearshore is characterized by series of terraces of late and postglacial basins. The contemporary shape of the investigated coastal zone was formed about 2,500–3,500 years ago, when Ladoga Lake waters burst out to the Baltic Sea, forming the Neva River.

The upper part of the geological sequence in the shoreline consists of relatively soft Quaternary deposits. Landward from the shoreline they are represented by Late-Pleistocene glacial, glacio-fluvial, glacio-lacustrine and Holocene lacustrine, marine, alluvial and aeolian deposits. In the nearshore bottom surface, glacial till, glacio-lacustrine clay and Holocene sand dominate (Spiridonov *et al.*, 2007). All these deposits are easily erodible and the relief of both land and bottom is rather plain and smooth. The average depth of the Gulf bottom is *circa* 38 m; depths increase westward from 1–5 m in Neva Bay to 105 m in the open part of the Gulf near Gogland Island.

As the coasts reached their contemporary shape only a few millennia ago, they apparently have not reached equilibrium. From a line from Cape Flotsky to Cape Grey Horse, the littoral drift is eastwards, stopped by headlands at the entrance to Neva Bay. Figure 2.11 represents the different types of coastal morphologies, but the intensity of coastal processes varies within different coastal typologies. Some parts of the eastern Gulf of Finland coastal zone are stable, e.g. the skerries of the north-western section, which mostly consist of stable bedrock formations, but shoreline configuration here is very complicated with lots of islands and narrow bays.

Within the large bays (Luga Bay, Koporsky Bay) of the southern coast, capes are severely eroded with formation of boulder-pebble beaches with a maximum rate of coastal retreat of 1.5–2 m/yr[1]. In the inner parts, stable sand beaches are observed, the largest being located in Narva Bay, with a length exceeding 20 km and a width of 40–50 m. Recent studies have shown that since 1990, the sand beach in Narva Bay has been stable, but during some severe erosion events, significant beach damage has been reported, e.g. after a storm surge in 1977, the width of the beach decreased from 50 to 20 m and coastal dunes were damaged (Ryabchuk *et al.*, 2009). To the

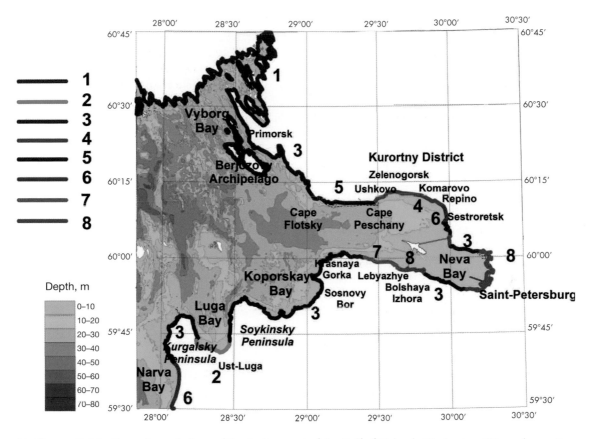

■ **Figure 2.11** *Coastal morphology of the Eastern part of the Gulf of Finland. (1) skerries; (2) sand accretion alluvial coast; (3) moraine erosion (boulder) coast with sand beaches within bays; (4) erosion and accretion moraine (boulder) coast with sand accretion zones; (5) moraine erosion (boulder) coast; (6) sand accretion coast; (7) sand accretion coast with cuspate foreland; (8) technogenic coast (embankments, seawalls, harbours and other hydrotechnical constructions)*

east of Cape Grey Horse, the highest cliffs of the region (up to 32 m high) can be found.

Differing from other coastal sections, beaches of the northern coast between Cape Flotsky and Cape Dubovskoy (Kurortny District) are largely open to the west. The magnitude of sediment flux along the northern coast tends to diminish in an easterly direction and in the vicinity of Zelenogorsk it is close to 30,000 m³/m/yr; while near Repino village, the longshore drift reaches exceeds 20,000 m³/m/yr whilst in the vicinity of Solnechnoye it almost ceases (Leont'yev, 2008).

Remote sensing data show that average rates of shoreline retreat within the Kurortny District are 0.5 m/yr, while the maximum reaches 2 m/yr (Sukhacheva and Kildjushevsky, 2006; Ryabchuk *et al.*, 2009). Total beach recession from 1990–2005 reached high values, e.g. 25–40 m in the vicinity of Serovo, Ushkovo and Komarovo village and the situation became dangerous, as the average beach width between Cape Lautaranta and Solnechnoye village is just 30–40 m. From the total length of about 45 km of this coastal zone, some 19 km have boulders both on shore and in the nearshore zone.

The major sand accretion section, with a total length of 15.5 km, is located between Capes

■ **Figure 2.12** *Drastic erosion of the coastal dune in Komarovo village after storms and floods during autumn 2006–winter 2007*

Solnechnoye and Dubovskoy. They have high recreational value and therefore their evolution is essential for the district's economic development, but they have a limited sand amount in the active body. The beaches are 30–40 m wide in sections that contain dunes, but boulders are usually found at short distances offshore. At some places, two to four sand bars are observed interspersed with boulders or till outcrops in the runnels.

Recently some extreme erosion events have been observed. A series of severe storms accompanied with high surges, in autumn 2006 to winter 2007, caused abrupt erosion, and these events have changed the appearance of many coastal sections much more than decades of previous, relatively slow development. During these events, the water level rose more than 2 m above the long-term mean in St Petersburg (Ryabchuk *et al.*, 2009). Extensive dune erosion occurred along the entire 3 km of the shore where they exist at Komarovo village and along some sections of the Solnechnoye coast (Figure 2.12).

In the vicinity of Fort Krasnaya Gorka, the highest cliffs (up to 30 m) can be observed with a rate of shoreline retreat up to 0.6 m/yr. According to aerial and satellite photo analyses from 1975–1976 and 1989–1990, sand beaches to the west of Lebyazhye village were eroded by up to 30 m, and near Bolshaya Izhora village, up to 70 m (Orviku and Granö, 1992). To the east, because of longshore drift, sand spits of different shapes and sizes have been formed since the middle Holocene. Sand cusps (of some hundreds of metres in length and some tens of metres wide) and hooked spits are among the most remarkable morphologic coastal forms and represent longshore sand waves. They move eastward along the shoreline in the vicinity of Bolshaya Izhora, causing alteration of erosion/accretion phases of coastal development along the shoreline accompanied by sand spit growth and degradation. Annual shoreline GPS surveys have shown that the western cusp has moved 33 m eastward from 2007 to 2009. Retrospective analysis of remote sensing data since 1989 to 2004 gives shifts of 315 m and 200 m, *i.e.* the average annual shifts of cusps were 20 m/yr and 13 m/yr (Ryabchuk *et al.*, 2011). During the last 30 years, because of longshore sand drift, a large accretion body has formed to the west of the Tchernaya River mouth. To the east a complicated hooked spit was formed, which at its distal part is still growing, while the attached part of the spit is intensely eroded with outcroppings of relict lagoon mud on the shoreline (Figure 2.13).

South-east Baltic

In general, the whole of the Kaliningrad Oblast coastline, except some segments on the Curonian and Vistula spits, can be attributed to an eroded shoreline (Figure 2.14; Boldyrev and Bobykina, 2008). From comparisons of coastline positions in the 1960s and 2000s, it was estimated that the long-term average rate of erosion was 0.6–0.7 m per year. Regular monitoring by ABSIORAS showed intensification of erosion rates up to 0.8–1.0 m per year during recent years, the eroded shore length increasing from 50 per cent in 1995 to 63 per cent in 2005 (Figure 2.15). Erosion in the Oblast region at present amounts to a physical loss of 70 ha per year at a financial cost of 60–100 million euros per year (Gilbert, 2008).

Coastal protection

Gulf of Finland

Within Neva Bay, huge clumps of water plants protect certain coastal sections, however more than

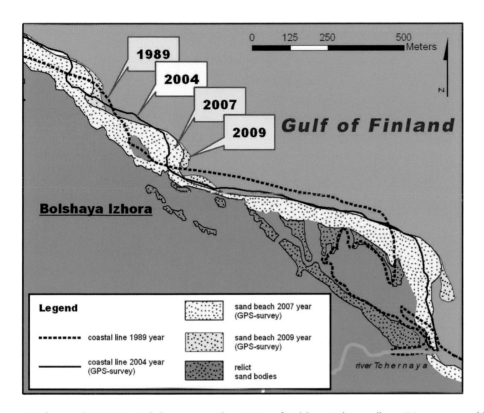

Labels on map: 1989, 2004, 2007, 2009, Gulf of Finland, Bolshaya Izhora, river Tchernaya

Legend
coastal line 1989 year
coastal line 2004 year (GPS-survey)
sand beach 2007 year (GPS-survey)
sand beach 2009 year (GPS-survey)
relict sand bodies

■ **Figure 2.13** *Coastal dynamics in the vicinity of Bolshaya Izhora village (Map prepared by A. Sergeev)*

40 per cent of the northern shore is heavily eroded and significant coastal protection measures have been carried out over the past three centuries along the eastern flank. Together with kilometres of embankments and many artificial islands, these measures should help to protect St. Petersburg against floods and storms. After the catastrophic flood of 1824, when a major part of historical St. Petersburg's centre was damaged, it was decided to protect the city from floods by engineering constructions. One proposed project was to build a stone dam from Cape Lisiy Nos to Kotlin Island, through the western part of the island and to Oranienbaum (Tillo, 1893). In 1979, a complex of Flood Protection schemes was commenced by building two dams between Kotlin Island and Neva Bay coasts (near Gorskaya in the north and Bronka in the south) with two navigation passes and several gates for water discharge. In the

early 1990s dam construction was interrupted due to economical and ecological problems, but in 2004 construction continued and flood protection was completed in 2011 and worked during two floods.

At the end of the nineteenth century, coastal protection structures were started along the northern Gulf coast, with the aim of stopping erosion and protecting sand beaches. Unfortunately, the hard engineering structures utilized (mostly groins) without sand nourishment were not effective due to a sediment deficit. In the case of groins, parts of the shore to the west of the construction can be stabilized, but to the east (the direction of sediment flow), erosion becomes more intense. For example, near Zelenogorsk (formerly Terijoki) since the construction in 1910 of a yacht-club harbour extending to about 100 m offshore, sand beaches east of the

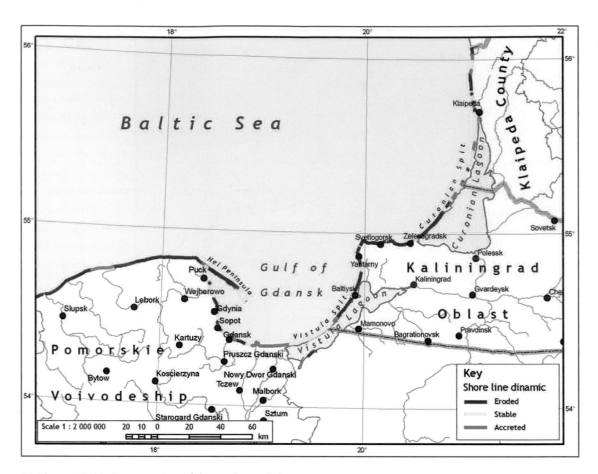

harbour have degraded and shoreline retreat has reached 90 m during the last 100 years (Figure 2.16). Currently, beaches east of the construction are completely washed out, shoreline recession is stopped by seawalls and other hard engineering constructions and its recreation value has drastically diminished (Ryabchuk *et al.*, 2009).

The only successful recent attempt at coastal protection was beach nourishment, realized in 1988 as a part of the First Coast Protection Plan (FCPP).

■ **Figure 2.15** *Temporal dynamics of segment lengths at Kaliningrad shores (Modified from Gilbert, 2008)*

Length of dynamic coastline

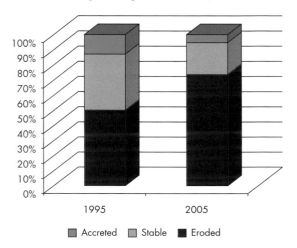

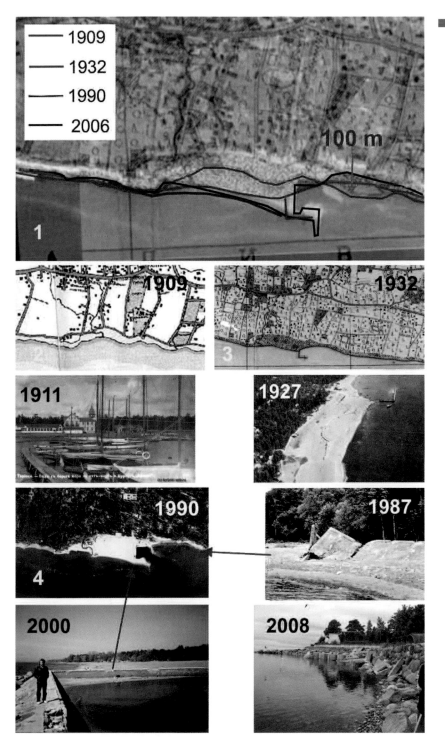

■ **Figure 2.16**
*Development of
the shoreline near
Zelenogorsk (former
Terijoki). 1: beach
plans; 2: 1909; 3:
1932 (www.terijoki.ru);
4: air-photo 1990
(Ryabchuk et al., 2009)
(Prepared by D.
Ryabchuk; published
with permission from
Estonian Journal of
Engineering)*

Along a 430 m long section of the Komarovo coast, an artificial beach 50 m wide and 2 m high was created, with a total sand volume of 32,250 m³. This was an experimental type of coastal protection, which was successful as it prevented even larger shore damage for two decades, but now is completely washed away.

Problems of coastal erosion are now becoming more and more important for regional authorities in St. Petersburg and Leningradskaya Oblast. The key roles of the FCPP were to consider further beach nourishment together with engineering structures, such as T-form groins or submarine breakwaters parallel to the shoreline and usage of artificial reefs as breakwaters. The programme has now been postponed due to lack of funding and absence of coastal legislation. The latter is needed, as Regional and Federal authorities have different responsibilities with respect to coastal protection and the offshore portion. An absence of coastal legislation leads to an increase of negative anthropogenic impacts, e.g. submarine sand exploration, unsustainable development of recreation facilities, etc.

South-east Baltic

In 1743, the first attempt of planting pines in order to fix moving dunes was made at Curonian Spit. Later, many different methods were used for stabilizing foredunes on the Curonian and Vistula spits. A first protective wall was built in 1887–1902 at Cape Taran, the most eroded coastal segment on the Kaliningrad shores. During the 1930s, protective walls (with a total length of 2.5 km) were constructed at the core part of the Curonian Spit, Svetlogorsk and Pionerskiy and the first groins were constructed near Baltiysk and Zelenogradsk at the end of the nineteenth century. During the first half of the twentieth century, groins were constructed around Svetlogorsk, Zelenogradsk, at the coastal segment from Svetlogorsk to Pionerskiy.

From the end of the nineteenth century, mining pulp was deposited into the sea near Yantarniy and up to the late twentieth century, 85–110 million m³ was deposited on the west shore. Such mass dumping

■ **Figure 2.17** *Old coastal protection constructions: permeable and segmented sea wall (partly destroyed) east of Pionerskiy, constructed in the 1960s*

in the sea affected coastal zone morphodynamics – new beaches were formed into two near-concave segments of the shoreline and slope wash moved sediment a distance of 200–800 m into the sea. This material migrated southwards, and stabilized the 25 km length of the coast uptil Baltiysk. After the cessation of dumping in the late 1990s, erosion started at this part of the shoreline (Burnashov *et al.*, 2010).

Kaliningrad's shore is characterized by longshore sediment drift; all coastal protection methods protect the shore immediately behind them but cause more intensive erosion further downstream (Figure 2.17). The first attempts to stop landslides from eroded cliffs were applied near Svetlogorsk in 1920–1930, but no coastal protective measures were applied during the 1945–1960 period and this continues. Activities to support foredunes at the Curonian and Vistula spits were made during the 1960s–1970s and a few coastal protection constructions and promenades were installed in Svetlogorsk, Pionerskiy and Zelenogradsk during the 1980s and 1990s.

Conclusions

Today, Russian coasts are under intensive development, especially in the Black Sea where the recreational industry has grown extremely quickly. There is a great deficiency of coastal areas suitable for recreational construction because of the mountainous relief of the Black Sea coasts. In former times, the greater part of the coast and part of the Russian Baltic Sea coast was protected by means of reinforced rigid concrete structures without paying attention to the resulting beach widths and to preservation/improvement of the coastal landscape. Installation of protective constructions in many limited segments stimulated even greater erosion processes downstream.

The value of a natural landscape has became well known, especially beaches – sand or gravel – as they are invaluable for any seaside region; their presence significantly increase recreational potential, and beach shortages limits resort capacity. Therefore, the common target of shore protection and recreationally oriented economic growth is to protect the land with beaches and not with concrete. However, on practically all Russian coasts there is a lack of beach-forming materials. The creation and broadening of beaches is possible only by artificial replenishment with borrow pit material, but constant replenishment is economically unprofitable. Artificial beaches under the protection of stone backfill retaining structures protect beaches from wave impact. Away from urban areas, the shoreline should be left in its natural condition.

In essence, the main problem currently is that constructing and sustaining artificial beaches, as well as structures for beach protection, costs too much even for state budgets and Russian coasts belong to the state.

References

Boldyrev, V. L. and Bobykina, V. P., 2008. Coasts of Curonian and Vistula Spits as transboundary territories. In B.V., Chubarenko (ed.), *Transboundary waters and basins in the South-Eastern Baltic*, 225–236, Kaliningrad: Terra Baltica.

Buhanovskiy A. V., Divinskiy, B.V., Kosyan, R. D., Lopatuhin, L. I. and Rozhkov, V. A., 2000. Black Sea wind-induced wave typification based on instrumental data. *Oceanology*, 40, 2, 289–297.

Burnashov, E., Chubarenko, B. and Stont, Zh., 2010. Natural evolution of western shore of a Sambian Peninsula on completion of dumping from an amber mining plant. *Archives of Hydro-Engineering and Environmental Mechanics*, 57, 2, 105–117.

Divinskiy, B. V., Kosyan, R. D., Podymov, I. S. and Pushkarev, O. V., 2003. Extreme waves in northeast part of the Black Sea in February 2003. *Oceanologiya*, 43, 6, 948–950.

Gilbert, C. (ed.), 2008. *State of the Coast of the South East Baltic: An indicators-based approach to evaluating sustainable development in the coastal zone of the South East Baltic Sea*, 160 pp, Gdansk: Drukamia WL.

Gurova, E., Chubarenko, B. and Sivkov, V. 2008. *Transboundary coastal waters of the Kaliningrad Oblast. / Transboundary waters and basins in the South-East Baltic*, ed. B. Chubarenko. Kaliningrad: Terra Baltica.

Kosyan, R. and Krylenko, M., 2009. Recent coastal processes of the Taman shore of the Black Sea. In *Proceedings of the 41th International Liege colloquium on Ocean Dynamics: Science-based management of the coastal waters*, 40–43, University of Liege, Belgium.

Kosyan, R., Krylenko, M., Petrov, V. and Yaroslavsev, N., 2005. Study of beach state and coastal protection in the neighborhood of Sochi ëity. In E. Ozhan (ed.), *Proceedings of the Seventh International Conference on the Mediterranean Coastal Environment*, 1007–1016, Turkey: METU.

Kosyan, R., Zhindarev, L., Chubarenko, B., Lukyanova, S. and Khanukayev, B., 2009. Modern coastal researches in Russia. In E. Ozhan (ed.), *Proceedings of the Ninth International Conference on the Mediterranean Coastal Environment*, 189–194, Ankara: METU.

Krivosheya, V. G., Ovchinnikov, I. M. and Titov, V. B., 1998. Meandering of the Basic Black Sea current and formation of whirlwinds in a northeast part of the Black Sea in summer 1994. *Oceanology*, 38, 4, 546–553.

Krylenko, M. V., 2009. Relief of the Black Sea Coastal Zone between Anapa and Novorossiysk as recreational factor. In E. Ozhan (ed.), *Proceedings of the Ninth International Conference on the Mediterranean Coastal Environment*, 463–468, Ankara: METU.

Leont'yev, I., 2008. Nearshore morphodynamics and forecast of coastal changes. In *Proceedings of the Second International Conference on the Dynamics of the Coastal Zone of Non-tidal Seas,* 96–103. Kaliningrad: Terra Baltica [in Russian].

Orviku, K. and Granö, O., 1992. Contemporary coasts. In A. Raukas and H. Hyvärinen (eds), *Geology of the Gulf of Finland*, 219–238, Tallinn: Valgus [in Russian].

Peshkov, V. M., 2003. *The Coastal Zone of the Sea*, Krasnodar: Lakont, 350 pp.

Ryabchuk, D., Sukhacheva, L., Spiridonov, M., Zhamoida V. and Kurennoy, D., 2009. Coastal processes in the Eastern Gulf of Finland: Possible driving forces and connection with the nearshore zone development. *Estonian Journal of Engineering*, 15, 151–167.

Ryabchuk, D., Kolesov, A., Chubarenko, B., Spiridonov, M., Kurennoy, D. and Soomere, T., 2011. Coastal erosion processes in the eastern Gulf of Finland and their links with geological and hydrometeorological factors. *Boreal Environment Research*, 16 (suppl. A), 117–137.

Sorokin, P. E., Gusentsova, T. M., Glukhov, V. O., Ekimova, A. A., Kulkova, M. A. and Mokrushin, V. P., 2009. Certain results of the research of the settlement Okhta 1: The Neolithic – the Early Metal Epoch. In, *Archaeological Heritage of Saint-Petersburg,* vol. 3, 205–222, St. Petersburg: Russian Research Institute for Cultural and Natural Heritage; North-Western Institute for Heritage Institute; Institute of the History of Material Culture, Russian Academy of Sciences. [In Russian].

Spiridonov, M., Ryabchuk, D., Kotilainen, A., Vallius, H., Nesterova, E. and Zhamoida, V., 2007. The Quaternary deposits of the Eastern Gulf of Finland. In, *Geological Survey of Finland, Special Paper* 45, 5–17.

Sukhacheva, L. and Kildjushevsky, E., 2006. Study of the eastern Gulf of Finland coasts on the basis of retrospective analysis of airborne and space data. In *Abstract of the VII International Environmental Forum: Baltic Sea Days*, 254–256, St Petersburg: Dialog Publishers [in Russian].

Tillo, E. I., 1893. Project of protection of Saint Petersburg against floods. In, *Reports of the Russian Geographical Society*, 25, 2, 63 pp [in Russian].

Zhindarev, L. and Kulakov, V., 1996. The Baltic Sea water level in Holocene. *Proceedings of the Russian Academy of Sciences, Geographical Series*, 5, 55–56 [in Russian].

3 Sweden

Magnus Larson and Hans Hanson

Introduction

The Swedish coast is dominated by rock shorelines (Bird, 1985; Bird and Schwartz 1985). Sand beaches are primarily found in the southern part (Lindh *et al.*, 1970; Rydell *et al.*, 2004), mainly in the county of Scania (Figure 3.1) but also to some extent in Halland, located on the Swedish west coast just above Scania. In Scania, sand constitutes the major source of material along about 200 km of shoreline, where stretches with sand beaches are mixed with beaches having a wide range of grain sizes, typically extending from very fine sand to boulders. Pocket beaches are common, where little material exchange occurs across boundaries, as well as straight, open coasts

with a predominant alongshore net littoral drift. Furthermore, beaches are often backed by sand dunes or cliffs of varying material, including silt and clay, that tend to erode when exposed to high waves and water levels. Although sand beaches are mainly found in the southern part of the country, the large islands in the Baltic Sea, Öland and Gotland, as well as the archipelagos found along several portions of the coast, also have such beaches. Furthermore, only limited sand shoreline stretches exist in northern Sweden and few protection structures have been built. Therefore, the focus will be on Scania where most sand beaches are found and where problems with sediment transport frequently occur, which call for different coastal protection types.

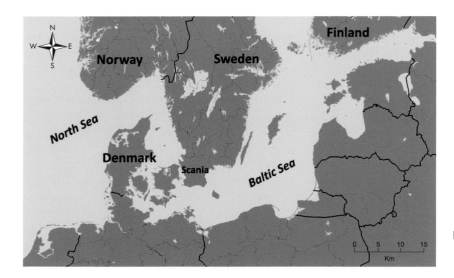

■ **Figure 3.1** *Location of Scania county in Sweden*

Presently, over 30 per cent of Swedish property is located within 100 m of the coast, implying that over 100,000 buildings are within a coastal zone defined by this width (Rydell, 2007). Therefore, the Swedish coast is developed along some 30 per cent of its total length and it has a total length of approximately 11,500 km. Along roughly 15 per cent of this length erosion is possible with regard to the geological conditions. Besides the potential threat to near-coast properties, erosion may significantly reduce the tourist value that many coastal areas have, implying a negative impact on the local economy of these areas. In a climate-change perspective, estimates show that 10–15 per cent of development along the coast may have to be protected in the future (Rydell, 2007). Although this may seem large, the cost of such protection is estimated as *circa* 2 per cent of the total value of development at risk, which is relatively low.

Coastal erosion

Rydell *et al.* (2004) performed a country-wide survey to identify shoreline stretches where erosion has been observed (for earlier surveys, see Lindh *et al.*, 1970; Hanson and Jönsson, 1978). Based on geological maps, stretches where erosion can potentially occur were also presented. Locations of observed erosion were mainly determined from questionnaires sent to coastal municipalities. Information obtained was primarily qualitative and the result was presented in terms of coastal maps where shoreline stretches suffering from erosion were marked. It should be pointed out that the reliability of available information may vary from one location to another depending on the person providing the information. Also, it was not possible to discern from the information, as to how long erosion has been occurring or its main cause. For Scania, a number of detailed studies have been performed that allow for detailed discussion of the erosion process and, in many cases, the typical rate.

Coastal erosion occurs primarily in three geographical areas with different geological and geomorphological settings, namely:

1. Along the Scanian coast and adjacent coastal areas (e.g., Halland to the north-west of Scania), where open beaches are prevalent (such beaches may also be found to a limited extent on the islands of Öland and Gotland in the Baltic Sea);
2. In the Swedish archipelagos formed close to the cities of Gothenburg, Stockholm, Piteå, and Luleå, where rock shorelines dominate but occasional pocket beaches or soft cliffs of erodible material may be found;
3. On the island of Gotland, where high erodible cliffs are often fronted by loose deposited material.

The conditions along the Scanian coast will be discussed in detail in the following section, so only the two latter erosional areas will be elaborated on here.

Several portions of the Swedish coast are fronted by archipelagos consisting of a multitude of islands of varying sizes. For example, the Stockholm archipelago, the largest one in Sweden and extending close to 60 km from the city itself, consists of about 30,000 islands. Along some of the island shorelines, sand beaches may be found where erosion has been observed, particularly in more sheltered inner parts of the archipelago. Limited wave energy can reach such areas and the possibility is greater for finer material to deposit there and form beaches. In certain locations, impact from wind-generated waves may even be of minor significance, whereas vessel-generated waves can be the main erosion cause. Within the archipelagos, there are often important shipping lanes frequented by large and fast-moving vessels.

Both in the Stockholm and Gothenburg archipelagos, erosion related to vessel-generated waves has been reported (Erikson *et al.*, 2000). Also, with the advent of high-speed vessels travelling at speeds that imply a Froude number above the critical value, the problem with erosion due to vessel-generated waves has increased, often making it necessary to introduce speed limits along sensitive shoreline stretches. Inventories from the archipelagos in Stockholm and Gothenburg show that the ratio of shoreline stretches along the shipping lanes that include erodible material is approximately 30 per cent and 15 per cent, respectively, the remainder being mostly rock shoreline (Allenström et al., 2003). These values include shoreline stretches with fine material, such as silt and clay, often deposited in cliff-type features of low elevation. The archipelagos around the cities of Piteå and Luleå also suffer from erosion, although at present erosion only occurs in limited areas.

Erosion has also been observed on the western side of Gotland island, particularly for shoreline stretches along the north-western portion where high cliffs are prevalent (Rudberg, 1967). The cliffs can reach up to 40–50 m above mean sea level, but in most cases the height is 20–30 m. Platforms, formed by waves that may extend 50–200 m in the seaward direction, typically occur seaward of the cliff (Philip, 1990). Loose material with a wide range of grain sizes is often found on the platform, and this material may form pocket beaches that are subject to erosion. The annual cliff retreat varies, but Rudberg (1967) reported 0.4–0.6 cm/yr as minimum values.

Overview of coastal erosion in Scania

Background

Scania's geological and geomorphological conditions are to a large degree determined by several glaciation periods. The last ice cover disappeared from Scania approximately 13,000 years ago and much beach material originates from glacial deposits. These deposits (moraine) are often poorly sorted with a wide range of material fractions containing a significant percentage of sand. However, deposits of fluvio-glacial well-sorted material are also found in the vicinity of the southern coast and supply beaches with material. Although the main portion of Sweden experiences a positive isostatic rebound, reaching almost 10 mm/yr in the northern parts, the Scanian beaches are at present exhibiting a relative sea level rise that vary from close to 1 mm/yr in southern parts to about 0 mm/yr in northern parts.

Scania is situated in the temperate zone with mild winters and cool summers. Severe storms with strong winds occur mainly in fall and winter, with a predominant direction from a sector south to west. Water level variations at Scanian beaches due to wind and differences in air pressure are pronounced because of the limited Baltic Sea water mass. Northerly winds tend to transport water to the southern Baltic Sea where Scania is located, which could increase the water level 1–1.7 m along the coast. Similarly, westerly winds transport water away from the southern part of the Baltic Sea, and low water levels occur that could be 1–1.5 m below Mean Sea Level (MSL). The tidal variation is negligible.

Short-period wind-generated waves are in general the main agent for determining beach change at an engineering time scale, which encompasses changes over days to centuries. Currents due to the large-scale Baltic Sea water movement only have minor influence on nearshore material movement, even though strong currents may occur in the Öresund Strait between Sweden and Denmark. At some beaches, wind-blown sand is of major importance for shaping the morphology, even causing problems because of excessive accumulation (VBB/VIAK, 1991). Nearshore wave conditions are in general fetch-limited with an average significant wave height of 0.4 m (including all wind directions) along the southern Scanian coast, as determined from a 16-year time series of wind data (Larson and Hanson, 1992). Significant wave heights above 2.0 m occurred about 2.5 per cent of the time and the maximum calculated wave height was 4.5 m. Dahlerus and Egermeyer (2005) obtained similar results in their wave climate hindcast for a 43-year-long wind time series. Larger maximum waves appear on the eastern side of Scania because of the increased fetch towards the Baltic States, whereas a milder wave climate prevails on the western side where the fetch towards Denmark is small.

Observed beach erosion

Figure 3.2 illustrates stretches of shoreline around Scania and adjacent areas where beach erosion has been documented during the latest decades (Lindh et al., 1970; Lindh, 1977; Lindh, 1979; Blomgren and Hanson, 1993; Blomgren, 1999; Rydell et al., 2004;

Dahlerus and Egermeyer, 2005; Brännlund and Svensson, 2005; Karlsson-Green and Martinsson, 2010; De Mas and Södergren, 2010). In general, long-term differentials in the longshore transport induce this erosion, especially in combination with storm events that shift material offshore from the upper part of the profile to the deeper part, making it available for longshore transport. Thus, erosion is generally not a continuous process in time but is often closely associated with such storm events. Furthermore, high water levels are required in combination with pronounced wave action for storms to cause significant erosion. For example, along the southern coast of Scania the most severe beach erosion typically occurs when at first northerly winds increase the water level in the southern Baltic Sea and then strong winds from the south-west induce major wave action along the coast. In addition, coastal erosion downdrift of obstacles blocking the longshore

transport – in some cases natural but typically man made – is quite common.

Therefore, beach erosion along the Scanian coast may be classified into one of the three following categories: (1) beaches where man-made structures disturb equilibrium conditions, typically in a pocket beach; (2) stretches of open coast that are locally out of equilibrium with the prevailing wave climate, and where the influence by man is secondary; and (3) dunes or cliffs that are attacked during storms when high wave and water levels prevail. The problems occurring in connection with shoreline adjustment to the local wave conditions are the classical ones involving accumulation of material in navigation channels and harbour entrances, erosion and loss of recreational beaches downdrift of harbours and man-made structures, and property loss due to shoreline retreat.

Examples of pocket beaches where man-made structures have affected the equilibrium of beach

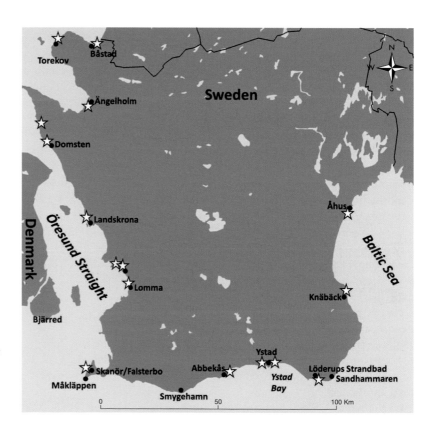

■ **Figure 3.2** *Examples of beaches around the study area of Scania where significant beach erosion is occurring*

conditions, causing significant downdrift erosion, are Båstad, Torekov, and Ängelholm (for locations, see Figure 3.2). The direction of net longshore transport varies between the different beaches depending mainly on shoreline local orientation and wave conditions. For example, beach change indicates a transport direction to the south at Ängelholm, whereas material is transported to the east at Båstad (Figure 3.3). The beach at Ängelholm has lost several 100,000 m³ since the middle of the last century on the southern side of two jetties constructed at a river mouth. Båstad and Torekov beaches contain material considerably coarser than sand that is not transported by the waves, which locally could produce an armouring effect as exposed finer material is transported away.

Beach erosion in the Öresund Strait between Denmark and Sweden, for example Domsten and Landskrona, is in general of a smaller magnitude because the fetch is short (Karlsson-Green and Martinsson, 2010; De Mas and Södergren, 2010). In Bjärred and Lomma, a main concern is dune and cliff erosion. At the southern end of the Öresund Strait there is a sand spit formation (Skanör/Falsterbo) that experiences very active sand transport with marked net transport to the north along the western side of the tip. The harbour at Skanör, constructed in 1860, acts as a barrier to north-going transport and has induced severe downdrift erosion. On the southern

side of the spit the transport is to the west, and the sand island Måkläppen, a bird sanctuary, is slowly becoming permanently attached to the spit (Blomgren and Hanson, 2000).

The most severe beach erosion cases are found along the southern coast of Scania, with Löderups Strandbad as the worst example. Along this beach, the shoreline has retreated 1–2 m/yr during the last 200 years, causing severe loss of property. Erosion at Löderups Strandbad is primarily induced by a local imbalance between the prevailing wave climate and the main shoreline orientation, although improperly designed shore protection structures have made the erosion worse. Ystad Bay is a good example of a crescent-shaped bay that constitutes a closed system with respect to sand transport (Larson and Hanson, 1992; Dahlerus and Egermeyer, 2005). The bay is not in complete equilibrium and there is an eastward transport in the western part of the bay, causing erosion along bay stretches that is quite sensitive to shoreline retreat. In the 1990s the main sewage treatment plant pipe in the western part of Ystad Bay was seriously threatened by erosion and about 200 m of pipe exposed, since remedied by a series of revetments. Figure 3.4 shows the beach in Ystad just after the dunes have been eroded by waves and high water levels. Beaches at Smygehamn and Abbekås are examples of eroding beaches with a wide range of grain sizes, and where coarser material functions as

■ **Figure 3.3** *Båstad marina with accretion occurring updrift (right) and erosion downdrift (left) (Photograph courtesy of Johan Stenlund)*

■ **Figure 3.4** *Coastal dunes in Ystad eroded by waves and high water levels*

armouring as exposed sand has been transported away.

On the eastern Scanian coast, marked erosion occurs primarily at Knäbäckshusen beach where shoreline retreat at present is 1–2 m/yr. This beach is unique for Scania, since the eroding cliffs consist mainly of cohesive fluvio-glacial deposits with no tendency to build beaches. Eroded material is lost into deeper water and no pronounced beach exists that can protect against wave attack. The glacial material has been deposited in the form of vertical cliffs with a typical height of 3–4 m, which implies little resistance to erosion.

Overview of coastal protection

Measures to combat coastal erosion in a systematic manner are recent. An important reason for this was that wide-spread recreational use of coastal areas started to develop only in the 1950s, which then accentuated the conflict between natural shoreline evolution and restrictions imposed by man-made activities and infrastructure. In order to resolve this conflict, various measures were introduced to control shoreline evolution. So far, the implemented measures have almost exclusively been hard structures, but soft solutions such as beach nourishment are now being carried out at several locations.

In the 1950's, a system of many short groins were built in Ystad (Figure 3.5), but since they were not very effective, a system of four longer groins was constructed in the 1960s. This has proved to better stabilize the shoreline, although local areas of erosion and accretion have occurred adjacent to individual groins (Dahlerus and Egermeyer, 2005), as well as downdrift of the groin system. Figure 3.4 shows a small groin in the foreground with a longer groin behind it. The longer groins are also used as summer swimming piers, when a superstructure is temporarily placed on the pier. An additional groin was built in the 1990s. Groins at a small scale have also been employed at some locations around the Skanör-Falsterbo Peninsula and in the north-east part of Scania.

Seawalls and revetments were used early on to stabilize the shoreline along stretches suffering from erosion because of gradients in the longshore

■ **Figure 3.5** *Ystad beach in 1952 where the many small groins are shown in the background. Some of the groins are still there on the previous figure, 55 years later! (Photograph source unknown)*

transport or direct wave attack on dunes and cliffs at high water levels. Some of the first efforts to employ such structures were made by private property owners who dumped stone illegally without any design considerations (Figure 3.6). Seawalls constructed in this manner suffered from excessive subsidence of the stones, making it necessary to carry on with dumping until all sand had disappeared and an extensive wall of stone existed that could dissipate incoming wave energy. Simultaneously, the water depth seaward of the wall increased so that larger waves could directly attack the wall. The protective structure created provided a coastal environment that was quite unattractive and had little access for people. Although stone dumping was illegal, the authorities did not enforce the law but quietly accepted the approach taken by owners to protect their property.

Revetments have been used with considerable success in areas with a milder wave climate and where

a wide range of grain sizes is present. In these areas the upper foreshore, often backed by a low cliff, contains fine material that easily erodes when exposed to waves during high water levels. This material is carried offshore by currents and will not be deposited in the nearshore area. By introducing a revetment, the sensitive parts of the foreshore are protected and erosion prevented. The revetment is typically not solid, but provided with drainage holes where vegetation can establish itself and hydraulic pressure may be released, which contributes to a pleasanter environment since the hard surface is covered by vegetation. Problems with seawalls and revetments are related to boundary effects, *i.e.* erosion at the foot of, or downdrift of, the structure and, in some cases, overtopping.

Gabions, cages filled with stones, are occasionally used to strengthen the dunes against erosion. Under normal wave and water level conditions the gabions

■ **Figure 3.6** *Löderups Strandbad in 1984 with sections of rock revetments built shortly before this date.*

remain buried, but during storms they may become exposed as sand is eroded away. They act as a kind of seawall under such conditions, maintaining the shoreline position and protecting the dune behind the structure. After a storm, the gabions typically need to be covered by sand again, unless they are located sufficiently close to the swash zone so that accretionary waves may cover them. In some cases the cages have not been properly sealed and during wave impact have opened and stones fallen out. When this happens, the gabions will not provide their protective function. In Ystad, a coastal section where the main sewage treatment plant pipe was exposed was protected with gabions in the early 1990s and has worked satisfactorily since that date.

Detached breakwaters (DBW) have only been used in a few cases in Sweden for the purpose of coastal protection. In the early 1990s three DBWs were constructed close to Borstahusen north of Landskrona, in order to reduce longshore sediment transport in the area. The DBWs have been quite effective in trapping sand; however, the amount of sand transported in the area is limited, and most sand has been trapped immediately behind the DBWs with limited shoreline impact.

Beach nourishment has rarely been applied in Sweden as a measure to combat erosion. The city of Ängelholm, located at the mouth of the Rönne River, which is stabilized with two jetties, has undertaken bypassing of sand from the northern to the southern side. The net drift is southwards and erosion occurs some distance downdrift of the river mouth. The sand moved from the area north of the river mouth is periodically placed in the eroding area. Some municipalities occasionally move sand from the outer part of the profile to the foreshore or dune region, typically in the spring after winter storms have finished. From an overall sediment budget perspective, this may not be an effective method, but it could help maintain the dunes. Several projects involving more extensive beach nourishment placement are under way, and approximately 100,000 m^3 is presently being placed on beaches in Ystad and Löderups Strandbad. However, the permitting process has been excessively complicated since permitting authorities have lacked basic knowledge on these types of projects.

Selected case studies of beach erosion

The beaches of Skanör, Ystad, Löderups Strandbad and Lomma illustrate well the previously discussed categories of Scanian beaches suffering erosion as typically encountered in south Sweden.

Skanör

Skanör is located on the western side of a sand spit in the south-west corner of Scania (Figure 3.2), where net sand transport is to the north (Hanson and Larson, 1993). A harbour was constructed on this beach around 1860, but initially access was through a causeway that did not markedly block sand transport. However, after the causeway was turned into an embankment, in about 1913, serious downdrift erosion started to occur north of the harbour and substantial accumulation south of the harbour (Blomgren, 1999). Another much discussed activity that may have affected erosion at Skanör previously was marine sand extraction performed at the northern end of the spit. Extraction stopped in 1992 when the

area became a natural park. Erosion on the northern side of the harbour currently seriously threatens a beach stretch that is of considerable recreational value and also offers protection to the infrastructure behind. Simultaneously, pronounced accumulation of sand occurs south of the harbour requiring periodic dredging of the entrance channel.

The spit formation is of post-glacial origin and a detailed description of the geological history and conditions of the spit is given by Davidsson (1963). The approximate net transport of sand to the north along the western side of the spit is estimated to be 40,000 m^3/yr and about half of this amount is transported past the harbour (VBB/VIAK 1991; Hanson and Larson, 1993). There is an offset in the shoreline position between the harbour south and north side of about 300 m at present (Figure 3.7). Sand transport is mainly induced by waves from the south and south-west, where fetches are longest, but the wave height seldom exceeds 2 m at Skanör because of limited fetch and shallow water. Since a large portion of the spit is at a low elevation, high water levels could create considerable flooding with property damage and overwash of the low-relief sand

■ **Figure 3.7** *Offset in shoreline position on either side of Skanör harbour*

dunes. A water level increase of 122 cm is exceeded on average every ten years and an increase of 145 cm has a return period of 50 years. In connection with climate change and expected sea level rise, grave concern has been expressed with regard to future spit erosion and the potential flooding damage (Lindh *et al.* 1989; Blomgren and Hanson, 1997; Pakkan, 2006).

Hanson and Larson (1993) studied beach evolution at Skanör and carried out profile measurements along four survey lines from 1987 to 1992, where two lines are located south of the harbour (#1: 2,500 m and #2: immediately updrift) and the other two lines north of the harbour (#3: 400 m and #4: 1,000 m). The same lines were measured again in 2006–2009. These lines were surveyed approximately every spring and fall, and manual observations of wave conditions were done regularly. Lines 1 and 2 displayed slight accumulation during the entire measurement period 1987–2009 with an average movement of the shoreline of +2.2 m/yr and +0.6 m/yr, respectively, whereas lines 3 and 4 displayed shoreline recession of −2.0 m/yr and −3.2 m/yr, respectively. Figure 3.8 displays four of the surveyed profiles for the northern-most line 4, and a marked retreat of the low-crest dune is noted. Because of the low crest dune, elevation overtopping frequently

occurs and the area behind the dune floods frequently.

Ystad

The city of Ystad is located on the western side of Ystad Bay, which constitutes the largest littoral cell on the south Swedish coast (see Figure 3.2). Larson and Hanson (1992) investigated the large-scale sediment transport pattern in the bay and identified areas of erosion and accumulation based on gradients in the derived transport pattern. The study focus was the longshore transport and associated shoreline evolution. Calculations indicated the most severe erosion should occur in the western part of the bay, which agrees with observations. In this area several different measures have been employed to combat erosion, including groins and revetments. Long-term data on shoreline positions from the 1950s to the present show that the eastern part of the bay is quite stable and most changes are found in the western part.

The calculated transport direction in the eastern part of the bay is westwards, whereas the opposite prevails in the western part. This implies a convergence point for sediment in the central part of the bay (*i.e.*, zero transport) about 8 km east of Ystad, close to the mouth of the small Kabusa River, where

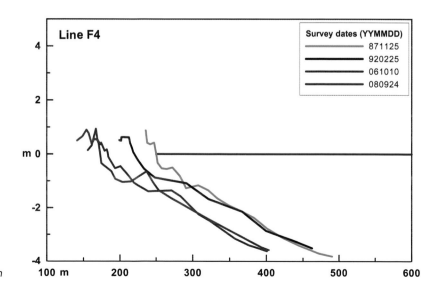

■ **Figure 3.8** *Profile measurements north of Skanör's harbour, where sand blocking contributes to the erosion*

the shoreline orientation is *circa* 30°. The point where no shoreline change occurs (the transport gradient is zero) was estimated to be located about 3 km east of Ystad. Thus, at the large scale, erosion is expected for less than 3 km whereas accumulation takes place for more than 3 km. Deviations from this pattern occur in areas with structures, for example, groins, which induce local erosion and accumulation in the vicinity of the structure. Based on the derived transport pattern and inferred shoreline change, together with information on land use, development potential, and regulations, Larson and Hanson (1992) developed a classification of the shoreline stretches in Ystad Bay with a focus on vulnerability towards erosion and flooding.

Several stretches of shoreline are backed by well-developed sand dunes, which may suffer severe erosion during storms with high waves and water levels (Figure 3.4). Dahlerus and Egermeyer (2005) examined dune response in Ystad Sandskog to storm impact based on available data and mathematical modelling (Larson *et al.*, 2004). These dunes are among the most exposed in Ystad Bay and if breaching occurs low-lying areas behind the dunes will sustain major damage. Predictions were also made regarding the effects of climate change on dune erosion, where the focus was on sea level rise estimated from available scenarios. For the present situation, dune breaching is not likely to happen, but with a rising sea level, such events will have a significantly higher probability of occurrence. Dune erosion was calculated to increase with about 25 per cent until 2100 for the average sea level rise scenario (+38 cm), and with 75 per cent for the most extreme scenario tested (+85 cm). Recently the permitting process was successfully concluded for a beach nourishment project, where the objective is to restore the 1993 shoreline location by placing 80,000 m^3 of materials on the beach.

Löderups Strandbad

This is an example of a case where the local shoreline orientation is out of equilibrium with the prevailing wave climate. This disequilibrium has existed for at least 150 years and erosion will most likely continue until shoreline orientation becomes similar to the orientation in the eastern part of Ystad Bay (Larson and Hanson 1992), which approximately corresponds to a further shoreline retreat of 300 m. Erosion has caused severe losses of property and several houses have fallen into the sea. Between 1950 and 1975 the shoreline retreated on average 2–2.5 m/yr (Lindh, 1979), and during the three first months of 1990 almost 20 m of beach was lost. In the eastern part, shoreline retreat between 1968 and 1993 was on average 4 m/yr. Although erosion is continuous at Löderups Strandbad, large storm events severely accelerate the shoreline retreat rate. During these events, waves in combination with high water levels induce offshore transport from parts of the profile typically not exposed to waves, making material available for longshore transport. Poorly designed shore protection measures have contributed locally to an increased erosion rate.

The marked shoreline retreat at Löderups Strandbad is clearly illustrated in Figure 3.9. Net sand transport direction is to the east and material eroded at Löderups Strandbad is deposited on the southeastern tip of Scania (Sandhammaren). The average net sand transport has been estimated to about 66,000 m^3/yr from calculations based on hindcasted waves (Holmberg and Åhnberg, 1977), and average yearly transport rates derived from measured shoreline retreat varies from 30,000–60,000 m^3/yr (Larson and Hanson, 1992; Blomgren and Hanson, 1993). Larson and Hanson (1992) computed transport rates from a 16-year time series of hindcasted waves with an eight-hour time resolution using the CERC formula (SPM, 1984). The transport rate coefficient K in the formula was calibrated to yield a mean yearly net transport rate of 60,000 m^3/yr, implying K= 0.30. The yearly net transport rate varied from −55,000 to 115,000 m^3/yr, and the average gross transport was 215,000 m^3/yr.

Rock revetments started to be built in the western part in the mid 1970s by private property owners with only limited knowledge about coastal processes or protection. As erosion developed downdrift of the structure, the revetments were gradually extended

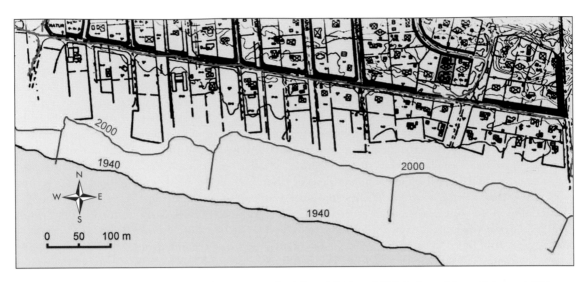

■ **Figure 3.9** *Coastal erosion at Löderups Strandbad from 1940 to 2000 (adapted from Holte and Ohlsson, 2000)*

and reached in the early 1990s about 1.5 km in length (Figure 3.10). The depth in front of the revetment had increased drastically and the municipality decided in 1993, in an attempt to restore the depth and the integrity of the revetment, to build a series of six groins along the structure. This project has been successful with respect to its objectives, but has at the same time resulted in dramatic erosion inside the National Park just downdrift of the protected coast. As a next step, the municipality will perform beach nourishment in front of portions of the revetment.

Lomma

The main problems concerning coastal erosion in the municipality of Lomma are retreat of cliffs at Bjärred and of dunes near the town of Lomma. Erosion takes place when the water level is high in the Öresund Strait, which typically occurs for northerly or easterly winds. Such winds push water from the northern to the southern part of the Baltic Sea, inducing an increase in the water level. Waves on the west Scanian coast however, are low for these wind directions, but if the wind direction is rapidly turning south or south-west, larger waves may be generated,

creating a situation that causes dune and cliff erosion. In order to combat erosion at Bjärred, the municipality has mainly constructed different types of revetments to stabilize the foreshore and the lower cliff portions. These revetments have been rather successful, except for minor boundary effects with erosion downdrift of the structures and at their toe.

Brännlund and Svensson (2005) investigated the cause of erosion in Bjärred and quantified the effect of wave and water level variations with respect to cliff retreat. An analytical model (Larson *et al.*, 2004) was employed to estimate the retreat rate, using historical aerial photos to determine the optimal value of the main transport coefficient in the model. The average retreat rate between 1960 and 2004 was about 0.2 m/yr. Relatively few events were calculated to affect the cliff in a year (about four), but the impact of these events was marked. The net longshore transport at Bjärred is towards the south with a rate of approximately 20,000 m³/yr (De Mas and Södergren, 2010). This number assumes that the material transported is mainly sand, whereas in reality the area exhibits a wide range of grain sizes, both finer and coarser than sand. The cliff contains a substantial amount of silt and clay that is transported offshore

■ **Figure 3.10**
*Rock revetment – 1.5 km
long in Löderups
Strandbad in 1990*

■ **Figure 3.11** *Concrete slab revetment in Bjärred
(Photograph by Johan Södergren)*

after being eroded from the cliff. De Mas and Södergren (2010) carried out detailed modelling of the longshore sediment transport and shoreline evolution, both at the regional and local scale, using a one-line model (Hanson, 1989). Furthermore, using additional data on the evolution of the shoreline and beach profiles in the area, cliff erosion rates were revised and estimates developed for a changing climate until 2070. Depending on the scenario studied, the number of cliff erosion events may increase substantially in the future with a mean eroded volume that is more than 200 per cent larger for the most severe scenario. De Mas and Södergren (2010) also reviewed all previously utilized coastal protection measures in Bjärred, mainly different types of revetments (Figure 3.11), and investigated a number of possible future measures (*e.g.*, detached breakwater, revetments, beach nourishment), assessing their potential from a functional, economical, and environmental point of view.

North of Lomma there are well-developed dunes that typically erode during fall and winter storms. The municipality periodically moves sand from nearshore areas to the dune face to build up the dunes again in the spring. Along the most exposed areas, gabions have been placed at the dune foot to stabilize the dune line. Behind the dunes there is a small road followed by a large dam, which was created through commercial extraction of clay but today functions as a bird sanctuary. If the dunes are breached, the road will be damaged and the dam will be affected by intruding sea water. The water level in the dam is 1 m below MSL and it is maintained at this level through pumping in order to protect a main road located shoreward of the dam. The potential net longshore transport in the area is small, but directed northwards (De Mas and Södergren, 2010).

Conclusion

Remedial measures have recently being implemented or are being discussed for several Scanian beaches that experience erosion. A groin field in combination with a minor beach fill was selected as the most cost-effective solution to halt erosion at Löderups Strandbad. In Ystad a fifth groin was added to the existing groin field. VBB/VIAK (1991) suggested bypassing of sand to reduce downdrift erosion at Skanör harbour, but this was never implemented. Different types of revetments have been constructed to protect the sewage treatment plant west of Ystad Bay and detailed field measurements are made of the beach evolution in the vicinity of the revetments to evaluate their performance. Blomgren and Hanson (1993) evaluated the efficiency of groins for reducing erosion at various beaches around Scania.

Therefore, beach protection in Sweden has up until now been very structure oriented. Much work carried out in the 1950s through the 1970s was actually done by private property owners and local municipalities, generally without any permits. From the 1980s and onwards the permitting process has been more rigorous, but still focused on structures. A number of beach nourishment projects have been submitted to the permitting authorities, but until recently without success. Authorities have been very conservative regarding 'new' methods such as nourishment. However, though a persistent effort from the coastal engineering community – researchers as well and practitioners – this attitude is now about to change. An ongoing application from the city of Ystad for a major nourishment project was recently approved and the authorities seem to have finally accepted beach nourishment as a technically and economically efficient measure.

A general problem concerning shore protection projects in Sweden is the unclear responsibility regarding activities in the nearshore zone. There is no special agency to coordinate and direct such activities, but responsibility is divided between several government agencies which tend to reduce the possibility of carrying out efficient measures against erosion. Furthermore, financing of shore protection project becomes complicated with difficulties in dividing costs between the private owner, local community and the government. As a general rule, the government is not allocating any money for coastal protection. Some local communities would only provide financial support for projects on their own properties, whereas others would support also private property owners if the project is regarded as 'politically motivated'.

Acknowledgements

The hard work of many dedicated Master's students contributed significantly to the above. The work of the authors was mainly supported by the Swedish Research Council Formas under Contract No. 214-2009-802. The photographs provided by Johan Stenlund and Johan Södergren are greatly appreciated.

References

Allenström, B., Bergdahl, L., Erikson, L., Eskilsson, C., Forsman, B., Hanson, H., Johansson, J. Johansson, L. and Svensson, U. 2003. *The interaction of large and high-speed vessels with the environment in archipelagos: Final report*, unpublished project report, Research Council VINNOVA, Stockholm, Sweden.

Bird, E. C. 1985. *Coastline Changes: A Global Review*, John Wiley and Sons, New York.

Bird, E. C. and Schwartz, M. L. 1985. *The World's Coastline*, Van Nostrand Reinhold Company, New York.

Blomgren, S. 1999. *Hydrographic and Morphologic Processes at Falsterbo Peninsula: Present Conditions and Future Scenarios*, Report No. 1027, Department of Water Resources Engineering, Lund University, Lund, Sweden [in Swedish].

Blomgren, S. and Hanson, H. 1993. *Coastal Erosion in Nine Scanian Communities: Damages and Remedial Measures*, unpublished report, Department of Water Resources Engineering, Lund University, Lund [in Swedish].

Blomgren, S. and Hanson, H. 1997. Future High Sea Levels in South Sweden, *World Resource Review*, 9(4), 476–490.

Blomgren, S. and Hanson, H. 2000. Coastal Morphology at the Falsterbo Peninsula, Southern Sweden, *Journal of Coastal Research*, 16(1), 15–25.

Brännlund, I. and Svensson, L. 2005. *Coastal erosion in Bjärred, Lomma Municipality: An Investigation of Impacts from Waves on Water Level*, Master's Thesis LU:2005-3, Department of Water Resources Engineering, Lund University, Lund [in Swedish].

Dahlerus, C.-J. and Egermayer, D. 2005. *Runup and Dune Erosion along the Coastline in Ystad Bay: Present Situation and Future Scenarios*, Master's Thesis LU:2005–11, Department of Water Resources Engineering, Lund University, Lund [in Swedish].

Davidsson, J. 1963. *Littoral Processes and Morphology on Scanian Flat-Coasts*, Dissertation XLII, Department of Geography, Lund University, Lund.

De Mas, C. and Södergren, J. 2010. *Modeling Coastal Erosion in Bjärred, Lomma Municipality. Long-Term Evolution and Protective Measures*, Master's Thesis TVVR 10/5024, Department of Water Resources Engineering, Lund University, Lund.

Erikson, L., Larson, M. and Hanson, H. 2000. *Coastal Erosion from Waves Generated by Large and Fast Vessels Traveling in Restricted Areas: Background and Preliminary Findings*, Report No. 3232, Department of Water Resources Engineering, Lund University, Lund, Sweden.

Hanson, H. 1989. GENESIS: A Generalized Shoreline Change Numerical Model, *Journal of Coastal Research*, 5(1), 1–27.

Hanson, H. and Jönsson, L. 1978. *Nearshore Sediment Transport Problems: Methods of Investigation and Analysis*, Report No. R39:1978, Swedish Council for Building Research, Stockholm, Sweden [in Swedish].

Hanson, H. and Larson, M. 1993. *Sediment Transport and Shoreline Evolution at Skanör/ Falsterbo*, Report No. 3166, Department of Water Resources Engineering, Lund University, Lund, Sweden [in Swedish].

Holmberg, R. and Åhnberg, H. 1977. *Beach Erosion at Löderups Strandbad*, unpublished report, Department of Water Resources Engineering, Lund University, Lund, Sweden [in Swedish].

Holte, S. and Ohlsson, K. 2000. *Löderups Strandbad 100 years*, unpublished report, Department of Water Resources Engineering, Lund University, Lund, Sweden [in Swedish].

Karlsson-Green, M. and Martinsson, S. 2010. *Flooding and Coastal Erosion in the City of Landskrona. An Overview with Regard to Climate Change*, Master's Thesis TVVR 10/5002, Department of Water Resources Engineering, Lund University, Lund [in Swedish].

Larson, M. and Hanson, H. 1992. *Analysis of Climatologic and Hydrographic Data for the Bay of Ystad*, Report 3159, Department of Water Resources Engineering, Lund Institute of Technology, Lund University, Lund [in Swedish].

Larson, M., Erikson, L. and Hanson, H. 2004. An analytical model to predict dune erosion due to wave impact, *Coastal Engineering*, 51, 675–696.

Lindh, G. 1977. *Aspects of the Beach Erosion Problem in South Sweden*, Report 3009, Department of Water Resources Engineering, Lund University, Lund, Sweden.

Lindh, G. 1979. *Erosion Processes and Restoration Attempts on the Scandian Coast*, Report 3027, Department of Water Resources Engineering, Lund University, Lund, Sweden.

Lindh, G., Andren, Y. and Bodlund, K. 1970. *Beach Erosion Problems on the Scanian Coast*, Report Series B, No. 9, Department of Water Resources

Engineering, Lund University, Lund, Sweden [in Swedish].

Lindh, G., Hanson, H. and Larson, M. 1989. Impact of Sea Level Rise on Coastal Zone Management in Southern Sweden, *Conference on Climate and Water*, Vol. 2, World Meteorological Organization, Helsinki, Finland, 128–147.

Pakkan, M. 2006. *Extreme Water Levels and Wave Run-Up in Falsterbo Peninsula, Sweden*, Master's Thesis LU:2006-2, Department of Water Resources Engineering, Lund University, Lund, Sweden.

Philip, A. L. 1990. Ice-Pushed Boulders on the Shores of Gotland, Sweden, *Journal of Coastal Research*, 6(3), 661–676.

Rudberg, S. 1967. The Cliff Coast of Gotland and the Rate of Cliff Retreat, *Geografiska Annaler, Series A, Physical Geography*, 49(2/4), 283–298.

Rydell, B. 2007. Adapting Developed Areas to a Future Climate in order to Reduce Natural Disasters, *Civil Engineering*, 5, 16–19 [in Swedish].

Rydell, B., Angerud, P. and Hågeryd, A.-C. 2004. The Extent of Coastal Erosion in Sweden: Overview of erosional conditions, *SGI Varia 543*, Swedish Geotechnical Institute, Linköping, Sweden [in Swedish].

SPM. 1984. *Shore Protection Manual*, Vol. I, Coastal Engineering Research Center, U.S. Army Waterways Experiment Station, Vicksburg, MS, USA.

VBB/VIAK. 1991. *Artificial Bypassing of Sand at Skanör Harbor*, Consultant Company VBB/VIAK AB, unpublished report, Malmö, Sweden [in Swedish].

4 The Baltic States

Estonia, Latvia and Lithuania

Hannes Tõnisson, Kaarel Orviku, Jānis Lapinskis, Saulius Gulbinskas and Rimas Žaromskis

Overview

The three Baltic States (Estonia, Latvia and Lithuania) are among the smallest countries in Europe, but the shoreline length is remarkable – approximately 4,500 km. Moreover, due to their geographical location between major geological structures, the Fennoscandian Shield and East European Platform, they are rich in different shore types and valuable coastal ecosystems. All shores have been significantly influenced by changing sea levels as a result of different Baltic Sea levels during the last 10,000 years. Postglacial isostatic rebound since the Ancylus Lake period has resulted in up to 75 m of land uplift on the northern coast. The zero isobase of this land uplift runs SW–NE through Riga in Latvia and as a result, beaches formed north of that line are frequently backed by series of beach ridges and dunes often reaching tens of kilometres inland.

The composition of most Lithuanian and Latvian coastal sediments range from pebbles to silt. The variability is mostly in glacial till and the silt shores increase northwards towards Estonia. In addition to till shores along the whole of the Estonian coast, many limestone cliffs can be found on its northern coast and at Saaremaa Island.

The Latvian and Lithuanian Baltic Sea (exposed to the Baltic Sea proper) coast has been straightened by prolonged marine erosion and accumulation and favoured by the non-existent land uplift. The Gulf of Riga Latvian coast is generally straight but includes some gently sloping forelands, separated by shallow 4–6 km long bays. In contrast over two-thirds of the Estonian coast is crenulated in outline, with capes and bays being either hard bedrock or unconsolidated Quaternary deposits, notably glacial drift with the cliffed north-eastern coast being straightened by erosion. The north-western shore in Hiiumaa Island, and also the Gulf of Livonia, has been straightened by a combination of erosion and deposition whereas the beach-fringed Narva Bay has a smooth outline as a result of sediment accumulation.

Prevailing winds, and therefore most waves on the eastern coast of the Baltic Sea, approach from the south-west, which is reflected in sediment transport patterns from south to north (except the western coast of Cape Kolka). Those patterns are clearly visible from ancient coastal formations which can be found along the Baltic States coast. The longshore sediment drift is also known as the Eastern Baltic Longshore Sediment Flow and has existed since the late-Littorina period.

Coastal erosion, a global phenomenon, is also currently one of the most important problems of the Latvian and Lithuanian coast, but a much smaller problem in Estonia due to the continuous land uplift and diverse geological structure. The fact that all three countries were considered as a border zone for the old Soviet Union, decreased the importance of erosion to human activities, as for about 50 years it was not permitted to live or construct non-military buildings close to the shore. However, all three countries have some significant similarities as to the main agents causing coastal erosion. There are no tides but wind-induced storm surges can still range over 4 m in Estonia, 3 m on the southern coast of the Gulf of Riga and over 2 m on the open Baltic Sea coast (Eberhards, 2003), with the most drastic erosion events taking place during those extreme storm surges. The effect of winter storms is affected by cold season climate warming and amelioration in ground frost and ice conditions both in the open sea and nearshore zone.

The northern part of the Lithuanian coast currently experiences rapid erosion, reaching 60–70 m during the last 70–80 years in some sections (Gudelis, 1998). Erosion is also evident on the lagoon side of the Great Curonian dune ridge. As a result of a longer vegetation period, stronger westerly winds and forestation, the height of many active dunes has decreased and the Curonian Spit has shifted lagoonwards (Povilanskas *et al.*, 2009). The southern part of the Lithuanian coast is stabler and accumulation has reached up to 80–90 m during the last 100 years (Gudelis, 1995).

In stretches of low depositional Latvian coast which have coastal dunes, major storms, occurring once or twice a decade, can cause erosion of up to 30 m^3/m of sand material, while along certain stretches of high cliffs, this figure can reach up to 70 m^3/m. Increased erosion is connected directly or indirectly with human activities, e.g. harbour and coastal sea defences, and reduced river sediment discharge as a consequence of hydroelectric power plants.

There are no recent major projects along the Estonian coast concerning coastal erosion. The few that exist are privately funded or project based and exhibit localized measures. Similar situations can be found on the Latvian coast, most protection measures being established during the Soviet Union period. For example, there have not been any major foredune restoration projects in Latvia in recent decades – one of the most useful measures to protect the shores (Eberhards, 2003) – but only small localized

protection measures, such as dune stabilization. The largest recent coastal engineering projects are related to the big harbours (Ventspils, Liepaja and Riga) and Jurmala beaches. Coastal erosion in Lithuania has been severely aggravated by human intervention such as construction of hydro-technical works, deepening of the Klaipéda harbour and recreational activities. The biggest problems appear to lie in popular tourist resorts in Palanga, where the sand beach has been almost completely eroded in the last ten years. To fight coastal erosion, all forests and dunes of the coastal zone have been classified as 'protected and preserved' according to the *Lithuanian Law of the Coastal Strip* since 2002. Furthermore, coastal forests cannot be cut down unless they are situated more than 1 km away from the coastline. Various soft protection measures have been carried out during the last decade including beach nourishment and dune stabilization. However, those measures suffered from a lack of funding until 2007, after which EU funds have been available to help extend soft protection measures.

Estonia

Introduction

Estonia is located in a transition zone between regions, having a maritime climate in the west and a continental climate in the east, and is a relatively small country (45,227 km^2), but its geographical location between the Fenno-scandian Shield and East European Platform and its comparatively long coastline (nearly 3,800 km) due to numerous peninsulas, bays and (>1,500) islands, results in a variety of shore types and ecosystems. The western coast is exposed to waves generated by prevailing westerly winds, with NW waves dominant along the north-facing segment beside the Gulf of Finland, contrasting with southern relatively sheltered sectors located on the inner coasts of islands and along the Gulf of Livonia (Riga).

The coastline classification is based on the concept of wave processes straightening initial irregular outlines via erosion of capes/bay deposition, or a combination (Gudelis, 1967; Orviku, 1974; Orviku and Granö, 1992). Much of the coast (77 per cent) is irregular with the geological composition of capes and bays being either hard bedrock or unconsolidated Quaternary deposits, notably glacial drift. The cliffed north-eastern coast around Ontika has been straightened by erosion, whereas beach-fringed Narva Bay has an outline smoothed by deposition (Orviku and Romm, 1992). Coasts straightened by a combination of erosion/deposition can be found on the northern shore of Kõpu Peninsula, the western part of Hiiumaa Island, and around the Gulf of Livonia. Orviku and Sepp (1972) have illustrated the stages in coastal landform evolution of the west Estonian archipelago.

Historical coastal evolution

Coastal evolution has been influenced by changing sea levels (Kessel and Raukas, 1967): the Ancylus transgression (9,500–8,000 years ago), followed by the Littorina transgression (8,200–7,000 years ago) and an ensuing regression (7,000–5,000 years ago). During the Littorina transgression there was extensive erosion, producing cliffs and shore platforms which generated large amounts of sand and gravel, with residual boulders occurring where glacial deposits

(till) had been dispersed. During the ensuing regression sand, gravel and boulders on the emerging sea floor formed beaches, beach ridges and dunes by wave and wind activity.

Postglacial isostatic movements since the Ancylus Lake period resulted in land uplift ranging from *circa* 45 m in southern Estonia up to 75 m on the northern coast. Beaches formed are often backed by beach ridges and dunes and tilting has continued on either side of a zero isobase that runs SW–NE through Riga in Latvia, with land uplift of about 1 mm/yr at Pärnu, 2 mm/yr at Tallinn and 2.8 mm/yr on the north-western coast (Vallner *et al.*, 1988).

Main agents causing erosion

Waves and currents (especially during heavy storms), longshore drift, onshore winds, and human activity are the main agents of coastline evolution. Tides are negligible on the Estonian coast (<5 cm), with sea level generally higher in winter than summer. Rivers are small and alluvial deposition is very limited. Westerly and south-westerly winds predominate, producing waves from these directions on west-facing coastal sectors, but wave energy is low, especially in places where the nearshore is shallow and boulder-strewn. During storm surges, onshore

■ **Figure 4.1** *Sea ice hummocks in a coastal forest at Paaste village, Saaremaa Island in 1997. (a) Hummocks of ice moved about 100 m inland within a 0.5 km wide sector of the shore as a result of north-westerly winds resulting in severe damage to a pine forest; (b) The results of the ice assault became evident in summer after the final ice melt: some boulders and sea bottom sediments have been transported to the damaged pine forest*

winds and low barometric pressure may raise sea level by nearly 3 m above the Kronstadt zero (benchmark for the eastern Baltic Sea), when major coast changes occur, with cliffs undercut at a higher level, high beach ridges formed, and large quantities of sand moved alongshore. During winter (November to April), a shore ice fringe develops which prevents wave activity, but on spring break up, waves may drive ice onshore, piling it up as 10–15 m high hummocks (Orviku 1965; Orviku *et al.*, 2011). Ice driven onshore scours sea floor sediment, displacing shoreward sand and gravel, and even boulders, damaging trees and buildings (Figure 4.1).

The coastline

Shoreline classification

The following shore types (Figure 4.2) have been distinguished based on the initial relief slope, geological character of the substrate and dominant coastal processes, (Orviku, 1974; 1992; 1993):

- cliffed (approximately 5 per cent of shores) – an abrasion bluff in resistant Palaeozoic rocks (limestone, dolomite, sandstone);
- scarp (very short sections between other types) – an abrasion bluff in brittle unconsolidated Quaternary deposits (sand, gravel, till, etc.);

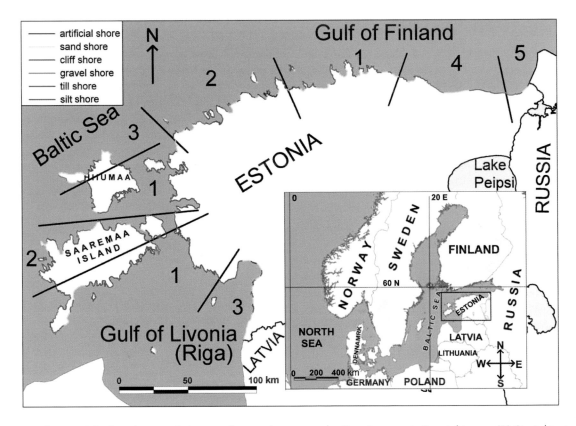

■ **Figure 4.2** *Distribution of shore and coastal types on the Estonian coast. Coastal types: (1) Straightening abrasion-accumulation coast in Quaternary deposits; (2) Straightening abrasion-accumulation coast in Pre-Quaternary bedrock; (3) Straightened abrasion-accumulation coast; (4) Straightened abrasion coast; (5) Straightened accumulation coast*

- till (35 per cent) – an abrasion sloping till;
- gravel (11 per cent) – depositional with beach ridges formed of gravel and pebbles;
- sand (16 per cent) – depositional with sand ridges often backed by foredunes or dunes;
- silt (31 per cent) – depositional with fine-grained (silt) sediments; usually it has a very flat nearshore and a tendency to become overgrown;
- artificial (2 per cent) – natural dynamics altered by anthropogenic constructions (breakwaters, protecting walls, berms).

For the above-listed types, the first three are erosional and the remainder mainly depositional shores (erosion might still occasionally occur); artificial shores may be either erosional or depositional (Figure 4.3 top). Classic examples are cliff formations outcropping in areas of carbonate rocks/sandstones, where initial topography was steep and waves sufficiently strong and till scarps subject to active erosion (Figure 4.3 centre; Orviku, 1982). As a result an erosion platform with boulders and cobbles formed on the sea floor or nearshore (Figure 4.3 bottom), acting as a natural breakwater, which considerably decreases storm wave influence on the scarp foot.

Explaining the formation of sand shores (Figure 4.4) is more difficult (Bird, 1985; Orviku *et al.*, 1995; Raukas *et al.*, 1994). The vast majority formed and developed because of alteration and re-sedimentation of old sand area deposits (buried valleys, landforms generated by glacier melt water, etc.). Small stretches of sand shores may also occur near artificial constructions, e.g. ports, where quays hinder natural sand movement. Sand origin on these shores is sometimes erroneously associated with erosion of neighbouring till shores, which is practically impossible because the till sand content is too low to enable formation of such shores.

Coastal types

Five different coastal types exist, which might include several different shore types and various coastal sediments but their evolution, geology and morphology follow similar patterns.

■ **Figure 4.3** *Cliff evolution. Top: Pakri cliff, northwestern Estonia. Centre: Erosional scarp formed in till in Mõntu, Sõrve Peninsula, Saaremaa Island. Bottom: Finer-grained sediments eroded by waves and remaining boulders, provide a good natural protection against further erosion (Abraded till shore on the western coast of Aegna Island, Tallinn Bay)*

■ **Figure 4.4** *(a) Accumulative sand beach with foredunes north to Lehtma harbour, Hiiumaa Island; (b) intensive erosion occurs south of the harbour jetty*

TYPE 1

Typical northern Estonian coastlines are large peninsulas separating deep bays. At the bay head there is usually a sand beach, backed by beach and dune ridges that rise landward. Erosion of glacial deposits has resulted in bay-head accumulation of sand and silt, with reeds and bushes colonizing silt flats. The coast consists of low scarps cut into glacial and fluvioglacial deposits, varved clays and gravel. Shore platforms cut into these soft sediments are strewn with erratic boulders fringed by reed beds. Bay-head beaches consist of sand and gravel, the finer silt and clays having been withdrawn seaward in suspension and deposited on the bay floor (Orviku, 1988).

Sections of Hiiumaa and Saaremaa Island are characteristically representative of type 1. The eastern and southern coasts of these large islands are low and flat, locally marshy, with pebble and cobble beaches. Large irregular inlets, open to this Baltic Sea sector sheltered by Hiiumaa Island, include many small rock islands. The southernmost bay (Matsalu) at the Kasari River mouth is bordered by extensive reed beds, which have spread rapidly seaward with the emerging nearshore area.

A sheltered low embayed coast made up of Quaternary deposits is also located between Saaremaa and the mainland where the offshore zone is very

shallow. There are low marshy areas and occasional shingle beaches on the shores lined with boulders derived from glacial drift along with some low scarps, cut into till and fluvio-glacial deposits. Dunes are either poorly developed or missing (Orviku, 1974). Exposure to wind and wave activities increases southwards and reed beds occupy sheltered bays and have even spread onto sand beaches (Bird *et al.*, 1990).

TYPE 2

This coast is irregular, consisting of valley-mouth bays with sand beaches and reed beds with limestone headlands and cliffs. Between the capes, wide beaches have formed, backed by dunes and fronted by sand bars. The sand has been derived from Lower Ordovician sandstone in bordering cliffs. On Cape Pakri, a cliff has been cut into relatively soft Lower Ordovician sandstone, overlain by harder limestone. The limestone upper cliff is undermined by wave-cut notches (Figure 4.5a), which causes frequent cliff failure resulting in talus slopes (Figure 4.5b), subsequently reduced by wave activity and eventually gravel beaches.

The north-western coast of Saaremaa Island has up to 20 m high cliffs cut into Ordovician and Silurian limestone and the nearshore area is a bare rock shore platform with pebble and boulder beaches widespread near the cliffs. Sand and gravel beaches, derived from eskers and end moraines, are distributed by longshore drifting (Orviku *et al.*, 1995). Erosion of the Silurian limestone cliffs has yielded gravel deposits, which drifted into intervening bays to accumulate as bayhead shingle beach ridges. Because of continuous land uplift, cliffs are usually eroded only during heavy storms (Orviku, 1974). In recent decades, there has been erosion on parts of the shingle beaches and spit prolongation. The islands were much smaller during the Ancylus Lake stage and successively enlarged at Littorina Sea and Limnea Sea stages, with subsequent land uplift of the coastal plain. Each of the earlier coastlines is marked by bluffs and beach ridges, particularly on Saaremaa and Muhu Islands.

Type 3

As an example, a long and wide gently curving beach of fine sand rich in quartz and backed by low dunes has formed at the head of Pärnu Bay. Reed beds have spread over parts of the shore. Emergence following the Littorina transgression (up to 7 m above present sea level), resulted in formation of a coastal plain and

shore fringed by boulders, sand beaches and segments of reed beds. Further south, as exposure to wave activity increases along the coast of the Gulf of Livonia, beaches are backed by lagoons overgrown with reeds. Similar processes and morphological features can be observed on the northern part of Hiiumaa Island.

Type 4

The Baltic klint (Ordovician limestone cliff) becomes higher (56 m) and westwards is located closer to this coastline type. It is cut into Cambrian sandstone, clay and shale and covered by Ordovician limestone, which is more resistant to erosion. There are accumulations of basal rock talus and beaches of cobbles and boulders derived from cliff erosion and the nearshore area is a boulder-strewn shore platform. The klint is generally stable and is now only locally and episodically attacked by waves.

Type 5

Close to the Russian border (Figure 4.2), this is epitomized by a long gently curving sand beach backed by numerous parallel beach ridges covered

■ **Figure 4.5** *(a) The upper part of the limestone cliff is undermined by deep wave-cut notches in non-resistant rocks (e.g. glauconite sandstone), which crop out on the coast of Osmussaar Island; (b) wave-cut notches collapsed and forming a talus slope in front of Pakri cliff*

with dunes. The sand has come partly from cliff erosion to the west; partly from glacial and fluvioglacial deposits on the Kurgalovo Peninsula to the north; and partly from the Narva River, whose sediment yield has diminished after upstream dam construction. With a reduced sand supply, the seaward margin of these dunes is now in recession forming abrupt scarps. In the 1980s, the beach was re-nourished, and a breakwater built at the river mouth (Orviku and Romm, 1992).

Recent trends and changes

Estonia is sensitive to climate change manifestations such as an increase in cyclonic activity, westerly circulation and a northward shift of the Atlantic storm track over the past decades. There has been a storm increase (ten major storms from 1965–2010) with an intensity that had previously occurred only once or twice a century (Orviku, 1995; Orviku *et al.*, 2003; 2009; Tõnisson *et al.*, 2008; 2011). Changes in meteorological conditions have changed wave climate and sea-level conditions, as well as the rate at which shore processes occur. Frequent storm surges and a general absence of ice cover on the shore with unfrozen shore sediments in milder winter conditions, allow waves to attack the coast and shape beaches even in winter. Despite a tectonically uplifting coast, beach erosion attributable to increased storminess has become evident in recent decades.

For example the Kiipsaare study site, located on the north-westernmost tip of Harilaid Peninsula, NW Saaremaa Island is exposed to the Baltic Sea proper. This is a small (4.3 km²) peninsula connected by a tombolo to the larger Tagamõisa Peninsula. The primary landform is a glacio-fluvial ridge. A coastal scarp borders the 50 m wide sand beach, reaching 3 m a.s.l. in the NW part where shore processes are strongest. The area is most influenced by waves from the SW, W, NW and N, and the sea bottom is particularly flat and shallow, north-west of Kiipsaare, the 5 m isobath being 4 km from the shoreline. The steepest underwater shore slope occurs north-east of the peninsula where the 5 m isobath is 350 m offshore.

In 1933, the Kiipsaare lighthouse was erected

approximately 150 m from the shoreline and in the middle of the Cape. Estimates of the shore retreat speed between 1933 and 1955 are not possible, but the mean annual velocity of shoreline retreat at Cape Kiipsaare is about 2 m/yr from 1955–1981. More rapid retreat is revealed when comparing aerial photographs from 1981 and 1990 (Figure 4.6). During that decade, the lighthouse 'shifted' from a safe distance near the erosion scarp. Shoreline retreat during these ten years was up to 30 m and by 1995, the lighthouse was in the middle of the beach. Even more severe storm damages occurred from 2001– 2010, when the shoreline receded approximately 50 m. Today the lighthouse stands over 50 m offshore. Former beach ridges explicitly show the position of earlier shorelines along the western coast. These shorelines are not parallel to the current one, as would be typical of sand beaches, but a 45° intersection of the current shoreline with the axis of former beach ridges is clearly visible in freshly formed scarps – evidence of rapid changes in the direction of shore processes. Additional evidence of the north-eastward migration is given by shipwrecks on the western coast of the Cape (Figure 4.6), probably wrecked on the north-eastern coast of the Cape 150 years ago. Sand later buried the wrecks, which were re-exposed on the western coast and consequently the Cape has 'rolled over' the shipwrecks.

■ **Figure 4.6** *Shoreline changes during the period from 1900–2010 (shipwrecks found in 2000 and 2010 are also indicated).Key: black = 1900; Purple = 1955; Yellow = 1981; Green = 1988; Lilac = 1998; Light blue = 2002; Red = 2008; Dark blue = 2010*

THE BALTIC STATES: ESTONIA, LATVIA AND LITHUANIA

Coastal defence

The Baltic Sea is a relatively shallow water body and its small dimension restricts wave size and reduces the erosion risk to shores. Major damage takes place during extreme events once or twice a century. Previously, the main coastal zone economical activities were fishing and agriculture, which did not need much shoreline infrastructure, and therefore all bigger infrastructures were planned further away at higher elevations. Lightweight buildings built close to the shoreline were easy and cheap to rebuild after extreme events. Humans lived in harmony with nature and there was no need for coastal protection.

Everything changed as a result of World War II. The Soviet occupation began at the end of the war in 1945 and continued for nearly half a century, during which time the coast was the western border of the Soviet Union and different degrees of restriction were established depending on location. On the mainland, east of the west Estonian archipelago, restrictions were not as stringent as on islands exposed to the Baltic Sea proper. Strict limits were established on sea-borne navigation and even land movement. Local people on small islands and in many villages situated on larger islands, were deported from their homes and many older buildings and infrastructure were taken over by Soviet border guards, with the result that many unused buildings collapsed after 50 years (Figure 4.7). Due to the restricted border zone regime, no developments were permitted close to the sea, so no coastal protection existed and natural processes reigned free.

After regaining independence in 1991, the coast became a popular recreation area, which affected coastal land use; it is now a popular place for summerhouses and even residences. According to legislation it is declared as a *force majeure* (Government has absolved itself for possible losses caused by coastal erosion) when erosion endangers houses or private infrastructure and this has more or less kept developers away from the shoreline. Therefore, most of the coast is akin to a natural laboratory and natural processes can be allowed with no threat of financial or human loss.

Moreover, the country is experiencing land uplift of up to 2.8 mm/yr and the shoreline is shifting slowly seaward. However, a few examples of recent coastal protection still exist. In Tallinn Bay, there are 4 km of breakwaters and jetties, and in 1980, a 2.5 km long sea wall was built purely to enclose and reclaim a shallow bay-head area for highway construction and the 1980 Olympic Yachting Centre (Martin and Orviku, 1988). Nearly 30 years later, because of relatively low wave energy, this construction is still in good shape, apart from a few ice attack events that bent some steel bars on the construction top (Orviku *et al.*, 2011).

Very close to the above is Pirita, where a seawall protects the roadway (Figure 4.8) and an artificial sand beach has been created. Pirita River sediments once fed the beach but at present have more or less stopped because of river mouth marina construction. Pirita beach has suffered repeatedly due to extreme storms (1967 and 1975/76). Approximately 30,000 m³ of additional sand was brought from inland sources to nourish the beach for the 1980 Olympic Games (Orviku, 2010). Many more extreme storms have taken place since, including the November 2001 storm and January 2005 storm (Gudrun) which caused extensive erosion. In addition to wave

■ **Figure 4.7** *The former lighthouse keeper house has been destroyed by various factors, including Soviet army destruction and purely natural processes. Eventually, it will be completely destroyed by coastal erosion*

■ **Figure 4.8**
*Seawall protecting
Pirita roadway in
Tallin Bay*

erosion, some sand was carried inland by wind to the pine forest. Pirita beach needs to be urgently renourished in the near future, as it is approximately 30 years since the last nourishment and the beach has lost much of its natural protection – every storm attack makes the beach more vulnerable.

The last of the coastal protection measure examples in Estonia is the area surrounding a former nuclear waste depository near Sillamäe town, north-east Estonia (Figure 4.9). At the beginning of the twentieth century, boat houses and summer cottages were the most common buildings here (according to a 1934 map), and the shoreline was stable. The first artificial changes appeared when the so-called 'Swedish harbour' was established at the end of the 1930s on the eastern side of Cape Päite. The harbour jetty became a sediment trap for gravel-pebble longshore transport from west to east. Sediments trapped behind the jetty accumulated on the northern shore of Cape Päite and the sediment amount in front of Sillamäe (east of the harbour) and the character of shore processes changed. Former accumulation gave way to erosion and the shoreline moved landwards,

especially in the area of the current Sillamäe town. The town itself and a highly specialized chemical and metallurgy plant were constructed in 1946 (Orviku *et al.*, 2008). It was forbidden to outsiders in 1947– 1991, as fuel rods and nuclear materials for the Soviet nuclear power plants and weapons were produced. Uranium enrichment finished in 1989, but enrichment of rare metals (such as niobium) still exists. On the immediate shore of the Gulf of Finland, is a 50 ha nuclear waste depository established since 1959, where some 1,200 tonnes of uranium, 800 tonnes of thorium enrichment residuals and other hazardous substances are buried.

The depository did not cause much change in the shoreline or coastal processes; accumulation was still dominant in front of the depository due to the Swedish harbour jetty. The most drastic changes, at the beginning of the 1980s, were related to construction of an ash depository whose northern tip extends up to 200 m into the coastal sea and therefore interferes with the longshore sediment transport pathway. As a result, the longshore transport of gravel-pebble from west to east has been artificially impeded forming a 200 m long and more than 20 m wide complex system of beach ridges. The process is similar to one caused by the harbour jetties. Simultaneously, the northern shore of the waste depository started to erode, caused by a lack of sediments from the west, which intensified longshore transport. Due to security concerns, the jetty at the Swedish harbour was demolished in 1984, which triggered massive amounts of formerly trapped sediments to move eastward. These sediments quickly formed a spit in the south-eastern direction. Rapid spit development was probably caused by intensive storm periods (the estimated average speed of the accumulation of the distal part of the spit was 700–800 m³/yr). Sediments formerly trapped behind the jetty, which protected the shores in front of the nuclear waste depository, were transported to the spit and left the north-eastern depository area without natural protection. By 1996, gravel pebble ridges formerly fronting the northern shore of the nuclear waste depository were completely eroded, giving storm waves free access to the gravel terrace built at the nuclear waste base, allowing subsequent erosion

to reach the nuclear waste's protective dam. It has been known only since 1989 that annually, some 4,000 tonnes of ammonia has leaked into the Gulf of Finland through the broken protective dam (Orviku *et al*, 2008).

Intensive wave activity caused the gravel terrace of the Sillamäe nuclear waste depository to erode, which, in turn, forced emergency repair work to be carried out to the foundation using whatever materials were available. These repairs were not effective and erosion continued until closure of the waste depository in 1998–2008. It was one of the highest priority environmental projects in the whole Baltic Sea basin. The dam has been reinforced and leakage through to the sea should be negligible. Currently, the northern shore of the nuclear waste depository is well protected from wave activity, as a result of conserving and construction of Sillamäe harbour in front of the dam (Figure 4.9). Intensive accumulation is current on the west of the western jetty. Unexpectedly, erosion has not intensified east of the harbour and accumulation has dominated during the last decade. This might be caused by changes in

■ **Figure 4.9** *Location of Sillamäe and directions of sediment transport (Base image with permission from the Estonian Landboard, compiled by H. Tõnisson)*

hydrodynamic conditions, as the longest fetch at Sillamäe is for waves coming from north and north-east. As a result of changes in wind climate, such as increases in westerly and decrease in northerly winds, over the last few decades, wave activity has decreased near Sillamäe (Tõnisson *et al.*, 2011).

Conclusions

The annual maximum sea-level on the Estoninan west coast has increased during the last decade. Fewer storms seemingly affect the coast but in western Estonia, they have become significantly more intense, resulting in an escalation of shore processes. Measurements at various study sites in western Estonia show that the current rate of coastal change is many times higher than in the 1950s. These factors are responsible for acceleration in the rate by which such coastal processes occur. Each subsequent storm reaches an already vulnerable beach profile. In addition, higher sea levels during storms have also caused the erosion area to move further inland with subsequent storms. Even accumulative shores (sand shores, gravel shores) in normal conditions have now become erosional in many locations (Tõnisson, 2008; Tõnisson *et al.*, 2011).

Latvia

Introduction

Latvia, with a coastline of about 496 km in length, is located on the eastern Baltic Sea shore. The Baltic Sea open coastline is approximately 183 km; the Irbe Strait coastline 57 km; while the Gulf of Riga total seashore length is 256 km. Because of Atlantic autumn/winter storms, high waves and storm surges occur especially on the Kurzeme coastline, which faces the open Baltic Sea. River flood plains, as well as lagoon remnant lakes specific for this particular area, created suitable environments for settlements to form and trade to develop. Today, three major ports exist (Riga, Ventspils and Liepāja), as well as eight smaller ports, which have had a significant effect

on the intensity of coastal erosion rates. Isostatic processes began at the commencement of the postglacial period and have slowed considerably, so that at present it has little effect on coastal processes (Ulsts, 1998). In terms of a morphogenetic classification, the coast generally belongs to a class of coasts straightened by erosion and deposition (Gudelis, 1967).

Coastal morphology and geological history

The coastline is comparatively straight with no explicit headlands and gulfs, but at some headlands the orientation changes by more than 30°, e.g. at the Capes of Akmeņrags, Ovīsrags, Kolka and Ragaciems. The Late Pleistocene relief has been modified through alternating periods of accumulation and erosion during former stages of development of the Baltic basin and also during the last 2,800 years, when sea levels have been comparatively stable. During this period, the Earth's vertical crustal movements have been activated and catastrophic events relating to basin water exchanges caused by de-glaciation process have occurred, all resulting in comparatively frequent and sharp water level changes. The intensity of coastal processes during the Baltic Ice Lake period (12,600–10,300 radiocarbon years BP) was significantly higher than today (Veinbergs, 1986).

Because of gradually increasing water levels during transgressions, modification of very flat coastal underwater slopes occurred in numerous places, mobilising and moving shorewards significant sediment amounts. Later, with a lowering of water levels, material on the upper coastal slope accumulated as barrier spits and accumulative terraces (Ulsts, 1998). Currently, underwater slope areas eroded during the Littorina Sea (7,000–5,000 BP) transgression phases, contain almost no fine sediment fractions, but gravel and boulders, which play significant roles in seashore dynamics. In shallow, sediment-active areas, sand bars are present (up to four sub-parallel bars in the Irbe Strait coast). Along major parts, the seaward depth increase is comparatively steep, but beyond a 4–8 m depth the slope becomes flat.

Along the major part of the coastal areas, the bedrock surface lies beneath sea level; hard rock cliffs outcrop only in Devonian sandstone and clay on the eastern shore of the Gulf of Riga. The bedrock is covered with Quaternary sedimentary deposits up to 60 m thick. These consist, mainly, of moraine and outwash deposits covered and in some places, replaced by silt, sand and gravel from the Baltic Ice Lake (average thickness of 2–8 m) and Yoldia Sea (10,300–9,500 BP); Ancylus Lake (9,500–8,000 BP) and Littorina Sea sediments cover them. The most widespread (almost along the entire coastline) are Littorina Sea sediments, often exposed at the surface

and involved in present day coastal processes. Significant volumes of Post-Littorina period coastal area deposits are found in northern Kurzeme on the shoreline of the Strait of Irbe, where these, along with more ancient sediments (in particular, those of the Littorina period) form accumulation terraces (Ulsts, 1998; Figure 4.10). A major depositional headland occurs at Kolka on the Kurzeme peninsula's tip, whose morphology has changed numerous times; today, coastal retreat occurs at the headland.

Littorina Sea stage and Post-Littorina dunes are present along the major part of the coastline (approximately 80 per cent), reaching up to 30–40

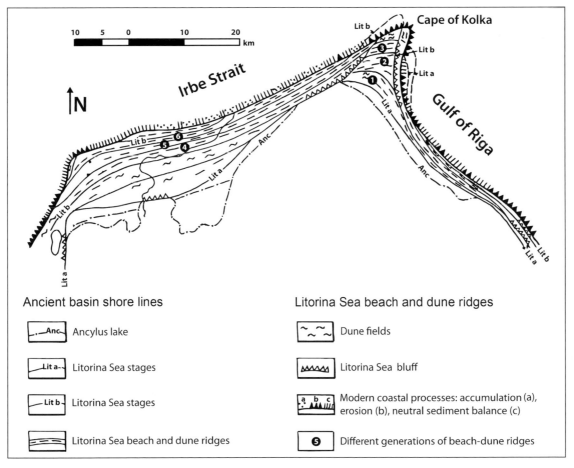

■ **Figure 4.10** *Generalized palaeogeographical map of the Littorina Sea accumulation terraces and Cape Kolka (Ulsts, 1998)*

m above sea level (Figure 4.14). Lagoons, separated from the sea during the Littorina Sea stage, have been transformed into shallow, actively overgrowing remnant lakes (Pape, Liepāja, Babīte, Engure, etc.), or have been partially eroded while the coastal system adapts to new conditions. The water level in several lagoon remnant lakes has been artificially lowered for agricultural development. At the same time, at the southern end of the Gulf of Riga, a depositional coast developed because large sediment volumes obtained from the major rivers (Daugava, Gauja and Lielupe) were moved by longshore sediment drift. Wave-dominated deltas were also built at the mouths of the rivers Daugava and Gauja.

All Latvian coastal sections possess several common peculiarities in that the:

- majority are formed of unconsolidated Quaternary sediments, except an approximately 8 km long section in the eastern part of the Gulf of Riga;
- seashore is comparatively low and flat (height mainly does not exceed 15 m);
- underwater coastal area has a gentle slope, covered with an amount of 'active' sediment material;
- entire coastline, with few seasonal exceptions, has a beach;

■ **Figure 4.11** *Soft cliff (bluff) in Quaternary deposits on the Baltic coast central section*

- longshore sediment drift reaches one million m³/yr in the open Baltic Sea coast.

Irrespective of the comparatively homogeneous geologic structure, the coastal geomorphology is changing and as a result, areas can be found where continuous accumulation of deposits has occurred on flat shorelines (about 140 km of total length) together with coastal erosion, where cliffs/bluffs (about 150 km in length), with different heights and different geological structures (Figures 4.11, 4.12 and 4.13).

■ **Figure 4.12** *Sediment-deficient slowly retreating coast on the northeastern side of the Gulf of Riga*

■ **Figure 4.13** *Accumulative coast with sand beach and growing foredunes on the central part of the Irbe Strait coast*

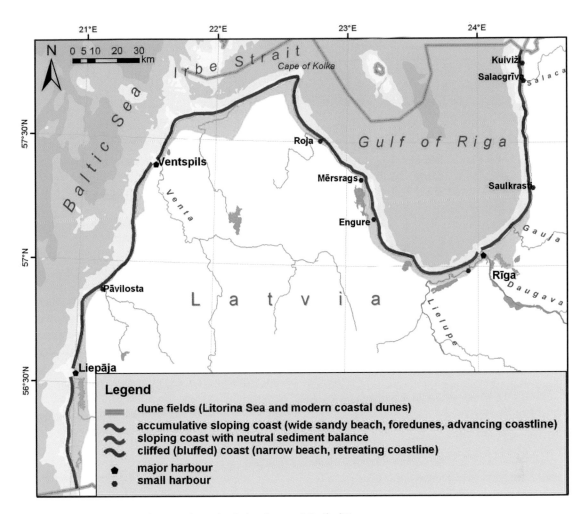

■ **Figure 4.14** *Coastal types along the Baltic Sea and Gulf of Riga*

Along other sections of the coast (about 200 km), conditions are stable. In some places, sediment movement occurs at a very slow rate, e.g. the northeast and north-west of the Gulf of Riga.

In general, coastal types can be defined by differences in sediment volume distributions, which, in its turn, is a function of many factors, the main being the Littorina Sea period heritage, coastline orientation against prevailing winds and anthropogenic factors (Eberhards, 2003). Artificial coasts (harbours, 8 km and shore protection structures, 4 km) represent 2.4 per cent of the entire coastline (Lapinskis, 2010).

Wave climate

The main Baltic Sea storm period is the autumn/winter season, when maximum wind waves in the open Baltic Sea can in exceptional cases reach heights of 13 m. Maximum wave height in the Gulf of Riga is 8.5 m in its southern part away from coastal areas. Waves arrive at the coast predominantly from a south-western and western direction and underwater sand bars, usually found on the flat coastal underwater slope, significantly affect wave transformation.

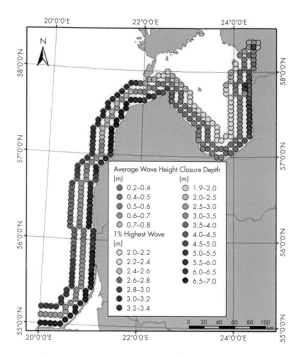

Figure 4.15 *Variation of the average significant wave height, threshold for the highest wave height and the closure depth for wave conditions between 1970–2007 (Soomere et al., 2011)*

Longshore variation in the simulated closure depth (Figure 4.15) largely coincides with similar variations in the average significant wave height and the threshold for the 1 per cent of highest wave conditions. The highest wave values are found along the western coast of the Kurzeme Peninsula (about 5.4 m); the smallest ones at the western Gulf of Riga coast (3.5 m). Wave climate differences in the Gulf of Riga evidently reflect the anisotropic nature of wind fields in this region.

The average breaking zone width is from 600 to 1000 m starting from a depth of 7–8 m. In the Baltic Sea and, in particular the Gulf of Riga, sea level changes are mainly caused by wind pressure (storm surges). Maximum storm surge levels along the open Baltic seashore were observed in 1967, when +174 cm was registered near Liepāja. In the Gulf of Riga, the highest water level (+247 cm) was registered in 1969 in Skulte. Wind speeds of 28 m/s have a return period of five years, and ten years is the return period for winds exceeding 32 m/s (Ulsts, 1998).

There is an increasing trend for long-term sea level changes, but this has not exceeded 10–15 cm since the end of the nineteenth century (Eberhards and Purgalis, 2008). Changes caused by gravitational tides do not exceed 15 cm and have no significant role in coastal area evolution.

Coastal settlement evolution

Within a coastal area of about 20 km in width (15 per cent of Estonia's area) live more than one million inhabitants, usually located outside the initial sea adjacent to a 100–200 m area/belt. They form approximately one half of the whole population, mainly because the capital, Riga, is located here, a function of Latvian history. Due to various social and political restrictions, there are no densely developed areas in the *direct* seashore vicinity and there exists a comparatively small number of civil infrastructures. During the Soviet occupation, significant parts of the western Gulf of Riga coast and almost all Baltic Sea seashores outside of large towns, were a 'closed area', for which access to a civilian population was not easy. Fishing villages existing before the occupation gradually disappeared. Within non-developed and poorly urbanized coastal areas in western Latvia, the only construction allowed was in relation to protection of the outer USSR borders. Therefore, paradoxically, the Soviet occupation 'saved' Latvia from various coastal management problems that seem specific to Western Europe, and the 255 km of protected natural areas located along the coast are partially a 'present' of the Russian occupation.

In the beginning and middle of the sixteenth century, demand for timber and firewood in all areas sharply increased (urban construction, shipbuilding, lighthouse fire, industry, warfare) and as a result, extensive areas of forests were harvested. The degraded areas were exposed to aeolian processes and inland dune migration followed, especially along the Kurzeme coast. Dune migration continued for centuries, causing significant damage and requiring much effort for replanting but sand movement stopped in the mid twentieth century (Eberhards, 2003).

Until the end of the nineteenth century, coastal areas, except river mouths, were very sparsely populated because of the coastal lowland's low fertility. The main source of subsistence for people living in coastal villages and farmsteads was fishery. At the end of the nineteenth century, along with industrial development, the first significant anthropogenic intrusion into coastal processes occurred when construction of large ports commenced. During the last century, because of a coastal retreat of 50 to 200 m, many small rural population centres in areas originally located a safe distance from the coast are now at risk. During recent decades, Latvia has been affected by a common worldwide trend: inhabitants gradually moving closer to the coast mostly to previously undeveloped areas. This rapidly increasing human impact on sensitive, but in most areas still intact, coastal belt ecosystems, is such that the typical natural landscape is being transformed and degraded.

However, significant coastal areas are still vacant. The total coastline length with developed areas closer than 500 m from the shore is *circa* 11 per cent. The total length of sections with a low or very low building density forms about 28 per cent (mainly in the Gulf of Riga); the total length of sections with no building, but with significant seasonal recreation load forms about 10 per cent, and about 51 per cent are those minimally affected by people (except erosion caused by harbours).

Coastal erosion

The most significant source of new material in the sand, gravel, shingle coastal system is erosion of cliffs/bluffs and sediments from the coastal underwater slope. These are moved laterally by longshore currents, but dredged sediment from harbours is lost to the system because of being transferred to offshore dump areas away from the active coastal zone.

Analysis of available materials show that since 1935 (the oldest qualitative cartographic material for the entire seashore; Eberhards, 2003), to 1992, shoreline retreat at an average rate of 0.5–2.0 m/yr has occurred in wide areas in the western part of Kurzeme, but only for some specific short sections in the Gulf of Riga (Figure 4.16; Eberhards and Lapinskis, 2008). The impact of Ventspils and Liepāja harbours is especially important, since ship entrance channels reach large depths, e.g. at Ventspils it is −19 m and these have completely stopped longshore sediment drift.

For over the past 20 years, more than 35 million m³ of sediment has been buried in offshore dumps or inland. To the north from both Liepāja and Ventspils harbours, shoreline retreat has been of the order of 200 m whilst south, accretion has exceeded 350 m at Liepāja and 800 m at Ventspils. The total length of seashore formed due to harbour excavation is significantly less than the length of the eroded retreating sections (35 km). For example, the approximate sediment deficit volume north of Ventspils harbour exceeds 50 million m³. In the Gulf of Riga, the role of ports affecting coastal dynamics must be considered significant, though the ports affect significantly shorter coastline lengths.

From 1992–2011, some 120 km of coast has been affected by erosion. For some 60 km withdrawal occurs at a rate >0.5 m/yr. Much higher rates (>2 m/yr) may be observed in 6 km of total length, e.g. at Bernāti headland, Šķēde, Jūrkalne, Melnrags, Staldzene, Cape of Kolka and Gauja delta (Figure 4.17). The retreat rate has probably remained constant for the central shoreline, but eroded lengths seem to have been extended (Eberhards *et al.*, 2009)

In the long term, certain adverse impacts have been caused by boulder removal from beaches and underwater slopes, these being used to fortify the subaerial coast (in the central part of the western Gulf of Riga and the south-east part of the Gulf of Riga). Additionally foredune trampling has resulted in deflation because of recreation overload, especially in the southern part of the Riga Gulf.

In the second half of the last century, extraction of, for example, sand-gravel for economic needs, hydropower dam construction, etc. has led to a decrease in alluvial sediment input from the major rivers (Daugava, Lielupe, Venta and Gauja), resulting

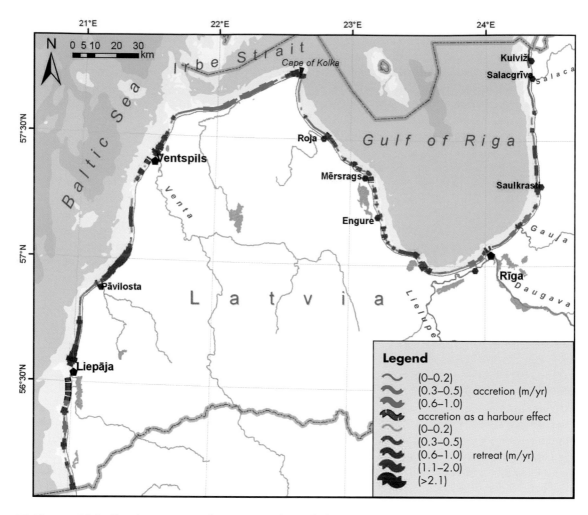

■ **Figure 4.16** *Shoreline retreat and accretion in the Gulf of Riga and Baltic Proper in 1935–1992*

in coastal erosion acceleration of some 6 km (Eberhards and Lapinskis, 2008). During recent decades, due to climate change, there has been a sharp reduction in the sea ice duration time (Meier *et al.*, 2004) along with a reduction in the number of days with sub-aerial frost on the shore (Briede, 2006). The result, particularly on the Gulf of Riga shore during winter storms, is that ice is not acting as a natural 'protective barrier'. For instance, during the hurricane-strength winds of 8–9 January 2005, when there was no ice cover, the volume of material washed into the sea reached 3.1 million m³ and about 90 ha of territory was lost (Eberhards *et al.*, 2006).

Coastal protection

According to estimates of coastal erosion, some 160 households, municipal or industrial (including cemeteries, 6 km of roads, electro transmission lines and several historical and cultural objects) are located within the 50 year risk zone, of which about 40 are located within the 15 year risk zone (Eberhards *et al.*, 2009).

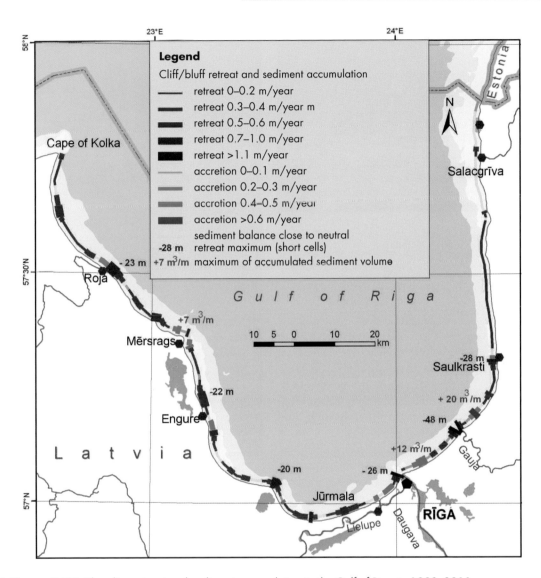

■ **Figure 4.17** *Shoreline retreat and sediment accumulation in the Gulf of Riga in 1992–2011*

Protected shoreline sections are comparatively short, their total length is small (about 4 km), and usually built from local natural material, e.g. many Gulf of Riga fishing villages have boulder revetments (Figure 4.18). Similar in function, though more solid, are structures formed of reinforced concrete prisms or tetrapods, placed near several lighthouses on the Kurzeme coast (Figure 4.19). In many cases, revetment section installations placed parallel to the shoreline for only a few 100 m have caused beach disappearance and erosion intensification of nearby sections. One major coastal protection structure with a length of 500 m was installed in 2006 at the Liepāja wastewater treatment plant (Figure 4.20). The chosen solution, an inclined wall of gabions, has still preserved its original function but caused significant

■ **Figure 4.18** *Boulder revetment at Valgalciems (western coast of the Gulf of Riga)*

intensification of coastal erosion to the north away from the protected section.

The massive military fortifications formed in the beginning of the twentieth century and located north of Liepāja port are unique in the context of coastal erosion. Despite being partially destroyed, they continue to function as coastal protection units, maintaining the coastline and further north, cause typical consequences of beach instability and intensification of erosion (Figure 4.21).

No soft shore protection projects have ever been executed in Latvia, except in episodic cases when material excavated from small harbours and shipping channels with volumes exceeding several thousand m³ have been transferred to the nearest beach vicinity to the port. No groins, detached or submerged structures have been constructed along the Latvian coast. The shoreline length covered by coastal protection structures is very small and all belong to passive structural types (seawalls and revetments). The end result is that scour, which increases reflected wave energy and end wall erosion, has occurred at all sites. Around 40 per cent of the existing coastal protection structures can ensure only short-term or partial protection of the endangered objects, since they are inappropriate for local situations and are in critical conditions (Lapinskis, 2009).

In several comparatively densely populated sections experiencing high anthropogenic stress and

■ **Figure 4.19** *Tetrapods around a lighthouse on the Kurzeme coast (aerial photograph by Metrum Ltd.)*

■ **Figure 4.20** *Liepāja wastewater treatment plant revetment*

■ **Figure 4.21** *Old military fortification north of Liepāja port*

where significant amounts of sand are available (e.g. Ventspils, Jūrmala), dune planting of osier and marram grass has been practised since the 1960s (Figure 4.22). It was carried out to enhance aeolian foredune accumulation and to accelerate depositional processes after erosion episodes, as well as to enhance any recreational qualities of the beach.

Conclusions

Coastal sections with landscape-valuable bluffs, as well as broad sand or pebble-gravel beaches, are significant objects for a coastal tourism industry. By artificially restricting natural processes, the intrinsic value of these objects may decrease or even disappear.

■ **Figure 4.22** *Dune planting close to the southern jetty of the port of Ventspils*

Negative effects of artificial protection features are seen adjacent to structures. Excavation related to port shipping channels has stopped longshore drift and large amounts of excavated sediment has been lost to the system by being dumped offshore. Transfer of excavated uncontaminated sediments from shipping channels and ports to sediment deficiency areas, should be mandatory. If existing harbour management practices continue, an extension can be expected with respect to the lengths of coastal erosion sections.

Lithuania

Introduction

Lithuania has one of the shortest coastlines (90.6 km) among all European countries, only three other European countries – Montenegro, Slovenia and Belgium – have shorter marine coastlines (Gudelis, 1967; Boldyrev *et al.*, 1976). With respect to geology and geomorphology the Baltic coast, which is shaped by wave-induced processes, can be divided into two

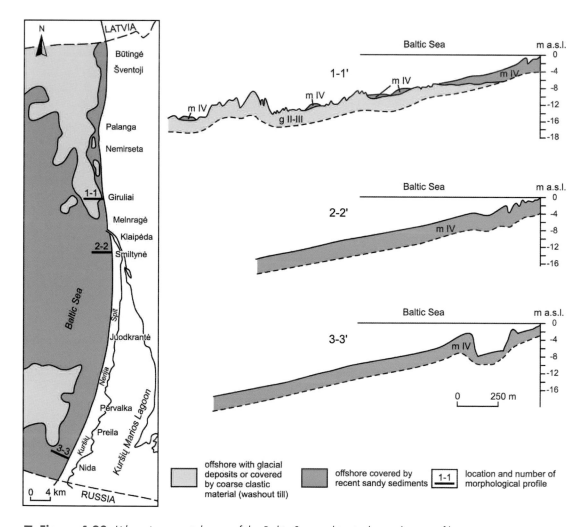

■ **Figure 4.23** *Lithuanian coastal area of the Baltic Sea and typical nearshore profiles*

different segments: a sand peninsula (the Curonian Spit), and the continental coast (mainland). They are separated by the narrow Klaipėda Strait that connects the Curonian Lagoon to the Baltic Sea, stretching north from Klaipėda (Figure 4.23), the third biggest Lithuanian city and one of the biggest Baltic Sea ports. The coast is important for nature conservation with the Curonian Spit protected as a UNESCO World Heritage Site. Intensifying use of coastal resources, especially port development and increase in recreational activities, are some of the most important factors causing coastal erosion resulting in application of coastal protection measures.

Coastal morphology

The Curonian Spit is the largest accumulative coastal landform in the Baltic Sea, being 95 km in length. The northern 51 km is in Lithuania, while the remaining 44 km is part of Kaliningrad Oblast, Russia. The spit width varies from a minimum of 380 m in Russia to a maximum of 3.8 km in Lithuania. Spit development started some 8,500–8,300 years ago at the end of the Ancylus Lake formation stage of the Baltic Sea and took its present shape in the Littorina Sea phase, 6,900–6,300 years ago (Bitinas and Damušytė, 2004; Damušytė, 2009). The barrier sand spit formed on the remnant of a glacial moraine, as a result of sand accumulation by longshore sediment transport. Historically, the Curonian spit has continued to evolve by dramatic changes in natural processes which took place during the 1700s and 1800s, when ancient parabolic dunes, which had prevailed since Holocene times, were completely destroyed by drifting sand and replaced by barchan dunes (Gudelis, 1995, 1998).

Different relief sections occur along the spit: beach, foredune ridge, blown sand plain, blow-out dune remnants, great dune ridges and the shore of the Curonian Lagoon (Gudelis, 1995). The most dominant element of Curonian spit relief is the ridge of high drifting dunes (Figure 4.24 top), stretching for some 80 km. The ridge width varies from 0.3 to 1 km and average dune height is 30 m, with some reaching 50–60 m. The 31 km long ridge of 40–60 m

■ **Figure 4.24** *Main landscape forms of the Curonian Spit: great dune ridge (top) and foredune (bottom) (Photograph by G. Gražulevičius)*

high migrating dunes is the longest coastal drifting dune ridge in Europe (Povilanskas and Chubarenko, 2000).

Another important element is the foredune ridge, a man-made protective beach dune, whose construction started 200 years ago. It stretches along the entire spit length (Figure 4.24 bottom) with heights varying from 7–8 m at Juodkrantė to 15 m at Kopgalis (in the northern spit) the width ranging from 50–60 m to 90–100 m. The presence of ravines and pits on the foredune top provides evidence of intensive blowout processes. Behind the protective foredune is a vegetated, blown sand plain. The Curonian Spit nearshore is characterized by very specific underwater topography and the presence of two or more 4–6 m high sandbars in water depths of up to 8–9 m (Figure 4.25; Gelumbauskaitė, 2003).

Three morpholithodynamic sectors may be distinguished on the Curonian Spit coast:

1. southern – erosional and accumulative (in the Russian part);
2. central – transitory (stable);
3. northern – accumulative (Kirlys and Janukonis, 1993).

The Lithuanian section includes both transitory and accumulative shoreline sectors.

The current state of the Curonian Spit depends very much on the functioning of the longshore sediment transport which occurs in a 3–4 km wide zone. Prior to the 1950s, the sediment flow intensity was distributed as follows: $350-400 \times 10^3$ m^3 of sand was transported along the spit; $500-700 \times 10^3$ m^3 of sediments reached Klaipėda and only $150-200 \times 10^3$ m^3 continued northwards from Klaipėda towards the Latvian border (Knaps, 1969). After the 1950s the amount of sand in the nearshore zone has decreased considerably (Žaromskis, 2007b), mainly through stabilization of the Sambian peninsula shores and depletion of erosional sediment sources on the Baltic Sea bottom (Zhamoida *et al.*, 2009). This resulted

in an increased length of abrasive spit segments, particularly in the Russian sector.

The Curonian Spit and mainland coast are separated by the narrow mouth of the Curonian Lagoon known as the Klaipėda Strait. The Curonian Lagoon (surface area 1,584 km^2) is the biggest shallow semi-landlocked freshwater basin located in the south-eastern part of the Baltic Sea. The Nemunas River (catchment basin of 97,864 km^2) drains into the lagoon on its way to the Baltic Sea and the Klaipėda Strait water area is occupied by one of the biggest sea ports in the Baltic region. Breakwaters and a permanently deepened port entrance channel act as a trap for longshore sediment transport, thereby affecting the state of the mainland coast.

The continental coast consists of alternating erosional and depositional zones. Several different lithologies occur, but sand or till deposits are dominant. While sand sediments, formed mainly in the Littorina and Post-Littorina phases of the Baltic Sea development, prevail in the northern part (Šventoji, Palanga), the southern part of the mainland

■ **Figure 4.25**
Beach dunes between Palanga and Šventoji

coast (Nemirseta, Giruliai) is characterized by glacial (till) deposits (Bitinas *et al.*, 2005), that mainly form eroded cliffs. A natural foredune stretches along the entire continental coast. This varies from a height of 4–6 m and a width of 50–60 m at Būtingė or Melnragė to a height of 9–10 m and a width of 100–130 m to the south of Šventoji (Figure 4.25). In several places, especially in the northern part near the Latvian border during the last 30 years, the foredune has been heavily destroyed and intensification of foredune erosion is related to an increasing frequency of storm events and recreational loadings.

Evolution of coastal settlements and coast usage

The first coastal settlements were created from 3–4,000 BC (Rimantienė, 1999). A convenient trading road connected Semba and Kuršas (Strakauskaitė, 2004) and in the thirteenth to sixteenth centuries AD the biggest settlements were created along this road. During spit deforestation due to overgrazing, timber harvesting and boat-building in the eighteenth century, dunes were activated and entire villages buried. The prevailing westerly winds reworked the parabolic dunes to form the single dune ridge which was supplemented by sand from the western part of the spit (Gudelis, 1998).

Between 1706 and 1846, 14 villages were buried beneath the sand (Gudelis, 1998). The massive sand movement was mitigated by large scale revegetation and reforestation efforts. Since the middle of the nineteenth century, the Curonian Spit has become known as a famous Lithuanian holiday resort with unique landscapes (Armaitienė *et al.*, 2007) and the whole territory of the Curonian Spit (including the Russian portion) has protected area status. Since 2000, the Curonian Spit has been on UNESCO's World Heritage List under cultural criteria "V", an outstanding example of a traditional human settlement, land-use, or sea-use which is representative of a culture or human interaction with the environment especially when it has become vulnerable under the impact of irreversible change (WHC-2000/CONF.204/21). In addition to its conservation

value, the Curonian spit is an important national and international tourism destination. In the Lithuanian part there are four resorts (Nida, Preila, Pervalka, Juodkrantė), integrated into one town.

Klaipėda's history began in 1252, when Memelburg castle was founded by the Order of Crusaders along the Klaipėda strait and the Danė river. In 1808, Klaipėda, for a short period, even became the capital of the Kingdom of Prussia. The territory of Klaipėda, which was densely populated by Lithuanians, was officially connected to Lithuania in 1923 and after this date the port developed rapidly. After World War II, the port fulfilled the primary needs of the Soviet Union and during Soviet times, most of the coastal zone was a closed military area. Sea access was possible only during the day and in narrow coastal strips designated for recreational purposes. When Lithuania regained its independence in 1990, the coastal area became fully opened to the public.

Currently the coastal region is characterized by the fastest economic growth in the country. The GDP per person is 1.5 times higher than in other regions (Žaromskis, 2006), one of the main reasons being the successful operation of Klaipėda port. In 1995, cargo turnover at the port was 12.7 million tonnes and in 2010 it had increased to 31.3 million tonnes (Figure 4.26). Moreover, Klaipėda developed as an attractive tourist destination. The northern part of the Curonian spit (Smiltynė and Kopgalis), as well as recreational settlements on the mainland coast (Melnragė, Giruliai), belongs to the municipality of Klaipėda.

A large part of the continental coast belongs to Palanga, which also includes Šventoji settlement. Palanga is the most popular resort on the Lithuanian coast, famous for its high quality sand beaches. At the end of the nineteenth century there was an attempt to establish a port in present-day Palanga and a breakwater was constructed (Žaromskis, 2006). Unfortunately, the quay rapidly filled with sediment, while the breakwater eventually became the promenade pier (Figure 4.26 centre). Dam construction had a major impact on further developments of the Palanga coast.

Ten kilometres north from Palanga, at the mouth of the Šventoji River, there was an ancient port,

■ **Figure 4.26** *Hydrotechnical constructions on shore: Klaipėda sea port breakwaters (top); Palanga promenade pier (centre); old Šventoji port filled with sand (bottom) (Photographs by V. Karaciejus, Klaipėda State Sea Port Authority and A. Tirlikas, PC ORLEN Lietuva)*

which is no longer used. Intensive port construction took place during 1924 to 1939, but stopped suddenly because of World War II. Afterwards construction works were discontinued and the port was covered with sand (Figure 4.26 bottom). After much discussion it was decided to reconstruct the port in 2011 for recreational fishing purposes.

Since 2000, the Būtingė Oil Terminal, near to the Šventoji settlement commenced operations with an annual turnover of 10 million tonnes per year. The terminal is the only one of its kind in the Baltic Sea, because tanker moorage, as well as cargo handling, takes place using a single point mooring (SPM) buoy, installed 7 km from the coastline in 20 m of water, crude oil being pumped to inland facilities via a pipeline.

Most of the coast is now under some form of protected area designation, which influences coastal zone usage possibilities; e.g. the entire Curonian Spit and part of the continental coast (Seaside Regional Park) are NATURA 2000 territories.

Beach erosion

This was already serious some 40 years ago (Žilinskas, 2005) and since then awareness has increased yearly, the length of eroding coastline moving from 18 per cent to 27 per cent between 1990 and 2003. By contrast, the proportion of accreting coast fell from 40 per cent to just 12 per cent over the same period, while the stable coastline rose from 42 per cent to 61 per cent (Figure 4.27). There are several reasons for the above. First, there was a significant reduction of sediment supply because of changes in long-shore sediment transport in the SE Baltic Sea region. There was a dramatic decrease of sand availability in the northern nearshore of the Sambian Peninsula, where sediment movement was blocked after shore stabilization involving hard structures (Zhamoida *et al.*, 2009). Geomorphological development was also influenced by human activity, particularly the operation of Klaipėda port. Breakwaters and the deep entrance channel intercept more than half the longshore sediment drift, resulting in the continental coast suffering an additional sand deficit (Žaromskis, 2007a).

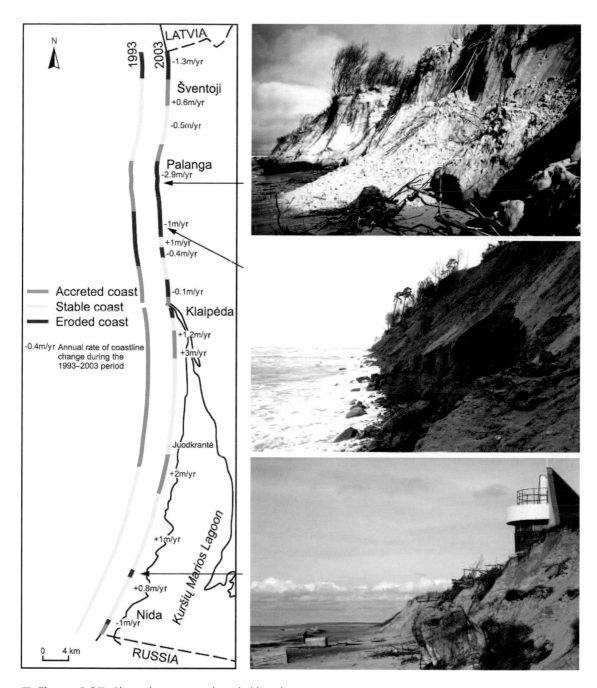

■ **Figure 4.27** *Shore dynamics and eroded beach segments*

Climate changes also affect shoreline stability and acceleration of relative sea level rise has been observed on the Lithuanian coast (Jarmalavičius *et al.*, 2007). Between 1970 to 2000, the Baltic Sea level rose by more than 15 cm (Johansson *et al.*, 2001; Dailidienė *et al.*, 2006) and long-term investigation of sea level rise in the Lithuanian Baltic Sea shows an annual increase of 6,5 mm. A rise of sea level up to 60 cm would cause significant problems for inhabitants and the land infrastructure (Žaromskis, 2001).

More frequent storms (wind speed >24 m/s) have caused sand loss from the coastal zone (Žaromskis and Gulbinskas, 2010). Formerly, deep cyclones occurred every six to eight years, but recently this frequency has changed to every two to three years, so that the coast has no time to restore its equilibrium profile (Žilinskas, 2008).

Shore protection

Coastal protection has a long history, even though the first coastal protection measures were not aimed to stop erosion, but rather to mitigate impacts of aeolian processes (Žaromskis, 2007b). In the Lithuanian part of the Curonian Spit, such works were first initiated in Klaipėda in 1810 and later (1825) activities were also undertaken in Nida. Continuous protective beach foredunes were also created at the beginning of the twentieth century. Foredune formation principles, developed in the nineteenth century were successfully applied in modern coastal management practices and are used today (Gerhardt, 1900). Until the mid twentieth century the protective foredune was created along almost the whole coastal length, except where several sectors had moraine cliffs (north from Klaipėda Strait).

Shore protection regulation

Coastal management is based on laws, long-term coastal management programmes and implementation of coastal protection projects. According to Lithuanian Law on the coastal strip, the coastal zone is that part of the coast including the nearshore and extending not less than 100 m inland, extending from the border of Latvia to the northern breakwater of Klaipėda port, including part of Curonian Spit up to the Russian border and the Baltic Sea part of territorial waters up to 20 m in depth (Seimas of the Republic of Lithuania, 2002).

The main principles of coastal protection are stated in the Lithuanian Baltic Sea coastal protection strategy (Ministry of Environment of the Republic of Lithuania, 2001) and are as follows:

- coastal protective measures can be applied only in strips where coastal erosion endangers human activities, protected areas and natural or cultural heritage values;
- priority should be given to the preservation of natural landscape and natural coastal formation processes;
- there is a need to ensure complex coordination of coastal protection and coastal use.

Long-term coastal management programs specify application of coastal protection measures according to functional priorities and natural coastal dynamics. Coastal segments are distinguished by their functions, coastal erosion rates, recreational capacities, etc. in order to identify the problem sectors, before agreeing to any final coastal protection and management measures (Gulbinskas *et al.* 2009).

Shore protection by elementary measures

During recent years, coastal protection works have expanded. The protective foredune is maintained by a brushwood flooring of foredune slopes and installation of brushwood fences, which help catch and accumulate sand, which builds up faster when surface roughness is increased and wicker fences at the dune base facilitate this process. Other measures include construction of wooden stairs and footpaths in recreational areas, planting of greenery in dunes, as well as maintaining natural processes (Figure 4.28).

■ **Figure 4.28** *Foredune maintenance by elementary measures near Palanga*

Hard constructions

Hard constructions are limited and used only for protection of a small segment at the centre of Palanga. For over 100 years, the impermeable Palanga promenade pier acted as a sediment trap and was dismantled in 1998 when a new permeable pier was erected. This changed the coastal dynamics and stimulated intensification of coastal erosion. In order to restore the former situation it was decided to build a stone groin in place of an old sea pier. A semi-permeable groin was built in 2005 (Figure 4.29a), but due to a relatively low height (*circa* 1 m above water level) it does not disturb longshore sand drift during strong storms, but traps sand brought by small waves.

Nearshore and beach nourishment

Beach and nearshore nourishment works commenced in 2001, sand extracted offshore and during the dredging of Klaipėda's port entrance channel being used for beach nourishment. Large volumes of sand are dredged from the Klaipėda sea port entrance channel (Figure 4.29a) and this sand is proposed to be used as a compensation measure to restore the sand balance disturbed by the port breakwaters. As a result the nearshore strip, north from the port entrance channel is regularly supplemented with dredged sediments. During the last ten years the nearshore has been nourished by a sand volume of 850×10^3 m³.

Palanga resort is the most problematic coastal sector with regard to coastal erosion. At the end of the nineteenth century, the width of Palanga beaches was 150–180 m but in the mid twentieth century, because of decreasing sand supply, beach width declined to 60–80 m. The most intensive erosion took place between 1993 and 2007, when average sand loss was calculated at 13.0 m³/m/yr (Žilinskas, 2008). At the beginning of 2005, the beach width in some places decreased to 10–15 m.

Palanga beach nourishment was implemented in several stages. In 2006 the central part (800 m length segment) was replenished with 40×10^3 m³ of sand, brought from an inland quarry (Figure 4.29b). As a result beach width increased up to 40 m. In 2008 the beach (1,200 m length segment) was nourished with ~111×10^3 m³ of sand, brought from the seabed (Figure 4.29c). This time the coastal sector width increased by up to 70 m. The third stage of beach nourishment was supplemented in 2011–2012 and the 2.4 km long Palanga beach sector nourished with 424×10^3 m³ of sand. Monitoring confirmed that beach nourishment had a positive effect on reducing erosion, resulting in an increase in beach width, stabilization of the coastline and nearshore sand accumulation.

■ **Figure 4.29** *Palanga beach regeneration area (right): building of groin (a); beach nourishment (b, c)*

Conclusions

Early coastal development was characterized by a large sand supply and a south–north longshore sediment drift, which created a dominance of accumulative sand coasts. However, during recent decades degradation and further disappearance of sand beaches was observed, mainly because of intensive human activity. Lithuanian coastal management gives priority to environmental protection and preservation of natural coastal dynamics. Coastal protection strategy and measures are selected according to functional coastal usage. Seeking to ensure harmonized nature and human interaction, the following soft coastal protection measures have been applied on sensitive coastal areas in Lithuania: 1) maintenance of the protective foredune using traditional 'soft engineering' measures; and 2) nearshore and beach nourishment.

References

Armaitienė, A., Boldyre, V. L., Povilanskas, R. and Taminskas J., 2007. Integrated shoreline management and tourism development on the cross-border World Heritage Site: A case study from the Curonian spit (Lithuania/Russia). *Journal of Coastal Conservation*, 11, 13–22.

Bird, E. C. F., 1985. *Coastline changes: A global review.* John Wiley & Sons, New York. 220 pp.

Bird, E. C. F., Martin, E. and Orviku, K., 1990. Reed encroachment on Estonian beaches. *Proceedings of the Estonian Academy of Sciences*, 39, 7–12.

Bitinas, A. and Damušytė, A., 2004. The Littorina Sea at the Lithuanian Maritime Region. *Polish Geological Institute Special Papers*, 11, 37–46.

Bitinas, A., Žaromskis, R., Gulbinskas, S., Damušytė, A., Žilinskas, G. and Jarmalavičius, D., 2005. The results of integrated investigations of the Lithuanian coasts of the Baltic Sea: Geology, geomorphology, dynamics and human impact, *Geological Quarterly*, 49(4), 355–362.

Boldyrev, V. L., Gudelis, V. K., Shuiski, and Yu. D., 1976. Baltic Sea coasts and their role in sediment supplying. In: *Geology of the Baltic Sea*, V. Gudelis and E. Emelyanov Eds., Mokslas, Vilnius, 141–158. [In Russian]

Briede, A., 2006. Latvijas klimata mainības tendences./Tendencies of climate change in Latvia. *LU 64. zinātniskā konference. Ģeogrāfija. Ģeoloģija. Vides zinātne. Referātu tēzes,* 19–21. [In Latvian]

Dailidienė, I., Davulienė, L., Tilickis, B., Stankevičius, A. and Myrberg, K., 2006. Sea level variability at the Lithuanian coast of the Baltic Sea. *Boreal Environment Research*, 11, 109–121.

Damušytė, A., 2009. Late glacial and Holocene subfossil mollusc shells on the Lithuanian Baltic coast. *Baltica*, 22, 111–122.

Eberhards, G., Lapinskis, J. and Saltupe, B., 2006. Hurricane Ervin 2005 coastal erosion in Latvia. *Baltica*, 19(1), 10–19.

Eberhards, G., 2003. *Latvijas jūras krasti. [Coasts of Latvia.]* Latvijas Universitāte, Riga, 259 pp. [in Latvian with English summary]

Eberhards, G. and Lapinskis, J., 2008. *Baltijas jūras Latvijas krasta procesi: Atlants. [Processes on the Latvian Coast of the Baltic Sea:. Atlas.]* Latvijas Universitāte, Rīga, 64 pp. [In Latvian]

Eberhards, G. and Purgalis, I., 2008. Pieaugošo Latvijas jūras krastu eroziju sekmējošie faktori[Aggravating factors of coastal erosion on former accumulative coasts of Latvia]. *Klimata mainība un ūdeņi*. Latvijas Universitāte, Rīga, 40–48. [In Latvian]

Eberhards, G., Grine, I., Lapinskis, J., Purgalis, I., Saltupe, B. and Torklere, A., 2009. Changes in Latvia's seacoast (1935–2007). *Baltica*, 22(1), 11–22.

Gelumbauskaitė, L., 2003. On the morphogenesis and morphodynamics of the shallow zone of the Kurš Nerija (Curonian Spit). *Baltica*, 16, 37–42.

Gerhardt, P., 1900. *Handbuch des deutschen Dünenbaues*, Berlin, 656 pp.

Gudelis, V. K., 1967. Morphogenetic types of coasts and shores of Baltic Sea. *Baltica*, 3, 123–145. [In Russian with English summary]

Gudelis, V. K., 1995. The Curonian barrier spit, south-east Baltic: Origin, development and coastal changes. In: *Coastal conservation and management in the Baltic Region* (Proceedings of the EUCC-WWF

Conference, 3–7 May, 1994, Riga–Klaipeda–Kaliningrad), V. K. Gudelis (Ed.). Klaipėda, Lithuania: University Publishers, 11–13.

Gudelis, V. K., 1998. *The Lithuanian offshore and coast of the Baltic Sea.* Lietuvos mokslas, Vilnius, 444 pp. [In Lithuanian]

Gulbinskas, S., Milerienė, R. and Žaromskis, R., 2009. Coastal management measures in Lithuanian Baltic coast (South Eastern Baltic). In: *Coastal Engineering 2008*, Proceedings of the 31st International Conference, Vol. 5, Jane McKee Smith (Ed.). World Scientific Publishing Co. Pte. Ltd., Singapore, 4042–4052.

Jarmalavičius, D., Žilinskas, G. and Dubra, V., 2007. Long-term dynamic peculiarities of water level fluctuations in the Baltic Sea near the Lithuanian coast. *Baltica*, 20 (1–2), 28–34.

Johansson, M., Boman, H., Kahma, K. and Launiainen, J., 2001. Trends in sea level variability in the Baltic Sea. *Boreal Environment Research*, 6, 159–179.

Kessel, H. and Raukas, A., 1967. *The deposits of the Ancylus Lake and Litorina Sea in Estonia.* Valgus, Tallinn. 134 pp. [In Russian, with English summary]

Kirlys, V. and Janukonis, Z., 1993. Dynamical characteristics and classification of the coastal zone in the South-East sector of the Baltic Sea. *Geography*, 29, 67–71.

Knaps, R., 1969. Sediment transport in the offshore of East Baltic. In: *Coastal development under the conditions of tectonic movements of the Earth crust*, G. A. Orlova and K. K. Orviku, Eds., Valgus, Tallinn, 36–44. [In Russian]

Lapinskis, J., 2009. Preterozijas pasākumi Baltijas jūras Latvijas krastā [Coastal protection measures on the Baltic Sea coast of Latvia]. *LU 67. Zinātniskā konference. Zemes un vides zinātņu sekcija.* Rīga, 78–80. [In Latvian]

Martin, E. and Orviku, K., 1988. Artificial structures and shoreline of Estonian SSR. In: *Artificial structures and shorelines*, H. J. Walker, Ed. Kluwer Academic Publishers, Dordrecht, 53–57.

Meier, H. E. M., Broman, B. and Kjellström, E., 2004. Simulated sea level in past and future climates of the Baltic Sea. *Climate Research*, 27, 59–75.

Ministry of Environment of the Republic of Lithuania, 2001. Strategic principles of Lithuanian Baltic sea coast management. *Valstybės Žinios*, 103–136.

Orviku, K., 1965. On the accumulation of boulders along the seacoasts of Estonia. *Oceanology*, 2(5), 316–321. [In Russian, English summary]

Orviku, K., 1974. *Estonian seacoasts.* Academy of Sciences of Estonia, Tallinn. 112 pp. [In Russian with English summary]

Orviku, K., 1982. Klint coast. In: *The encyclopaedia of beaches and coastal environments*, M. L. Schwartz, Ed. Dowden, Hutchinson & Ross, Stroudsburg, PA, 502–503.

Orviku, K., 1988. Some aspects of shallow water sedimentation along the Estonian coast. In: *The Baltic Sea,* Papers prepared for a colloquium on Baltic Sea marine geology in Parainen, Finland, 27–29 May 1987, B. Winterhalter and H. Ignatius, Eds., 73–77.

Orviku, K., 1992. *Characterisation and evolution of Estonian seashores.* Tartu Ülikool, Doctoral thesis, 20 pp. [English summary]

Orviku, K., 1993. Nüüdisrandla. In: *Geology of the Estonian shelf.* J. Lutt and A. Raukas, Eds. Estonian Geological Society, Tallinn, 29–39.

Orviku, K., 1995. Extensive storm damage on the Estonian seashore: Sharpening conflict between man and nature? In: *Coastal Conservation and Management in the Baltic Region: Proceedings of the EUCC – WWF Conference*, 3–7 May, Klaipeda, V. Gudelis, R. Povilanskas, and A. Roepstorff, Eds., 19–22.

Orviku, K., 2010. Tallinna rannikuala geoloogia. *Tallinna Geoloogia [Geology of Tallinn]*, A. Soesoo and A. Aaloe, Eds. Noria Books, Tallinn, 202–229.

Orviku, K. and Granö, O., 1992. Contemporary coasts. In: *Geology of the Gulf of Finland Tallinn,* A. Raukas and H. Hyvarinen, Eds., 219–238. [In Russian, English summary]

Orviku, K. and Romm, G., 1992. Litho-morphodynamical processes of Narva Bay. *Proceedings of the Estonian Academy of Sciences.*

Geology. 41(3), 139–147. [In Russian, English summary]

Orviku, K. and Sepp, V., 1972. Stages of geological development and landscape types of the islets of the West-Estonian Archipelago. *Geographical Studies*, 15–25.

Orviku, K., Bird, E. C. F. and Schwartz, M. L., 1995. The provenance of beaches on the Estonian islands of Hiiumaa, Saaremaa and Muhu. *Journal of Coastal Research*, 11, 96–106.

Orviku, K., Jaagus, J., Kont, A., Ratas, U. and Rivis, R., 2003. Increasing activity of coastal processes associated with climate change in Estonia. *Journal of Coastal Research* 19(2), 364–375.

Orviku, K., Suursaar, Ü., Tõnisson, H., Kullas, T., Rivis, R. and Kont, A., 2009. Coastal changes in Saaremaa Island, Estonia, caused by winter storms in 1999, 2001, 2005 and 2007. *Journal of Coastal Research* SI 56, 1651–1655.

Orviku, K., Tõnisson, H. and Jaagus, J., 2011. Sea ice shaping the shores. *Journal of Coastal Research*, SI64, 681–685.

Orviku, K., Tõnisson, H., Aps, R., Kotta, J., Kotta, I., Martin, G., Suursaar, Ü., Tamsalu, R. and Zalesny, V., 2008. Environmental impact of port construction: Port of Sillamäe case study (Gulf of Finland, Baltic Sea). In: 2008 IEEE/OES US/EU-Baltic International Symposium: *Ocean Observations, Ecosystem-Based Management and Forecasting*, IEEE-Inst Electrical Electronics Engineers Inc, 350–359.

Povilanskas, R. and Chubarenko, B. V., 2000. Interaction between the drifting dunes of the Curonian barrier spit and the Curonian lagoon. *Baltica*, 13, 8–14.

Povilanskas, R., Baghdasarian, H., Arakelyan, S., Satkunas, J. and Taminskas, J., 2009. Secular morphodynamic trends of the Holocene duneridge on the Curonian Spit. *Journal of Coastal Research*, 25(1), 209–215.

Raukas, A., Bird, E. and Orviku, K., 1994. The provenance of beaches on the Estonian Islands of Hiiumaa and Saaremaa. *Eesti Teaduste Akadeemia toimetised. Geoloogia*, 43(2), 81–92.

Rimantienė, R., 1999. *Die Kurische Nehrung aus dem Blickwinkel des Archäologen*. Vilniaus Dailės Akademijos leidykla, Vilnius, 110 pp.

Seimas of the Republic of Lithuania, 2002. Law of the Coastal Strip of the Republic of Lithuania. *Valstybės Žinios*, No 73–3091.

Soomere, T., Viška, M., Lapinskis, J. and Räämet, A., 2011. Linking wave loads with the intensity of coastal processes along the eastern Baltic Sea coasts. *Estonian Journal of Engineering*, 17(4), 359–374.

Strakauskaitė, N., 2004. *The Curonian Spit: The old European postal road*. Versus Aureus, Vilnius, 141 pp.

Tõnisson, H., 2008. Development of spits on gravel beach types in changing storminess and sea level conditions. Tallinna Ülikool, Ph.D. thesis. 245 pp.

Tõnisson, H., Orviku, K., Jaagus, J., Suursaar, Ü., Kont, A. and Rivis, R., 2008. Coastal damages on Saaremaa Island, Estonia, caused by the extreme storm and flooding on January 9, 2005. *Journal of Coastal Research*, 24(3), 602–614.

Tõnisson, H., Suursaar, Ü., Orviku, K., Jaagus, J., Kont, A., Willis, D. A. and Rivis, R., 2011. Changes in coastal processes in relation to changes in large-scale atmospheric circulation, wave parameters and sea levels in Estonia. *Journal of Coastal Research*, SI 64, 701–705.

Ulsts, V., 1998. *Baltijas jūras Latvijas krasta zona* [*Latvian coastal zone of the Baltic Sea*]. Valsts Ģeoloģijas Dienests, Rīga, 96 pp. [In Latvian]

Vallner, L., Sildvee, H. and Torim, A., 1988. Recent crustal movements in Estonia. *Journal of Geodynamics*, 9, 215–223.

Veinbergs, I., 1986. *Drevnye berega Sovetskoy Baltiki i drugih morei SSSR. /Ancient coasts of Soviet Baltic republics and other seas of the USSR*. Zinatne, Riga, 168 pp. [In Russian]

WHC-2000/CONF.204/21., 2000. *Report of the World Heritage Committee, twenty-fourth session*. Cairns, Australia, 27 November – 2 December 2000, 155 pp.

Žaromskis, R., 2001. Impact of climatic changes on the shores of the Baltic Sea and Kurši˜ marios lagoon. In: *The influence of climatic variations on physical geographical processes in Lithuania*, A. Bukantis, Ed., Institute of Geography, Vilnius, 122–164.

Žaromskis, R., 2006. *Lithuanian seaside and dunes.* Klaipėda County Governor's Administration, Klaipėda, 48 pp.

Žaromskis, R., 2007a. Impact of harbour moles and access channels on the South-East Baltic shore zone. *Geography*, 43(1), 12–20. [In Lithuanian]

Žaromskis, R., 2007b. The formation of protective dune ridge along the Southeast Baltic Sea coast: Historical and social aspects. *Journal of Coastal Conservation*, 11, 23–29.

Žaromskis, R. and Gulbinskas, S., 2010. Main patterns of coastal zone development of the Curonian Spit, Lithuania. *Baltica*, 23(2), 146–156.

Zhamoida, V., Ryabchuk, D. V., Kropatchev, Y. P., Kurennoy, D., Boldyrev, V. L. and Sivkov, V. V., 2009. Recent sedimentation processes in the coastal zone of the Curonian Spit (Kaliningrad region, Baltic Sea. *Z. Dt. Ges. Geowiss.*, 160(2), 143–157.

Žilinskas, G., 2005. Trends in dynamic processes along the Lithuanian Baltic coast. *Acta Zoologica Lituanica,* 15(2), 204–207.

Žilinskas, G., 2008. Distinguishing priority sectors for the Lithuanian Baltic Sea coastal management. *Baltica*, 21(1–2), 85–94.

5 Poland

Kazimierz Furmańczyk

Introduction

With its population of some 38.2 million and an area exceeding 300,000 km², Poland is a Central European country bordering, in the north, the southern Baltic Sea; if the total population were to stand along Poland's >500 km coastline (excluding the shores of the Szczecin and Vistula Lagoons), there would be 76.4 people in each metre! Some 420 km of the coast is fringed with wide sandy beaches backed by dunes; about 80 km are cliff shores rising above sand-shingle-gravel beaches shaped during Pleistocene and Holocene times (Figure 5.1). The coast is a very popular tourist destination, which means that, in summer, its sandy beaches are crowded despite very unstable weather and water temperature

that rarely exceeds 20°C. Both the natural factors and, to some extent, the recreational pressure makes the Polish coast exceptionally susceptible to erosion.

The southern Baltic coast morphology is a result of the Scandinavian ice sheet retreat and the sea level changes that occurred during the Pleistocene and Holocene.

Throughout the recent 18,000 years of Baltic Sea history, its water level was also affected by the presence or absence of the connection with the North Sea. Other factors that significantly influenced the process of shaping coastal morphology over various time scales include storm surges and isostatic movements caused by ice sheet melting and retreat.

Land forms left in the coastal zone in the wake of the retreating ice sheet include frontal moraine hills,

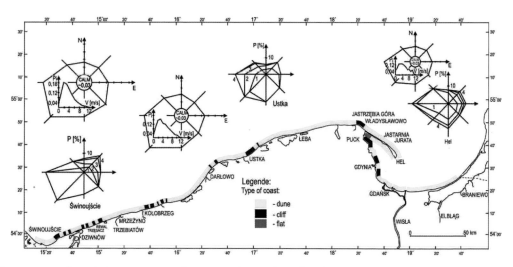

■ **Figure 5.1** *The Polish coast. Types of coasts and annual wind and wave roses (based on Zeidler et al., 1995; Zawadzka-Kahlau, 1999)*

moraine highlands built of boulder clay and sands as well as lowlands and moraine valleys of fluvial origin, built of clays and sands. Post-Pleistocene, the present coastal zone was gradually flooded whenever the Baltic Sea level rose; this formed cliffed shores at the edge of highlands and flat and low shores in the lowlands and ice-marginal valleys. In the Holocene, oscillations associated with the changing sea level gradually decreased. Whereas about 6,000 to 4,000 years ago the southern Baltic Sea shore was situated about 2–3 km offshore its present position, changes that occurred within the last 2,000 years – apart from some minor exceptions – did not exceed 600 m (Uścinowicz, 1999).

During the periodic sea level fluctuations, long-shore and cross-shore sediment transport resulted in the formation of spits and coastal lakes. With a sea level rise (sea transgression), cliff shores were eroded and the sand material created was transported along the shore and deposited in embayments and estuaries of ice-marginal valleys forming wide sandy beaches backed by dunes (Figure 5.2). With dropping sea level (regression), many accumulative forms emerged, such as spits or dunes covering lower cliffs (Musielak, 2006).

The coastal zone seabed is built of sand, shingle and gravel sediments, which form the constituents of many longshore bars, including single, double and multibar systems. Generally, single bars are usually present off cliff shores, double bars are to be found off cliff and dune shores, while multibars can be seen off some dune shores. Those bars are a form of natural underwater wave breaks, as they partially dissipate wave energy.

The Baltic Sea is microtidal (the tidal amplitude does not exceed several centimetres), therefore factors

■ **Figure 5.2** *A Holocene dune shore near Dziwnów (Photograph by P. Domaradzki)*

not related to tides, essentially storm surges and waves (their amplitude and duration), play an important role in shoreline formation.

Most storms occurring during low sea levels produce seabed and beach changes, but do not affect the dunes or cliffs, as dune or cliff erosion is a result of 'significant storms'; these usually happen once every 10–12 months and are associated with very high storm surges (Furmańczyk and Dudzińska-Nowak, 2009). The majority of winds in the southern Baltic are south-westerly, blowing from the offshore sector, but due to coastal exposure to the north-east and north-west, the most severe storms come from those directions (Figure 5.1).

Another important factor influencing coastal erosion is a result of North Sea water inflow to the Baltic Sea. Inflow occurs when a deep depression travels slowly above the Danish Straits, increasing the water volume that flows from the North Sea to the Baltic Sea. Such a situation occurs once every several years and high sea levels continue for many weeks. When an inflow is accompanied by a strong storm surge, the shore suffers catastrophic erosion; in such situations, the possibility of flooding is also very high. The highest storm surges associated with the North Sea inflows were recorded in 1874, when the Świnoujście tide gauge showed 1.96 m above mean sea level, and in 1913/14, when it indicated 1.95 m. This resulted in major erosion over almost the entire coast; dunes were torn apart and storm surge flooding ensued (Majewski *et al.,* 1983; Zeidler *et al.,* 1995; Wiśniewski and Wolski, 2009).

When sea level drops in post-storm conditions, the beach rebuilds, but the pace of shore rebuilding is five times slower than that of erosion (Musielak, 1980). Due to wind activity, beach sand accumulates to form foredunes or to enlarge an existing main dune. The extent of dune baseline changes caused by one major storm may be similar to the total of long-term changes occurring within a 40-year period (Furmańczyk, 1995). Therefore, if a series of storms occurs within a short time period, each shore reconstruction response time is short and total erosion changes are at their largest (Ferreira, 2005).

Historical evolution of coastal settlements

In ancient times, people settled at the sea coast mainly because they considered the sea to be a source of food as well as a transport route. Frequently, small fishing settlements were established on the dune coast; fishing boats were stored at the dune base (a practice still used by present-day fishermen), because, in the case of storms, the boats could be easily pulled behind the dune.

For the coast to be considered a transport route, at least an elementary form of port infrastructure is required, consisting basically of a properly secured quay that would not easily be destroyed during a storm, and a basin whose depth would be such that, with time, it could gradually receive larger ships. Such conditions were found at river mouths; the larger the river, the further from its mouth could quays be built, and these ports created opportunities for towns to develop in the hinterland.

In the thirteenth and fourteenth centuries, boat-building was embedded in naval transport, which stimulated development of seaside towns. The main development commenced at port towns belonging to the Hanseatic League, especially those located on the present Polish coast, such as Gdańsk, Kołobrzeg, Szczecin, Elbląg and Braniewo. Gradually, smaller ports including Puck, Łeba, Ustka, Darłowo and others developed (Figure 5.1).

In the first half of the twentieth century, the recreational values of the Baltic coast began to be realized and appreciated. Sea bathing came into fashion among people belonging to the 'upper classes' and this had a large influence on the conversion of fishing settlements into seaside resorts with extensive hinterlands. Many substantial developments came about and facilities offering therapeutic baths were set up in towns located in the vicinity of hot springs, natural brine pools, or peloid resources. Peloids consist of humus and minerals formed over a long time by geological, biological, chemical and physical processes, and are used therapeutically in spas. Typically, they mature for a period of up to two years in special ponds.

The breakthrough in coastal settlement came after World War I, in 1918, when – by the Treaty of Versailles – Poland gained independence and access to the Baltic Sea along a 140-km stretch of the coast extending from the Free City of Danzig to the mouth of the River Piaśnica. As there was only a minor fishing port in Puck and several fishing quays, this coastline was inaccessible to cargo ships. Even though Poland was granted the use of the Free City of Danzig port in the form of a single small stretch (used as the Polish Military Transit Depot known as Westerplatte), this was not enough for the needs of a country which, after more than a hundred-year-long occupation by foreign powers, had gained independence and sea access. In accordance with the then Polish maritime policy, a new trade and military port – and an accompanying town within a 20 km distance from Danzig (Gdańsk) – of Gdynia was built, as were a number of smaller ports used for military and fishing purposes, for example Władysławowo, Hel and Jastarnia and the two seaside resort towns of Jurata and Jastrzębia Góra (Figure 5.1).

After World War II, the Polish coastline was extended to 500 km (excluding the shores of the Szczecin and Vistula Lagoons). Much attention was paid to reconstruction of the destroyed country. The ports in Gdańsk, Gdynia, Szczecin and Świnoujście received a lot of attention and enjoyed major development.

In the early 1950s, sun and sea bathing once again came into fashion countrywide and many new seaside resorts, intended for mass recreation, were built. However, these developments very often encroached on the shore, for which reason the coastal protection costs for these areas are now high. On the other hand, most coastal towns are accessible by roads running perpendicular to the coast and joining the major motorway whose distance from the shore varies by up to several kilometres. It was fortunate that roads and railways were not built near the coast, as this later helped to avoid coastal protection problems.

After transition to a democratic form of government in 1989, these coastal towns have grown steadily and intensively.

Erosion process: natural vs. anthropogenic factors

The seashore erosion process may be studied on different time scales. Taking into consideration single events and years, observable changes are that during one single storm, the coast could change more than it had over several decades. For example, changes in the Hel Peninsula dune baseline provided evidence that such a major storm caused changes (beginning with erosion at the cliff/dune base) comparable to those that had occurred over 40 years (Furmańczyk, 1995), supporting the concept of an oscillatory nature of short-term coastline changes,

Short-term changes usually prove to be of an oscillatory nature, changing from erosion to accumulation. On the other hand, long-term changes involve mainly, with minor exceptions, shore erosion (Furmańczyk, 1994).

In addition to processes acting directly on the shore, coastal changes can be induced by rivers. Along the Polish coast, there are several rivers that discharge to the Baltic Sea, including two major ones: the Odra (Oder) and the Vistula. Generally, discharges should strengthen the shore by providing sand material, which would ease the effect of shore erosion. The Odra empties into the Szczecin Lagoon and affects the sediment budget of the lagoon's shores, but its influence on the open sea shore is negligible. Additionally, the river has been navigable and regulated for more than a hundred years. The Vistula discharges directly into the Gulf of Gdańsk. However, the sand material the Vistula carries has been diminishing, mainly due to construction of two dams in its lower reaches and some attempts at regulation, all of which was carried out within the last few decades. Most of the small coastal rivers are regulated and jetties were built in their mouths. This is in part responsible for elimination of river-borne sand material from entering processes in the nearshore zone, as sand is transported directly to sea rather than being retained close to the shore. Under such conditions, the amount of sediment that reaches the active part of the coastal zone is diminished, resulting in increased shore erosion (Furmańczyk and Basiński, 2006).

Within the last 50 years, due to the gradual increase in coastal protection activities and a significant increase in the strengthened coastline length, the volume of material delivered from both cliff and dune shore erosion has decreased. Moreover, wave action produces different effects on a protected shore compared to those on a natural sandy one. Although seawalls reduce cliff or dune erosion, they generate wave reflection, which results in increased beach erosion as well as the seabed and beaches fronting those constructions are usually narrower and lower. In addition, in their vicinity there is increased shore erosion the extent of which, while dependent on local conditions, can be up to several kilometres. The longshore sediment transport can be hampered by groins, thus providing material necessary to build up the beach, but unfortunately, groins cannot prevent shore erosion itself. On the contrary, at the groin field end increased erosion is the norm, just as it is in the case within the vicinity of seawalls (Basiński *et al.*, 1993).

The comparison of topographical maps drawn in the 1880s and 1980s with respect to the Polish coast (Zawadzka-Kahlau, 1999) indicates that the coastline has changed little (Figure 5.3). Although the number of erosion-prone areas has been, and is, higher than that of accumulation areas, most of the coastline is classified as stable, with only minor changes (+50 m or −50 m per 100 years). Furthermore, shores of the inner part of the Puck Bay (also known as the Puck Lagoon), partly flat and similar to shores of the Szczecin and Vistula Lagoons, are stable due to the reed belt they support. The only changes occurring on these shores are those caused by ice-related phenomena.

As mentioned above, although cliff shores in many places are strengthened, they are still prone to erosion. Wolin Island features natural cliff shores soaring up to 90 m (Figure 5.4). The material originating from cliff erosion contributes to the largest coastal accumulation area in the vicinity of Świnoujście, where annual accumulation rates exceed one metre per year.

Dune parts of the coast are very dynamic and bear witness to different trends in shore development. Dune growth can be observed in Świnoujście and at the mouth of the Vistula River, while the coast near Łeba is in dynamic equilibrium. Other parts of the coast show a stronger or a weaker tendency toward erosion, the tendency being somewhat modified by standard protection measures. The strengthened section of the coast, including cliffs and dunes, is altogether some 137 km (Przyszłość, 2006).

Recent research in the western part of the Polish coast demonstrates that, although erosion processes have diminished within the last six decades, the total length of coastal sections affected by erosion has

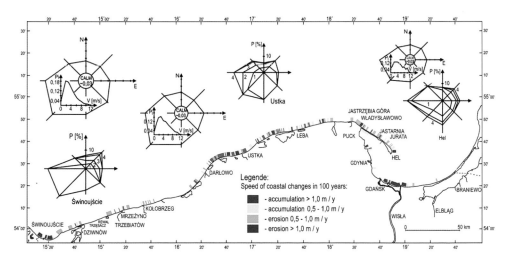

■ **Figure 5.3** *The Polish coast. The rate of coastline change in 100 years (1880s–1980s) and annual wind and wave roses (based on Zeidler et al., 1995; Zawadzka-Kahlau, 1999)*

■ **Figure 5.4** *The Pleistocene sand-clay cliff about 90 m high at Wolin Island shore (Photograph by P. Domaradzki)*

increased (Dudzińska-Nowak, 2006). Each human intervention, such as a hard protection measure, brought results contrary to expectations, *i.e.* erosion was observed to increase in the neighbourhood of the structure. In general, the process of erosion is stopped or reduced within a decade or so after the coast has been strengthened, but erosion increases in neighbouring areas. In the end, such a method of coastline protection appears to be problematic and expensive, mainly due to maintenance costs and the need for strengthening still new sections (Boxes 5.1 and 5.2).

Post-storm measurements carried out by the Maritime Offices and analyses of the data collected within the framework of Project MICORE (Morphological Impacts and COastal Risk induced by Extreme storm events (https://www.micore.eu) suggest that extreme storms, with high significant waves, together with large storm surges associated with North Sea inflows, produce erosion of beaches, dunes and cliffs (Furmańczyk and Dudzińska-Nowak, 2009). Results of such storms are very severe: wave action lowers the beach and the dune baseline can be moved landward up to several dozens of metres. After a storm, the beach and part of the dune become rebuilt either due to natural processes or human activities.

A history of shore protection activities on the Polish coast

Construction of ports and harbours on river banks proceeded in the Middle Ages, but such projects did not require any seashore protection measures other than periodic dredging of river mouths. The first coastal intervention to be carried out on a large scale occurred in the second part of the fifteenth century, when the dune shore near Mrzeżyno was breached in order to create both an extra channel for the Rega River and to build a port in its vicinity. As a result, the hydrography of the area changed and produced unexpected results such as flooding, which since then has affected the surrounding area, including the nearby town of Trzebiatów.

In the second half of the sixteenth century, to provide military protection of waterways leading to major ports such as Gdańsk, Kołobrzeg and Świnoujście, fortifications were built in the river mouth at which those ports were situated. It is believed that the first hydraulic structure on the Polish southern Baltic coast was a jetty built in the Vistula mouth near Gdańsk (Figure 5.5a; Wagener, 1589).

Jacobsz's Nautical Atlas (Jacobsz, 1644) shows jetties built near Gdańsk, on the Vistula and near

■ **Figure 5.5** *Jetties at the mouth of the Vistula River, from Wagener's Atlas (Wagener, 1589; left); jetties at the mouth of Parsęta River in Kołobrzeg from Jacobsz's Atlas (Jacobsz, 1644; right)*

Kołobrzeg, on the Parsęta River (Figure 5.5b). Then, not only were these ports protected by jetties, but they were also deepened to maintain navigability. Later jetties were built in the mouths of other rivers on which ports were located. The jetties limited sand accumulation and increased navigability; however, they became an obstacle to longshore sediment transport and contributed to coastal erosion in the vicinity of river mouths. As port and sea town developments were gradually moving towards river outlets, shore erosion soon became a direct threat to buildings, including fortifications, port facilities and support buildings. Similarly, fishing settlements were first located at a safe distance from the shore, and it was not until the nineteenth century that the inflow of tourists in summer seasons caused seaward town development, which resulted in large numbers of structures becoming endangered due to coastal erosion.

In the mid-1880s, anthropogenic intervention on the coast – especially dredging in river mouths to facilitate their navigability – was intensified. At this time, the mouth of the Vistula River channel was both dredged and straightened; as a result, sand deposition reduced in Gdańsk. Additionally, the Kaiser Canal (now called Piastowski) in the Świna Strait was also dredged, making the port of Szczecin more accessible. The Dziwnów Spit, in the vicinity of Dziwnów, was also cut through to shorten the Dziwna Strait transport pathway. Consequently, seashore erosion occurred in Dziwnów, which made it necessary to take steps to protect buildings situated too close to the beach. As a result, despite different protection measures, this stretch of shore is threatened by erosion.

In the 1870s, coastal protection structures were put in place to secure short stretches of the seashore in Niechorze, Jarosławiec, Darłówek, Mielno, Ustronie Morskie and Kołobrzeg. The structures included timber piles, groins and coastal seawalls, initially made of wood and later of stone and concrete (Basiński *et al.*, 1993). Erosion was a particular threat for people living in towns located on cliffs, or in places separated from the sea by a narrow dune belt, such as in the upper part of the Hel Peninsula. At each coastal town, there was a person responsible for dune conservation and restoration (Wunsche, 1904). Fascine fences at the dune baseline were put in place to stimulate sediment accumulation and dune accretion; such methods are still in use today (Figure 5.6).

Increased erosion caused by many factors, such as, protection of cliff-top lighthouses, sand dune enhancement, shore strengthening, river dredging and jetty construction and also urban development, makes it necessary to pay more attention to coastal protection. Prior to World War II, groins and seawalls were built in many localities where buildings were endangered; on the other hand, piers and bathing areas were established during the war and immediately after its end such constructions were left uncared for and many deteriorated with few survivors.

Already, in the wake of World War I, the state administration had established institutions whose mandate was to oversee coastal management. In 1920, the Merchant Navy Office (Urząd Marynarki Handlowej) and the Maritime Fishing Office (Morski Urząd Rybacki) were inaugurated. In 1927, they were relocated to Gdynia, and were fused in 1928

■ **Figure 5.6** *Fascine fences at the dune baseline on the shore near Dziwnów; the fences enhance sediment accumulation and dune accretion (Photograph by P. Domaradzki)*

to form the Maritime Office. The new maritime administration, established in 1945, involved three Maritime Offices: in Gdańsk (1945), Szczecin (1948), and Koszalin (1955), located in Słupsk; (www. umgdy.gov.pl). The mandate for these state administration institutions included coastal protection. A 10 m to 1 km wide 'technical belt' adjacent to the coastline was identified as belonging to the Polish state and managed by the Maritime Offices. Their task is to alleviate effects of erosion and strengthen the shore to protect developed areas. After each storm, erosion damage is assessed and recorded. In 2000, a seashore protection programme was developed to become a legal act of Parliament and was decreed in 2003. The programme, envisioned to be in force until 2050, describes the methods of protection to be used at different seashore sections and identifies sources of funding for the actions required (Przyszłość, 2006).

The following coastal protection measures are used: constructions (jetties, different kinds of seawalls, groins), dune maintenance and reconstruction (fences, vegetation planting), cliff slope stabilization and artificial nourishment of both the beach and inshore seabed. Other complex measures are also being used; on a smaller scale, e.g. submerged breakwaters.

Dune conservation and restoration

The oldest seashore protection measure, used at the end of the eighteenth century and maybe even earlier, involved dune strengthening and restoration by building fences at their base. Such actions resulted in aeolian sand accumulation and building up the dune height (Figure 5.6). Later dune slopes were stabilized by planting appropriate grasses and shrubs, which prevent the wind blowing sand away and reinforce the accumulation process.

Groins and seawalls

The most popular measures for strengthening the shores involves construction of different kinds of seawalls and groins. Double row timber pile seawalls were most widely used in the 1950s and 1960s. Two parallel rows of timber piles were dug into the base of a dune or cliff and the void between filled with concrete blocks placed on a fascine mattress. Unfortunately, these seawalls were rapidly destroyed by storms and even contributed to acceleration of erosion by basal scouring when water overtopped the seawall. Later, the placement of such constructions was discontinued and the remaining timber piles

■ **Figure 5.7** *A gabion seawall on the dune shore in Dziwnów. Gabions are located at the core of a dune as a second line of coastal defence. Permeable groins in the background*

have been recently removed to keep beaches safe for tourists. Those constructions were replaced with mound seawalls made of stones or tetrapods and pyramids placed at a dune or cliff base (see Box 5.2). These dissipate wave energy reaching dunes or cliffs and decrease erosion rates. The other type of protective constructions, which replaced the double row of timber pile seawalls, were gabions (Figure 5.7), *i.e.* wire containers filled with stones of a specific diameter and amount. However, gabions need to be filled very carefully. When not enough care is expended and if the amount of stones is not adequate and they are packed too densely, gabions act as a concrete wall and the fronting beach either becomes narrower or disappears.

On the other hand, if the structure is not adequately fixed and stones are allowed to move, they will destroy the wire container. Gabion seawalls are often placed inside a dune so that they are invisible and do not affect a coastal landscape; in the case of a strong erosion-causing storm, they protect the dune from breaking up. In 2000, over 41 km of the Polish coast was strengthened with various kinds of seawalls, including gabions (Przyszłość, 2006).

In some cases, when erosion is very severe and threatens very valuable buildings located on a cliff or dune, strong concrete seawalls are used, or different measures are combined to produce the best protection. Such examples can be found in Niechorze (Box 5.1), Trzęsacz (Box 5.2), Ustronie Morskie and Mielno. Usually, those seawalls are additionally reinforced with groins or mound seawalls. Until recently, groins were the second (after various kinds of seawalls) most widely used measure of shore protection. Groins consist of one or two rows of timber or concrete piles perpendicular to the water line. Occasionally the space in double rows is filled with stones or tetrapods and in 2000, almost 85 km of seashore were protected by such groins. They break up longshore sediment transport, but do not stop shore erosion. On the contrary, major shore erosion occurs in the area adjacent to a group of groins. Groin construction and placement has been gradually abandoned over the last 20 years (Przyszłość, 2006).

Beach nourishment (the dominant form of protection)

An increasingly popular shore protection measure is beach and/or seabed nourishment in the area near to a submerged longshore bar. The sand material is dredged from the seabed of an area where the sediment grain size closely matches that found in the beach (and the seafloor) to be nourished. The sand material is deposited underwater on the longshore bar or – more frequently – on the beach itself. Sand material that can be used in beach nourishment is also obtained from spoils resulting from dredging waterways and river mouths. Despite its repetitive nature (sand deposition has to be repeated every several years), this is currently the most popular protection measure, mainly because it is easily incorporated into the shore processes and is also tourist-friendly, as it results in wider beaches. Beach nourishment is often used in combination with other protection measures (Box 5.1 and Box 5.2). Since its introduction in Poland in 1985, the measure has been applied on nearly 33 km of shoreline (Przyszłość, 2006).

An example of large-scale beach nourishment is provided by the protection of the Hel Peninsula. Following construction of the port in Władysławowo, at the base of the Peninsula in the 1930s, the eastward longshore sediment transport was disturbed resulting in intensive erosive processes to the east of the port. To counteract erosion, groins were initially built, but they did not stop long-term erosion; the narrowest sections of the dune belt were almost entirely eroded and water overtopped into the hinterland, disrupting the road and railway tracks along the Peninsula. It was suggested that a concrete seawall be built, but as the area is a very popular tourist destination, it was eventually decided to introduce beach and dune nourishment. Where dune erosion happens most frequently, a gabion seawall was constructed to create a dune core. Even though nourishment has to be repeated every two or three years, the natural landscape and tourist value of the place was preserved.

Complex protection

A complex shore protection was applied in 1994–2000 to a 30 m high cliff location in Jastrzębia Góra. The cliff base was strengthened with a gabion seawall and the slope stabilized by using drainage and a stair-shaped technology termed 'Green Teramesh' (layers of clay sands interlaid with plastic nets and covered with stones of a specific diameter). Although cliff erosion ceased, the whole process had a detrimental effect on the landscape and the seawall resulted in the beach becoming narrower or even in its periodic disappearance. The future will show what effect this protection measure will have on beach preservation, and whether it will decrease the popularity of Jastrzębia Góra, therefore affecting the economy of the area.

Tourists do not always dislike heavy concrete structures, especially in towns. An example can be found in Gdynia, where a 1.7 km stretch of the shore features a heavy seawall which protects a cliff-base promenade. Currently, the promenade appears to be one of the most popular places in Gdynia (Figure 5.8).

In Gdynia Orłowo, three 70 m long submerged breakwaters were placed at about 4 m depth. Additional complex forms of protection are described in Boxes 5.1 and 5.2.

BOX 5.1 COMBINED CLIFF SHORE PROTECTION IN REWAL

Since 1950, the seaside resort of Rewal has expanded in size, and coastal protection measures have had to be applied to counteract erosion of the Pleistocene sand and clay cliffed shore. Protection of the most heavily eroded section involved a double row of timber-pile seawalls which was destroyed by several strong storms. In 1992–94, they were replaced by a steel wall about 15 m high, built across the cliff face. The space between the wall and cliff face was filled with sand and shingle and topped by a viewing terrace. A system of pumps and drains was installed to remove ground water. A scarp made of sand and shingle to support the wall from the sea side was also constructed, the scarp base being strengthened by a tetrapod-mound seawall. The construction is presently one of the tourist attractions in Rewal. Although the structure succeeded in its protective role, particularly with respect to the nearby buildings, the adjacent part of the shore is experiencing increased erosion

■ **Figure 5.8**
A Pleistocene cliff in Gdynia protected by a heavy seawall with a promenade (Photograph by P. Domaradzki)

(Figure 5.9). Consequently, in 2009, it was decided to protect the endangered cliff sections by constructing an 'up-shore mound' seawall made of natural stone. An almost vertical cliff wall was made gentler and natural stones put on the lower slope section. The construction is porous enough to not generate wave reflection, therefore avoiding its negative consequences; whilst supporting the slope, it does not block water flow and is more aesthetically pleasing than concrete. Additionally, beach nourishment has been applied to keep the beach at its natural summer width. Several seawalls of this type have survived heavy storms (Figure 5.10; Dudzińska-Nowak et al., 2005).

■ **Figure 5.9** *Cliff protection in Rewal. Erosion is noticeable in the neighbourhood (Photograph by P. Domaradzki)*

■ **Figure 5.10** *Rewal, the same section of the coast during the holiday season (Photograph by P. Domaradzki)*

BOX 5.2 PROTECTION OF THE CHURCH RUINS ON THE CLIFF IN TRZEØSACZ

The church in Trzęsacz was founded in the twelfth century at a distance of some 250 m from the 15 m high cliff composed of Pleistocene sand-clay. In 1890, the church was closed because of cliff erosion; in 1990, only a small part of the church wall still remained, despite construction of a mound seawall as a protection (Figure 5.11). In 1999, it was decided that the Trzęsacz church remnants should be protected, since they are a valued tourist attraction. The ruins were anchored to the ground, the cliff base strengthened with a gabion seawall and the slope stabilized by using drainage and the stair-shaped 'Green Teramesh' technology. Near the remnants, a viewing terrace was built, accessible from the beach by a flight of stairs. Additionally, the beach was nourished and today the Trzęsacz church ruins are one of the most popular tourist attractions in the area (Figure 5.12).

■ **Figure 5.11** *Church ruins in Trzęsacz on a cliff protected by a mound seawall in 1990 (Photograph by P. Domaradzki)*

■ **Figure 5.12** *Church ruins in Trzęsacz on a cliff protected by a mound seawall with a viewing terrace (Photograph by P. Domaradzki)*

Conclusions

The Polish coast was shaped during the Pleistocene
and Holocene from clay and sand, which makes
it exceptionally susceptible to erosion. Coastline
changes are induced by significant storm events,
heavy storm surges and extremely high sea levels. The
most popular protection measures include groins
(gradually being withdrawn), seawalls (different
kinds) and beach nourishment (recently the most
popular measure). In individual cases, combinations
of different measures are applied as well. A strategy
for coastal protection extending to 2050 was passed
by the Parliament; this is a good starting point for
developing Integrated Coastal Zone Management in
Poland.

References

Basiński, T., Pruszak, T., Tarnowska, M. and Zeidler,
R. B., 1993. *Ochrona brzegów morskich*. IBW PAN.
Gdańsk, 536 pp. [In Polish]

Dudzińska-Nowak, J., Furmańczyk, K. and
Łęcka, A., 2005. Ochrona brzegu na odcinku
Międzyzdroje-Niechorze, in K. Furmańczyk (ed.),
*Zintegrowane Zarządzanie Obszarami Przybrzeżnymi
w Polsce-stan obecny i perspektywy, Problemy erozji
brzegu*, 1, 96–105. [In Polish]

Dudzińska-Nowak, J., 2006. *Zmiany morfologii jako
wskaênik tendencji rozwojowych brzegu*, Ph.D. thesis,
University of Szczecin, Poland, 225 pp. [In Polish]

Ferreira, O., 2005. Storm groups versus extreme
single storms: Predicted erosion and management
consequences, *Journal of Coastal Research*, 42,
221–227.

Furmańczyk, K., 1994. Współczesny rozwój strefy
brzegowej morza bezpływowego w świetle badań
teledetekcyjnych południowych wybrzeży
Bałtyku, Uniwersytet Szczeciński. *Rozprawy i
Studia*, vol 161, 149 pp. [In Polish]

Furmańczyk K.. 1995. Coast changes of the Hel
Spit over the last 40 years. Polish Coast: Past,
present and future, *Journal of Coastal Research*, SI22,
193–196.

Furmańczyk, K. and Basiński, T. 2006. Rola
działalności człowieka w modyfikacji funkcjonowa-
nia systemu strefy brzegowej, in K. Furmańczyk
(ed.), *Zintegrowane Zarządzanie Obszarami
Przybrzeżnymi w Polsce-stan obecny i perspektywy,
Brzeg morski – zrównoważony*, 2, 36–46. [In
Polish]

Furmańczyk, K. and Dudzińska-Nowak, J., 2009.
Extreme storm impact to the coastline changes:
South Baltic example, *Journal of Coastal Research*.
SI56, 1637 –1640.

Jacobsz, A., 1644. *De Lichtende Columne ofte Zee-
Spiegel, inhoudente de Zee-Custen van de Noordtsche
. . .*, Schipvaert Vol. I–II, Amsterdam.

Majewski, A., Dziadziuszko, Z. and Wiśniewska,
A., 1983. *Monografia powodzi sztormowych 1951–
1975*. IMGW. Wyd. Komunikacji i Łączności,
Warszawa, 214 pp. [In Polish]

Musielak, S.. 2006. Geneza i funkcjonowanie systemu
przyrodniczego morskiej strefy brzegowej, in
K. Furmańczyk (ed.), *Zintegrowane Zarządzanie
Obszarami Przybrzeżnymi w Polsce-stan obecny i
perspektywy, Brzeg morski –zrównoważony*, vol 2,
21–23. [In Polish].

Musielak, S.. 1980. Impact of basic hydrodynamical
factors on beach dynamics, *Hydrotechnical
Transactions*, 41, 159–164.

Przyszłość. 2006. *Przyszłość ochrony brzegów morskic'*,
Dubrawski, R., Zawadzka-Kahlau, E. (eds),
Instytut Morski w Gdańsku, 302 pp. [In Polish]

Uścinowicz, S., 1999. Southern Baltic area during
the last deglaciation, *Geol. Quart.* 43, 137–148.

Wagener, L.J., 1589. *Dess Spiegel der Seefart, von
Navigation des Occidentischen Meers oder der Westseen
. . . In vielen See Carten . . . Durch . . . T.* I–II,
Amsterdam.

Wiśniewski, B. and Wolski, T., 2009. *Katalogi
wezbrań i obniżeń sztormowych poziomów morza oraz
ekstremalne poziomy wód na polskim wybrzeżu*.
Wydawnictwo Naukowe Akademii Morskiej,
Szczecin, 156 pp. [In Polish]

Wunsche, H., 1904. *Studien auf der Halbinsel Hela*.
Ph.D. Dissertation, University of Leipzig, Druck
Heinrich, Dresden, 79 pp. [In German]

Zawadzka-Kahlau, E., 1999. *Tendencje rozwojowe*

polskich brzegów Bałtyku południowego. GTN, IBW, Gdańsk, 147 pp. [In Polish]

Zeidler, R. B., Wróblewski, A., Miętus, M., Dziadziuszko, Z. and Cyberski, J., 1995. Wind, wave, and storm surge regime at the Polish Baltic coast. Polish Coast: Past, Present and Future, *Journal of Coastal Research*, SI22, 33–55.

6 Denmark

Per Sørensen

Introduction

Denmark has an area of about 43,000 km² and is situated in the northern part of Europe. Because Denmark consists of one larger peninsula and about 500 islands, of which 200 are inhabited, it has a very long coastline of 7,300 km, implying that no Dane lives more then 50 km away from the sea (CERC, 1996); consequently, most of the largest cities in Denmark are coastal (Figure 6.1). Some 80 per cent of the people live in urban areas connected to the coast and half of these within 3 km of the coast. Coastal usage can be divided into: urban 13 per cent, holiday homes 12 per cent, recreational 3 per cent,

nature 67 per cent, others 5 per cent (Kappel *et al.*, 2010).

Coastal morphology

Geologically, the majority of Denmark has been built from Neogene and Holocene sediments that was primarily a result of sedimentation and sculpturing effect of the last two glaciations, the Saale and the Weichsel, in combination with marine impacts during the Holocene period. The entire area was last covered during the Saale glaciations. During the Weichsel, several ice advances covered Denmark – except for the south-western parts of Jutland – and ice moulded the landscape so that a large variety of glacio-tectonic structures can be found in many coastal cliffs. A large part of the country is sand/littoral dunes and apart from moraine tills, areas of calcareous Cretaceous marine layers have been folded into 130 m high cliffs, e.g. on the Baltic Sea coast, and Palaeocene marine layers of molar (diatomite) are found in the Limfjord area. Only on the island of Bornholm, in the Baltic Sea, can hard rock coasts be found (Safecoast, 2008).

Coastal classification and sediments

The Danish coastline is divided into five classes, based on the three classes described by Nielsen and Binderup (1996), incorporating wave energy and extended to cover the Wadden Sea tidal flat/

■ **Figure 6.1** *Location of Denmark and its larger cities*

■ Table 6.1 Coastal classifications in Denmark.

Coastal classification	Properties/characteristics
Rock coast	Consists of hard rock stratum
Soft cliff coast	Consolidated sediments. Glacial tills, meltwater deposits, calcareous cliffs, etc.
Tidal flat/marsh coast	Dominated by tidal processes.
Protected coast	Dominated by water level variations and sheltered from wave impact
Sand/littoral dune coasts	Loose to unconsolidated sediments under the influence of waves

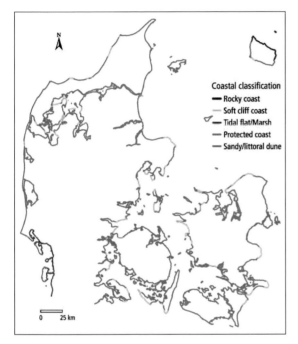

■ Figure 6.2 Coastal classes in Denmark

marsh coasts in south-western Jutland together with the rock coast of Bornholm (Tables 6.1; Figure 6.2).

Topography and depth

Denmark is surrounded by shallow water (Figure 6.3), with water depths close to the shore ranging between 0 and 20 m. The shallowest part is in the Wadden sea and the deepest lies close to the shore near the island of Bornholm, which mainly consists of rock. The west coast of Denmark is relatively deep, up to a depth of some 15 m.

Wave energy and direction

Denmark is primarily affected by low pressure weather systems from the west or north-west, *i.e.* a wave climate dominated by waves coming from westerly directions. Because wave energy is also affected by fetch, which is large at the North Sea coast, the largest waves ($H_{s,100}$ = 8.1 m) are found here. Wave impact on coastal erosion can be evaluated if the wave energy that reaches the coast can be determined. In the Safecoast (2008) project, this was carried out by calculating wave energy from the CERC (1996) formula based on wind speed, wind direction and fetch. Wind data was recorded at 16 stations around Denmark in the period from 1995 to 2005, and divided into 11 directional intervals. Total wave energy was calculated for each cell by multiplying by the corresponding fetch and by the share (in per cent) for that cell. Total wave energy was obtained by summing contributions and the calculated average yearly wave energy can be seen in Figure 6.4.

The highest wave energy levels occur on the Danish North Sea coast. Relatively high energy levels (red and grey colours), may be seen for parts of the coastline facing the Baltic Sea and the Kattegat, whereas sheltered conditions prevail in the belts and in fiords (Figure 6.4). As mentioned, wave energy calculations are based on fetch and, as duration limitation may occur with long fetches, some

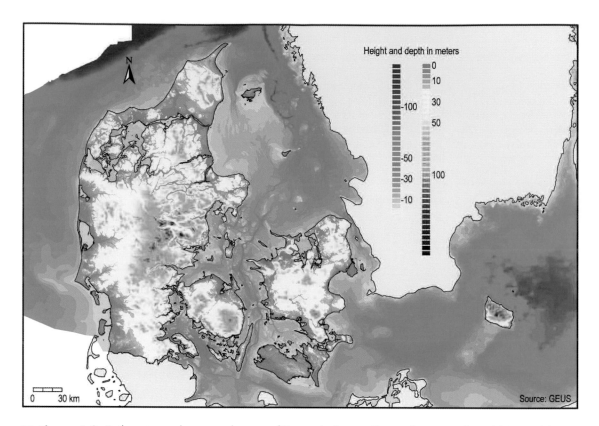

■ **Figure 6.3** *Bathymetry and topography map of Denmark. Source: Farvandsvæsenet, Danish National Survey and Cadastre*

overestimation may occur at the open coasts facing the North and the Baltic Seas.

Water level gauges and storm water levels

Water levels are measured continuously in more than 50 locations around the coastline by the local harbour authority, the Danish Meteorological Institute or the Danish Coastal Authority (DCA). If the time series for a water level gauge exceeds 10 years, the DCA calculates extreme water level statistics, used in Denmark for the planning and design of coastal structures. Calculated 50-year return periods are shown in Figure 6.5. The highest water levels are reached on the West coast of Denmark, which is due

to the large tidal range of *circa* 2 m, whereas the tidal range is less than 10 cm in the south-east. The highest water levels in the south-east are however influenced by seiching in the Bothnian Gulf area between Sweden and Finland, which can cause some very high water levels (up to 3.5 m), but they are quite rare.

Current erosion

Combining the coastal classification, the hydrodynamic impact and geographical conditions, the Danish coasts can be classified into two major coastal stretches and three distinct subgroups (CERC, 1996).

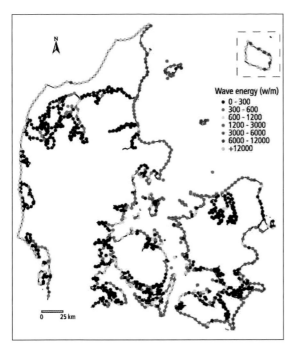

Figure 6.4 *Calculated average yearly wave energy in Denmark*

Figure 6.5 *Storm water levels (cm) in Denmark. 50 year return period shown in black (Kystdirektoratet, 2007)*

North Sea coasts

The northern Jutland headland coast erosion has been predominantly of the order of 2–4 m/yr over the last 20 years. Net littoral drift is northward due to the sheltering effect of Norway, going from zero in the south close to the inlet to Limfjord, to some 1 million m³/yr to the north of Skagen on Skaw spit (Figure 6.6).

The central west coast is dominated by barrier beaches, which separate three major areas – Limfjord, Nissum Fjord and Ringkøbing Fjord – from the sea. Erosion rates are high, of the order of 2–8 m/yr. Net littoral drift is towards the inlet to Limfjord in the north, where sediment is deposited in the flood tidal delta. The net drift is south in the southern part, zero at Ferring Sø to more than 2 million m³/yr in the southernmost part, where most sediment is deposited on a shoal north-west of Blåvands Huk (Kystinspektoratet, 1998). The Wadden sea coasts are in overall sediment balance despite the fact that there has been a relative sea level rise over many hundreds of years. The barrier islands found on the western coast, south of Esbjerg, Fanø, Mandø and Rømø are growing seawards because of the sediment supply from north and south due to their position.

Inner coasts

Apart from the North Sea coasts, all other coasts are defined as inner coasts. They account altogether for about 7,000 km of the total length of the Danish coastline and naturally show much variation (Eurosion, 2004). They fall into three major categories (CERC, 1996):

1. Northern medium exposed coasts along the Kattegat, with littoral drift rates of an order of magnitude less than along the North Sea, *i.e.* 10–75,000 m³/yr;
2. Eastern and southern medium exposed coasts in the Baltic, with littoral drift rates of 10–75,000 m³/yr;

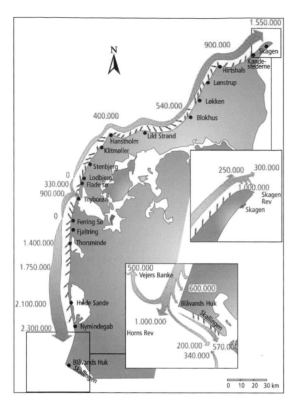

Figure 6.6 *Sediment budget (m³/yr) for the North Sea coasts (Kystdirektorat, 2001)*

3. All remaining coasts along straits, belts and fiords or on sheltered islands experience low exposure with a littoral drift of less than 10,000 m³/yr.

Coastal protection

Coastal Zone Management

Coastal Zone Management (CZM) has been in existence in Denmark for many decades and physical planning, as such, has been applied for more than 100 years. CZM has not been defined as a separate issue but has matured through gradual harmonization and coordination for administrative and legal frameworks (CERC, 1996). An example of an early CZM initiative provoked by necessity (and, at present, actually

an integral part of CZM), was the Dune Protection Law, compelled by coastal dune migration, which was aggravated in the sixteenth century and continued into modern times. The first decree was issued in 1539 and the first law enacted in 1792.

The Nature Preservation Act (1917) reinstated the right of public access to all Danish beaches. The Act was revised in 1937, when a definitive stop was enforced for placing buildings on the beach proper and on the adjacent 100 m of the hinterland. In a directive in 1978, the Ministry of Environment and Energy has provided a general stop to the planning of new vacation housing and hotels in a 3 km wide zone along open beaches. Finally, in 1994, protection of near shore areas in the whole country was extended to ban all construction, except coastal protection works in a 300 m wide zone along all Danish coasts.

Since 1874 and 1927, respectively, Denmark has had a Dike Protection Law and a Coastal Defence Law based on the landowner's individual responsibility for coastal protection. The integrated act of 2006 substitutes previous laws and takes into account new dimensions in the coastal zone, including natural development, recreational use and economic relations. In principle, the protection of land and property affects only front property owners, while coast erosion management relates to areas and interests in considerable depths in the hinterland, dependant on the proposed coastal protection scheme. Furthermore, beach nourishment schemes especially required conceptual thinking, planning and financing on a larger scale.

The local municipality has a vital role in the approval process and is therefore well suited to combine a good understanding of the local values at stake, with properly balanced problem assessment. In addition, regulatory procedures still require final permission from the DCA for all coastal protection works and other man made changes in the coastal zone, 100 m landward of the coastline as well as 12 NM seaward in territorial waters. Financing of coastal protection schemes remains the responsibility of individuals, municipalities and regional authorities, except along the exposed North Sea coast, where considerable governmental resources are allocated to schemes of general and regional importance.

History

The oldest well-established dike scheme in Denmark dates back from about 1550, when it was executed just north of the present Danish-German border in order to protect low-lying marshlands along the Wadden Sea coast (CERC, 1996). In 1868, the Harbour and Coastal Authority (today's Danish Coastal Authority) was established to deal with large erosion problems on the North Sea coast. The first problem solved was the Thyborøn barrier breach in 1862 and the consequent high erosion; the first groin was built at Ferring Sø in 1875 to counteract this problem. During the next 25 years, 77 large groins were built at the North Sea coast consisting of concrete blocks cast on a nearby site; to stabilize the groins, rows of timber piles were used, as can be seen in Figure 6.7.

On the Inner Danish coasts, improvements in welfare in the beginning of the nineteenth century and the corresponding decrease in weekly working hours, kick-started the building of holiday homes close to the coast. Often these were built disregarding the erosion or flooding risk, so coastal tension began, resulting in the building of not only groins, but revetments and shore parallel breakwaters using local

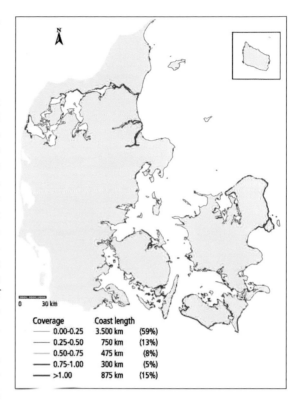

Figure 6.8 *Coastal protection coverage on the Inner Danish coast*

Coverage	Coast length	
0.00-0.25	3.500 km	(59%)
0.25-0.50	750 km	(13%)
0.50-0.75	475 km	(8%)
0.75-1.00	300 km	(5%)
>1.00	875 km	(15%)

material, typically small rocks, concrete and wood (Crumlin-Pedersen *et al.*, 1996). Very strong evidence of the coastal tension prevalent in Denmark today is revealed when observing how much of the coastline has been protected against the sea. Figure 6.8 shows protection in the Inner Danish coasts in 1999, red lines indicating that the entire coastline is protected (Kystinspektoratet, 1999). Values higher than 1 can occur if there are more than one coastal protection measure, for example, if there are both revetments and groins.

Sand nourishment

The only coastal protection measure that can solve the fundamental problem of an eroding coast without negative leeside effects is sand nourishment, where sand is brought to the coast to replace the sand that

Figure 6.7 *Groin 1, the first groin built on the Danish North Sea coast, at Bovbjerg*

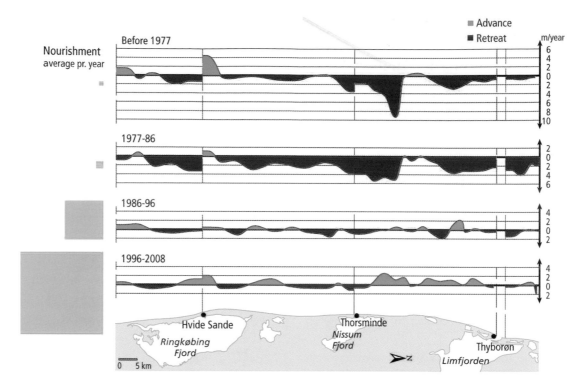

■ Figure 6.9 *Movement of the active coastal profile and corresponding sand nourishment volumes. The largest box equates to nourishment of 2.5 million m³/yr*

nature has eroded away. The first location where sand nourishment was used on a large scale was in 1976 at Thyborøn at the Danish North Sea coast, at the inlet to Limfjord. Here, coastal steepening had been very large since the inlet was kept artificially open with groins [this is the term used in Denmark but in most of Europe and the USA these would be called jetties]. A small amount (30,000 m³) of sand was pumped to the beach. Since then use of the soft erosion protection/sand nourishment technique has been used repeatedly, so now more than 2.5 million m³ of sand is pumped to the 110 km of coast each year. Sand nourishment has been able to stop coastal profile retreat, even on the most eroding locations where natural coastal retreat is approximately 8 m/yr, as shown in Figure 6.9.

Three different methods are used, the favoured being shoreface nourishment since it is the cheapest. Where the coastal profile is steep, nearshore nourish-

■ Figure 6.10 *Beach nourishment in a groin field south of Thyborøn*

ment is used, and beach nourishment is used where there is a need to keep dune flooding probability to an acceptable level. Sixty percent of the total volume is nourished, either on the shoreface or nearshore.

■ **Figure 6.11** *Nearshore nourishment (Rainbowing) over the dredger's hull on the coast at Thyborøn where depths close to the coastline are large*

A continuous Research and Development (R and D) programme, run by the DCA, constantly tries to optimize sand nourishment works at the North Sea coast. A beach nourishment example is shown in Figure 6.10; one of nearshore nourishment in Figure 6.11.

On the Inner Danish coasts, sand nourishment is seldom used even though smaller projects have been carried out and a very large one is been planned on the north coast of Zealand. The reasons are various, but the most important is a lack of understanding of the coastal processes and tradition. A test was carried out on the north coast of Zealand from 1984 to 1986 where 23,100 m³ of sand was pumped to the beach (Fællesudvalget for kystpleje og kystsikring af Nordkysten, April 1987: Haldstrand 1984–1986). Twenty percent of the nourished sand was still present after three years and the project was considered successful. Nourishment sand is typically dredged from designated areas. Some effort has been made to use sand that has certain characteristics, for example, coarser and more cube-shaped sand grains, but it is not always possible to do that in a feasible way.

Groins

Groins have been and are the most common coastal protection measure because they have an immediate effect of collecting sand on the updrift side; more than 12,000 groins have been built (Kystinspektoratet, 1999). They are cost effective because they utilize different cheap materials such as wood, concrete and stone/rocks. Typically, individual landowners built groins on their property in order to trap some of the material transported by longshore currents. For many years groins have been constructed of timber piles or from stone found nearby, the stone typically being rounded in shape because of glacier transport. Many

■ **Figure 6.12** *Groin group on a very exposed North Sea coast*

■ **Figure 6.13** *Groin group on a medium exposed North Funen coast*

newer groins are now composed of quarry rocks because they are cube-shaped, have a higher density so are stable, and are easy to obtain from Norway. Figures 6.12 and 6.13 show examples of a groin group from very highly and medium exposed coasts respectively.

Shore parallel breakwaters

Groins have some disadvantages because they make accessibility along the beach worse, so people sometimes have to climb over each groin when walking on the beach. Shore parallel breakwaters do not have these disadvantages and combined with the technological development in construction and design, shore parallel breakwaters have been used more and more on very exposed coasts in Denmark (Kystinspektoratet, 2000). At present, more than 500 shore parallel breakwaters exist (Kystinspektoratet, 1999). Most have been constructed as detached low crested emerged breakwaters made from either quarry or sea stone. Some of the very small shore parallel breakwaters are constructed without filter layers, based on the consultant's local experience. Figures 6.14 6.15 are examples of typical shore parallel breakwaters that can be found on very and medium exposed coasts respectively.

■ **Figure 6.14** *Shore parallel low crested emerged breakwaters on very exposed North Sea coast*

Revetments

When a coast suffers from erosion, a revetment is for most people the most obvious coastal protection measure, as they are easy to construct; some 700 km of revetments have thus been built (Kystinspektoratet, 1999). They have been placed mainly on the Inner Danish Coasts by private landowners in an attempt to stop erosion of their property. The seawalls and revetments are either made of concrete as vertical/sloping walls, or made of quarry/sea stone (Figures 6.16 and 6.17 respectively). Figure 6.17 gives a closer view of the

■ **Figure 6.15** *Shore parallel low crested emerged breakwaters on a medium exposed North Zealand coast*

■ **Figure 6.16** *Seawalls on a very exposed North Sea coast*

■ **Figure 6.17** *Revetments on a medium exposure coast, West Zealand*

revetment showing how close the house is to the water, as there is no beach left.

Coastal drainage

Another way to trap longshore sediment transport is by drainage, *i.e.* lowering the beach water table and actively pumping water away. There have been some tests with this method both on a very exposed coast and on less exposed coasts. Only at one location, Karrebæksminde on a less exposed coast, is the method currently in use. Success was not achieved at, for example, Hornbæk, as evaluation showed no effect with pumping. There have also been tests with passive beach drainage (tubes drilled into the beach), but after three years it was concluded that the method cannot be used as a coastal protection measure on a very exposed coast.

Climate change

It is self evident that climate will change in the future and sea level will rise. In Denmark it is estimated that the mean sea level will rise between 0.15 and 0.45 m until 2050 and between 0.30 and 1.00 m until 2100 excluding vertical land movement. It has been estimated that the maximum wind during storms

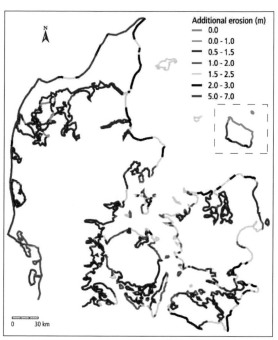

■ **Figure 6.18** *Estimated additional erosion until 2050 from climate change, based on the IPCC 2007 A2 scenario (Safecoast, 2008)*

would increase between 1 and 10 per cent until 2100 (Klimatilpasning, 2010).

Increased sea level rise and increased storminess will result in increased erosion rates. In the Safecoast (2008) project, the expected additional erosion until 2050 is based on the IPCC 2007 A2 scenario which has been estimated by expert judgement (see Figure 6.18). The estimated additional erosion can easily be counteracted by coastal protection measures, in that it is straightforward to increase sand nourishment volumes, or to put one armour layer on top of groins or shore parallel breakwaters.

Research and development

Coastal engineering in Denmark has a long history going back to establishing the Technical University of Denmark (DTU) in 1829 (CERC, 1996). Ever since, coastal engineers have strived to improve knowledge in the field of coastal engineering through national and international cooperation. Coastal engineers such as Helge Lundgren, Per Bruun, Frank Engelund, Jørgen Fredsøe, Hans Burcharth, Rolf Deigaard and many others, are internationally well-known coastal experts. Their experience and knowledge has been integrated in one of the world's leading coastal engineering software tools, the software package 'The Mike' developed by the Danish Hydraulic Institute (DHI).

The Research and Development community in Denmark is strong because of its long tradition of cooperation across any kind of border, *i.e.* between universities, coastal engineers and authorities. At present, the prime effort is on better understanding of coastal processes in both short and long term and on developing coastal protection measures in a sustainable way.

Conclusions

Coastal tension is increasing in Denmark and will continue to do so due to climate change and because more and more people want to live very close to the coast. This will also increase the need for erosion protection and spatial planning in a sustainable way. Many projects addressing these issues are already running, e.g. the EU project BLAST. There has been formulated a national climate adaptation strategy and a national coastal protection strategy was introduced in 2011, which stated that protection must be designed so that it does not have any impact on downstream coasts; protection has to be a sustainable solution to the problems. There are also three or four EU projects at present in Denmark that are investigating how to address erosion in a better sustainable way, e.g. SusCod and Baltica. Many municipalities also include the problem in their spatial planning. There is a Research and Development project currently in progress, CoAdapt, which aims to develop coastal protection structures designed to be able to adapt to climate change. It is obvious that erosion management will continue to cost much money; perhaps not every problem can be solved, due to

the general money shortage. To be able to prioritize resources, socio-economic models have been adapted so that they can be used in coastal protection cases. (One example is at the town of Løgstør, situated by the Limfjord, as Løgstør has been flooded several times.) Most are based on a risk approach where risk is defined as probability multiplied by consequence.

References

CERC, 1996. *History and heritage of coastal engineering*, ASCE, New York.

Crumlin-Pedersen, O., Porsmose, E. and Thrane, H. (eds), 1996. *Atlas over Fyns kyst i jernalde, vikingtid og middelalder*, Odense Universitetsforalg, Denmark.

Eurosion, 2004. *Living with coastal erosion in Europe: Sediment and space for sustainability, PART II – Maps and statistics*, available at: http://www.eurosion.org/reports-online/part2.pdf (accessed 11 September 2012).

Fællesudvalget for kystpleje og kystsikring af Nordkysten, April 1987, *Hald strand 1984–86, Sjællands Nordkyst, strandfodringsforsøg 1984–1986, Slutrapport*, Fællesudvalget for kystpleje og kystsikring af Nordkysten.

Kappel V., Rasmussen, T. and Waneck, J., 2010. *Danmarks kyster*, Politikens forlag, Copenhagen.

Klimatilpasning 2010. Forandringer i havniveau. Available from: http://klimatilpasning.dk/da-DK/service/Klima/KlimaaendringeriDanmark/vandstandihavet/Sider/Forside.aspx (accessed 9 August, 2012.

Kystinspektoratet, 1998. *Kystinspektoratet 1973–1998, Menneske, hav, kyst og sand*, Kystinspektoratet, Lemvig.

Kystinspektoratet, 1999. *Danmarks indre kyster, Kortlægning af kystbeskyttelsen*, Kystinspektoratet, Lemvig.

Kystinspektoratet, 2000. *Indre kyster-skitseprojekt 1, Indre kyster – Skitseprojekter Liseleje-Hyllingebjerg*, Kystinspektoratet, Lemvig.

Kystdirektoratet, 2001. *Sedimentbudget Vestkysten*, Kystdirektoratet, Lemvig.

Kystdirektoratet, 2007, *Højvandsstatistikker 2007* (or www.kyst.dk), Kystdirektoratet, Lemvig.

Nielsen, M. and Binderup, N. 1996. *Fysiske rammer for marine aktiviteter på Fyn*, Odense Universitetsforlag, Odense.

Safecoast, 2008. *Coastal flood risk and trends for the future in the North Sea region: Synthesis report,* Safecoast Project Team. The Hague, 136 pp., available at: www.safecoast.org/editor/databank/File/Safecoast geheel-lres.pdf (accessed 11 September 2012).

7 Germany

Jürgen Jensen and Klaus Schwarzer

Introduction

The 2,400 km of Germany's coastline are located along two different seas, the tide-dominated North Sea and the intra-continental non-tidal Baltic Sea. The North Sea is connected to the Atlantic Ocean and the only access the Baltic Sea has to the world's oceans is to the North Sea via the Skagerrak. Both seas are different in their hydrographic characteristics, sedimentological conditions and geomorphological features (Schwarzer et al., 2008). In the North Sea, storms from westerly directions can induce a water level rise of up to five metres above mean water level, especially if wind waves and spring tides interact. Those storm surges can cause tremendous coastal changes. In the non-tidal Baltic Sea, storms from north-easterly directions induce the highest water levels and have the strongest influence on coastal changes. For both seas, coastal erosion and flooding is a severe problem. Therefore protection of coastal areas has a long history in Germany, starting circa 1,000 years ago. Constructions, such as groins, walls of tetrapods, beach nourishment and/or detached breakwaters have been established during the last 120 years.

Coastal morphology

Coastal evolution and coastal processes depend to a certain extent on the geology and sedimentology of the ocean basin (Figure 7.1). Neotectonic activities and isostatic crustal movements occur in the coastal areas of the German North and Baltic Sea, controlling exposure and delineation of their coastlines. The thickness of sedimentary units, their texture and lithological composition have a strong influence on compaction and resistance against erosion. Additionally, sediment availability and transport conditions are essential for the formation of geomorphological structures composed of soft rock components, such as barrier islands, spits, bars and beaches.

After the last glaciations (Figure 7.2), flooding was at its highest between 15,000 and 7,000 BP, followed by pronounced deceleration during the last 7,000 years. Due to this relatively short time-span of water coverage, the continental shelves presently contain a lot of relict features inherited from past glacial times, e.g. river valleys, partly of subglacial origin and sometimes deeply incised in the shelf sequences (Stackebrandt, 2009; Lutz et al., 2009). The veneer of marine sediments is on many shelf platforms still relatively thin, sometimes not exceeding a few metres (Zeiler et al., 2000).

The North and Baltic Seas are separated from each other by the 60 to 80 km wide and 450 km long Jutland peninsula.

North Sea

As the southern North Sea was not covered by ice during the last glaciations (Figure 7.2), the seafloor underwent one regression but two transgressional phases, both shaping and levelling the seafloor (Figge, 1981). Accordingly, the sediment distribution pattern

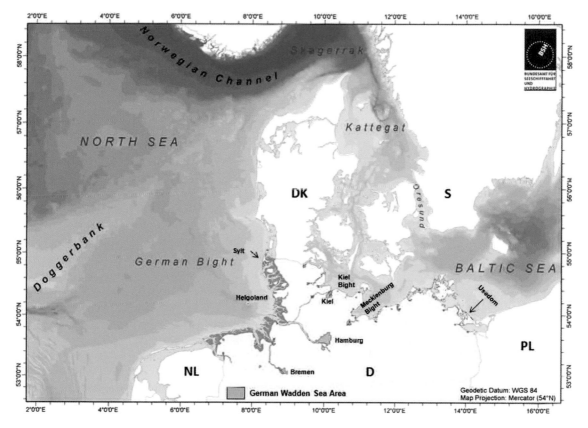

■ **Figure 7.1** *Seabed morphology of the North Sea and south-western Baltic Sea (Source: Federal Maritime and Hydrographic Agency; modified with permission from the German Coastal Engineering Research Council, KFKI)*

in the German Bight reflects to a high degree the reworking processes that have been acting since the Saalian ice shield retreat. Relict sediments ranging from coarse sand to boulders usually cover glacial till and protect the sea floor against further erosion while fine to medium sand is almost constantly replaced by waves and tidal currents.

The latest depositional evolution of the North Sea basin began between 8,600 and 7,100 BP, when sea level rose from −45 to −15 m below the present mean sea level (Figure 7.3). At that time the coastline position was approximately 5 to 10 km seaward compared to nowadays (Flemming and Davis, 1994; Behre, 2003) and tidal currents and waves were pushed further towards the hinterland eroding parts of the submerged landscape. Large quantities of

reworked sediment were shifted landward, building up a 10 to 25 km wide wedge-shaped body of Holocene coastal deposits striking parallel to the coastline and extending also into the Weser and Elbe estuaries. While the accumulation wedge tapers off towards the mainland, it can reach a thickness of more than 40 m at its relatively steep seaward slope (Ahrendt, 2006), where it mainly consists of fine to medium-grained sand.

With deceleration of the sea level rise (7,000 to 3,000 BP; when not explicitly mentioned all data are given in conventional 14C Data) coupled with an increase of tidal range from 1.3 m to 2.2 m (van de Plassche, 1995; Figure 7.3), the coastal zone developed, forming salt marshes, tidal flats and barrier islands (Figure 7.4). Due to the further decreasing of

■ **Figure 7.2** *Maximum extent of the inland ice-sheets during the Saale and Weichselian glaciations in middle Europe. The lower courses of the rivers Ems, Weser and Elbe formed a common spillway into the North Atlantic via the North Sea (Source: © BGR Hannover; modified)*

sea level rise to 0.11 m/100 years over the past 3,000 years BP (Figure 7.3), the Wadden Sea area formed. It is characterized by a complex pattern of marine and terrestrial deposits and reworked postglacial drainage material. Basal peat layers, directly overlying Pleistocene deposits, often mark the beginning of the Holocene depositional sequence (Streif, 2004).

Besides the Elbe palaeo-valley and Helgoland Island, the North Sea seabed shows no distinctive morphological variation. In its south-western part the 20 m isobath is located up to 50 km offshore. The upper subsurface is composed of soft-rock deposits (Zeiler *et al.*, 2008), except Helgoland Island, which

is built up of Triassic strata, uplifted due to an active subsurface salt dome structure.

The coastal area encompasses the North and East Frisian barrier island chains as the seaward border of the Wadden Sea, together with the funnel-shaped river-estuaries of the Eider, Elbe, Weser, Jade and Ems. The Holocene accumulation wedge surface is almost flat, as are the North Frisian Hallig islands, which are relics of a former much wider marshland partly destroyed by storm surges in the fourteenth and seventeenth century.

The Holocene transgression inundated the glacial drift topography. Reworking of glacial deposits by waves and tidal currents generated sands, silts, and clays. While sand was being deposited mainly on barrier islands the finer fractions accumulated in the Wadden Sea and estuaries.

Due to the presence of coastal dunes or elevated cores of Pleistocene (Amrum and Föhr Island) and/or early Neogene age (Sylt Island), the land elevation exceeds spring tide high water level or the water level of severe storm tides. Hence, these areas are not endangered by flooding. The typical elongated shape of the East Frisian barrier islands with a sand foreshore, beach and dune sediments, as well as fine-grained organic rich deposits at the sheltered backside, indicates that their formation is very much related to interaction of mobile sediments with currents and waves. Due to a rising sea level, the development of these islands is characterized by an almost continuous landward shift of the shoreline.

The Wadden Sea forms the interface between the open North Sea and the mainland, extending over a distance of about 450 km from Denmark via Germany to the Netherlands and reaching a width of up to 25 km along the German coast. The barrier island chains of East Frisia and North Frisia separate the tidal flats from deeper offshore waters, while in front of the high mesotidal to low macrotidal estuaries of the Elbe and Weser, there are no barrier islands.

Tides are semidiurnal, ranging from 2 to 3 m in the East Frisian Islands region, increasing to more than 4 m in the inner German Bight, but decreasing to 1.6 m in the northern North Frisian region. They

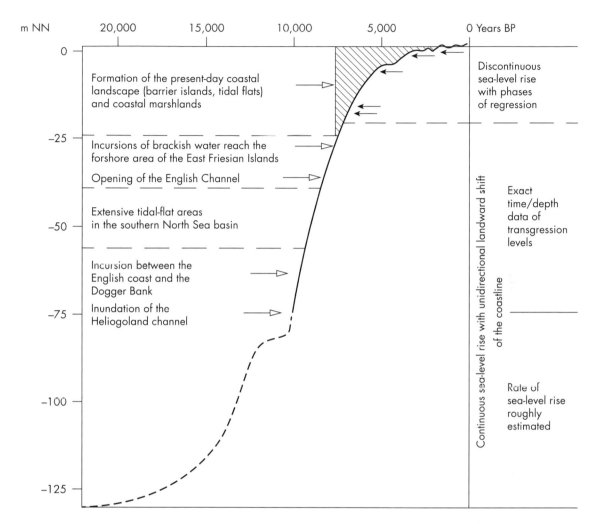

■ Figure 7.3 *Relative sea level rise since the last glacial maximum. The marine transgression is divided into three phases. The last phase, which began 7,100 years BP, led to the development of the present landscape (Streif, 2004, with permission from Elsevier)*

are strong enough to generate currents forming ripples and sand waves of various dimensions with a crest spacing up to 30 m in the tidal inlets, as well as in the flood and ebb deltas. At low tide, extensive sand and mudflats are exposed in the Wadden Sea area. The supratidal zone carries salt marsh vegetation of which large areas have been embanked and reclaimed in recent centuries.

The surficial sediments of the intertidal area show a typical zonation from predominantly sandy sediments in more exposed areas to mud deposits in more sheltered parts. Overall a dominance of sand flats with a mud content (mud: silt content >50 per cent) of less than 10 per cent is observed (Ricklefs and Asp-Neto, 2005). Eisma and Irion (1988) concluded that most fine-grained sediment deposition takes place in the tidal flats, whereas 3 to 7 × 10⁶ t/yr of suspended material is accumulated in the German Bight and 12

■ **Figure 7.4** *Reconstructed coastlines of the German Bight for 1AD, 800AD and 1500 AD (Behre, 2003)*

to 19×10^6 t/yr are transported into the Norwegian Channel (Dyer and Moffat, 1998).

On Sylt Island (Figure 7.1), Pleistocene deposits are exposed in cliffs up to 30 m in height, topped by coastal dunes. Through spit prolongation, Sylt has been growing northward and southward and the island was connected to the mainland by the Hindenburg Dam in 1927, which caused increased sedimentation and salt marsh formation.

In the shelter of Amrum Island, which like Sylt Island is underlain by Pleistocene deposits, exists the rounded island of Föhr, which is bordered by salt marshes and completely surrounded by man-made dykes. South of Amrum exist broad sand intertidal shoals. They are elevated locally to only 1 m above mean high tide level. Due to their broad shallow topography, wave action is reduced, thus protecting the islands of Pellworm and Nordstrand, which contain only embanked marshlands.

The Halligen, a group of small islets in the Wadden Sea with a total area of only 23 km², are remnants of an old populated marshland, which was destroyed mainly by two severe storm tides in 1326 and 1634. These small islands have been further reduced in size by later storm surges. In the innermost part of the German Bight there are a number of sand shoals, the largest of which is Trischen. West of the Elbe estuary, on a broad sand shoal that runs westwards out from Cuxhaven, the flat islands of Neuwerk and Scharhörn have developed, the latter seriously reduced by erosion during the past century. Alte Mellum shows a similar situation on the shoal extending seaward between the Weser and Jade estuaries (Figure 7.4).

The East Frisian Islands are a chain of elongated dune-capped barrier islands extending to the Dutch border. They have wide sandy beaches on the outer coast, backed by dunes. This barrier island chain was originally formed some distance to the north, and driven southward with the secular rise in sea level by a combination of dune migration and storm surge washover. Land reclamation and substantial catchment area reduction resulted in enlargement of the East Frisian barrier islands together with a decrease of mudflats and salt-marsh areas. During the recent past, along with the sea-level rise and increase in wave energy, fine-grained particles have been removed from the back barrier area (Flemming and Bartholomä, 1997).

For the bottom of the North Sea, coarse sand, drowned dunes and sand ribbons have been described as typical bedforms (Köster, 1974; Werner, 2004). Other geomorphological elements are elongated nearshore bars on the upper shore-face, which like Sylt Island, are supported on a regular basis with sediment from beach nourishments (LKN S-H, 2009). In shallow parts, offshore of the East Frisian barrier islands, nearshore bars occur down to a water depth of about 6 m NN (Normalnull = German ordnance datum). Further offshore, shore-face connected sand ridges, 2 to 5 m in height, are present. They occur in water depths between 12 to 18 m below NN, exhibiting a pronounced dynamic behaviour showing cross-shore migration rates of up to 100 to 200 m/yr at their maximum (Antia, 1996).

The thickness of the mobile sediment cover along the German North Sea coast down to 20 m below mean tidal water level has given rise to three distinct areas. Offshore, the Wadden Sea in water depths between 0 to 10 m has a mobile sediment cover up to 10 m thick; a relatively thin layer of mobile sand material is located between 10 to 15 m water depth; a slight increase in thickness of 2 to 3 m, with local maxima of about 5 to 6 m, can be observed in water depths from 15 to 20 m. A conceptual model of the net sediment transport regime comprises a zone of relatively small longshore sediment bypassing. This results in a substantial sediment supply in the inner-most German Bight and a shore-normal bedload transport shifting sand back and forth along the coastal profile with a net seaward transport component (Zeiler et al., 2000).

During storm surges sediment can be transported offshore from tidal flats and sand bars, down to water depths of 40 m below NN. Large-scale sedimentary structures are shifted during storm wave action mainly by foreset movement of fine sand. The mouths of the estuaries are tidally scoured, with distinct ebb and flow channel systems. In the Elbe and Weser Estuary the seabed is dominated by morphological features, such as channels and sand tongues, sand waves in channels or longitudinal sand ridges. Pronounced migrating bedforms occur on the sandy river bed of the Elbe and Weser estuary and migration of these bedforms shows a cyclic pattern correlating with the tidal cycle.

Due to an increasing sea level and/or mean tidal range, a deepening of inlets and tidal channels is expected (Hofstede, 2002). Another hint for adaptation of the depositional environment to changing hydrodynamic forcing is the landward migration of characteristic geomorphological elements.

Baltic Sea

Over the last 2.4 million years, the Baltic Sea basin was carved by several ice advances of which the latest, with distances between the different ice-marginal lines increasing from west to east, formed the specific current geomorphological shape and coastal areas. As meltwater deposits, composed of gravel, sand and silt, have formed between these lines, the amount of

this material in coastal areas increases from west to east.

During the Holocene development glacio-isostatic movements and climatically controlled eustatic sea level fluctuations caused transgressions and regressions causing four evolutionary stages: Baltic Ice Lake, Yoldia Sea, Ancylus Lake and Littorina Sea (Björk, 1995). The Baltic Sea was finally connected to the North Atlantic via the North Sea about 7,900 years BP (Björk, 1995; Lampe, 2005). This connection has continued until today (see Figure 7.1) The rapid water level rise of the Littorina transgression led to widespread inundation of the pre-existing glacial relief. All areas below −5 m related to the present mean sea level were flooded during this period. Typical sub-merged landforms are pronounced ridges of morainic material and embedded meltwater channels, all associated with glacier-tongue shaped troughs and fjords.

About 6,000 years BP, when the water level was close to its present position, waves and currents began to modify the coastal profile by intensive erosion, transportation and redeposition, resulting in an initial development of spits, hooks and beach ridges. Small islands, consisting of morainic material, were connected by growing spits, forming the famous Bodden bay-mouth coast. Rügen Island, as well as thr Fischland-Darss-Zingst peninsula are well-known examples of this kind of coastal configuration (Figure 7.5).

■ **Figure 7.5** *The coastline of the Darss peninsula, which is exposed to westerly winds, showing old beach ridges with cats-eye ponds inbetween (Photograph by B. Gurwell)*

In response to glacio-isostatic rebound, the northern part of the Baltic Sea is dominated by a still ongoing uplift with rates of up to 9 mm/yr relative to the present sea level. In the southern part, subsidence with rates of up to 2 mm/yr occurs (Meyer and Harff, 2005) and as this generally causes erosion, wide areas of the southern Baltic Sea coastline suffer from sediment loss.

Today, the German Baltic Sea outer coastline is about 724 km in length (Figure 7.1), while the inner coastline including the Bodden areas is twice as long. Cliffs, built up of morainic material, are eroding with recurrent landslides resulting in a boulder-strewn cliff foot. The intervening coast is low-lying and beach-fringed sometimes backed by dunes with an increasing west to east number of dunes being found. Beach and nearshore sediments are mainly composed of sand with some gravel and boulders, derived from cliff and sea floor abrasion. Along the low-lying sectors, lagoons are formed where valley mouths or depressions in the glacial drift topography were submerged by the Holocene transgression and enclosed by formation of spits or barrier beaches.

Besides the geological prerequisites, only waves and wind-driven currents are relevant for the seafloor conditions and sediment dynamics in the Baltic Sea. Wave conditions and sediment transport in coastal waters depend on exposure to the predominant wind and wave direction. The predominance of westerly winds in the southern Baltic Sea yields a prevalence of waves from the west and north-west, consequently inducing an eastwardly moving sediment transport system. A good example of these conditions is the Darss peninsula (Figure 7.5), a major cuspate foreland built up in stages indicated by up to 120 sub-parallel beach ridges whose development started 3,500 years BP. The pattern of these ridges indicates a north-east migration with rates up to 2 m/yr. They are truncated on the retreating western shore, clearly visible by 'cats-eye' ponds which are now under erosion. Accretion continues at the northern tip Darsser Ort and in Prerow Bay, east of Darsser Ort. The volume deposited amounts to 390 × 10^6 m^3 (Naumann and Lampe, 2010). North-easterly, easterly and south-easterly winds are also frequent resulting in sediment transport directed westward, especially in the most south-western part of the Baltic Sea. As strong storms can last for days, high water levels can last for the same duration (Schwarzer, 2003), which is in contrast to North Sea conditions.

Although sediment distribution is affected strongly by subsurface geology, a depth-dependent overall zonation of surface deposits is found (Seibold et al., 1971). In the south-western and southern Baltic Sea coarse-grained lag deposits form a thin veneer (few decimetres) on top of coastal till deposits and on submarine sills and shoals in water depths of 5 to 15 m. These sediments result directly from erosion of underlying till deposits, as sand is removed leaving coarser material behind. Apart from the immediate proximity of the coast and abrasion platforms, these sand veneers are relatively thin, e.g. only 0.5 to 2.0 m in Kiel Bay and inner Mecklenburg Bay (Schrottke and Schwarzer, 2006).

Typical features of wave-dominated coasts are highly dynamic nearshore bars of different orientation, such as longshore or crescentric bars (Figure 7.6). Their mobility and shape depends on waves, climate, sediment availability and seafloor gradient (Short, 1999; Schwarzer, 2003). They are mainly established in front of lowlands where several bars can exist down to 6 m below sea level. In the south-western Baltic Sea their thickness seldom exceeds 2.5 to 3.0 m. However, these features have a significant influence on coastal stability, as they induce wave breaking and energy dissipation offshore. In front of active cliffs nearshore bars are often missing.

Schwarzer et al. (2003) demonstrated that waves are the main controlling factor for seasonal variations on the upper shore-face. On the lower shore-face, stable sedimentological and morphological behaviour is observed; changes are only measurable on centennial and millennial scales, as long-term processes, i.e. sea-level fluctuations or neotectonics, are the main driving forces. Even on the upper shore-face offshore of Usedom Island, the sedimentological pattern remained stable after passage of a severe storm (Schwarzer et al., 1996). Comparison of near-bottom hydrodynamic forcing and experimentally derived critical shear velocities indicate a storm-controlled

■ **Figure 7.6**
System of crescentric bars in inner Lübeck Bay

particle transport in the Mecklenburg Bight. Statistical analysis of sediment parameters and hydrodynamic modelling show evidence that the preferred transport pathways for clastic material in the south-western Baltic Sea is based on the direction of the average current vectors (Bobertz and Harff, 2004).

Schwarzer and Diesing (2001) compared sediment dynamics and geo-morphological changes on seasonal and annual scales in two different sandy nearshore areas, Bottsand (Kiel Bight) and Tromper Wiek (Rügen Island). A geographical variation of seasonal maximum mobility exists, showing an offshore progradation of geomorphological structures during stormy winter conditions, but even intense sediment mobility did not invoke substantial morphological changes, especially with respect to seasonality.

Along cliff sections, slumping of soft sediments during winter and spring produces fans of shore debris which are consumed and dispersed by wave action during high water levels, distributing sand and gravel to adjacent lowland beaches.

Several offshore cliff sections measured by Schrottke (2001) indicated seafloor erosion rates of 2 to 5 cm/yr in water depths down to 6.5 m and up to 300 m offshore. The eroded sediment is partly moved onshore. In many cases the amount of sediment supplied by seafloor erosion is underestimated, sometimes completely neglected. Depending on mineralogical composition and exposure to wave attack, such sediment supply can be of the same order of magnitude as the supply from the exposed parts of retreating cliffs (Schrottke and Schwarzer, 2006). Along the coastline of Schleswig-Holstein, there is no cliff protection as these landforms are regarded as natural sediment sources supporting adjacent lowlands with sand material.

In Mecklenburg Bay and inner Lübeck Bay, a Bodden-type coast did not develop; here cliff sections alternate with lowlands sometimes with intensive nearshore sand bar systems in front of them. In inner Lübeck Bay, Hemmelsdorf Fjord and Trave Fjord were created, leaving the Brodten cliff (up to 20 m in height) area to separate them. This cliff has retreated some 6 km during the past 6,000 years, supplying adjacent lowlands with a huge amount of sediment, which about 1,000 years ago blocked Hemmelsdorf Fjord. Trave Fjord is kept open artificially, as it is the main navigation channel for Lübeck-Travemünde.

Spits have formed at several localities in Kiel Bay. Graswarder spit at Heiligenhafen is the best known structure, which has grown eastward since *circa* 3,000 years ago. In Kiel Bay another four fjords were created by glacier tongues, running west to east: Flensburg Fjord, Schlei, Eckernförde Bay and Kiel Fjord, all surrounded by cliffs with lowlands in between.

For the retreating southern and south-western parts of the Baltic Sea there might be an increasing future demand of sand for beach nourishment, which is related to increasing erosion and coastal retreat due to the predicted sea level rise (HELCOM, 2007). Unfortunately, due to their Pleistocene and Holocene development, these areas have the most limited amount of natural mineral resources and most of these deposits are of fossil origin and non-renewable.

Storm surges

Storm surges are natural events that have always threatened life at the coast and may cause severe damage due to increasing use of the coastal regions. Both coastal protection and natural disaster defence measures are based on analyses of extreme flood levels. One of the oldest reports of storm surges at the North Sea coast dates from 340 or 120 BC and a reliable early record of a Baltic Sea storm surge dates back to the year 1044 (Jensen and Töppe, 1990).

Extreme storm surge events causing major losses of land have shaped the North Sea coastline and islands for centuries. An example is shown in Figure 7.7, which depicts the development of the largest German North Sea island of Sylt since the fourteenth

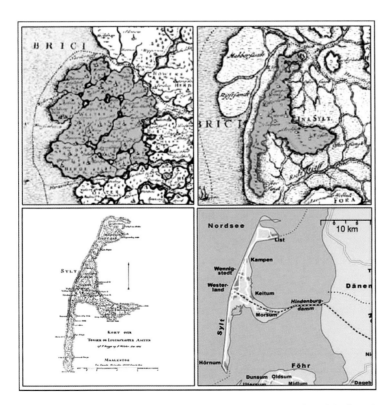

■ **Figure 7.7** *Development of the island of Sylt. Top left: Sylt before the severe storm surge of 1362. Top right: Sylt before a strong storm surge in 1634 (from Danckwerth, 1652). Bottom left: Sylt in 1793 (from Bugge and Wilster, 1805, in Jensen and Müller-Navarra, 2008). Bottom right: Sylt today (www.sylt.citysam.de/landkarte-foehr-amrum.htm; with permission from the German Coastal Engineering Research Council, KFKI)*

century. The figure illustrates the enormous land losses due to several storm surge events. The first studies on surge event frequency were conducted by Brahms (1754), Woebecken (1924) and Schelling (1952). Although storm surges have hit the North Sea coasts several times during the past centuries, since the disastrous flooding of the Netherlands in 1953, surges in the recent past have given rise to the question whether there has been a change in the pattern of storm surge occurrences in the North Sea region (Führböter 1976 and 1979; Siefert 1978; Führböter et al., 1988).

Petersen and Rohde (1977, p. 9) defined a storm surge as a "period of time during which water levels on the coasts and in estuaries are high, primarily due to strong winds". Studying storm surge generation

needs a clear distinction between the North Sea and the Baltic Sea, as they have different hydrological regimes.

Storm surges at the German North Sea coast result mainly from a build-up of water masses along the coasts (e.g. wind set-up), i.e. they are caused by stochastic impacts of meteorological origin, which are superimposed on astronomical tides (Figure 7.8). As a rule, extreme storm surges at the German North Sea coast occur when heavy storms from north-westerly directions reach wind speeds in excess of 25 m/s. In general two types of North Sea storm surges can be distinguished. The wind set-up type is characterised by winds blowing for a long time from a north-westerly direction, pushing water masses into the south-eastern North Sea. Storm surges of this type

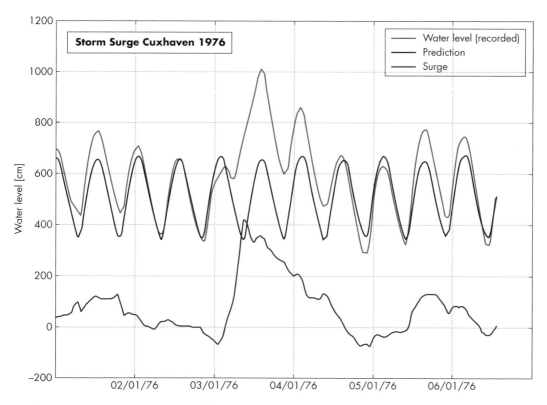

■ **Figure 7.8** *Storm surge event at the Cuxhaven gauge from 1976 and its build-up (water level, prediction (astronomical fraction) and surge (meteorological fraction)) generated by the programme T-Tide (based on Pawlowicz et al., 2002)*

can be reliably forecast, and warnings issued up to 18 hours and more in advance. In contrast, the circulation type is more difficult to forecast, because a small intense low-pressure system tracks across the British Isles at high speed, gaining strength over the North Sea. Consequently there may be situations in which no exact forecasts can be made until a few hours before the peak water level is reached (Müller-Navarra, 2005). The ratio of stochastic influences in relation to deterministic influences (e.g. astronomical tide) in water levels at the German North Sea coast is very high and this has to be taken into account when computing storm surge levels based on probability calculations.

Whereas the North Sea represents by definition a semi-enclosed sea, the Baltic Sea constitutes a (nearly) closed system (see Figure 7.1). It is connected to the North Sea only through narrow Belts and the Sound, leading to a complex system behaviour. The differences between the systems account for the different mechanisms of storm surge generation.

The Baltic Sea does not have one particular predominant weather pattern which creates particularly high water levels. Therefore a variety of weather patterns can potentially lead to storm high water levels. The principal difference among storm surge events in the Baltic Sea is their classification either as a wind set-up event, where wind is the only cause, or a storm high water where seiches involving the whole water body of the Baltic Sea can influence water levels in the western Baltic by a few decimetres. Moreover, the actual water volume in the Baltic Sea also affects the development and peaks of extreme water levels in this region.

The total number of storm surges with extreme water levels is substantially higher at the North Sea coast compared to that of the Baltic Sea coast with the main cause being the meteorological situation that triggers such events.

The storm surges of the last 30 years produced maximum water levels especially in the inner German Bight and along the North Frisian coast. This is evident, as the highest tidal high water level at Borkum and Emden gauge stations was recorded during the 1906 storm surge, whereas in the area between the Weser and Elbe estuaries the highest storm surge level ever recorded occurred on 16 February 1962. In Cuxhaven and on the west coast of Schleswig-Holstein, the historically highest storm surge level was measured on 3 January 1976. This in turn was exceeded by levels recorded at the Dagebüll and List/Sylt stations which are located farthest north.

The first records of storm surge events at the Baltic Sea coasts, including the maximum measured water levels, date back to the fourteenth century. The first precise measurement showed an elevation of 3.2 m above the Baltic Sea mean water level, measured at Lübeck during the storm surge of 1320 (Jensen and Töppe, 1990). In November 1872, a storm surge of unprecedented severity hit the Baltic coasts, with a maximum of up to 3.5 m above mean water level in Flensburg Fjord.

Flood events on the Baltic Sea and North Sea coasts usually occur during autumn and winter season from October to March, especially in the Schleswig-Holstein area of the Baltic Sea. The extreme flood of 12/13 November 1872, has to be considered a singular event with regard to its maximum water level, and thus will continue to serve as a design flood in the future.

Life at the coast has always been strongly influenced by recurring storm surges. In order to be able to assess future risks and to take early action to strengthen coastal defences, some detailed studies still have to be made. In particular, questions concerning the probability of occurrence of extreme storm surges and the impact of climate change on storm surge events still have not been satisfactorily answered.

Mean sea level of the North and Baltic Sea

One of the most pronounced effects of climate change has been a rising sea level over the last 150 years. A higher sea level increases relative heights of storm surges and thus the number of erosion areas along coastlines. Therefore, knowledge about sea level development is very important for coastal engineering and coastal zone management. Many studies have been accomplished in the last 30 years, especially on the German Bight, Führböter and Jensen (1985) presented the first results of analysing mean high water in the German Bight. Since then many different studies have been accomplished. Most projects dealt with mean high water, the mean tidal range or the mean sea level in the German Bight (e.g. Jensen, 1985; Wahl et al., 2010), but similar studies were conducted for the Baltic Sea area (Jensen and Mudersbach, 2004).

North Sea

Wahl et al. (2010 and 2011) analysed mean sea level changes in the German Bight. Records from 13 tide gauges covering the entire German North Sea coastline and from 1843 to 2008 have been used to derive relative mean sea level time series. Changes in mean sea levels are assessed using non-linear smoothing techniques and linear trend estimations for different time periods. Time series from individual tide gauges are analysed and then 'virtual stations' constructed which are representative of the German Bight and the southern and eastern regions of the Bight. It was found that an acceleration of sea level rise commenced at the end of the nineteenth century followed by a deceleration. Another acceleration with its starting point in the 1970s and intensification from the 1990s has been identified, but the rates of sea level rise during this period are comparable with rates at other times during the last 166 years. Higher rates of sea level rise have been detected for tide gauges covering the eastern part of the German Bight compared to those covering the southern part. This is mostly due to different rates of postglacial vertical land movement. In addition, different temporal behaviour of sea level change is found in the German Bight compared to wider regional and global changes, highlighting the urgent need to derive reliable regional sea level projections for coastal planning strategies.

Baltic Sea

Investigations of sea level development and variations in the Baltic Sea have a long tradition. Various authors (Hünicke et al., 2008) have studied sea level rise and its variations using tide gauge data and different diagnostic, statistical and harmonic approaches.

For the German part of the Baltic Sea coastline, Jensen and Mudersbach (2004) studied mean, low and high water levels in comparison to the German North Sea coast using the normalized arithmetic mean of the four gauges at Travemünde, Warnemünde, Wismar and Sassnitz. Results for the normalized 'Baltic Sea' mean sea level gauge show an increase of 1.1 mm/yr for the period from 1838 to 2001 and 1.3 mm/yr for the period from 1949 to 2001. The intensity of this rise is much smaller than in the North Sea. Interpretation of these results needs to take into account that the mean water level of the North Sea and the Baltic Sea are comparable but not equal in detail, as different definitions of 'mean water level' exist. In the last 40 years the increase of mean sea level is stronger than before and future investigations need to analyse trends of water level developments.

Beach erosion

Even if the North Sea and the Baltic Sea are different in their hydrographic characteristics and geological development, beach erosion is a severe problem along the coastline of both seas.

Baltic Sea

Large sections of the Baltic Sea coastline are retreating at an average rate of 0.2 to 0.3 m/yr with maximum rates of up to 1.5 m/yr (Schwarzer et al., 2003; Ziegler and Heyen, 2005; Niedermeyer et al., 2011). Approximately 70 per cent of the coastline of the state of Mecklenburg-Vorpommern, extending from Mecklenburg Bay to Odra Bay, is permanently under retreat (Harff et al., 2004). Erosion of active cliffs is controlled by storm events combined with high water levels. The rate of active cliff retreat of Schleswig-Holstein is 24 cm/yr on average (Ziegler and Heyen, 2005) and up to 30 cm/yr for the cliff-coast of Mecklenburg-Vorpommern.

Erosion is predominant along the coastline of Kiel Bay and Lübeck Bay where lowlands are protected by a combination of different shore protection measures. The strongest shore protection scheme can be observed in front of the Probstei area, at the eastern outer Kiel Fjord. A 14.3 km long dyke, 4.5 m in height combined with groins and breakwater elements, together with beach nourishments, have been constructed here (see 'Shore protection' later in this chapter and Figure 7.19).

East of Rügen Island the coastline turns towards a formation which looks as if it is in equilibrium but even here coastal retreat dominates. This response is related to glacio-isostatic/neotectonic sinking, sea level rise, frequency of storm surges and the geology of the coastal areas. As such, it requires continuous replacement of the eroded material to maintain a stable coastline in those areas where settlements, different kinds of infrastructures and/or industrial use predominate.

The North Sea

Waves reaching the outer shores of the barrier islands vary in height depending on the season. The average annual significant wave height at the North Sea Buoy II location (55° 00' N, 006° 20' E) is 1.6 m (Klein and Frohse, 2008). Because of the dominant westerly winds, longshore drift is eastward along the East Frisian Islands beaches, while along the North Frisian Islands it is alternately northward (south-westerly waves) and southward (north-westerly waves). The coastal region is subject to occasional storm surges produced by strong winds accompanying the passage of depressions across the North Sea. Storm waves during such periods accomplish substantial erosion of the barrier island chain sand shores, some reshaping the intertidal morphology and damaging the dike system that protects the hinterland. Major storm surges occurred during the years 1164, 1219, 1287, 1362, 1436, 1532, 1570, 1634, 1717, 1756, 1792, 1825, 1904, 1909, 1911, 1953, 1976, 1981,1990 (five hurricanes in three days), 1999, 2000 and 2007 (Jensen and Müller-Navarra, 2008). The 1362 surge inundated 1,000 km² of land, part of which remained permanently below high tide level; the 1634 surge broke the island of Strand into the islands of Pellworm and Nordstrand, which are now separated by the largest tidal channel in the North Frisian area, the Norderhever. In recent centuries the building, enlargement and repair of dykes has maintained the mainland coast and much of the barrier island's inner shoreline. However, the outer beaches and tidal inlets between barrier islands are subject to continuing changes, especially during stormy periods.

The entire west coast of Sylt Island is subject to erosion (Figure 7.10) and for more than 110 years, several shore protection schemes have tried to preserve the shape of the island. Since 1972 mainly beach nourishments have been carried out, partly combined with the usage of geotextiles. Since 1972, the beach has been nourished with *circa* 40 million m³ of sand (Hofstede, 2008). The southern end of Sylt, Hörnum Odde, is characterized by severe erosion due to the construction of cross-shore and longshore jetties of tetrapods, which have been partly removed during the past years.

Shore protection

Historically inhabitants of coastal areas began to protect themselves against floods due to rising sea level and storm surge impacts. Different measures have been applied to the local problems, but sea and estuarine dikes are the main coastal defence structures in Germany (Schüttrumpf, 2008).

Dikes

Dikes protect low-lying areas in Lower Saxony, Schleswig-Holstein, Bremen, Hamburg and Mecklenburg-Vorpommern. More than 2.4×10^6 people and an area of more than 12×10^3 km² are protected by more than 1,200 km of sea dikes and estuarine dikes in Germany. Sea dikes have a long tradition in Germany. The first references to sea dikes are found in 10 BC (Garbrecht, 1985). Due to continuous impacts through storm surges, the development of sea dikes is now quite advanced.

Fatalities after storm surge disasters in the Middle Ages were high and the consequences severe for those who survived the flood. Large regions were flooded, houses and farms destroyed and many areas remained useless for agriculture and stock farming. These fatalities, damages and economic losses were caused by flooding through breached dikes. The resistance of dikes in the Middle Ages and in later centuries against wave attack and high storm surge water levels was low and many dikes overflowed. The first dike breaches along the German North Sea Coast were reported from a storm surge in 1164 with about 20,000 fatalities. Even if the number of fatalities is uncertain, the importance of this event is obvious for that time. More historical storm surge disasters were reported in 1362, 1634, 1717 and 1825 for the Belgian, Dutch, German and Danish coasts. These storm surge disasters were responsible for development of sea dikes along the coastlines in these countries. Essentially based on experience, sea and estuarine dikes became higher and broader over the centuries. The first dikes – called Stackdikes (Figure 7.9) – were very steep and sometimes even vertical, and consisted of a wooden front face on the seaward slope.

The coastal disasters (a summary of dike failures is given in Schüttrumpf and Oumeraci, 2002) of recent times changed the design philosophy of sea dikes in Germany. Before 1950, the crest level and slopes of sea dikes were designed based solely on experience. From *circa* 1950, the crest levels of sea dikes were designed deterministically based on a statistically designed water level and corresponding wave run-up height. Experimental investigations were applied to determine the wave run-up height (the first experimental investigation in Germany was by Hensen, 1954) and the wave overtopping rate (the first experimental investigation in Germany was by Tautenhain, 1984). Today, the objective is to improve scientific knowledge concerning the probabilistic design of sea dikes (Kortenhaus, 2003), but as yet, probabilistic design has not found its way into practice.

In general, two different types of sea dikes can be distinguished in Germany (Figure 7.10). Type 1 has a wide foreland above mean high water to protect it. A high foreland reduces incoming energy by wave breaking at storm surge water levels and its width can reach several hundred metres. At normal tides no water, waves or currents affect the dike toe, so revetment construction can be avoided to protect the dike toe. If a smooth slope is not possible due to areal constraints, a light revetment is recommended. If no foreland protects the dike, a heavy revetment is recommended at the dike toe (Type 2). These revetments are often constructed with a slope of 1:3. The toe is embedded in the sea bed to avoid scouring and a crest is built, reaching a height of usually about 1.50 m to 2.0 m above mean tidal high water. A berm of asphalt or concrete is located landward of the revetment. Berms are constructed 1.0 m to 2.0 m above mean tidal high water with a width of up to 3.0 m.

The dike seaward slope can differ between 1:3 (some estuarine dikes and Baltic sea dikes without heavy wave loads) and 1:7 (at very exposed locations along the North Sea coast). In general, a 0.5 m (Baltic Sea) up to 2.0 m (North Sea dikes) thick clay layer is preferred to avoid erosion and scouring due to wave loadings. The quality of the clay is defined in EAK (2002). Some dikes are covered by asphalt or concrete, but grass-covered dikes are preferred. Currently, the dike core mostly consists of sand with a drainage system towards the landward foundation trench.

■ **Figure 7.9** *'Stackdike' circa 1600 AD at the dike museum in Büsum (Photograph by D. Meier, 2006; with permission from the German Coastal Engineering Research Council, KFKI)*

The crest of a sea dike or an estuarine dike in Germany has a width of 2.0–3.5 m to allow vehicles or pedestrians to drive or walk. The crest is slightly sloped to enable overtopping water or rain to flow landward or seaward and to avoid infiltration. The landward slope has to fulfil geotechnical aspects (no sliding), erosion or infiltration due to wave overtopping should be avoided and harvesters should be able to drive on the landward slope. Therefore, landward slopes range between 1:2 and 1:5, but most of the slopes are constructed with a gradient of 1:3.

A landward berm is situated at the landward slope toe with a width of up to 10 m with a 3–4 m wide road to allow heavy vehicles to drive along the dike during even very severe storm situations. The landward berm is located about 0.5–1.0 m above mean tidal high water to ensure vehicle passage even when low-lying areas are flooded during extreme situations. Finally, an inner ditch is constructed at the landward berm toe to collect drained water or rain.

Possible future climate change represents an important issue in coastal engineering at the moment and, based on predicted increasing sea levels, dikes should be resistant against future floods. To avoid too large investments in more resistant dikes for the far future, they are nowadays shaped in such a manner that dike enlargement can be easily done at any time without requiring major effort.

Detached breakwaters

In Germany, detached breakwaters for shore protection have been mainly constructed at the Baltic Sea coast at Mecklenburg-Vorpommern, with just a

(a) Dike without foreland

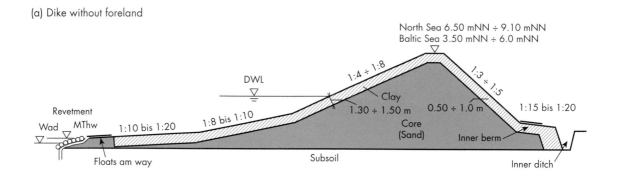

(b) Dike with foreland

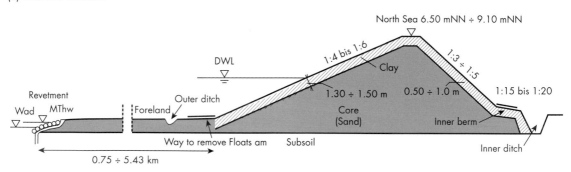

■ **Figure 7.10** *Typical dike profiles in Germany (with permission from the German Coastal Engineering Research Council, KFKI)*

few in Schleswig-Holstein (Schilksee, Stakendorf). They are used for protection of eroding sand from coastal stretches as well as cliff sections (Schilksee, Ahrenshoop, Scllin, and Streckelsberg). Currently, eight detached breakwaters have been installed along the Mecklenburg-Vorpommern coast.

Detached breakwaters for coastal protection are mainly constructed using natural rocks, but in Schilksee some concrete has also been used. The single breakwater elements arc nearly homogeneous with no core or filter layer below. The weight of the natural rocks, used for construction, ranges between 2 and 7 tonnes. The heaviest blocks are used for the armour layer and the construction crown, since these are the positions where wave energy release is concentrated. A typical cross-section of a German Baltic Sea Coast detached breakwater is shown in Figure 7.11.

The breakwaters are used as single breakwaters or in systems with normally up to four elements, but in Schilksee nine breakwater elements have been built. The length of breakwaters along the German Baltic Sea coast are between $L = 50$ m and $L = 400$ m (Stakendorf), the distance of the detached break-waters from the shore is between 70 m and 200 m and the crest height is between $H_c = 0$ m and $H_c = 2$ m above mean sea level. In systems, the gap between the breakwater elements ranges from $L_g = 50$ m up to $L_g = 100$ m (see also Carstensen *et al.*, 2004) or breakwaters are constructed in two lines with staggered gaps (Schilksee).

The ratio between length, distance to shore and gap width between the single elements normally allows development of a tombolo, while a complete tombolo attached to the breakwater is normally not desired. To support the development of stable salient constructions these are normally combined with initial beach nourishments. The breakwater system in front of Streckelsberg/Usedom Island (Figure 7.12) serves as a good example of such a construction.

The function and influence of detached breakwaters on the sediment budget and on the morphological development in a coastal area is shown in Figure 7.13, assuming a sand coast and longshore sediment transport from the indicated direction. Coastline development shows the desired accretion in the protected area behind the breakwater and downdrift erosion on the construction lee-side. For longshore sediment transport in the other direction, the development of the coast would be *vice versa*, respectively.

Typical configuration of a detached breakwater

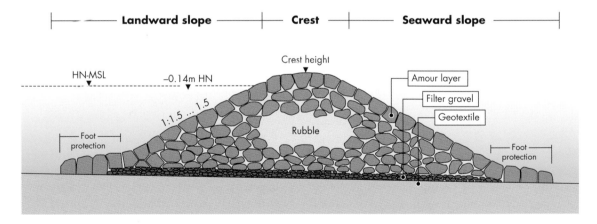

■ **Figure 7.11** *General layout of detached breakwaters in Mecklenburg-Vorpommern (STAUN Rostock; with permission from the German Coastal Engineering Research Council KFKI)*

■ **Figure 7.12**
Detached breakwater at Ahrenshoop (Photo by B. Gurwell, 2000)

OFFSHORE – BREAKWATER

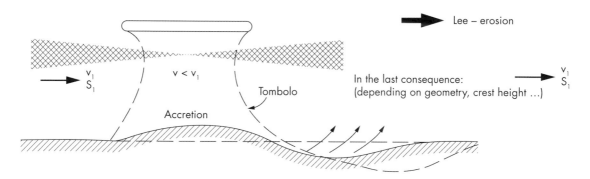

■ **Figure 7.13** *Influence of a detached breakwater on sediment transport and coastline development (schematic, after Kohlhase, 1991; Kohlhase, 2004; with permission from the German Coastal Engineering Research Council KFKI)*

The design of detached breakwaters is normally separated into functional and constructional tasks. The functional design ensures performance, namely the development of a salient or a tombolo, of the breakwaters/breakwater systems. The constructional design ensures that the designed cross section of the breakwater is statically or dynamically stable for selected design parameters.

The functional design of detached breakwaters is in general based on results of sediment transport numerical simulations in the project area. This is done to determine in detail the influence of the structure on sediment transport on a local scale for the surrounding area and therefore future morphological development of the area. The numerical simulation includes assessment of the long-term behaviour of the project area. In order to assure the results of numerical simulations, results are double-checked based on nomograms and empirical and/or analytical solutions (Pope and Dean, 1986; Silvester and Hsu, 1993). The structural design of breakwaters is performed based on the stability of rubble mound structures (Hudson, 1953; van der Meer, 1993). Since breakwaters are used to steer the morphological development of a coastal stretch, the probability of occurrence of the design parameters (wave height, etc.) is selected to be comparatively high, e.g. in the range of p = 0.05 to p = 0.02.

Streckelsberg is located in the middle of the north-eastern coast of the Island of Usedom, an area which is very exposed to wave action from North-East to East. Streckelsberg, with a central part reaching a height of up to 56 m, is an erosive cliff consisting of outwash deposits and glacio-limnic deposits topped by coastal dunes. The length of the protected area is approximately 500 m. South of Streckelsberg, the coast is formed by meltwater deposits underlain by boulder clay. North of Streckelsberg the coast changes from an active boulder clay cliff to the typical low-lying sandy coast with dunes and comparatively wide beaches. This low-lying area forms a narrow (in some areas <50 m) border between the open Baltic Sea and the so called Achterwasser (Bodden Coast). For this reason it is very vulnerable to breaching caused by high water levels combined with strong wave attack.

The vulnerability of the low-lying areas in conjunction with breaching is the main reason for protection of the Streckelsberg cliff, since it is seen as the focal point of development for the complete coastal area near the Island of Usedom. Erosion at this point, which happened severely in the past, might also cause retreat north and south of the Streckelsberg (see Figure 7.15).

For more than a century the Streckelsberg was protected by a sea wall (Streckelsberg Wall), which was destroyed (Figure 7.14) and partly rebuilt several times during the past after the beach in front had been completely eroded due to wave reflection by the vertical wall superimposed by the negative sediment budget in the area (see Figure 7.15).

The breakwater system design was optimized using a numerical model, where a wide variety of criteria, such as transmission-coefficients and shoreline distances, have been analyzed. Assessment of the situation was based on long-term wave information

■ **Figure 7.14**
Destroyed Streckelsberg Wall

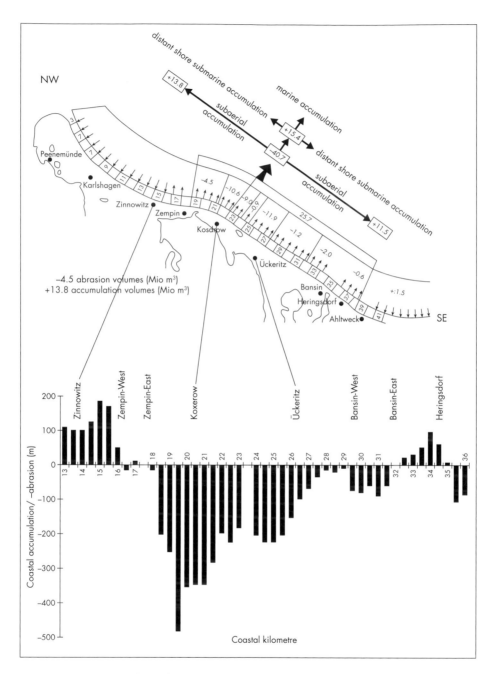

Figure 7.15 Top: Sediment fluxes for the open (outer) shoreline of Usedom Island for the period 1692–1986 reflecting almost equal longshore accumulation volumes for the beach barrier systems located in the northwest (Peenemünde Spit) and southeast (Svina Gate). Calculations are based on quantitative interpretations of historical and modern maps. Below: Shoreline displacement showing four centres of erosion (most intensive off Koserow) and two centres of accumulation (highest off Zinnowitz) (Source: from Schwarzer et al., 2003; modified)

■ **Figure 7.16** *System of detached breakwaters offshore of Streckelsberg and Usedom (Photograph by B. Gurwell, 1997)*

covering a period of five years. Criteria for the layout were mainly the formation of a stable salient in front of the Streckelsberg cliff without the tendency of growing a complete tombolo, mainly to minimize negative downdrift effects.

As the result, a system of three detached breakwaters with the following attributes was constructed: length, approx. 190 m; gap width, approx. 50 m; distance offshore, approx. 200 m; considering construction costs, this was the maximum distance possible from the shore. In addition to the detached breakwaters a wall was built directly at the cliff foot to protect it against extreme high water levels. South and north of Streckelsberg groin systems were re-established to minimize negative effects (see Figure 7.16).

Coastal groins

Coastal groins represent one of the oldest coastal protection structures in Germany and still represent an important part of coastal protection schemes. Groins can be found at the North and Baltic Sea coast in significant numbers and they will continue to be important elements of coastal protection schemes in Germany, especially in Mecklenburg-Vorpommern, where more than 800 groins exist. Even though their efficiency and value is controversial, groins have been used to increase the retention period of sand from

eroding and artificial nourished shorelines. Table 7.1 displays a review of the total number of groins along the coastlines of Mecklenburg-Vorpommern, Schleswig-Holstein and Lower Saxony.

Referring to their function, there is a differentiation between beach groins influencing mainly wave-induced currents and sediment transport and groins located at islands and tidal inlets mainly acting against tidal currents. In the German "Recommendation for Shore Protection Measures" (EAK, 1993) the latter are named 'stream groins'.

At the German Baltic Sea coastline, groins are mostly deployed as entire groin systems (Figure 7.17) to influence shoreline sediment transport, as a result of wave-induced longshore currents. They are mostly built as permeable groins consisting of wooden piles. Others have been made from natural stones or constructed as box-type groins. In the Probstei area (Kiel outer Fjord) 100 m long groins have each been combined with a 40 m long breakwater element at its most offshore part (see Figure 7.19). At the German North Sea coast, groins are often installed as stream groins, especially at islands and estuaries, to protect sandy coasts against erosion caused by tidal currents. These groins are mostly constructed using rock and concrete due to usually heavy loads (Figure 7.18).

Between 1815 and 1821, the first simple groins were built at the Island of Wangerooge (North Sea) for the purpose of protecting a lighthouse. These groins were built of brushwood material without additional rubble protection and were seriously damaged in the winter of 1821/22 (Fülscher, 1905). Nevertheless, more groins of this simple design were built at Wangerooge

■ **Table 7.1** *Number of groins in Germany*

Federal state	Number of groins
Lower Saxony	262 (North Sea)
Schleswig-Holstein	1,245 (North Sea)
	1,143 (Baltic Sea)
Mecklenburg-Vorpommern	1,129 (Baltic Sea)

■ **Figure 7.17** *Wooden single-row pile permeable groins at low water at the Baltic coast near Warnemünde (Photo by F. Weichbrodt; with permission from the German Coastal Engineering Research Council, KFKI)*

■ **Figure 7.19** *T-groin system in the Probstei area, Kiel Outer Fjord. Each groin is 100 m long; at its offshore end it is connected to a 40 m breakwater element*

■ **Figure 7.18** *Rock groin at Hallig (holm) Südfall (Photograph by H. Hinrichsen; with permission from the German Coastal Engineering Research Council, KFKI)*

in 1832 and 1834, but had to be abandoned around 1850 because of destruction by currents and waves. Starting in 1843, further simple bush groins were constructed on the Island of Norderney, but were destroyed shortly after completion.

Using this experience, coastal engineers began to construct various types of groins, composed of rubble or broken rocks, on some North Sea island coasts (Figure 7.19). First, seven solid groins were installed at the western part of the Island of Norderney between 1860 and 1867. They were built with a length of between 190 m and 215 m using layers of

fascines, wooden piles and a cover of rock or rubble (Figure 7.20). Little damage was detected after this modification of groin construction. Until 1877, five more of these groin constructions were placed on the western side of Norderney Island. Following the positive experience of groin construction at Norderney, more groins of this type were built on other German islands in the North Sea, e.g. Borkum, Baltrum, Spiekeroog, Wangerooge and Sylt.

Wooden pile groins using various designs were also established on German North Sea islands, constructed both as permeable and impermeable groins. As a special construction, wooden pile boxes filled with rubble or concrete blocks (called box-type groins), were introduced first in 1925. However, wooden pile groins became generally unpopular at the German North Sea islands although they have seldom sustained damage by currents and waves. This may be also due to the fact that wooden constructions in sea water are susceptible to infestation by the naval shipworm (*Teredo navalis*).

Starting in the mid 1920s, different construction materials such as steel, concrete, reinforced concrete and asphalt were increasingly used for groin construction. In the 1930s single wall groins made of steel sheet piling were built. Later on, box-type groins made of two rows of sheet piling filled with sand rubble or broken rock with or without a concrete

Sectional drawing of groin head

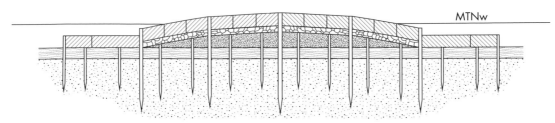

MTNw

Width: approx. 15m

Groin head

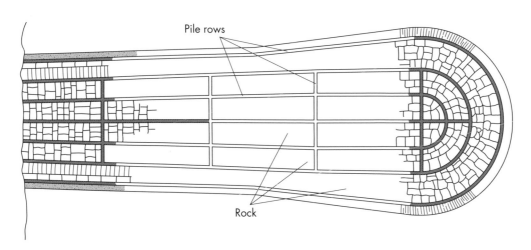

Pile rows

Rock

■ **Figure 7.20** *First rock groins on Norderney, 1861 (Tritt and Altemeier, 1987; after Fülscher, 1905; with permission from the German Coastal Engineering Research Council, KFKI)*

armour layer became common along the coast. Combinations of various construction methods and materials in one groin were tested and very long submerged groins were built. Today, natural stone (e.g. basalt columns), concrete, steel and asphalt as construction material for groins, is common along the German North Sea coast.

Following recommendations by Hagen (1863), the first groins at the German Baltic coast were installed in 1843 at the island of Ruden (located by Usedom, near the border of Poland). Similar to those first ones

constructed at the North Sea coast, these groins were very simple constructions, consisting of two rows of piles entwined with pine branches. Compared to present-day groins they were very small. With a length of 15 to 30 m, they were built in water depths not deeper than 1.0 m. They were destroyed by waves and scouring some years after completion. Later, advanced structural design and construction methods were introduced. Figure 7.21 and Figure 7.22 show some examples of groin construction in the nineteenth century using fascines and rock armouring.

Pile groin at Warnemünde 1895

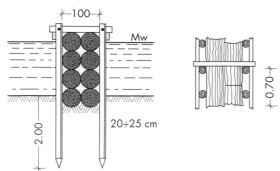

■ **Figure 7.21** *Wooden groin construction on the Baltic Sea coast near Warnemünde, 1895 (Tritt and Altemeier, 1987; after Fülscher 1905; with permission from the German Coastal Engineering Research Council, KFKI)*

Pile groin at Baltic Sea 1897

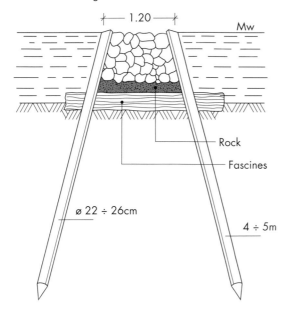

■ **Figure 7.22** *Wooden groin construction on the Baltic Sea coast, 1897 (Tritt and Altemeier, 1987; after Fülscher, 1905; with permission from the German Coastal Engineering Research Council, KFKI)*

With the invention of pile drivers, the embedded pile length in the ground could be significantly increased making groins more stable and so increasing the resistence against wave and ice forces. The groin length was increased and groins were built up to a water depth of 3.0 m. Before World War II, some groins were built with steel sheet piling. However, substantial corrosion problems were encountered. In the meantime the tourist industry had developed and steel sheet piling groins were causing a risk for tourists at sand beaches. Since they do not merge well into the environment, as do the more natural-looking stone or wooden groins, they were not very popular and are not accepted today.

At the German Baltic Sea coast and that of Mecklenburg-Vorpommern in particular, the wooden single-row pile groin became readily accepted and has been the favoured type of groin since World War II. This is due to lower costs compared to double-row pile or stone groins, as well as the option to build the groin with a varying permeability depending on its intended function.

It is not possible to describe the design details of all groins which were established along the German coast during the last 200 years. Numerous examples can be found in Kramer and Rohde (1992). The first papers on the experience with the functional and structural design of groins were published by Plener (1856), Hagen (1863) and Franzius (1884). An overview of German publications on groin construction, with a focus on the effectiveness of beach groins, from the beginning until 1961 is given in Petersen (1961). Detailed recommendations for the functional design of beach groins in Germany may be found in EAK (1993).

The crest elevation for North Sea groins is designed to be more than 0.5 m above mean tidal high water at the groin root and 0.5 m above mean tidal low water at the head. Along the Baltic Sea coast, the crest elevation of wooden groins is determined to be 0.5 m above mean water level. This ensures their functioning even during events with elevated water levels.

Groin length, spacing within a system and its permeability, are the key parameters. Groin length can be determined as a function of local wave parameters (wave length) and beach slope. The spacing depends on its length, permeability and category of site wave load. Usually permeable single-row pile groins with a permeability >20 per cent are used at the German Baltic coast of Mecklenburg-Vorpommern for water depths >1.0 m at the seaward part of the groin. In a groin system, permeability is often increased towards the end of the groin field in order to avoid downdrift beach erosion.

Today groin constructions functionally are designed based on practical knowledge, field and model investigations, recommendations, design formulae or manuals for the design of groins, such as those offered by CERC (1984), CIRIA (1990) and EAK (1993 and 2002). The first concept for the functional design of a single groin or a groin system is often obtained by numerical model tests. Experiences with existing groins near the proposed construction site are included in the functional and structural design. At the moment, no generally accepted tool exists to predict long-term morphological changes near groins or groin systems in detail. There are still various questions considering both hydraulic and numerical modelling. Cost-benefit analysis will be an important aspect in planning and design in the future.

Independent from the design of a groin field or materials used for groin construction, the purpose of each groin construction along coastlines is to trap sediment or increase the retention time of sand artificially put into the system. As sand availability increases from west to east in the Baltic Sea, the functioning of groin systems is much more effective in Mecklenburg-Vorpommern compared to Schleswig-Holstein, where a tremendous lack of mobile sand material exists. There have been several beach nourishments carried out in the Probstei areas since 1987 to stabilize beaches in between groins; material was taken from the German part of the Baltic Sea only for the first project. Due to the lack of proper material in German offshore waters, for the last three nourishments (each in the range of about $2 \times 10^5 m^3$) sand has been imported from Denmark.

Sand nourishment

A sustainable method of coastal protection is sand nourishment and especially at the islands of Sylt and Föhr, sand nourishment has been one of the major coastal defence strategies. Since 1972, more than 40 million m³ of sand have been transferred to these islands' beaches (Hofstede, 2008) resulting in costs of about $163 \times 10^6 €$. The first artificial beach nourishments were carried out in 1950 at the island of Norderney, located in the western part of the German Bight (Stefan and Kunz, 2000). The replacement aim was to compensate storm surge caused sand erosion at beaches, with sand from offshore areas. Nourishment included an enlargement of beach profiles giving higher protection against storm surge impacts and to the present time, sand nourishment is one of the most effective techniques against shoreline retreat. Sylt is the northernmost island in Germany (Figure 7.2), located west of Schleswig-Holstein. Its coastal zone is composed of sand beaches backed by dunes. A nearshore bar system and the beaches serve as natural wave breaks in front of the dunes. As a result of storm surges, beaches on Sylt Island lose more than one million m³ of sediment per year (Hinrichsen, 2009). Though nourishment cannot stop erosion processes, it stops coastal withdrawal and at Sylt it is considered to be the most effective measure against coastal retreat. The advantage compared to 'hard' coastal measures is the absence of negative side effects, such as lee-erosion and the morphologic and ecological alterability in respect to climatic changes (Hofstede, 2008; Reise and Buschbaum, 2007). In contrast the disadvantages are the high cost of expensive technical equipment and the negative aspect of long-term changes in biodiversity of the extraction areas (Reise and Buschbaum, 2007).

Several studies have been carried out to assess the impact of dredging activities on the North Sea and the Baltic Sea seafloor sedimentology. Zeiler et al. (2004) showed that refilling of pits on the shore-face, especially in front of Sylt Island, is very slow and may last for decades. The same is true for pits dredged in Tromper Wiek offshore Rügen Island. Diesing et al. (2006) showed a high stability of dredging effects.

Regeneration depends on the adopted dredging technique and the nature of the dredged material. Pits remain more stable than furrows, indicating higher impact of anchor suction dredging in comparison with trailer hopper suction dredging (Manso et al., 2010).

The amount of sand and gravel resources located in the coastal waters of Mecklenburg-Vorpommern amounts to 25.5 million m^3 of which, up to 2004, 31 per cent have been exploited (Harff et al., 2004). There is no estimation of sand and gravel resources for the Baltic Sea part of the state of Schleswig-Holstein. Furthermore, first studies have showed that sand abstraction from offshore areas will imply higher wind wave potential for the future, with consequent increasing impacts for the coastal zone, including higher erosion rates. Therefore, there have also been attempts at foreshore sand nourishment (Hofstede, 2008).

Acknowledgements

We gratefully acknowledge the very valuable contributions from Peter Fröhle, Holger Schüttrumpf and Frank Weichbrodt to this work. We are also grateful for the help we received from Sönke Dangendorf in preparing this article.

References

Ahrendt, K., 2006. Ein Beitrag zur holozänen Entwicklung Nordfrieslands, *Die Küste*, 71, 1–33.

Antia, E. E., 1996. Shore-face-connected ridges in German and U.S. mid-Atlantic bights: Similarities and contrasts, *Journal of Coastal Research*, 12, 141–146.

Behre, K.-E., 2003. Eine neue Meeresspiegelkurve für die südliche Nordsee, *Probleme der Küstenforschung im südlichen Nordseegebiet*, 28, 9–63.

Berner, U. and Streif, H. J., 2000. *Klimafakten. Der Rückblick* – ein Schlüssel für die Zukunft, Schweizerbartsche Verlagsbuchhandlung, Stuttgart, 238 pp.

Björk, S., 1995. A review of the history of the Baltic Sea, 13,0–8,0 Ka BP, *Quaternary International*, 27, 19–40.

Bobertz, B. and Harff, J., 2004. Sediment facies and hydrodynamic setting: A study in the south western Baltic Sea, *Ocean Dynamics*, 54, 39–48.

Brahms, A., 1754. *Anfangs-Gründe der Deich- und Wasser-Baukunst,* 2. Auflage, Aurich.

Bugge, W., 1805. *Aufnahme der Königlich Dänischen Gesellschaft der Wissenschaften aus dem Jahre 1793*.

Carstensen, D., Fröhle, P., Jäger, B. and Sommermeier, K., 2004. Bemessung von Küstenschutzbauwerken in Mecklenburg-Vorpommern [Dimension of Coastal Defence Structures in Mecklenburg-Vorpommern], *Jahrbuch der Hafenbautechnischen Gesellschaft,* 54, Verlag Hansa, Hamburg.

Coastal Engineering Research Center (CERC), 1984. *Shore Protection Manual,* Department of the Army, US Army Corps of Engineers, Washington DC.

Construction Industry Research and Information Association (CIRIA), 1990. (CIRIA), *Guide on the Uses of Groins in Coastal Engineering,* C. A. Fleming (Editor), Report 119, London.

Dankwerth, C. and Meyer, J., 1652. *Newe Landesbeschreibung der zweij Hertsothümer Schleswich und Holstein,* Atlas.

Diesing, M., Schwarzer, K., Zeiler, M. and Klein, H., 2006. Comparison of marine sediment extraction sites by means of shore-face zonation, *Journal of Coastal Research*, SI 39, 783–788.

Dyer, K. R. and Moffat, T. J., 1998. Fluxes of suspended matter in the East Anglian plume Southern North Sea, *Continental Shelf Research*, 18, 1311–1331.

EAK, 1993. *Empfehlungen für die Aufführung von Küstenschutzwerken,* Editor: Kuratorium für Forschung im Küsteningenieurwesen, Heide i. Holstein, p. 541.

EAK, 2002. *Empfehlungen für die Aufführung von Küstenschutzwerken,* Editor: Kuratorium für Forschung im Küsteningenieurwesen, Heide i. Holstein, p. 589.

Eisma, D. and Irion, G., 1988. Suspended matter and sediment transport, In: W. Salomons, B. L. Bayne, E. K. Duursma and U. Förstner (Editors), *Pollution of the North Sea: An assessment*, Springer Verlag, Heidelberg, 20–35.

Figge, K., 1981. Sedimentverteilung in der Deutschen Bucht, *Deutsches Hydrographisches Institut,* Karte Nr. 2900 (mit Begleitheft).

Flemming, B. W. and Davis, R. A., 1994. Holocene evolution, morphodynamics and sedimentology of the Spiekeroog Barrier Island System (Southern North Sea), *Senckenbergiana Maritima,* 24.

Flemming, B. W. and Bartholomä, A., 1997. Response of the Wadden Sea to a rising sea level: A predictive empirical model, *Deutsche Hydrographische Zeitschrift,* 49, 343–353.

Führböter, A., 1976. Über die zeitlichen Veränderungen der Wahrscheinlichkeit von Extremsturmfluten and der Deutschen Nordseeküste, *Mitteilungen des Leichtweiss-Instituts für Wasserbau der TU Braunschweig,* 51.

Führböter, A., 1979. Wahrscheinlichkeiten und Häufigkeiten von Extremsturmfluten, *Küste,* 34, 40–52.

Führböter, A. and Jensen, J., 1985. Säkularänderungen der mittleren Tidewasserstände in der Deutschen Bucht, *Die Küste,* 42.

Führböter, A., Jensen, J., Schulze, M. and Töppe, A., 1988. Sturmflutwahrscheinlichkeiten an der deutschen Nordseeküste nach verschiedenen Anpassungsfunktionen und Zeitreihen, *Die Küste,* 47, 163–186.

Fülscher, J., 1905. *Über Schutzbauten zur Erhaltung der Ost- und Nordfriesischen Inseln,* Ernst und Sohn, Berlin.

Franzius, L., 1884. Wasserbau am Meere und in Strommündungen, *Handbuch der Ingenieurwissen-schaften, III.* Band, Franzius, L. und Sonne, E., Hannover.

Garbrecht, G., 1985. Wasser: Vorrat, Bedarf und Nutzung in Geschichte und Gegenwart, *Deutsches Museum Kulturgeschichte der Naturwissenschaften und Technik,* Rororo Sachbuch.

Hagen, G., 1863. *Handbuch der Wasserbaukunst,* Teil III, Band 2, Verlag von Ernst und Korn, 398 pp.

Harff, J., Bobertz, B., Granitzki, K., Lemke, W. and Wehner, K., 2004. Sand and gravel deposits in the south-western Baltic Sea, their utilization and sustainable development, *Zeitschrift für Angewandte Geologie,*111–123.

HELCOM, 2007. Climate change in the Baltic Sea area: HELCOM Thematic Assessment, *Baltic Sea Environmental Proceedings,* 111, 49 pp.

Hensen, W., 1954. Modellversuche über den Wellenauflauf an Seedeichen im Wattengebiet, *Mitteilungen des Franzius-Instituts,* 5.

Hinrichsen, A., 2009. *Morphologischer Zustand Westküste Sylt 2008: Untersuchungszeitraum 1984–2008.* Bericht 02/2009, Landesbetrieb für Küstenschutz, Nationalpark und Meeresschutz Schleswig-Holstein (LKN), 58 pp.

Hofstede, J. L. A., 2002. Morphologic responses of Wadden Sea tidal basins to a rise in tidal water levels and tidal range, *Zeitschrift für Geomorphologie,* 46, 93–108.

Hofstede, J. L. A., 2008. Coastal flood defence and coastal protection along the North Sea Coast of Schleswig-Holstein, *Die Küste,* 74, 134.

Hudson, R. Y., 1953. Wave forces on breakwaters, *Transactions of the American Society of Civil Engineers,* ASCE, 118, 653.

Hünicke, B., 2008. *Atmospheric forcing of decadal Baltic Sea level variability in the last 200 years: A statistical analysis,* Ph.D. dissertation, Hamburg University, 118 pp.

Jensen, J., 1985. Über instationäre Entwicklungen der Wasserstände an der deutschen Nordseeküste, *Mitt. d. Leichtweiss-Instituts f. Wasserbau der TU Braunschweig,* 88, 1–14.

Jensen, J. and Töppe, A., 1990. Untersuchungen über Sturmfluten an der Ostsee unter spezieller Beachtung des Pegels Travemünde, *Deutsche Gewässerkundliche Mitteilungen,* 34, 1/2.

Jensen, J. and Mudersbach, C., 2004. Zeitliche Änderungen in den Wasserstandszeitreihen an den Deutschen Küsten, In: G. Gönnert, H. Graßl, D. Kelletat, H. Kunz, B. Probst, H. von Storch, and J. Sündermann (Editors), *Klimaänderung und Küstenschutz,* Proceedings, University of Hamburg.

Jensen, J. and Müller-Navarra, S. H., 2008. Storm surges on the German coast, *Die Küste,* 74, 92–124.

Kohlhase, S., 1991. The need to monitor the coastal response to structural intervention, *Proc. Seminar on Causes of Coastal Erosion in Sri Lanka,* Colombo, Sri Lanka.

Kohlhase, S., 2004. Zur Konzeption von technischen Maßnahmen zur Küstensicherung [Examples of technical measures for Coastal Protection)], *Jahrbuch der Hafenbautechnischen Gesellschaft*, 54, Verlag Hansa, Hamburg.

Kortenhaus, A., 2003. Probabilistische Bemessungsmethoden für Seedeiche, *Promotion am Leichtweiss-Institut für Wasserbau*, p. 154; http://opus.tu-bs.de/opus/volltexte/2004/525/.

Köster, R., 1974. Geologie des Seegrundes vor den Nordfriesischen Inseln Sylt und Amrum, *Meyniana*, 24, 27–41.

Kramer, J. and Rohde, H., 1992. *Historischer Küstenschutz: Deichbau, Inselschutz und Binnenentwässerung an Nord- und Ostsee*, Deutscher Verband für Wasserwirtschaft und Kulturbau, Wittwer Conrad Verlag, Stuttgart.

Lampe, R., 2005. Late glacial and Holocene water-level variations along the NE German Baltic Sea coast: Review and new results, *Quaternary International*, 121–136.

Landesbetrieb für Küstenschutz, Nationalpark und Meeresschutz Schleswig-Holstein (LKNS), 2008. Morphologischer Zustand Westküste/Sylt, *Untersuchungszeitraum 1984–2008*, Bericht 02/2009, 58 pp.

Lutz, R., Kalka, S., Gaedicke, C., Reinhardt, L. and Winsemann, J., 2009. Pleistocene tunnel valleys in the German North Sea: Spatial distribution and morphology, *Z. Dt. Ges. Geowiss.*, 160(3), 225–235.

Manso, T., Radzevicius, R., BlaǏauskas, N., Ballay, A. and Schwarzer, K., 2010. Nearshore dredging in the Baltic Sea: Conditions after cessation activities and assessment of regeneration. *Journal of Coastal Research*, SI 51, 187–194.

Meyer, M. and Harff, J., 2005. Modelling palaeo coastline changes of the Baltic Sea, *Journal of Coastal Research*, 21(3), 598–609.

Müller-Navarra, S. H., 2005. Sturmfluten: Land unter an Nord- und Ostsee, In: *Entfesselte Elemente: Der Mensch und die Kräfte der Natur*, Wissen-Media-Verl., Gütersloh, 92–99.

Naumann, M. and Lampe, R., 2010. The Fischland-Darss-Zingst peninsula: Late Pleistocene and Holocene evolution of the southern Baltic and its coastal zone, In: R. Lampe and S. Lorenz (Editors), *Eiszeitlandschaften in Mecklenburg-Vorpommern*, Geozon Science Media, Greifswald, 34–49.

Niedermeyer, R.-O., Lampe, R., Jahnke, W., Schwarzer, K., Duphorn, K., Kliewe, H. and Werner, F., 2011. *Die deutsche Ostseeküste, 2. Völlig neu bearbeitete Auflage*, Gebrüder Bornträger, Stuttgart, 370 pp.

Pawlowicz, R., Beardsley, B. and Lentz, S., 2002. Classical tidal harmonic analysis including error estimates in MATLAB using T_TIDE, *Computers and Geosciences*, 28, 929–937.

Petersen, M., 1961. Das deutsche Schriftum über Seebuhnen an sandigen Küsten, *Die Küste*, 9, 1–57.

Petersen, M. and Rhode, H., 1977. *Sturmflut: Die großen Fluten an den Küsten Schleswig-Holsteins und in der Elbe*, Karl Wachholtz Verlag, Neumünster.

Plener, E. van., 1856. Bemerkungen über die ostfriesischen Inseln in geognostischer und hydrotechnischer Beziehung, *Hannov. Z. Arch. U. Ing. Ver.*

Pope, J. and Dean, J. L., 1986. Development of design criteria for segmented breakwaters, *Proc. 20th International Conference on Coastal Engineering*, ASCE, New York, 2144–2158.

Reise, K. and Buschbaum, C., 2007. Mehr Sand statt Stein für die Ufer der Nordseeküste, *Meeresbiolog. Beitr.* 17, 77–86.

Ricklefs, K. and Asp-Neto, N. E., 2005. Geology and morphodynamics of a tidal flat area along the German North Sea coast, *Die Küste*, 69, 93–127.

Schelling, H., 1952. Die Sturmfluten an der Westküste Schleswig: Holsteins unter besonderer Berücksichtigung der Verhältnisse am Pegel Husum, *Die Küste*, 1, 63–146.

Schrottke, K., 2001. Rückgang schleswig-holsteinischer Steilküsten unter besonderer betrachtung submariner Abrasion und Restsedimentmobilität. *Ber.-Rep., Inst. f. Geowiss.*, 16, 168.

Schrottke, K. and Schwarzer, K., 2006. Mobility and transport of residual sediments on abrasion platforms in front of active cliffs (Southern Baltic Sea). *Journal of Coastal Research*, SI, 39, 459–464.

Schüttrumpf, H., 2008. Sea dikes in Germany, *Die Küste*, 74, 189–199.

Schüttrumpf, H. and Oumeraci, H., 2002. Schäden

an See- und Stromdeichen [Failures of sea and estuary dikes], *Mitteilungen des Leichtweiss-Instituts für Wasserbau*, H. 149, 129–172.

Schwarzer, K., 2003. Beeinflussug der Küstenmorphodynamik durch Wasserstandsänderungen, *Die Küste*, 66, 223–243

Schwarzer, K. and Diesing, M., 2001. Sediment redeposition in nearshore areas: Examples from the Baltic Sea, Coastal Dynamics: Proceedings of the fourth conference on coastal dynamics, 808–817, ASCE, Reston, VA.

Schwarzer, K., Ricklefs, K., Schumacher, W. and Atzler, R., 1996. Beobachtungen zur Vorstranddynamik und zum Küstenschutz sowie zum Sturmereignis vom 3./4. 11.1995 auf den Vorstrand vor dem Streckelsberg/Usedom, *Meyniana*, 48, 49–68.

Schwarzer, K., Diesing, M., Larson, M., Niedermeyer, R.-O., Schumacher, W. and Furmanczyk, K., 2003. Coastline evolution at different time scales: examples from the Pomeranian Bight, southern Baltic Sea, *Marine Geology*, 194, 79–101.

Schwarzer, K., Ricklefs, K., Bartholomä, A. and Zeiler, M., 2008. Geological development of the North Sea and the Baltic Sea, *Die Küste*, 74, 1–17.

Seibold, E., Exon, N., Hartmann, M., Kögler, F.-C., Krumm, H., Lutze, G. F., Newton, R. S. and Werner, F., 1971. Marine geology of Kiel Bay, In: G. Müller (Editor), *Sedimentology of Parts of Central Europe,* 8th International Sedimentology Congress Heidelberg, 209–235

Short, A., 1999. *Handbook of beach and shore-face morphodynamics*, John Wiley & Sons, Chichester, 392 pp.

Siefert, W., 1978. Über das Sturmflut Geschehen in Tideflüssen, *Mitt. des Leichtweißinstituts der TU Braunschweig, (Die Küste)*, 63.

Silvester, R. and Hsu, J. R. C., 1993. Coastal stabilization: Innovative concepts, PRT Prentice-Hall, Englewood Cliffs, NJ.

Stackebrandt, W., 2009. Subglacial channels of Northern Germany: A brief review, *Z. Dt. Ges. Geowiss.*, 160(3), 203–210

Stefan, H.-J. and Kunz, H., 2000. Fifty years of experience with the implementation of artificial sand nourishment techniques on the East Frisian barrier islands, Germany, *Abstracts for the ICCE 2000*, Sydney.

Streif, H.-J., 2004. Sedimentary record of Pleistocene and Holocene marine inundations along the North Sea Coast of Lower Saxony, Germany, *Quaternary International*, 112, 3–28

Tautenhain, E., 1981. Der Wellenüberlauf an Seedeichen unter Berücksichtigung des Wellenauflaufs, *Mitteilungen des Franzius-Instituts*, 53, 1–245.

Tritt, W. P. and Altemeier, G., 1987. *Buhnen im Küstenschutz,* Franzius-Institut für Wasserbau und Küsteningenieurwesen der Universität Hamburg, Sonderübung.

van de Plassche, O., 1995. Evolution of the intracoastal tidal range in the Rhine-Meuse delta and Flevo Lagoon, 5700–3000 yrs cal BC, *Marine Geology* 124, 113–128.

van der Meer, J. W., 1993. Conceptual design of rubble mound breakwaters, *Delft Hydraulics Publications*, 483, 221–315.

Wahl, T., Jensen, J. and Frank, T., 2010. On analysing sea level rise in the German Bight since 1844, *Nat. Hazards Earth Syst. Sci.*, 10, 171–179

Wahl, T., Jensen, J., Frank, T. and Haigh, I. D., 2011. Improved estimates of mean sea level changes in the German Bight over the last 166 years, *Ocean Dynamics*, 701–715.

Werner, F., 2004. Coarse sand patterns in the southeastern German Bight and their hydrodynamic relationship, *Meyniana*, 56, 117–148.

Woebecken, C., 1924. *Deiche und Sturmfluten an der Nordseeküste,* Friesenverlag, Bremen.

Zeiler, M., Schulz-Ohlberg, J. and Figge, K., 2000. Mobile sand deposits and shore-face sediment dynamics in the inner German Bight (North Sea), *Marine Geology*, 170, 363–380.

Zeiler, M., Figge, K., Griewatsch, K., Diesing, M. and Schwarzer, K., 2004. Regenierung von Materialentnahmestellen in Nord- und Ostsee, *Die Küste*, 68, 67–98.

Zeiler, M., Schwarzer, K., Bartholomaä, A. and Ricklefs, K., 2008. Seabed morphology and seabed sediments, *Die Küste*, 74, 31–44.

Ziegler, B. and Heyen, A., 2005. Rückgang der Steilufer an der schleswig-holsteinischen westlichen Ostseeküste, *Meyniana*, 57, 61–92.

8 The Netherlands

Frank van der Meulen, Bert van der Valk and Bas Arens

Introduction

The Netherlands coast covers *circa* 350 km along the North Sea (Figure 8.1) and is primarily a sand barrier coast, backed by low-lying land. These low-lying areas are polders, locally extending to 8 m below Mean Sea-Level (MSL). Close to the coast, the North Sea bottom is mostly sand, so that the Netherlands coast (beach grain sizes range between 270 and 330 mµ) has a large sand reservoir on its doorstep. It is probably the only European country in such a situation. The major part of the coast consists of dunes, which together with the beach and shoreface, represent a natural sand defence against the sea. The dune areas vary in width from 100 m to a maximum of 5 km. This chapter focuses on dynamic preservation of the sand coast and the adaptations made, as a result of environmental and climate change, rather than on hard structures.

For centuries, rivers did not carry sediments to the sea and relative sea-level kept rising (last century by some 20 cm), hence most of the coast is erosive. For hundreds of years, an important dune function (amongst others) was coastal defence, especially for protection of low-lying hinterland areas. Dunes represent about 75 per cent of the defence line, the remainder consists of sea dykes, storm surge barrier dams and beach flats in the north of the country. In general, the coast is receding and as such, counteracting erosion is a main management activity. Van Koningsveld *et al.* (2008) have given a short account of man's interference with this coast. Although these activities started early in the ninth century, only in

■ **Figure 8.1** *The coastal foundation (dark yellow) defined as the area in the shallow North Sea between MSL and a 20 m water depth. The dunes themselves (a narrow band of light grey) are also part of the coastal foundation. Physiographic regions from north to south are: 1: the Wadden barrier island and lagoon coast, 2: the Holland barrier coast, 3: the Zealand estuarine coast. Dark blue = North Sea and Wadden Sea; light blue = Lake IJssel. (Figure drawn by Rob van der Laan, Rijkswaterstaat, Road and Hydraulic Engineering Division)*

the nineteenth and twentieth centuries has the profound influence of man's actions become apparent and effective measures put in place to halt coastal retreat.

The year 1953 marked an important turning point in coastal defence management. A disastrous flood struck large parts of the south-western delta, causing much damage and over 1,850 casualties. As a result, the Netherlands government established the Delta Plan. Dams and dykes closed the large estuaries in the south and the entire coastal defence system was raised to a delta height at a minimum of 11 m above MSL. The year 1990 marked another important turning point in coastal defence management. Before 1990, hard constructions were generally used to counteract erosion (see e.g. van Koningsveld *et al.*, (2008); see also www.eurosion.org). In 1990, the government decided on a new policy, termed 'dynamic preservation'. In principle, the coastline was to be kept at the 1990 position (the average position from 1980 to 1990) and where necessary, since 1996, soft measures have been applied to counteract erosion for most of the sand coastline, while 'abandonment' was preferred at some parts deemed unimportant for the direct safety of the inhabitants. According to policy, these areas are allowed to erode/prograde naturally, e.g. the eastern parts of the Wadden barrier islands. A scheme of intermittently repeating sand nourishments, based on systematic monitoring, has been carried out by the Rijkswaterstaat (the National Authority responsible for safety against flooding, which includes coastal protection), sand being taken from the North Sea below the 20 m depth contour.

Most nourishment works were placed on the beach or, later, the foreshore. This method for coastal defence is regarded as being much more in line with the dynamic natural character of the soft coast than the hard measures usually taken before 1990. Therefore, the term 'building (or working) with nature' was coined for this (see also Waterman, 2010). The Dutch have now nourished their coast for almost 20 years and much management and monitoring data experience has been gained, which is a unique situation in Europe.

Dynamic coastline management was not practised until 1995. Even now, the water board organizations that recently became responsible for day-to-day as well as long-term coastal maintenance (the latter previously being a central government task) have difficulties with sand 'being moved by nature': it should remain in place! Nature conservation and Natura 2000 targets, however, aim for an even more 'dynamic coastline management' than is already possible.

The sediment management perspective is clearly the key issue, as coastal evolution on a longer timescale is a function of demand and supply of sediments. Understanding the pathways of sediment along a coast is the start for proper management (Van Koningsveld *et al.*, 2008), and the basis used in the Netherlands is the concept of a 'coastal foundation', which encompasses all areas between MSL and a 20 m water depth together with the landward coastal dune boundary (Figure 8.1). The entire area cannot be used for sand mining, as coastal defence law prohibits this. With a rising sea-level it will be a major task to increase land levels while maintaining the natural and dynamic character of the coast.

A geomorphological overview of the Netherlands coast

General

The central Dutch barrier coast developed approximately 6,000 years Before Present (BP) and was situated some 8 km east of the current coast. Coastal retreat prevailed until 6,000 years BP, as sediment supply lagged behind availability of sediment accommodation space. After this date, sea-level rise reduced from *circa* 0.6 m/century to 0.3–0.4 m/century, and the sediment supply drastically changed, leading to an aggrading sediment wedge behind the oldest sand barrier and a prograding coastline in front of the oldest barrier along the North Sea. The progradation (8 km at maximum) lasted until the period 2,500 to 2,000 years BP, the rate gradually reducing until coastline stabilization occurred. This was especially apparent on the Holland and part of the Zealand coastline. Areas further south, as far as the current border with Belgium and the Wadden area in the north, showed continued erosion, as sediment supply lagged behind the increase in coastal lagoon size located behind the narrow coastal barrier in Zealand and along the Wadden Sea. In the meantime, coastal erosion continued, counteracted by using sand sediment of the previously prograded coastline for filling in gaps in other parts of the coast (Van Koningsveld *et al.*, 2008). Additionally, continuing coastal erosion resulted in formation of a new coastal dune, around 700 years BP, in some parts of the country and some centuries later elsewhere. Decrease of available sediment was the main reason for this due to depletion of readily available sea-bed sand together with reduction of transport capacity in some areas. With large-scale river canalization, sluicing and regulation, sand no longer reached the sea. Before that, the fluvial origin of the coastal zone was already low, never contributing more than an estimated, on average, 10 per cent to the coastal sand budget.

Physiographic regions

The Wadden barrier islands and Wadden Sea

The Wadden Sea probably existed for some 5–6000 years, more or less in its current form although especially towards the west, it expanded considerably in size (Figure 8.1). A rapid sea-level rise during the early Holocene drowned the pre-existing Pleistocene landscape, enhanced by human excavation of peat areas for fuel. Embayments and smaller estuaries were first eroded (removing extensive peat covers), which were subsequently filled in by tidal sedimentation. From the twelfth century, land reclamation started and large parts of the Wadden Sea became available as arable land (Middelzee, Fivel area and Dollard area). Over time, polder dykes reclaimed very large areas, this activity continuing well into the twentieth century. The 'endikement' contributed to the currently prevailing 'coastal squeeze' (see for instance Van Koningsveld *et al.*, 2008), together with fixation of the inhabited part of the Wadden Islands.

Over time, barrier islands shifted landward as sediment supply to the coast was not enough to fill embayments and keep the islands at the same location. Between barrier islands, tidal inlets always existed and some inlets showed a steady development and are still relatively stable; others have shown a remarkable increase in importance such as the Marsdiep in the west (Elias and Van der Spek, 2006). Closing the Zuiderzee in 1932 (now Lake IJssel, see Figure 8.1) meant a profound change in boundary conditions, which caused a solid re-orientation of the Marsdiep tidal inlet (lying between the mainland and the first Wadden island). It is not in equilibrium with the environment and appears to expand towards the east at the cost of other inlet discharge areas.

Barrier islands, inlets with ebb-tidal deltas and tidal flats inside the Wadden Sea, form parts of a 'sand-sharing system', with mud playing a secondary role, as when the coastline changes, or changes occur in a tidal basin, other parts of the system react. For instance, the combined effects of enlargement of the Marsdiep and of coastal erosion means that the ebb-tidal delta is diminished in size, as sand has been eroded at a steady pace over the last few decades (Elias and Van der Spek, 2006). In view of a potential sea-level rise that may drown the Wadden Sea tidal flats, the sediment issue is a shared problem for the entire Wadden Sea area, stretching all over the northern Netherlands, Germany and Denmark. Nourishment may have to be designed to counter-effect such drowning, but without damaging nature values in the area.

The Holland barrier coast

The slightly curved coastline from Den Helder to Hook of Holland remained virtually untouched by man until well into the nineteenth century (Figure 8.1). A number of coastal cross cuts were made in the 1860s for harbour entrances at Amsterdam and Rotterdam. This caused coastal compartmentalization (Short, 1992; van Koningsveld *et al.*, 2008). For as long as measurement records have been kept, the Holland coast has known a relatively moderate (~1 m/yr) steady retreat throughout the nineteenth

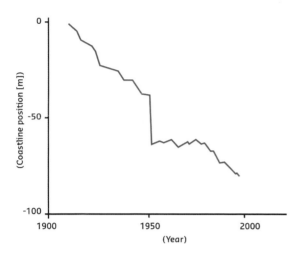

Figure 8.2 *Typical example of coastline retreat at the mid-Holland coast during the past century (before the start of the nourishment programme in 1995). The impact of the major storm surge of 1953 is clearly visible: in one night, some 20 m of coastline regression occurred due to wave attack and erosion of the coastal dune front. After the storm surge, the coastline stayed more or less in place, but the long-term regressive trend picked up once more after a few decades*

and twentieth centuries. The coastal dunes are largely fossilized because they are covered with dense vegetation, and mobile dunes have become rare over the

past century. The main reasons are: widespread and routine sand stabilization; inaccessibility of the dune area to humans (protected area for drinking water); and atmospheric NOx deposition acting as a fertilizer, all leading to dense vegetation growth and cover. Occasional major storms, such as the 1953 storm, create a rapid retreat during the storm period itself, but almost no erosion in the post-storm period after which erosion gradually comes back at the rate of 1 m/yr (Figure 8.2).

In the 1990s, the national beach nourishment programme commenced. It is remarkable that along stretches of coast that have not been nourished since 1990, coastal regression continued by pushing the coastal dune landward (Bakker *et al.*, 2012). Apparently, the current coastal dune needs aeolian nourishment to remain in place (or if this becomes too much, progradation occurs).

With regards to development, the Holland coast is somewhat 'over-nourished', as over the years and over large dune stretches, has occurred development of fresh new embryo dunes in front of the long-standing foredune ridge. The relative over-nourishment is possibly also due to execution of the so-called 'weak-link' programme (see below). These embryo dunes are often eroded during storms but new dune growth was facilitated recently (since 1995) because of relatively low north-westerly storm

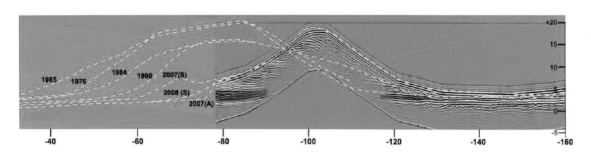

Figure 8.3 *Combined image of GPR profile (200 MHz, unshielded) across the foredune at beach pole 49.500 near Heemskerk aan Zee and the Jarkus profiles (yearly coastal profiles) from selected years (1965–2007) at this position. A distinct landward shift, narrowing and over steepening of the foredune can be observed. The dune crest has migrated ~0.5 m/yr and the dune foot (beach side) ~1.0 m/yr. In the current foredune the groundwater table (in blue) and historical storm surge beds (in red) are imaged by the GPR profile. A set of strong and parallel reflections and inclined strata at the leeward side, both in aeolian records are clearly shown (in yellow) (Figure drawn by Marcel Bakker, in Bakker et al., 2012.)*

intensity and frequency together with an ample availability of sand (nourished beaches are wider, and hence allow a longer fetch to develop with higher potential for sand transportation). Despite a more lenient policy since 1995 towards coastal dune dynamics, relatively few blow-outs have formed.

The Zealand (former) island-and-estuarine coast

The 120 km long Delta coast has wide estuaries and islands, which have sand dunes at their tips (Figure 8.1). Long-term development of this coast is governed by sea-level rise and sediment supply and the actual demand for sediment is in the order of some 3×10^6 m^3 per year (Mulder *et al.*, 2010). This coast, which includes the Zuid-Holland islands of Goeree and Voorne, was an open estuarine coast until quite recently. The 1953 storm surge disaster led the government to install the previously mentioned Delta Plan; adoption of this law implied a large re-alignment of the Delta coast. The most northerly exit of the Rhine estuary has to stay open to keep free access to Rotterdam harbour. Likewise, the most

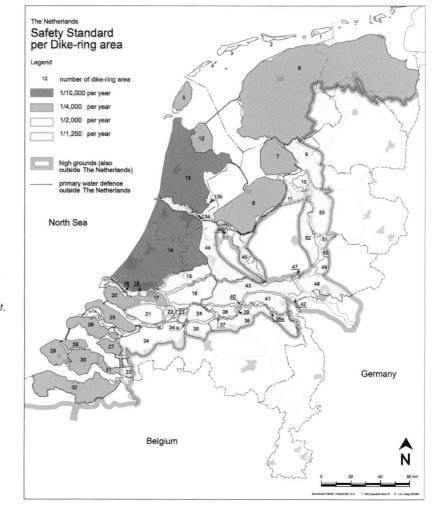

■ **Figure 8.4a**
Safety standards and dike-ring areas in The Netherlands. Note highest safety standard 1/10,000 in the west along the Holland coast. Compare with Figure 8.4b. A dike-ring area is an area (polder) surrounded by a dike, the height of which offers a certain safety standard to the people living in that area protected by that dike (Figure drawn by Rob van der Laan, Rijkswaterstaat, Road and Hydraulic Engineering Division)

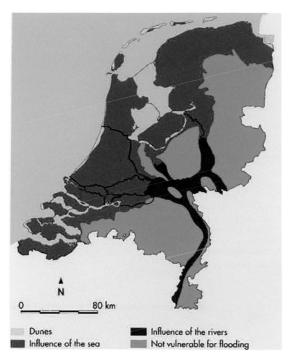

Dunes
Influence of the sea
Influence of the rivers
Not vulnerable for flooding

0   80 km

N

■ **Figure 8.4b** *Flood-prone areas in the Netherlands. The highest socio-economic activity is in the western part, so this area has the highest safety standard. Compare with Figure 8.4a (Figure drawn by Rob van der Laan, Rijkswaterstaat, Road and Hydraulic Engineering Division)*

southern estuary, the Scheldt River, must remain open in order to allow Antwerp free sea access. All other estuary branches were either closed off, or made semi-permeable, as in the case of the Eastern Scheldt where a storm surge barrier reduced tidal volumes by some 10 per cent.

As to be expected, large-scale structural landscape changes altered current and wave refraction patterns. These, in turn, caused some re-arrangement of the so-called Voordelta area, the shallow sand banks seaward of the island heads. Re-alignment caused the distribution of bed forms over parts of the ebb-tidal deltas into more wave-dominated forms.

Close to the coast, erosion occurs by the activity of channels alongside the island heads of Walcheren, Schouwen, Goeree and Voorne. A regular nourishment programme has been carried out since 1996,

both along channel-sides, as well as on the foreshore. In exceptional cases, dune re-enforcements have taken place in front of coastal dunes, e.g. Walcheren, Schouwen and Voorne, or landward re-enforcement (Goeree and Voorne). Both Goeree and Voorne are additionally so-called 'weak links' and therefore in some recent years received more extensive and massive nourishments than the regular coastal maintenance programme. On average, the Zealand part of the coast is the least nourished. Nourishment equals the volume that is being lost on a yearly basis and repetitive measurements are needed. Currently, a more regular monitoring programme is applied, leading to better knowledge of the fate of the fore-shore and channel-side nourishments. In some wider dune areas, such as Schouwen, programmes for increasing the level of coastal dune dynamics are being investigated. In most other Zealand areas, the coastal dunes are so narrow that increased coastal dune dynamics is considered to be a direct threat to safety (Figure 8.4a and b) or to other dune area functions (e.g. the production of drinking water).

Towards a nourished coast

Before 1990, hardly any sandy coastal area was nourished and protection was by groins (Figure 8.5),

■ **Figure 8.5** *Coastal defence by groins along the North Sea coast, Texel island (Photograph courtesy of www.beeldbank.rws.nl)*

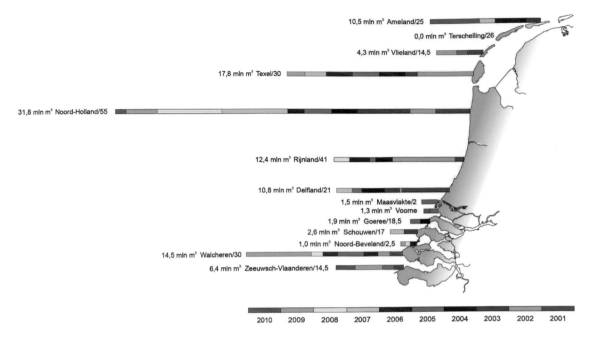

10,5 mln m³ Ameland/25
0,0 mln m³ Terschelling/26
4,3 mln m³ Vlieland/14,5
17,8 mln m³ Texel/30
31,8 mln m³ Noord-Holland/55
12,4 mln m³ Rijnland/41
10,8 mln m³ Delfland/21
1,5 mln m³ Maasvlakte/2
1,3 mln m³ Voorne
1,9 mln m³ Goeree/18,5
2,6 mln m³ Schouwen/17
1,0 mln m³ Noord-Beveland/2,5
14,5 mln m³ Walcheren/30
6,4 mln m³ Zeeuwsch-Vlaanderen/14,5

2010 2009 2008 2007 2006 2005 2004 2003 2002 2001

■ **Figure 8.6** *Total volumes of sand nourishments for coastal maintenance purposes 2000–2010, per coastal division. Legend bar: each unit of a colour bar indicates 1Mm³ (Figure drawn by authors, after Anonymous, 2010)*

but some nourishment took place locally, usually civil engineering works or harbour extensions. During the 1990s, the nourishment type changed from beach – and/or dune foot – reinforcements, to shallow foreshore nourishments (Figures 8.6 and 8.7) as, during the 1990s, an extensive evaluation showed that shallow foreshore nourishments dispersed rapidly and adequately over the entire foreshore and adjacent beach (Nourtec, 1997). This offered an equal safety perspective in terms of sand availability for combating erosion, whilst using natural processes to spread sediment, which is much better for shallow water and foreshore ecology and beach fauna.

Over the past few years, a scaling up of nourishment schemes has taken place due to various reasons:

• More confidence has been gained in applying this particular method of coastal defence, as many evaluations showed its success (Figure 8.9).

• The market showed growth, dredging ship size and transportable volumes of sand increased, requiring deeper water for ships to offload their temporary cargo whilst reducing the average price per cubic metre. Modelling results showed that the large Rotterdam harbour works using enormous amounts of sand (220 Mm³), will have no consequences for the coastal sand budget and later requirements for maintenance would only be of the order of $1–3 \times 10^6$ m³/yr.

• A programme taking care of so-called 'weak links' (in the coastal defence) has added a considerable volume of sand to the coastline (some 120×10^6 m³, not incorporated into Figure 8.7) and does not lead to a higher requirement for maintenance. Simultaneously, it combines coastal safety measures with economical development, notably where nourishment was carried out in front of coastal resort towns, so that coastal communities benefit greatly from such measures.

A further step in scaling up coastal nourishment is the 'sand-engine' (*Zandmotor*), currently being placed between the coastal resorts of Kijkduin and Terheijde (Delfland coast). Some 21.5×10^6m³ of sand has been

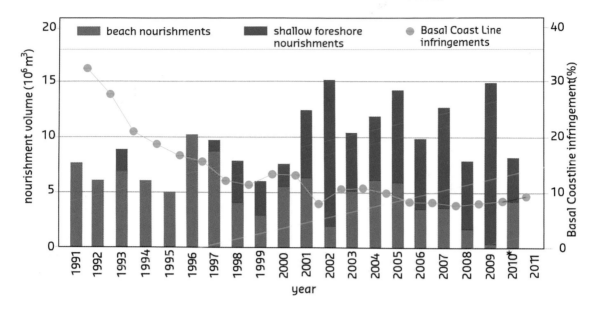

■ **Figure 8.7** *Nourishments executed for coastal maintenance since 1991 (beach nourishments in brown, foreshore nourishments in blue). Since 2001, the average volume of sand nourished is circa 12 Mm³/yr. The dotted line indicated the number of exceedance cases per year, showing a stabilization over the years, indicating the success of the method (Figure drawn by authors after Anonymous, 2010)*

deposited in the shape of a sand hook. The purpose of this mega-nourishment is to:

- stimulate natural coastal dune development for safety, nature and recreation;
- generate knowledge development about the above processes including added value combining these functions;
- add a nature conservation and recreational area to the Delfland coast.

Nourishment has had specific effects when placed in the various areas, both in the shallow North Sea and on land in dune areas. In the Wadden area, they are predominantly placed along the extended island coast, or around island heads and occasionally in the past, dune reinforcements were made. In general, the Wadden Islands have received less sand on average along the coast, the Holland coast traditionally receiving most of the nourishment budget. The Zealand area has received the 'right' amount of sand, and has its own regime of channel-side nourishments, as well as beach nourishments.

Effects of nourishment

The Netherlands are unique in Europe in the sense that there is now more than 20 years of experience with nourishments and a large amount of monitoring data exists. As this chapter shows, to assess properly the effects of nourishments, one has to consider the entire active (that is, sediment-sharing) coastal zone. In addition, one has to look at aspects of civil engineering as well as geomorphology (e.g. sand budgets and coastal morphology) and ecology (effects on nature values). In doing so, a more integrated approach to coastal defence management develops (from coastline management to coastal zone management).

The nourishment policy to counteract beach and dune erosion works well, as ongoing coastal erosion is generally prevented. Apparently, it is possible to interfere with large-scale coastal processes in such a way that the coastline position can be maintained and stabilized. Between 1990 and 2011, around 300 nourishments were executed at more than 60 locations, many locations being nourished several times. Differences also exist between the main coastal regions. On the Wadden Islands and Holland coast, the first nourishments were on the beach, and currently there is a shift towards shoreface nourishments, which are cheaper and more efficient. In the Delta region, most nourishment is still placed on the beach, because of the presence of gullies close to the beach, but increasingly these gulley sites are nourished with larger volumes, but with lesser frequency, as this has proven to be an efficient method.

Because of nourishment, the coastal sediment budget has adapted to yielding more sand to the beach and dunes. The policy of 'dynamic coastline preservation' adopted around 1995 has abandoned historically prevailing strict foredune management, expressed by planting marram (*Ammophila arenaria*) grass at any spot where vegetation cover was temporally lacking. For centuries, routine fixation of foredunes by planting marram grass and the placing of sand fences was the dominant management strategy to counteract erosion. A decrease in management intensity in many places also resulted in important changes in the system, mainly because natural processes of sand mobility are no longer suppressed and a more uncontrolled movement of sand is allowed within the system. After 20 years, it is evident that the system is changing (Arens *et al.*, 2010). On the other hand, concern is growing about the ecological effects of nourishment on the shoreface, due to periodical shoreface/beach burying by huge quantities of sand and the possible changes in marine processes. Consequently, several research projects have started to investigate the changes in the system and its possible effects on geomorphological and ecological processes (Arens *et al.*, 2010). There is a need of information regarding how this coastal defence strategy interferes with the requirements for nature conservation imposed by Natura 2000.

Sediment budget foredunes: Trend breaks

It has become clear that the amount of sand transported from beach to foredunes has increased considerably over past decades (Arens *et al.*, 2010). Calculation of foredune sand volumes revealed many cases having trend breaks in the volume development of foredunes (for an example, see Figure 8.8). In some cases, the trend break clearly coincides with the start of local nourishments; in others, this correlation is less clear. There are cases with beach nourishment where a trend break became clear only after the first shoreface nourishment. In addition, there are examples of locations where the trend break is earlier than the start of the nourishments, as seems to be the case in Figure 8.8. The trend break is thought to be related to previous nourishments 'upstream', *i.e.* in the south-west, but these relationships have not yet been studied and much remains unclear about the spatial extent of nourishment influences. The case of Figure 8.8 shows that there is a very clear tendency to increased foredune volumes from the start of the nourishments. Also the change of volume in time becomes more consistent, since R^2 has been rising from the first to the second to the third period.

There is still little insight as to the exact movement of the sand: what is the relationship between processes on the shoreface, supply of sand to the beach and aeolian withdrawal, and what is the exact trigger for dune development? What is the relationship between the sand budget of shoreface, beach and foredunes? Table 8.1 shows theoretical combinations of shore/beach volumes and foredune volumes; it appears that along the Dutch coast most combinations exist (Arens *et al.*, 2010). Studies of a number of sites showed that after nourishment, in most cases the sediment budget of both shoreface and dunes became positive. There was only one area where a negative trend became less negative. Before nourishment, the situation was much more complicated. In the studied sites, three different scenarios exist:

1. Foredunes erode and sand accumulates on the shoreface. A negative correlation exists between

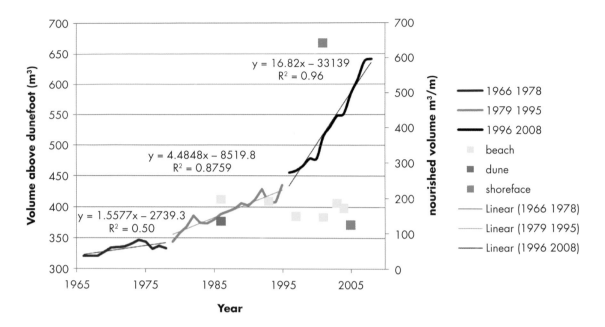

■ Figure 8.8 *Development of foredune volume over time. Example: Delfland Coast (Zandmotor location)*

■ Table 8.1 *Possible combinations of sediment budgets on shoreface/beach/foredunes*

		shoreface / beach		
		+	**0**	**–**
Fore-dunes	**+**	sedimentation, accretion, coastal progradation, sand supply from elsewhere	shoreface is transfer zone, or sand losses to foredunes are compensated by import of sand from elsewhere	erosion of shoreface provides sand supply to beach for dune development
	0	supply of sand on shoreface without transfer to the beach, adaptation to steep shore profile?	stable	erosion of shoreface has not reached the foredunes, adaptation to shallow shore profile? *(no examples exist)*
	-	cliff development, erosion foredunes, transfer of sand from dunes/beach to shoreface	shoreface is transfer zone, or sand supply from foredunes to beach/shoreface is discharged to elsewhere; possible loss of sand from foredunes by secondary blowouts	erosion, cliff development, coastal retreat, sand losses to elsewhere

volume development on the shoreface (positive trend) and in the foredunes (negative trend).

2. In one area the correlation is negative as well, but here the volume of the foredunes (positive trend) increases while the shoreface volume decreases (negative trend). Apparently here dunes are built at the cost of the shoreface.

3. Cases with positive correlation before nourishment: both volumes decreased in time.

The relationships between shoreface and foredune development seem to be an essential link in coastal development. But they are still poorly understood. It is important with respect to coastal defence management to properly assess the effects of nourishments.

Changing foredunes: Classification

On a large scale, the coastal sediment budget compartments are changing because of nourishment. However what is the effect on dune morphology, in particular the foredunes? Arens and Wiersma (1994) published a classification of foredune types (reflecting the situation as at 1988), based on their development, management interference and activity of natural processes. Application of this classification to the current situation indicates that foredune state has changed considerably. On average, the percentage of stable foredunes (S, Figure 8.9) is equal over this

period (40 per cent), but the occurrence of regressive foredunes (T, Figure 8.9) has become rarer, and has decreased from 21 per cent to only 8 per cent. Progressive foredunes (U, Figure 8.9) have become the dominant type (27 per cent in 1988, over 50 per cent now). Obviously, the change of sediment budget has resulted in a completely different distribution of foredune types.

Morphology and dune building

Since nourishments began, the intensity of management efforts in the foredunes has decreased considerably. The sediment budget has become mostly positive, foredune types changing to more progressive types, but also aeolian activity rate has increased (Figure 8.9). Two major types of development can be distinguished, depending on local factors, which are not yet completely understood.

1. Along large parts of the coast, the available sand surplus from nourishment was deposited in front of the foredunes, resulting in new, embryonic dune types. In some areas, these have developed into new foredune ridges and all these dunes became progressive. In 1988, comparable types of progressive foredunes were mainly found on the Wadden Islands; in 2008, they are important in all regions. These dunes have a natural

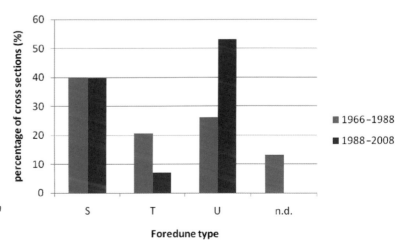

■ **Figure 8.9** *Changes in foredune types due to new management (Figure drawn by Bas Arens from Arens, van Puijvelde and van der Valk (in prep.))*

morphology, although locally disturbances occur due to recreation. Their surface areas are mostly limited, but typically cover a zone of some 10–50 m in front of the former foredunes.

2. Along other parts of the coast, an increasing sand input is distributed over existing foredunes, either on the seaward slope, the crest or even the landward slope. Also locally, the foredunes sand is remobilized due to development of blow-outs and formation of embryo parabolic dunes, being induced by either small-scale wave erosion, or pure wind erosion. These foredunes may become very dynamic and their shape transformed into a more or less natural foredunes with blow-outs and other signs of wind erosion and deposition, all within 20 years (after many years of strict management, foredunes have taken the unnatural shape of a 'sand dike').

Response types and implications for ecology

These types of changes have important consequences for dune ecology. In a natural dune system, the foredune is the area of exchange of sand (containing carbonates), salt and wind (*i.e.* stress factors) between the beach on the one side and the inner dunes on the other side. Depending on the rate of exchange, the inner dunes profit from fresh calcareous sand. The type of coast decides the rate of exchange and roughly, this is determined by the foredune response to the (increased) sand supply from the beach. Five so-called response types can be distinguished (Figure 8.10).

1. Supply of sand is limited and the foredune remains static. Hardly any exchange between beach and dunes.

2. Supply of sand is completely stored in front of the foredune in the form of new embryonic dunes or ridges. Although substantial, the sand supply is not enough to bury foredune vegetation, the original foredune does not receive fresh sediments and the dynamic processes zone gradually moves seaward. As in 1 above, hardly any exchange exists between beach and dune (apart from a new zone of embryonic dunes).

3. Supply of sand is completely stored in foredune seaward slope so the supply is limited and not enough to bury vegetation. The dynamic processes zone is limited to the seaward slope and there is no exchange with landward slope and landward dunes.

4. Supply of sand stored in entire foredune and there is little further transport to the inner dunes (quantities are being studied). Although insignificant in terms of volumes, the ecological impact might be significant (especially with respect to calcareous sand inputs and vegetation depending on such sand-spray).

5. Supply of sand is stored in entire foredune and interaction with locally remobilized sand in blow-outs. There is full exchange between beach and dunes and an important gradient of landward transport from the foredunes.

Some questions arise about the sand mobility processes to which answers would help better to assess the effects of the nourishments. For example, are differences between types 2 and 3 forced by processes alone, or (also), by the shoreface sediment budget and is this budget decisive for the rate of dune

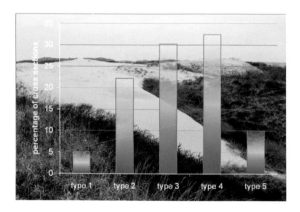

■ **Figure 8.10** *Distribution of foredune response types along the Dutch coast. For type key, see text below (Figure drawn by Bas Arens after Arens, van Puijvelde and van der Valk (in prep.))*

■ **Figure 8.11** *Back of the foredune buried by extreme sand input, Terschelling (Photograph courtesy of Evert Jan Lammerts, State Forestry Organization)*

development? How? Types 4 and 5 may also coincide with embryonic dune development in front of the foredunes. This happens in some locations with an excessive sediment supply when the foredune back is buried, such as on Terschelling island (Figure 8.11).

Currently, there is so much sand available that vegetation is buried and sand transported further landward. Some spectacular examples of type 5 occur in the North Holland Dune Reserve, where it is expected that new parabolic dunes will be formed (Figure 8.12).

With types 4 and 5, there is a very small transport of sand to some hundreds of metres landward of the foredunes, especially during storms (Arens *et al.*, 2010). Implications for geomorphology are minor, since the transport amount results in non-detectable elevation changes of the soil substratum. However, with respect to ecology, even these small sand amounts may contribute, for example, in cases where soil acidification could be counteracted by the input of small amounts of calcareous sand. There is hardly any data on this kind of transport. Currently a research project is being undertaken, which aims at quantification of the landward transport and of the

geochemical consequences in the soils (Stuyfzand *et al.*, 2010; Arens *et al.*, 2010).

Natura 2000

The Dutch dunes are important because of the grey dunes, an important dry short grassland habitat in the European Natura 2000 network (type 2130, EC 2007). These dunes need some dynamics, usually shallow burial with sand. They seem to benefit when situated behind types 4 and 5. This is the subject of current studies (Everts and de Vries, 2010). In the case of types 2 and 3, the new dunes belong to the 'embryonic dunes' habitat type (2110, EC 2007), and partly to the 'white dunes' with *Ammophila arenaria* (2120, EC 2007). There are signs that the occurrence of types 1, 2 and 3 in front of the grey dunes might be a threat to them, because these types block *all* sand dynamics. Although development of new dunes is positive with respect to nature values (an increasing surface of embryonic dunes), a possible threat to the existence of the grey dunes might result in a negative overall effect of a decreasing surface of this type (Arens *et al.*, 2010). This is also the subject of current studies (Everts and de Vries, 2010).

■ **Figure 8.12**
Development of a new parabolic dune from a blow-out in the foredune, North Holland Dune Reserve, Mainland Coast (Photograph courtesy of Tycho Hoogstrate, PWN - Drinking Water Company North-Holland)

In the past, at least before 1850, management of the foredunes was performed at a low level and large parts of the coastline were regressive. As a result, sand remobilization from dune cliffs was an important process and resulted in development of mobile, parabolic dunes. Although no systematic information is available on the exact state of the foredunes, it is likely that most of them belonged to types 4 and 5. The exchange of sand between beach and foredunes was probably very large, resulting in an increasing width of the dune belt, even with a regressive coastline.

Possible implications for management strategies

The new type of management, with large-scale nourishment and an increasing importance of natural processes to redistribute the extra supply of sand, offers new chances for future coastline and nature management. It is clear that the state of the coast is adapting, with potential negative and positive effects on ecological development. The challenge is to adapt current management in such a way that negative effects are limited while positive effects are optimally exploited. This can be done either by different ways of nourishment (frequency, time of intervention), or by active interference in foredune dynamics, for example by stimulating sand mobility (Box 8.1).

The future and climate change

What makes the coast of the Netherlands special in the European coastal context?

This coast is almost completely sand (beaches, dunes, estuaries, barrier islands) and protects nearly half the country, which lies below sea level and is the socio-economic heart of the country. The coast has an ample sand reservoir in the North Sea that can be tapped apparently without limit and/or without sizable environmental damage. Because of coastal defence reasons for the densely populated hinterland, the coast is well monitored and studied. Much data exists and is available for evaluation of coastal

BOX 8.1   A SPECIAL CASE OF NOURISHMENT, THE DELFLAND COAST

with contributions by Frank van der Meulen, Bas Arens, Bert van der Valk and Rien van Zetten (Rijkswaterstaat)

A unique situation

The Delfland Coast between the Hook of Holland and The Hague is designated as one of the 'weak links' along the coast, *i.e.* the coast is safe now, but not when future climate change is taken into account. A large nourishment was carried out from 2009–2011 and as part of these protection works, a new dune area was constructed to compensate for the predicted loss in ecological values in existing dunes due to the use of the new harbour extension of Rotterdam: Maasvlakte 2 (Figure 8.13). Some 6.5 Mm^3 of sand was used for this new area. Coastal defence and development of high quality dune nature will go hand in hand at this place and recreation will also benefit from the new developments.

Project Mainport development Rotterdam (PMR)

Already in the early 1990s, studies from various sources showed a need for a new industrial area in the Rotterdam harbour region. Public debate led towards a double aim of the final project: to provide new

■ **Figure 8.13** *Southwestern part of the coast near Hook of Holland, with Rotterdam harbour (dark grey), new extension Maasvlakte 2 (yellow), and compensation areas: marine reserve (dark blue), dune compensation Delfland (dark blue strip, on top of figure), new recreation areas near the city (hatched green) (Figure drawn by authors after Project Mainport Rotterdam, 2007)*

possibilities for both economic and *ecological* improvements by working on both fronts in a balanced way. This led to the birth of the PMR-project (Figure 8.13), which consists of three independent parts:

1. The redesign of (mainly older and smaller) existing harbour facilities near the city of Rotterdam, in order to create 200 ha for various new purposes;
2. The increase of recreational areas in the neighbourhood of the city, totalling 750 ha of nature; and
3. The realization of 1,000 ha of a new industrial area, beside the latest existing harbour extension from the 1960s, the so-called Maasvlakte 1 near Hook of Holland. The new area, Maasvlakte 2, had to be reclaimed from the sea.

All the PMR activities took place in a Natura 2000 area, which has protected values regarding birds, fish and habitats based on European Directives. According to EU regulations, the predicted damage to Natura 2000 areas has to be compensated. In the case of PMR, this means that both marine and terrestrial nature (dune) habitats have to be compensated. This complicates the situation, but it is also a reason for careful project design. By reclaiming land from the sea, part of the marine ecosystem is completely destroyed. Since it is not possible in this region to 'make a new sea', it was decided, as compensation, to improve the quality of part of the existing shallow sea in the mouth of the southwestern delta. For the dune nature area, it was decided to create nearby new dune environments at the Delfland coast. This coast had to be re-enforced to guarantee its safety under conditions of future climate change and the new dune area would be an extra addition to these protection works.

The EIA that preceded the PMR project estimated damage to nearby dunes, due to the use of Maasvlakte 2 (from 2013 onwards). Maasvlakte 2 will be used for deep-sea-bound harbour activities, *i.e.* for container handling and distribution activities, and for chemical activities, all of which will lead to increased air pollution. Realization of Maasvlakte 2, its operational start date and development of compensation areas will take decades. To make Maasvlakte 2 possible, a part of the nearby sea has to improve in quality and a new dune area has to be developed, both as part of the same project and at approximately the same time. This is a unique development, both for planners, constructors and ecologists. Construction of the new dune area took place from 2008 to 2009, whilst construction of Maasvlakte 2 started in 2010.

Estimated dune damage and required compensation

The EIA estimated that the use of Maasvlakte 2 would damage certain important dune habitats and plant populations in the immediate surroundings that are under protection of Natura 2000 (Table 8.2). The damage is expected to be caused by NOx emissions from traffic (boats, cars, etc) that use the new harbour. Airborne deposition of N in fact acts as an extra nutrient load to the rare dune ecosystems, which are (extremely) nutrient poor.

According to Natura 2000 regulations, the damage has to be compensated. Using some multiplying factors, it was calculated that *circa* 35 ha of new dune was needed. This was created by adding an extra 6.5×10^6 m^3 of sand to nourishments already planned along this coastal stretch. However, sand for this special extra area had to be of a suitable grain-size composition in order to create the best ecological soil situation for rare vegetation to develop in this area, as after nourishment was completed, aeolian sand transport and deposition was planned.

■ **Table 8.2** *Expected damage to nature values in dunes near Maasvlakte 2, as a result of the use of this area (after 2013, causing extra emission of NOx) and total compensation required (Table drawn by authors after: Vertegaal and Arens, 2008)*

Habitat type/species Natura 2000	Estimated maximum damage/loss	Total compensation required
	according to worst-case scenario	*including multiplier factor because of exceptional nature value*
H 2130 grey dune (short dry dune grassland)	4.5 ha	10 ha
H 2190 wet dune valley grassland	1.5 ha	6 ha
H 1903 *Liparis loeselii* (orchid) population (Loesel's Twayblade)	1 locality	1 locality
Other (scrub)	15 ha	15 ha
		32 ha + 1 locality

The new dune area will develop in four phases:

1. Extra nourishment (as part of larger coastal defence works along the Delfland coast), to lay down the basic terrain form (Figure 8.14).
2. Slight remodelling of this terrain form by aeolian activity (local sand deflation and accumulation).
3. Spontaneous development of vegetation on slopes and in depressions, with the compensation requirement (Figure 8.15), as an ultimate management goal.
4. Fine tuning by means of specific management actions in case development deviates from the management goal.

Figure 8.15 a and b show the desired lay-out of the dune compensation area. Basically, the area consists of a new dry foredune ridge with a narrow (maximum 100 m) elongated valley behind it, where dune slack

■ **Figure 8.14** *Dutch coast near Hook of Holland, looking north towards The Hague. Construction of the dune compensation area as extra part of coastal protection works between Hook of Holland and The Hague. Phase 1: beach nourishment (Jan. 2009). Note the first layout of the narrow, elongated wet dune valley in the centre. Compare with Figure 8.15a and b, section A–B (Photograph courtesy Nico Bootsma, Rijkswaterstaat)*

■ **Figure 8.15a**
Design of the desired dune compensation area near Hook of Holland. Compare with Figure 8.14 and Figure 8.15b. For transect A–B see Figure 8.15b (Figure drawn by Jasper Fiselier from Veeken et al., 2007)

existing beach
new beach
existing dune
new dry dune
new wet dune valley

North Sea

A

B

P

N

0   1 km

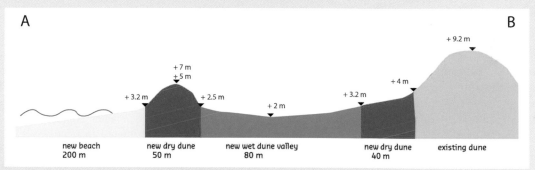

A   B

+ 9.2 m

+ 7 m
+ 5 m

+ 3.2 m   + 2.5 m   + 4 m

+ 2 m   + 3.2 m

| new beach | new dry dune | new wet dune valley | new dry dune | existing dune |
| 200 m | 50 m | 80 m | 40 m | |

■ **Figure 8.15b** *Transect showing the design of the new dune compensation area from A (beach) to B (inner dune ridge). Numbers indicate metres above MSL. Compare with Figure 8.15a (Figure drawn by authors after Veeken et al., 2007)*

vegetation can develop. It is clear that the desired vegetation elements and plant populations need time to develop, as well as the fresh groundwater level. Figure 8.16 shows expected times for development of the main nature elements.

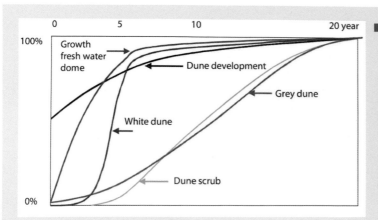

■ **Figure 8.16** *Estimated time for development of new natural elements in the first 20 yrs as % after 20 years. It is expected that the desired situation will develop in the course of some 20 years (Figure drawn by authors after Veeken et al., 2007)*

First developments and management

After two years of development, the first monitoring results showed that wind activity is doing its work and the phreatic level is slowly developing. Most beach sand is transported by wind over the first dune ridge across the valley into the old foredune ridge, where vital *Ammophila arenaria* (marram grass) captures it and supports new dune soil development. The new area has the official status of a nature reserve and is managed by a nature conservation organization (Zuid-Hollands Landschap), working towards Natura 2000 status. A monitoring program will show if the desired natural development is indeed occurring and if extra supporting management activities are required.

development, although scaling up nourishments, as would be the case under the new Delta Programme's recommendations (Deltacommissie, 2008), could make the environmental issue predominant.

From the start of the regular nourishment programme, it was decided that sand could not be dredged from within a continuous depth of up to 20 m, as this was firmly regulated and supervised. Environmental concerns have generated monitoring programmes that keep track and investigate resulting effects and the strategic importance of managing this resource carefully has been established.

Climate change and the Netherlands coast

Coastal erosion is a long-standing issue, especially in densely populated areas and under conditions of climatic change, coastal erosion rates may be enhanced. The European Commission sponsored the *Eurosion* (2001–2004) project (www.eurosion.org). Eurosion tackled this issue and established a reference for coastal erosion as was prevailing then. Netherlands coastal erosion was estimated to occur over some 134 km. In view of occurring and expected sea-level rise, this figure is likely to increase. The European Commission (DG-ENV, 2004) investigated the potential of incorporating coastal erosion into a project Strategic Environmental Assessment (SEA).

Until now, coastal protection policies have proven to be efficient (and cheap). However, in anticipation of climate change, the Netherlands Government has decided to embark upon a pro-active policy. It has installed the so-called Delta Commission (nicknamed the 'commissie Veerman') in 2007, which reported a year later (Deltacommissie, 2008). The committee's recommendations included, amongst others, that the Netherlands should be protected against the ongoing rise of sea-level by keeping up its coastal defences. Two of the methods are the obvious choices: rigid 'hard' defences (e.g. storm surge barriers) and 'soft' defences (nourishments). It is also obvious that the relatively small nourishments, as had been practiced until now, will not be enough: very large volumes of sand will be needed, up to possibly sevenfold of the volumes nourished today. The challenge is to decide if these nourishments are really needed and if 'yes', how they should be applied and where they should be placed. Currently these questions are under study (Rijkswaterstaat/Deltares). The 'Zandmotor' ('sand engine') is a large nourishment in front of the Delfland coast which is currently under construction.

The three different physiographic regions require a differentiated approach. The Holland coast cannot simply be widened all over its 130 km and choices must be made where trial projects will be placed. One is currently being established (the Zandmotor). Likewise, it is difficult to see the Wadden coast islands extend seaward, as it is the Wadden Sea area itself which will be under threat of drowning, especially its intertidal flats. The challenge here will be how to raise the surface of these tidal flats without nourishing them directly. The Delta programme for the Wadden is currently starting to design methods and trial projects (requiring an inclusive monitoring programme for plugging knowledge gaps). Likewise, in Zealand, nourishments are evaluated with more attention than previously. In the Wadden, the island of Ameland has been chosen as a pilot site to study the environmental effects on nearshore ecological effects of nourishments in the shallow North Sea. It will take some years of monitoring and evaluation to understand the effects of large-scale nourishments along the coast and the Zandmotor site is specially designated for this purpose.

An artificial supply of sand, in combination with free spreading by natural processes, provides several advantages, benefiting both shoreline consolidation and nature development. In the past, sea defences and nature management had contrasting goals, leading to many conflicting situations. Since sea defence was the most important function, this usually resulted in declining natural values, increasing artificiality of foredune shape and vegetation and a complete block of the sand exchange between beach and dunes. The current strategy favours both safety and nature. Recently, the Zandmotor, consisting of 21.5 Mm^3 of sand, was constructed in front of the South-Holland coast, heralding a new age and a new dimension of coastal management. It is thought that the Zandmotor will safeguard the coast for at least 20 years by spreading sand along the coast, and strengthening coastal dunes (www.dezandmotor.nl).

There are more advantages as, in a fixed system with sea level rise, the height difference between sea level and the average dune surface becomes smaller. Using natural processes to release foredune sand to move into the inner dunes, ensures a gradual adaptation of the surface: the dunes can grow with the sea level. In this way, natural processes are essential for long-term dune safety. Meanwhile, habitat types like the grey dunes will benefit, because they need a light sand spray. Given the need to apply sand nourishment, a sound coast (ensuring safety against flooding in the long term), dynamic dunes and good opportunities for the maintenance of grey dunes and other important nature values using sustainable nature management, all seem to be a viable ambition.

Conclusions

Apart from a limited stretch of strong sea dikes and storm surge barriers, Dutch coastal defence management largely deals with extensive sand foreshores, beaches and dunes.

In the past half century, this management evolved from traditional hard measures (*i.e.* with groins) to soft solutions (nourishments). The latter is also designated as working with nature, because it fits better into the dynamic character of the soft sandy dune coast. Along with this development, the focus widened to a more integrated approach. The various, often conflicting, functions of the coast (*i.e.* flood defence, nature conservation, recreation) are taken into account. In addition, the focus shifted from strict dune stabilization to providing room for mobile dunes. In this way, coastal defence management is attempting to give positive support to all these functions. This means that not only some parts, but also the whole coastal zone needs to be considered in its full width: foreshore, beach, foredune and inner dunes (together called the 'coastal foundation'; see Figure 8.1). As can be seen in this chapter, aspects of civil engineering, geomorphology and ecology are needed and all will contribute in the future. Based on the evaluation of many nourishments carried out so far, it is likely that such an approach is a better solution to make the coast more robust and adaptive to the consequences of climate change. It will build a safe and healthy coast for the millions of people that live there.

Acknowledgements

Research for the section on the effects of nourishment was funded by Rijkswaterstaat (the former Ministry of Public Works, Transport and Water Management) and by OBN (a programme of the former Ministry of Agriculture, Nature and Fisheries). Hans van Bergem (Deltares) assisted with the preparation of several figures.

References

Anonymous, 2010. *Kustlijnkaarten 2011.* Rijkswaterstaat/Ministry of Infrastructure and Environment, The Hague, 161 pp. [in Dutch]

Arens, S. M. and J. Wiersma, 1994. The Dutch foredunes: Inventory and classification. *Journal of Coastal Research*, 10, 189–202.

Arens, S. M., S. P. van Puijvelde and C. Brière, 2010. *Effecten van suppleties op duinontwikkeling: Geomorfologie. Rapportage fase 2.* Arens Bureau voor Strand- en Duinonderzoek en Deltares RAP 2010.03 in opdracht van Directie Kennis, LNV, 141 pp + annexes. [in Dutch]

Arens, S. M., van Puijvelde, S. P. and L. van der Valk, in prep. *Large-scale response of Dutch foredunes to changes in management.*

Bakker, M. A. J., van Heteren, S., Vonhögen, L. M., van der Spek Ad J. F. and L. van der Valk, 2012. Recent coastal dune development: Effects of sand nourishments. *Journal of Coastal Research*, 28(3), 587–601.

Deltacommissie, 2008. *Working together with water.* Findings of the Deltacommissie, Hollandia Printing, 138 pp. Available at: www.deltacommissie.com (last accessed 16 August 2012).

DG-ENV European Commission, 2004. *Development of a Guidance Document on Strategic Environmental Assessment (SEA) and Coastal Erosion.* Final Report, 68 pp. Available at: www.ec.europa.eu/environment/iczm/pdf/coastal_erosion_fin_rep.pdf (last accessed 16 August 2012).

Elias, Edwin P. L. and A. J. F. van der Spek, 2006. Long-term morphodynamic evolution of Texel Inlet and its ebb-tidal delta (the Netherlands). *Marine Geology*, 225, 5–21.

Everts, F. H. and N. P. J. de Vries, 2010. *Plan van aanpak ecologische effecten van zandsuppletie op de duinen langs de Nederlandse kust.* Rapport 890EGGev, Groningen. [in Dutch]

Mulder, J. P. M., Cleveringa, J., Taal, M. D., van Wijsenbeeck, B. K. and F. Klijn, 2010. *Sediment perspective on the SW delta area, the Netherlands.* Deltares report 1203404, 62 pp. [in Dutch]

Nourtec, 1997. *Innovative nourishment techniques evaluation, Final report.* Coord. Rijkswaterstaat, National Institute for Coastal and Marine Management/RIKZ, The Hague, the Netherlands, 105 pp, with figures.

Project Mainport Rotterdam, 2007. *Map, Havenbedrijf Rotterdam N.V.,* 1 p.

Short, A .D., 1992. Beach systems of the central Netherlands coast: Processes, morphology and structural impacts in a storm driven multi-bar system. *Marine Geology*, 107, 103–137.

Stuyfzand, P. J., Arens, S. M. and A. P. Oost, 2010. Geochemische effecten van zandsuppleties langs Hollands kust *[Geochemical effects of sand nourishments along Hollands coast].* Bosschap Report OBN141 DK, Ministry of Agriculture, Nature Management and Food Quality, The Hague. [in Dutch]

Veeken, L., Hoeven, J. ter. and J. Fiselier, 2007.

Ontwerpplan Duincompensatie Delflandse Kust. 2007. Kustvisie ZuidHolland. DHV/H+N+S/Alterra, 141 pp. [in Dutch]. Available at: www. delfland-sekust.nl (last accessed 16 August 2012).

Van Koningsveld, M., Mulder, J. P. M., Stive, M. J. F., van der Valk, L. and A. W. van der Weck, 2008. Living with sea-level rise and climate change: A case study of the Netherlands. *Journal of Coastal Research* 24, 367–379.

Vertegaal, C. T. M. and S. M. Arens, 2008. *Natuurbeheerplan Duincompensatieproject Delflandse Kust 2009–2029.* ZHL–Vertegaal–Arens, 44 pp. [in Dutch].

Waterman, R. E., 2010. *Integrated coastal policy via building with nature.* Ph.D. Thesis TUDelft, 71 pp. Available at: http://repository.tudelft.nl (last accessed 16 August 2012).

9 Belgium

Roger H. Charlier

Introduction

The coast of Belgium stretches for approximately 67 km, between the borders with France (near De Panne) and the Netherlands (Zwin inlet; Figure 9.1). It is intensively developed, with several examples of the movement pattern of former inland settlements, as a result of tourist pressures (Figure 9.2), to new sites which are now coastal conurbations having coastal lengths of several kilometres, as from Blekkaard to De Panne (35 km). The Shoreline 'promenade' is lined by multi-storeyed constructions that replaced Belle-Epoque and Victorian villas that German occupation forces had soldered together in 1940 as part of the Atlantic Wall (World War II).

Tides range is from 4 to 5 m and longshore currents of 1 m/s can develop causing significant impact on coastal dynamics (Magnusson *et al.*, 2003). Although sheltered by Great Britain in the West, the Belgian coast can be impacted by severe storms of significant wave heights of 4 m. Sandbanks run parallel to the coast, but currents differ between two major segments: from De Panne to Wenduine, and Wenduine to Het Zwin. The Belgian coast has been extensively studied by major research centres such as those belonging to the Universities of Ghent and Louvain.[1] A recent review (2007) of the Belgian sand beach ecosystem was made by a group of researchers, many belonging to the marine biology department of the University of Ghent. Their findings were

■ **Figure 9.1** *Location map: Belgian Coast – Zwin Inlet and the Scheldt and Yser river estuaries*

■ **Figure 9.2** *Coastal 'Riminization' at Niewpoort (photograph by E. Pranzini)*

published in a volume edited by Speybroeck (2007).

Geology and geomorphology

Stratigraphy

Quaternary surface sediments on coastal zones are well known, but knowledge of deeper Quaternary units remain fragmentary. Fieldwork carried out by the University of Ghent has provided new data on Tertiary substratum relief and the lithostratigraphy of Pleistocene and Holocene sediments in the eastern part of the coast (De Moor and De Breuck, 1973; Behre, 2001). Several lithostratigraphic units have been identified, from the gravel–sand deposit on the bottom (Ostend formation, from Eemian age, 125,000 years BP) to more recent deposits (Zuienkerke formation; Nieuwmunster peat; Dunkirk formation, 3,000–500 years BC).

The Belgian coast of West Flanders presents a gently sloping, fine sand beach on its western end, near Koksijde, which is bordered by a ribbon of dunes, of extremely variable width from 100 m to a few kilometres (De Moor *et al.*, 2010) and moderate heights (*circa* 20 m). Beyond the dune field, a 5 to 10 km wide coastal plain is cut by sand creeks and clay pits; its landward limit corresponding to the limits of the Holocene transgression. Clay pits often serve as ornithological refuges and those near De Panne and Heist have been converted into protected areas (Figure 9.3). Quaternary deposits are a succession of marine and continental deposits.

■ Figure 9.3 *Zwin inlet (photograph by E. Pranzini)*

Stratigraphic layers on the Belgium coast consist of: Upper Clay (Polder Upper Clay Layer); Coquina marine sand (containing the cockle, *Cardium edule*); marine clay presenting *Cardium*; peat layers showing evidence of modern flora (such as birch, hazelnut, oak and poplar); marine sand containing roots of trees from the overlying peat layer; and sand containing *Corbicula fluminalis* (an organism currently present in the Nile River estuary). Sea level is located either at the base of Lower Polder Clay or the base of the upper limit of the peat layer (where Roman coins and cut silex from the Neolithic Period have been found).

Coastal barriers and dunes were formed in the Holocene: young tidal and lagoon deposits are overlaid by 'Younger Dunes', which occasionally top 'Older Dunes'. After the Gallo-Roman historical period, when barriers were destroyed, the shoreline migrated landward and peat deposits were either eroded or covered by tidal flat sedimentary deposits. However, retrieval of coins and artefacts from peat layers of the former Sincfal Channel, near the Zwin area, demonstrated that several changes in sea level in this area occurred very recently. Peat layers

continue under the present sea surface for a distance and reappear near the English coast.

The western coast presents several ephemeral pans which eventually join the coast. These are usually depressions or shallow lakes, salt or brackish – the levels of salinity depending on precipitation, runoff and evaporation rates. In Belgium, rivers do not discharge into coastal pans, although the Aa River criss-crosses the area on the French side. The border town of Adinkerke–De Panne developed in an area of numerous dried-up pans (De Breuck *et al.*, 1969; Anonymous, 1995; Anonymous, 2010).

Geomorphology

Coastal geomorphology in Flanders has been highly influenced by the formation and migration of tidal inlets, together with anthropogenic factors that interfere with natural processes. Some of the major associated issues are:

- Tidal inlets are believed to be the largest sediment sink along the coast and responsible for beach erosion (De Moor *et al.*, 2010), but may foster accretion in some areas. They play a critical role due to their cumulative effect on regional orientation and composition of coasts, and of barrier island migration.

• In the event of exceptionally intense storms and floods the physical aspects of the Flanders coast and adjoining Zealand seem to 'conform' to the natural conditions that caused changes to occur (such as sea-level rise, tidal variations, wave energy and its seasonal variations, sediment budget and geology). Anthropogenic factors that interfere with coastal morphology include the building originally of hard protective structures, and in the twentieth century periodic beach nourishment, polder creation (now discontinued), dredging, mining and hydraulic engineering projects.

• In addition to the well known Zwin inlet and the lesser known Schipgat, it should be noted that other inlets and waterways were present in Flanders before the French King Louis XIV (1638–1715) amputated the westernmost segment of Flanders County, a medieval vassal state of France (Figure 9.4).

The Flemish plain gradually extended further from the North Sea because of sedimentation, development of nearshore littoral barriers and anthropogenic

■ **Figure 9.4** *Belgii Veteris Typus map by Ortelius (Claes Jansson Visscher excudebat, 1624) (Source: E. Pranzini)*

impacts. The oldest barrier is that of Ghyvelde which lies across the border with France and is now the site of anchored dunes. Seawards, another elevation (Great Moërs dune ribbon) forms a barrier against seawater.

The Herm River (which today is French territory) used to flow through Mardijke[2] and reaches the sea as a single estuary – such as the Enna River estuary at Grevelingen.[3] Contemporary clay soils were originated by *slikkes* and *schorres* that dried up. *Slikkes* are the lower part of the coastal 'muddy' wetlands that are covered by the sea at high tide. *Schorres* are the upper part of these wetlands, often covered by halophytic meadows.

Of *wateringen, möeres* and *wadden*

These terms are widely used in studies of French, Belgian and Dutch flatlands near the North Sea (Postma, 1961; Van Straaten, 1951). Philippe of Alsace, the ruling count of Flanders in the twelfth century, set up a multipurpose system of drainage, reaching in some cases up to 100 km inland, which tempered potential intrusion of saline waters into the phreatic freshwater network, and irrigated when necessary.

A *moër*[4] designated a polluted lake (pan) in Flanders whose waters eventually ran into a channel – using windmills – and to the sea. *Moërs* were repeatedly dried up (and still are); they were also integrated into defensive schemes of military protection in Dunkirk.[5] *Moërs* were flooded for military purposes, in the seventeenth and eighteenth centuries, and then dried in the early nineteenth century. During World War II, German occupation forces halted the flow of freshwater at first, but later reversed the policy and allowed saltwater to flow in.

An unusual consequence of the sophisticated system of hydraulic drainage developed after World War II, was the disappearance of *wadden*[6] in French Flanders and the Belgian Westhoek. *Wadden* is the name for tidal marshes and include several types of environments as described by e.g. Van Straaten (1951) and Postma (1961). Mud is deposited at virtually all depths and distances from the shore and is highly concentrated on these flats. They have been recently associated with the asymmetrical nature of tidal currents, since low velocity lasts longer at high tide than it does at low tide, therefore allowing deposition of very fine particles. The *wadden* off the French-Belgian area fall into the category of tidal marshes.

Off- and on-shore changes

Sandbanks lie parallel to the recent shoreline (Figure 9.5) at Middelkerke, Wenduine and Paardenmarkt and coastal erosion appears to be caused by marine currents flowing from east to west; beaches are under direct attack and sand is carried to a sand bank located *circa* 0.2 km off the beach. The Paardenmarkt Bank is split in two by the Appelzak Channel/Deep. Incoming tides bring sand to the shore, but in smaller quantities than those taken by outgoing tides; these are deposited mainly in the Appelzak, which has been gradually filling up. This process has led to discussions regarding the outward limit of the Flanders Region and associated territorial claims.

Close to Appelzak lies Paardenmarkt Sandbank, which expanded while the Appelzak Deep was being filled. Approximately 30 years after the end of World War II, this uninterrupted process of high magnitude allowed the bank to emerge at low tide, when it could be reached on foot; this is still the case today. The original Paardenmarkt Bank gradually sank to below 10 m depth, and the name was then reused to identify this contemporary bank, located offshore Knokke Heist at less than 4 m below low tide level (Charlier, 1955). The area has acquired an infamous reputation as a resting place for some 30,000 metric tons of ammunition and discarded military equipment (Anonymous, 2010). The shifting and disappearance of sandbanks along the coast have an influence on faunal behaviour. Seals have been found in offshore waters and sandbanks are their usual resting and sun-bathing area. Deprived of them, they have been lately seen resting at low tide on groins, locally known as 'wave breakers', which were constructed decades ago to halt beach erosion along the entire coast (except at Oostduinkerke) and that have now re-emerged.

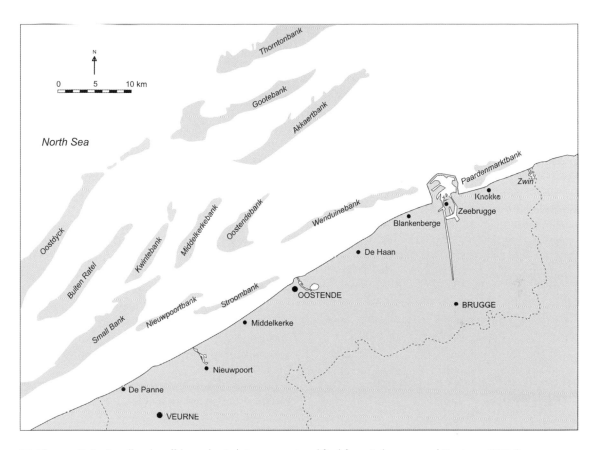

■ **Figure 9.5** *Sandbanks offshore the Belgium coast (modified from Belpaeme and Konings, 2004)*

Therefore, the coast and offshore area have undergone numerous changes during the 'Anthropocene' (Houthuys and De Moor, 1993). Many events may be accounted for, such as the Zwin Inlet case, the filling up of channels (such as Sincfal), and the disappearance of towns drowned by the North Sea (such as Harendyke, near contemporary Wenduine and Walraverside, near Ostend and Middelkerke). Readjustments have also occurred on the continental shelf at relatively short distances from current coastline positions. Sincfal Channel dates back to the Middle Ages; other changes occurred on river estuaries, *schorres* and sandbanks. Originally an inlet, Westernscheldt became an estuary between theninth and twelfth centuries. A few islands also underwent modifications: Wulpen gradually disappeared between 1377 and 1513, Koezand disappeared completely by 1570, Zuidzande, Cadzand (now a small coastal harbour and resort) and Schoneveld (spread out from where the mouth of the Westernscheldt is presently located).

Storm floods periodically affected the Rhine (De Moor and De Breuck, 1973), Meuse and Scheldt estuaries and had significant effects on the coastal lands of Flandrian Zealand (De Moor, 1991). Dunes protect only the western sides of the islands whereas coastal inlets and outlets are protected artificially by dykes; the development of dunes appears to be contemporary to construction of early man-made defence structures (twelfth to thirteenth century). The hard structure approach yields benefits but also has negative outcomes. Larger inlets and outlets were closed by dams, as part of the Dutch Delta Plan, implemented after the catastrophic storm and flood of 1953, which caused many casualties.

Man-made polders dominate in the eastern coastal zone (Oosthoek). The first 'real' polders date back to the eleventh or twelfth century; they are found in the Flandrian North Sea coastal zones of France and Belgium and in Zealand. Mud flats, a type of pan that often forms in front of polder dykes, constitute an enticement to further polder incorporation. However, 'polderization' and reclamation are often looked at with suspicion, as dykes seem to favour coastal erosion and, in a reversal of approach, in an application of the new maxim of 'Let Nature have its way', tidal waters are allowed to regain access to old channels. The policy has been tested in Northern France and near De Panne in the Westhoek dune areas. Erosion rates in experiments at La Panne area (Westhoek) have shown encouraging signs of abatement (Charlier and De Meyer, 2000).

Historical context of coastal protection

The relative sea levels varied from proto-, pre- and historic times to the late Gallo-Roman period, and again from about 1250 onwards. As a result, Flanders and Zealand coastal areas in France, Belgium and Netherlands were repeatedly covered by North Sea waters (De Moor and Mostaert, 1993) and subject to both silting and erosion during that time (De Moor, 1981).

The most spectacular processes were the silting of the Zwin (Charlier, 2011) and first the decline, then disappearance, of Bruges as the wealthiest port and city of medieval Northern Europe. A thirteenth-century storm (Houthuys and De Moor, 1993) broke through continental barriers and opened a channel which allowed ships to sail to towns such as Sluis, Damme and Hoek, sites that for several centuries were outer-harbours of Bruges and they became to some extent harbours in their own right. Damme – a 'pre port' for cargo trans-shipment to shallow-bottomed boats, and military strongholds – originated

a new trade, which developed the wealth of Bruges[7] and gave Zwin Inlet the nickname 'Golden Inlet' (Charlier, 2011). However, silting caused occlusion of the Zwin, to the benefit of Antwerp, which became the largest city in the world. Today, the Zwin Inlet is merely a natural 'swimming pool' where at high tide a rivulet of water allows some canoes to manoeuvre. On the other hand, the region became an internationally known nature reserve for birds and plants (Lanckneus, 1992). Erosion and sedimentation induced new modifications on the shoreline, some of which impacted the fauna.

On the other end of the short Belgian coast, an unusual variety of dunes, *pannes* and *möers*, and silted-in sea channels are found. The west coast is also the site of a natural reserve that includes the oldest dunes of historical Flanders.[8] On the limit between Oostduinkerke and Coxyde,[9] a dune break indicates where (in historical times) the North Sea Schipgat channel led, through the Doornduinen (thorn dunes), to Dixmude[10] – then a thriving inland harbour for the clothing trade.

Groins and dykes were eventually built for protection against recurrent floods in polders lying below sea level. However, these polders were voluntarily flooded more than once. Sea-locks were opened on certain well-known occasions: in 1383, when the Bishop of Norwich, who had sworn allegiance to the Avignon-seated papacy, laid siege to Blankenberge, Ostend and Newport,[11] as they had chosen allegiance to the Rome-seated Pope; in 1600, when Spanish defenders of Newport needed to stop Dutch invaders, while the Gueux[12] opened an inlet in the dunes near Ostend which had been captured from the Spaniards; and during World War I, when Albert I, King of the Belgians, gave an order to stem the advance of German troops to Furnes,[13] by drowning the Flemish plain of the Yser[14] river mouth. His son, King Leopold III, vetoed a new version for this strategy during World War II (Leopold III, 2003) so as not to endanger refugees fleeing the German advance.

In the fifteenth century, Count John, Duke of Brabant, constructed a dyke along part of what is today Belgian coast, as a response to the early fourteenth century floods that had extended over wide areas of land and drowned several villages. The remnants of John's engineering efforts are still traceable (Charlier *et al.*, 2005).

Approximately 6 km to the east lies the Yser estuary, which was the front line between German-Austrian and Allied forces in World War I. Once a major waterway, the only river entering the North Sea from Belgian territory has lost much of its commercial significance. It still works as a home port for some fishing trawlers, and scores of pleasure crafts. The regional government bought the area from the naval base (Westende, later changed to Lombartsijde) owned by the Military, *i.e.* the national government, and allowed the river estuary to return to its natural condition. Other harbours, located along the North Sea and Atlantic coasts (or not far inland), have also had their hour of glory. On the Yser River, Dixmude was an important river port in medieval times, though on a much smaller scale than Bruges. This area, serving as a harbour for cloth export, also hosted a large urban settlement, and benefited from its vicinity to Ypres,[15] then a world-famous trading place for clothing. Gradual silting of the river meant the death knell for Dixmude, which, along with Ypres, was practically demolished during World War I.

Mining for sand and gravel on the continental platform also exacerbated the natural erosion affecting beaches. Yet the practice still persists along the Belgian coast (Lanckneus, 1992). The dune belt, a few kilometres wide in the west and east, narrows to barely 100 m in the centre, where man-made infrastructures were built up between the ninth and thirteenth centuries. According to mediaeval documents, the sea pushed the coastline back for as much as 5 km and 'swallowed' several coastal communities, such as, Harendijke, situated between Wenduine[16] and Blankenberge (Houthuys and De Moor, 1993). Two other coastal communities were lost to the North Sea on 24 November 1334: Scarphout, seawards from Blankenberge, and Ter Streep, offshore Mariakerke near Ostend. On the other side of the channel numerous towns suffered a similar fate (Sheppard, 1912).

Coastal protection

Coastal protection is a concern dating back to the early Frisian settlements and is recorded in the writings of Plinius the Elder. Dykes (actually sand heaps) were the main approach used then, and reports on their construction and maintenance can be found from the thirteenth and fourteenth centuries. The ocean was envisaged for centuries as a new domain to be recovered, dammed, drained and transformed into agricultural and/or pasture land, especially by the Dutch, who started reclamation in the thirteenth and fourteenth centuries. Belgium and France also added a number of polders, *polderland* and *wateringen* to their territory with equal success.

Seawalls

Hollanders, Zealanders, Brabanters and Frisians were among the first in Northern Europe, as early as a thousand years ago, to protect their buildings from rising sea-level through the use of earth mounds. By the 1200s they had already performed reclamation and built coastal protection works. The eleventh-century methods for building dykes were adopted by Flemings and Hollanders for land protection and reclamation. Cistercian and Premonstratensian monks may also have contributed substantially to the creation of dykes which formed polders in Flanders, Zealand and Friesland.

Concerns regarding coastal protection in the Lowlands already dominated in the thirteenth and fourteenth centuries. William I, Count of Holland, who was probably the founder of the *Rijkswaterstaat*, the Water Administration of contemporary Netherlands (Van Veen, 1962), surrounded the coasts of his territory with dykes – and probably also the islands of Walcheren and Schouwen, although these belonged to Zealand County. William I may have traced a network of canals to drain the moors. Shipwrecks close to dyke breaches often formed the base for fill material which was secured with mats and/or brushwood. Vierlingh, a sixteenth-century researcher, strenuously opposed this method of repairing dyke breaches, pointing to the nonhomogeneity of the shipwrecks used in dyke structure but this approach remained in use in the Netherlands for a considerable time.

Breakwaters placed at Zeebrugge in 1870 were made of concrete, but around a hundred years later the preference shifted to stone blocks placed at the site (1980). Detailed descriptions have been provided by Charlier and Auzel (1961), De Moor and Blomme (1988) and Moller (1984). Hard defence structures made of stone/rock started to be constructed around 150 years ago, such as seawalls in Ostend (1885; Figure 9.6), Wenduine, Blankenberge and Zeebrugge (1870), Nieuwpoort (1897), Middelkerke (1898) and Heist (1899). Seawalls were close to the village/town centre only in Ostend, Mariakerke and Heist. Since the 1980s, the seawall in Coxyde[17] has stood landward from the local dune fringe. The concrete seawall and promenade between De Haan (Le Coq) and De Haan-Golf, built after 1912, were 'lost', buried under sand drifts and 'rediscovered' in 1976. Short stretches of seawalls[18] were constructed prior to World War II in Knokke and Oostduinkerke. The beaches were at times reduced to a few square metres, until artificial nourishment was later undertaken on a large scale.

The catastrophic flood of 1953 and subsequent extreme storm events led to seawall extensions, sometimes 'temporary', such as from De Panne to the French border; recreating tidal inlets was favoured as the solution instead of restoring the dune toe for protection (De Moor, 1991).

Groins

Groin-like structures are known to have been placed in the Lowlands (today part of the Netherlands, Belgium and Northern France) well before the sixteenth century, but there is not sufficient information on them. More comprehensive information is available on the history of the development of protective structures in the past 100–150 years. Even though groins have increased in size, engineering principles remained the same: stone (rubble) pitching on gravel and mattresses in the centre; stones on mattresses on the sides, with two or more pile walls as supports. However, hard materials were soon

■ **Figure 9.6**
Seawall at Ostende (photograph by E. Pranzini)

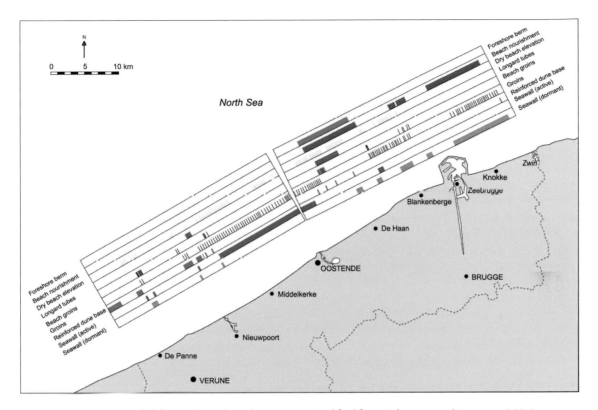

■ **Figure 9.7** *Coastal defence along the Belgium coast (modified from Belpaeme and Konings, 2004)*

brought in to reinforce dykes, and rock blocks were placed perpendicular or parallel (in some rare cases even at an angle) to the coastline. These hard structures became the groins, breakwaters and seawalls that are known today (Figure 9.7).

Beach erosion started to threaten budding coastal tourism in the late 1800s. Accurate warnings by field observers were ignored by high-ranking government officials; beaches shrank, and eventually groin fields were placed on and in front of beaches. Yet, in most areas the shoreline continued to migrate landwards to such an extent that beaches were entirely covered by the sea at high tide (as for instance in Heist), while only a few square metres of dry sand remained available for recreational tourists in other locations (e.g. Knokke). Groins built on the Northern part of the Belgian coast back then did not differ from those placed along adjacent French and Dutch beaches.

Neither solved the problem of erosion (Charlier and De Meyer, 1998). These eighteenth- and nineteenth-century structures proved to be site-specific remedies that usually transferred the problem to a neighbouring coastal sector. Beach nourishment was widely proposed as a new approach to coastal protection, but both hard structures and costly nourishment schemes have finally given place to the approach of 'let Nature have its way' (Charlier and De Meyer, 1998).

Coastal protection in the face of rising sea-levels has increasingly become a concern for governments during the past 150 years. Modern groins (Figure 9.8) started to be placed in the twentieth century and some backed existing seawalls. In addition to the 17 groins located in the Ostend area (Figure 9.9), some 75 were placed west of Wenduine in 1912. Little has changed since then, except for additional groins being added (Figure 9.10), artificial nourishment in Coxyde,

■ Figure 9.8 *Modern groin at Knokke Heist (Photograph by E. Pranzini)*

Ostend, Knokke and Heist, and berm nourishment in De Haan. In Belgium, as in The Netherlands, groins are believed to have a 'positive impact', but this view is not unanimous (De Moor and Blomme, 1988). It is still uncertain whether they are beneficial for renourished beaches or not.

In essence, the Belgian defence approach is best characterized by massive hard engineering, but artificial nourishment has started to be preferred in more recent decades – especially as cost assessments and technical improvements made it an attractive alternative. In addition, other alternative approaches to coastal protection have also been experimented with.

Layers of different materials have been used for building dykes, overlaying silt, silt-and-sand, coquina, shell, and willow mattresses, among other materials. Bottom willow mattresses are still used occasionally, but more sophisticated materials have gradually been replacing them. This method had already been proposed over a decade ago for artificial nourishment in Ostend. Over time, dykes grew in size, and it became necessary to reinforce them with hard surfaces such as basalt blocks and/or other structures parallel and/or perpendicular to the shore. Reinforcing or supporting structures were developed and modified as experience and exposure increased. Gradual reinforcement by structures such as seawalls and groins may have contributed to a (not fully justified) sense of security.

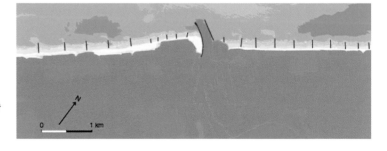

■ Figure 9.9 *Ostend beach, protected by groins. They are visible on the top left corner of the illustration. The artificially dug Port access channel is in the middle of the illustration*

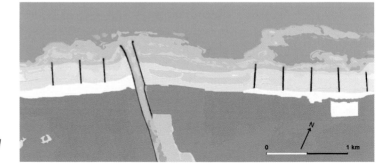

■ Figure 9.10 *Nieuwpoort aan Zee. Groins on beaches right and left of Yser River estuary*

Beach nourishment

Beach nourishment forms a well-established shore protection system in Belgium, and 22.1 per cent of the coast benefits from it (Eurosion, 2004). In the early 1980s, tonnes of sand were dredged due to the expansion of Zeebrugge harbour (Figure 9.11) which provided large amounts of free material. The Government's engineering department decided to use this sand for a mega beach nourishment project at Knokke (Charlier and De Meyer, 1998). At the time it was the largest undertaking of this kind in the world. Sand deposited kept being 'lost' to the Appelzak Deep, but the beach was maintained, partly due to periodic renourishment (Charlier and Beavis, 2000; Bruun and Gerritsen, 1959; Bruun, 1978, 1990, 1994; De Roeck, 1996; US NRC, 1995).

This nourishment project, the largest in Belgium, has now been completed as combined profile nourishment and berm feeding have begun to be implemented. A three-phase beach profile nourishment scheme was put in place at De Haan[19] in 1997 (Charlier and De Meyer, 1998) for restoring the beach, built with 600,000 m³ of sediment placed offshore and 800,000 m³ placed directly on the beach (Figure 9.12).

The cost for Phase 1 (1991–2) over 2.15 km; phase 2 (1994–5) over 3.2 km; and phase 3 (1996) over 4.2 km, was €1.25million/km, with an offshore bar (feeder bar) fill volume of some 250 m³/m, which was 2.5 times cheaper than direct beach placement of nourishment material. The effect was noticeable within six months. The seaward extension at high tide was some 70 m, with the refill schedule envisaging some 50 per cent of the 'design' fill every five years. Success was measured by comparisons of the area with both nourished/non-nourished sections further along the Belgian coast and the mayor of the area is extremely pleased with the stability of the beach.

New technologies and approaches, e.g. the above, to coastal defences developed slowly at first – though improvements and refinements were often made – but the pace of change accelerated considerably in the twentieth century, especially during the past few decades (Figure 9.6). In addition to profile nourishment and berm feeding, other tests included synthetic sand-filled tubes ('sausages'), which were environmentally acceptable though requiring anti-ultraviolet ray protection, and burlap sand-filled bags (similar to devices used by the Spanish in the seventeenth-century siege of Ostend), which have been used in the past decades at the foot of endangered dunes in Vosseslag and Bredene.

Additionally, in some places, e.g. from De Panne to Bray-Dunes, the toe of hard structures that protected dunes was not repaired and during the past few years the sea created breaches through bordering dunes, so seawater can now pass this barrier; results are said to be encouraging and local erosion reduced (Charlier and De Meyer, 1998). The experimental reopening of a sea inlet, with concomitant removal of dune toe protection, offers an opportunity to study

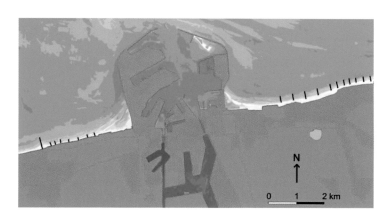

■ **Figure 9.11** *Zeebrugge harbour. To revive Bruges' economy, King Leopold II decided to create an artificial port in 1908, linked to Bruges by a sea channel. It has expanded since then and is now a leading port*

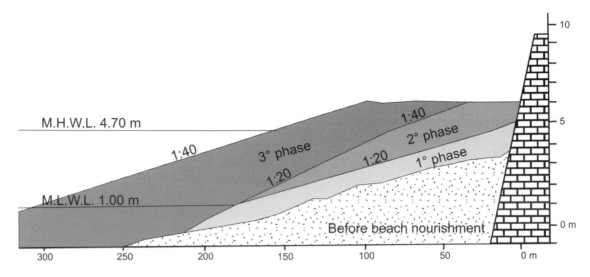

Figure 9.12 *Profile of beach nourishment scheme in Heist and Knokke, where 8.1 million cubic meters of sand were deposited. (Source: European Coasts: An Introductory Survey. Picture book of the European coasts, prepared by the Liverpool/Thessaloniki network as part of the Erasmus project (1996), Delft University of Technology, Hydraulic Engineering Group)*

Figure 9.13a *Dune stabilization between Wenduine and De Haan (photograph by E. Pranzini)*

Figure 9.13b *Plastic sand fences and groins south of Raversijde (photograph by E. Pranzini)*

the relationship between growth of the breach with time, discharge, water height and sediment. Related observations could encompass reciprocal effects of penetrating discharge and the volume involved, as well as assessment of inundation height (Bruun, 1978).

Dune stabilization

Dune stabilization is part of the present shore protection strategies and is achieved through vegetation planting, access limitation and fence construction (Figure 9.13a, b).

Notes

1 Important developments on coastal science in Belgium have been achieved by research groups in these universities, such as those headed by Guy De Moor, Roland Paepe and Cecile Baeteman (Geology and Physical Geography) and Adriaan Verhulst (Human Sciences applied to Stratigraphy, Sedimentology, Geomorphology and Geological Engineering).
2 Mardik today.
3 French Gravelines, Dutch-Flemish name Grevelingen.
4 French adaptation of the Flemish term *moer* (plural *moeren*).
5 In French, Dunkerque; in Flemish Duinkerke.
6 *Watten*, in German.
7 Bruges became the wealthiest and most powerful city in northwestern Europe during medieval times.
8 Historical Flanders includes the areas annexed by France and by Zealand.
9 In Flemish, Koksijde.
10 In Flemish, Diksmuide.
11 In Flemish, Nieuwpoort.
12 In Flemish, Geuzen.
13 In Flemish, Veurne.
14 In Flemish, IJzer.
15 In Flemish, Ieper.
16 Wenduyne.
17 Koksijde.
18 The term seawall generally means a structure put in place to hold back the sea; here it doubles as a promenade lined with villas and buildings, and remains separated from the actual sea by a beach.
19 In French, Le Coq.

References

Anonymous, 1995. Automatic beach accretion proves its worth in the UK. *Dredging and Port Construction* 22, 11, 18–19.

Anonymous, 2010. Zeewoorden. *De Grote Rede* 22, 26–28.

Bakker, W. T. and Joustra, D. S., 1970. The history of the Dutch coast in the last century. *Proceedings of the Twelfth Coastal Engineering Conference,* Washington, DC, ASCE, New York, pp. 709–728.

Behre, K. E., 2001. Holozäne Kùstenentwicklung, Meeresspiegelbewegungen und Siedlungsgescheben an der sùdlichen Nordsee. *Bamberger Geogr. Schriften* 20, 1–18.

Belpaeme, K. and Konings, Ph. (eds.), 2004. *The Coastal Atlas: Flanders–Belgium.* Publication of the Coordination Center for Coastal Zone Management, Ostende, 100 pp.

Bruun, P., 1978. *Tidal inlets and littoral drift.* Elsevier, Amsterdam, 510 pp.

Bruun, P., 1990. Beach nourishment: Improved economy through better profiling and back-passing from offshore sources. *Journal of Coastal Research* 6, 2, 265–277.

Bruun, P., 1994. *Coast stability.* Danish Society of Civil Engineering Press, Copenhagen.

Bruun, P. and Gerritsen, F., 1959. Natural by-passing of sand at coastal inlets. *Journal of the Waterways and Harbors* Division, 2301, 75–107.

Bruun, P. and Gerritsen, F., 1960. Stability of coastal inlets. *Proceedings of the Seventh Conference on Coastal Engineering,* Berkeley, CA, 23, 386–417.

Charlier, R. H., 1955. Belgian coastal erosion. *The Professional Geographer* 7, 2, 10–12.

Charlier, R. H., 2011. From golden inlet to nature refuge. *Journal of Coastal Research* 27, 4, 746–756

Charlier, R. H. and Auzel, M., 1961. Géomorphologie côtière: migration des sables sur la côte belge: *Zeitschr. für Geomorph.* 5, 181–184.

Charlier, R. H. and Beavis, A., 2000. Development of a nearshore weed-screen: A Nature coastal defense idea: *International Journal of Environmental Studies* 57, 4, 457–268.

Charlier, R. H., Chaineux, M. C. P. and Morcos, S., 2005. Panorama of the history of coastal protection. *Journal of Coastal Research* 21, 79–111.

Charlier, R. H. and De Meyer, C. P., 1998. *Coastal erosion: Response and management.* Springer Verlag, Heidelberg and New York, 194–222.

Charlier, R. H. and De Meyer, C. P., 2000. Ask nature to rebuild beaches. *Journal of Coastal Research* 16, 2, 385–390.

De Breuck, W., De Moor, G. and Maréchal,

R., 1969. Litostratigrafie van de kwartaire sedi-
menten in het Oostelijk Kustgebied (België).
Natuurwetenschappelijk Tijdschrift 51, 3–8, 125–137.

De Moor, G., 1981. Erosie aan de Belgische kust
[Erosion at the Belgian coast]. *De Aardrijkskunde*
1–2, 279–294.

De Moor, G., 1991. The February-1990 storms and
their impact on the beach evolution along the
Belgian coast. *De Aardrijkskunde* 15, 3, 251–316.

De Moor, G. and Blomme, E., 1988. Belgium. In:
Walker, H. J. (Ed.), *Artificial structures and shorelines.*
Kluwer, Dordrecht NL, 125–126.

De Moor, G. and De Breuck, W., 1973.
Sedimentology and stratigraphy of some Pleistocene
deposits in the Belgian coastal plain. *Natuurwet.
Tijdschr.* 55, 1–3, 3–96.

De Moor, G. and Mostaert, F., 1993. Eemian and
Holocene evolution of the eastern part of the
Belgian Coastal Plain. In: Baeteman, C. and De
Gans, W. (Eds), *INQUA Fieldmeeting 1993.
Quaternary shorelines in Belgium and the Netherlands,
September 18–25: excursion guide*, 94–108.

De Moor, G., Ozer, A. and Heyese, I., 2010.
Belgium. In *Encyclopedia of the World's Coastal
Landforms.* Springer Science, Dordrecht, 669–671.

De Roeck, J., 1996. *De stabiliteit van strandhoofden.*
Doctoral thesis, University of Ghent, Faculty of
Applied Sciences, Ghent, 346 pp.

Eurosion, 2004. *Living with coastal erosion in Europe:
Sediment and space for sustainability: Guidelines for
incorporating coastal erosion issues into Environmental
Assessment (EA) procedures*, prepared for European
Commission Directorate General Environment,
Service contract B4-3301/2001/329175/MAR/
B3.

Houthuys, R. and De Moor, G., 1993. The shaping
of the French–Belgian North Sea coast throughout
recent geology and history. In: Hillen, R. and
Verhagen, H. J. (Eds), *Coastlines of the World:
Coastlines of the Southern North Sea*, ASCE, New
York, 27–40.

Lanckneus, J., 1992. Zand- en grintwinning op het
Belgisch Continentaal Plat en monitoring van de
eventuele gevolgen voor de bodemstabiliteit. In:
Bolle, I. and Brijsse, Y. *et al.* (Eds), *Colloquium:
Oppervlaktedelfstoffen problematiek in Vlaanderen,*
Proceedings 24–25 Oktober 1991, University of
Ghent, 188–214.

Leopold III, 2003. *Pour l'histoire.* Racine, Brussels,
520 pp.

Magnusson, A. K., Jenkins, A., Niedermayer, A. and
Niedo-Borge, J. C., 2003. Extreme wave statistics
from time-series data. *Proc. MAXWAVE Final
Meeting*, Geneva, WP, 2, 6 pp.

Moller, J. T., 1984. *Artificial structures on a North Sea
coast.* Geoskrifter 19, Aarhus, Denmark, 25 pp.

Postma, H., 1961. Transport and accumulation of
suspended matter in the Dutch Wadden Zee. *Neth.
J. Sea Res. I*, 148–190.

Sheppard, T., 1912. *The lost towns of the Yorkshire coast.*
The Author, London.

Speybroeck, J., 2007. *Ecologie van macrobenthos als een
basis voor een ecologische bijsturing van strandsuppleties
[Ecology of macrobenthos as a baseline for an ecological
adjustment of beach nourishment].* MSc Thesis,
University of Ghent, Belgium, 189 pp.

U.S. National Research Council, 1995. *Beach
nourishment and protection.* National Academy Press,
Washington DC, 180 pp.

Van Straaten, L. M., 1951. Texture and genesis
of Dutch Wadden Sea sediments. In: van Andel,
Tj. (Ed.) *Proceedings of the Third International
Congress of Sedimentology*, Groningen-Wageningen,
Netherlands, 5–12 July, 225–244.

Van Veen, J., 1962. *Dredge, drain, reclaim: The art of
a nation.* Martinus Nijhoff, The Hague, 200 pp.

10 Great Britain

Simon J. Blott, Robert W. Duck, Michael R. Phillips, Nigel I. Pontee, Kenneth Pye and Allan Williams

An overview of coastal erosion in Great Britain

The Futurecoast project analysis of shoreline change indicated that some 28 per cent of the coast in England and Wales was experiencing erosion higher than 0.1 m /yr (Burgess, et al, 2004; Evans et al., 2004), with the highest rates being in eastern England where a great deal of the coastline is artificially protected. Later work by Masselink and Russell (2008) indicated that erosion on coasts with *hard* defences was also most prevalent in England (Table 10.1).

As outlined in Chapter 1 and many subsequent chapters, there are numerous causes of coastal erosion that are likely to vary in significance from place to place. However, a significant control on the large-scale spatial distribution of erosion in GB is the underlying geology (Figure 10.1). The prevalence of erosion on the east coast of England arises in part from extensive areas of soft or unconsolidated sediments, which are readily eroded by wave action. In contrast, the lesser extent of erosion on the west coast of Britain is partially explained by the more resistant geology that makes up extensive areas, although in

■ *Table 10.1 Coastal erosion and protection in the UK. Islands with a surface area <1 km² and inland shores (estuaries, fjords, bays, lagoons) where the mouth is less than 1 km wide are not included in the analysis (Masselink and Russell, 2008)*

Region	Coastal length	Coast length eroding		Coast length with artificial beaches and defence works	
	km	km	%	km	%
NE England	296	80	26.9	111	37.4
NW England	659	122	18.5	329	49.9
Yorkshire and Humber	361	203	56.2	156	43.2
East Midlands	234	21	9,0	234	99.8
East England	555	168	13.3	282	68.8
SE England	788	244	31.0	429	54.2
SW England	1,379	437	31.7	306	22.2
England	4,273	1,275	29.8	1,947	45.6
Wales	1,498	346	23.1	415	27.6
Scotland	11,154	1,298	11.6	733	6.6
Northern Ireland	456	89	19.5	90	19.7
TOTAL	17,381	3,008	17.3	3,185	18.3

the north-west of England onshore directed tidal currents drive onshore transport, producing wide intertidal areas that reduce shoreline wave energy.

In Scotland, isostatic uplift means that relative sea levels are falling in a number of areas and this, coupled with extensive areas of hard geology, means that erosion is less severe than elsewhere, but erosion is still the dominant coastal response. Based on a survey of nearly 650 sandy beaches between 1969 and 1981, Ritchie and Mather (1984) reported that approximately 40 per cent of beaches were eroding, 22 per

cent were stable, 11 per cent were advancing, 18 per cent showed evidence of both advance and retreat, and 9 per cent were protected or backed by some stable feature such as rocks. Advancing beaches tend to be located in areas where relative sea level is falling and/or there are contemporary supplies of sediment from fluvial inputs, alongshore littoral transport (from cliff/dune erosion systems) or biogenic carbonate production.

A National Coastal Erosion Risk Mapping (NCERM) project started in 2006 and a pilot project produced some results in 2011, but the main report is expected in 2012/13. This is mapping erosion/instability around the coastline of England and Wales to evaluate the risk to built assets, taking due account of the influence of current coastal defences and management activities (Rogers *et al.*, 2009). A key aspect is capturing local knowledge and expert opinion using web-based mapping techniques to allow local operating authorities to verify, interrogate and amend input data and provide a live visualization of the outputs generated. When completed, the project will complement the National Flood Risk Assessment (NaFRA), to provide a complete representation of flood and erosion risks along the coastlines of England and Wales (Rogers *et al.*, 2009).

Wales

The coastal setting

Wales, lying between latitudes 53° 43′ N and 51° 38′ N, is part of the United Kingdom and the island of Britain, bordered by England to the east and the Atlantic Ocean and Irish Sea to the west. It has a total area of *circa* 20,779 km² (8,023 square miles) with more than 1,200 km (746 miles) of coastline, including offshore islands, the largest being Anglesey. It is essentially a hard rock country and rock types heavily influence erosion rates. Cambrian and Ordovician volcanics and Carboniferous limestones show very little erosion, whilst softer rocks show higher rates. The relief consists essentially of a narrow coastal belt which fringes mountainous terrain in the north and central areas, shaped

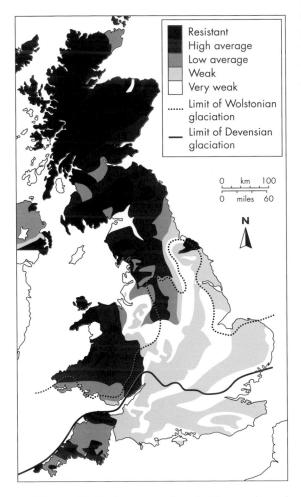

■ **Figure 10.1** *Resistance of underlying geological units across the UK (Source: Reprinted from Clayton and Shamoon, 1998, with permission from Elsevier)*

during the Devensian glaciation. The climate is maritime temperate. Inland, the mean annual rainfall exceeds 2500 mm/yr and average annual coastal temperatures are 10.5 °C (50.9 °F), due to the warm water of the North Atlantic drift. The prevailing winds are south-westerly and coastal areas are usually very windy, with gales occurring up to 30 days per year.

Some 70 per cent of the Welsh coastline is designated (under European Union Directives and UK law) for its environmental quality. It encompasses the only UK coastal park, the Pembrokeshire Coast National Park, together with five Areas of Outstanding Natural Beauty. Some 42 per cent of the coastline of South and West Wales is designated as Heritage Coast – an exceptionally high figure, with 13 specific designated strips of coastline maintained by the Countryside Council of Wales. The National Trust, a charity, owns a sixth of Wales' coast, some 230 km (143 miles).

A succession of peoples, from the prehistoric period to the present, have left their mark on the coastline, and today some 60 per cent of the *circa* 4 million population live on the coast supporting (directly and indirectly) 92,600 jobs, contributing £2.5 billion of GDP to the economy. Overnight visits to the coast brought in Sterling £601 million in 2009, which accounts for 43 per cent of the total tourism sector spending. The bulk of tourism (78 per cent) is seaside leisure oriented and is vital to the economy, the direct impacts covering 3.2 per cent of the whole economy added value in Wales (WAG, 2008). All major cities and many important towns are located on the coast. For example, ports such as Holyhead, Milford Haven, Swansea, Port Talbot, Barry, Newport and Cardiff – the capital – and myriads of small communities are dotted around the coastline, especially in the west.

Settlements in the south grew because of the nineteenth-century coal trade and the bulk of the population is located here. West Wales is essentially rural apart from the giant oil terminal at Milford Haven, around which many small coastal settlements, which started as fishing villages fronting an agricultural hinterland, have now grown. Most of these villages have breakwaters and/or groins associated with their growth, as historically, these forms of construction constituted the raison d'être for fishing village harbours. To the north, most settlements have agglomerated to form linear holiday resort coastal areas stretching from Rhyl to Caernarvon, with an emphasis on seawall construction. These areas owed their success to large-scale leisure tourism, which began in the nineteenth century with railway construction, which made this region accessible to the populous Midland and Lancashire conurbations.

Marine influences

Around the Welsh coastline, offshore wave conditions vary due to differences in local fetch and degree of shelter provided by Ireland to the waves generated further out in the Atlantic Ocean. Nearshore waves are influenced by the offshore wave climate plus nearshore controls provided by changes in bathymetry, islands and headlands (Figure 10.2). The largest wave heights occur between Worm's Head and St David's Head in the south-west where the 10 per cent exceedance significant wave height is 2.5 m. The smallest wave heights occur on the north-facing coast of North Wales where the 10 per cent exceedance wave height is 1 to 1.5 m. (Halcrow, 2002).

Low-pressure atmospheric systems can give rise to increased water levels as they travel over the UK from the south or westerly quadrants and the Irish Sea and Bristol Chanel are particularly influenced by these storm surges. Tidal range varies from a maximum in the Bristol Channel (12 m) to a minimum of around 3 m in the central portions of Cardigan Bay. Tidal currents are generally strongest in high tidal range areas or where there are bathymetric constrictions such as in the Menai Strait (greater than 0.4 m/s) and the Bristol Channel (greater than 0.3 m/s; Halcrow, 2002).

Coastal erosion and tidal flooding risks

The Environment Agency (EA) has carried out sea defence surveys and estimated that coastal erosion occurs along 23 per cent of the Welsh coastline, which is protected by various structures (EA, 1991;

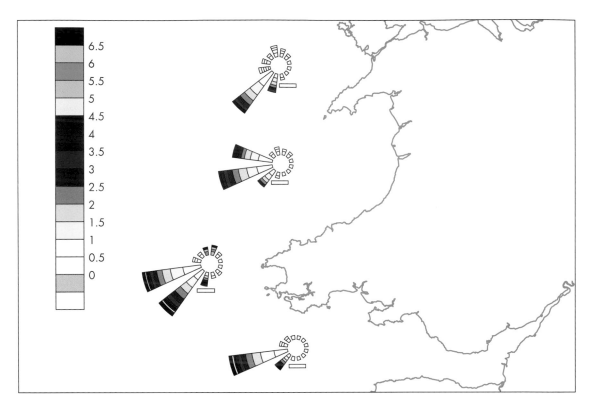

■ **Figure 10.2** *Offshore wave roses for the west coast region. The scale bar associated with each rose represents 10% of the data. Sector length represents the relative frequency of waves from that direction (RF:2,200,000) (Source: © Halcrow Group Ltd, 2002)*

2010). Current structures are estimated to delay erosion of properties valued at £1.5 billion, and some 39,000 properties are at risk from flooding compared with some 9,000 properties at risk from erosion. The National Trust stated that some 75 per cent of the Welsh coastline it owns (66 coastal sites covering 1,572 ha) could be adversely affected over the next century, arguing that policy-makers must plan for future rising sea levels (NT, 2007).

Some 415 km of man-made sea defence structures exist (Figure 10.3) and protect more than Sterling £8 billion of assets from coastal erosion and tidal flooding (WAO, 2009; Masselink and Russell, 2008). Sea defence structures are geared to the flooding problem, coastal protection to erosion.

The replacement costs of current defences, which are maintained by a number of organisations, are esti-

mated to be some Sterling £750 million. Maritime local authorities spend public money managing the potentially damaging effects of coastal erosion. The Environment Agency Wales also builds sea defences and undertakes other work that provides increased protection from tidal inundation in vulnerable low-lying coastal locations. The Agency's power over local authorities is, in the main, discretionary and usually progress is made through collaboration with the Welsh Government, public sector bodies and private stakeholders. This necessitates careful management due to the complex range of issues involved, hence the rise of ICM, which encompasses sectoral approaches so common in the past.

The extent of enhanced erosion due to climate change affecting sea levels and waves is uncertain and the current view is not to build higher defences, as was

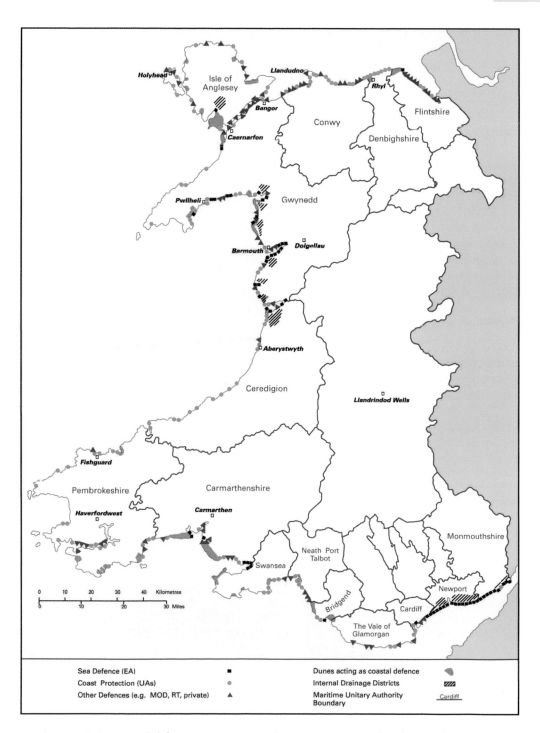

Figure 10.3 *Coastal defence structures in Wales (Figure courtesy of © The Welsh Government)*

common in the past, but to utilize risk management approaches and work with nature wherever possible. Establishment of entities such as the Environment Agency and Welsh Government has facilitated a holistic approach. Shoreline Management Plans, e.g. SMP (1999), SMP & HRW (1999) and Halcrow (2010), have been completed for the coast and policy options identified for future shoreline management. However, it must be stressed that site-specific factors, such as isostatic changes, wave/tide conditions and longshore sediment transport, solid/glacial/moraine/dune geology, as well as anthropogenic impacts, are extremely important in affecting any coastal response to global sea-level rise and long-term data sets are needed on all these parameters. Maintaining flood defences (including rivers) will cost some £135 million a year by 2035, with the Welsh Government providing the bulk of the funds, compared to approximately £44 million in 2010 (EA, 2010).

Coastal protection

Hard engineering approaches

BREAKWATERS, SEAWALLS, JETTIES, REVETMENTS AND GROINS.

A long history exists of these structures – e.g., relics of mediaeval breakwaters occur at Llantwit Major – but most were built in the nineteenth century and

are associated with harbours, both on a large scale, (e.g. Milford Haven, Swansea, Cardiff), and on a smaller scale (e.g. Aberaeron, Solva). Shore-connected breakwaters exist at all these locations. As an example, Pwllheli, located on the Lleyn Peninsula, is an important tourist destination with the harbour and is a centre for tourism.

■ **Figure 10.4** *Abererch frontage with narrow dunes fronting a railway and low-lying land (Source: © Halcrow Group Ltd, 2010)*

■ **Figure 10.5** *Rock revetment backed by dunes at Traeth Crugan (Source: © Halcrow Group Ltd, 2010)*

■ **Figure 10.6** *Jetty arm at Pwllheli harbour entrance (Source: © Halcrow Group Ltd, 2010)*

At Abererch, plastic piling and rock armouring protection has been used to protect the dunes in front of a nineteenth century railway (Figure 10.4) At Traeth Crugan, near Pwllheli, North Wales, a rock revetment has been constructed to protect an eroding dune system (Figure 10.5) in proximity to the Pwllheli harbour modification (lock gates, jetties, causeway, etc.; Figure 10.6).

The coastline of Wales is dotted with groin fields, especially along the western and northern coasts. For example, at Aberaeron,18 groins exist with lengths of ~50 m and a spacing of ~70 m extending to 125 m to the north. At Amroth 22 groins exist; two 75 m in length, two at 16 m, the rest at 36 m and with a spacing of *circa* 43 m. At Newton, Porthcawl, four rock groins constructed east of a central access slipway by the old National River Authority (NRA), were built in 1992 at a cost of Sterling £34,000 and a further three added to the western sea wall in 1994 (Figure 10.7). Their lengths are *circa* 25 m with spacings of: 30 m, 43 m and 57 m, and heights of *circa* 1.5 m above beach level).

With respect to beach users' perception of groins, surprisingly the general opinion was that they were very visible and well liked. Interestingly, surveys showed that groined beaches were perceived to be more attractive than ones without groins, metal groins were unpopular and the preference was for rock groins (Williams *et al.*, 2005). Age, sex, socio-economic status, and whether a visitor or local had no effect on findings.

SOME BRIEF CASE STUDIES:

Newton, located on the eastern edge of Porthcawl town In order to prevent flooding of lower stretches, the UK Environment Agency constructed a rubble mound rip-rap revetment extending for some 250 m adjacent to four rock groins (Figure 10.7; see above section). An adjacent dune system (Merthyr Mawr Warren) represents the eastern edge of soft sediments before the hard rock cliffs of the Glamorgan Heritage Coast. The sand dune system backing the beach is a relic of a much larger system that formally covered much of the South Wales coastline. Sand was blown into the system in the stormy period between 1450 and 1660 (Lamb, 1991). Much sand extraction for building purposes occurred, but ceased in the mid 1950s. Today, the system is well vegetated and very stable apart from a problem with *Hippophae rhamnoides* (sea buckthorn) that has been spreading at a rate of some 2 ha/annum.

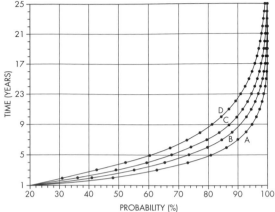

■ **Figure 10.7** *Newton revetment and groins (left) (photograph by Viv Griffiths) Cumulative exceedance probabilities of damage levels (DL) of revetment; A: DL= (0-5)%; B: DL= (10-15)%; C: DL= (20-30)%; D: DL= (40-50)% (right)*

The tidal flow reaches locally up to 3 m/s in front of Porthcawl fishing harbour breakwater where the mean high water spring level and the spring tidal range are 4.6 m AOD and 8.1 m respectively (HN, 1999). This is enhanced by storm surges in excess of $S_S= 1.5$ m (HRW, 1995), the theoretical maximum being 3.4 m. Dominant wave directions are the SW–W sector, which is exposed to Atlantic Ocean swell waves having maximum fetch distance of 6,500 km (IOS, 1980). Extreme value significant wave height and zero crossing periods were determined for a return period of Rp =50 years as Hs=6.1 m and Tz= 8.6 s., irrespective of direction, for an average storm duration of 12 hours (SMP and HRW, 1999).

Several nearshore sand banks, e.g. Scarweather, Hugo and Nash, provide shelter for Newton beach. The significant revetment design wave height was determined as Hs=1.8 m from wave transformation studies which included shoaling, refraction and breaking. The maximum non-storm tidal level was at the rip-rap structure's toe. A reliability-based risk assessment and design (REBAD) analysis utilizing Monte Carlo simulation was used to determine the damage probability. Revetment resistance parameters plus 30,000 trials of simulation characteristics of failure function were carried out which showed it to

be unstable (Balas *et al.,* 2000). Maintenance works are a necessity and periodically, from three per year to once every two/three years (depending upon storm activity, new rubble is placed over the mound by the EA (Pretty, *pers. comm.*). Approximately 500 tonnes of 20–25 cm clean stone is deposited each time, at a cost of some £16 a tonne.

Porthcawl 'paved' revetment   A sloping seawall was constructed in 1887 in this residential coastal town, which has a substantial tourism capacity due to its beaches, sand dunes, fishing harbour and amusement park. The aim was to reflect wave energy, as waves were causing problems with adjacent infrastructure (hotels, homes, road, etc.). Vertical seawalls superseded this seawall in 1906 and 1908, with a buffer-zone promenade being created – the Lower Promenade (Figure 10.8). The outer seawall exacerbated beach lowering, resulting in exposure of the base and pilings in 1932, which necessitated building a newer sea wall 450 m long. A concrete buttress was added to crumbling wall segments in 1942, as the mean high water mark intersected the seawall base. During the next *circa* 30 years erosion wore away the structure so that the base and pilings again became exposed (Figure 10.8). Proposed solutions included insertion of a rock revetment, beach nourishment, and off-shore breakwaters, but the final decision was to design and implement a bitumen-grouted revetment, work commencing in 1984/85. Asphalt was poured over the sand fill surface to create a 15 cm thick asphalt–concrete–cobble layer, the seaward end being a bitumen-grouted toe, 0.7 m thick at bedrock and 0.5 m thick at the asphalt–concrete junction. Bedrock to seawall base height is 3.0 m and the surface coated with a tar spray together with a light chipping veneer (Figure 10.8). Maintenance is cheap and easy and holes are filled before the spring/summer onrush of visitors, as the location is renowned for sun bathing, lying as it does close to the heart of the town. To the knowledge of the authors, this is the first and only location where this technique has been implemented.

Towyn A large-scale protection development (costing some £11 million) has recently been completed. Beach level dropped by more than 3 m over the past century and in 1995 a proposal was made for seven rock fishtail groins, but the project floundered due to political disagreement. In June 2001, Gwynedd Council conceived a plan of placing thousands of tonnes of boulders on the beach, to

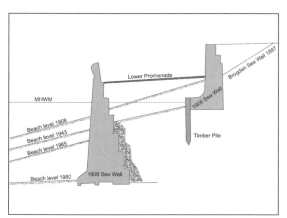

■ **Figure 10.8** *Clockwise from top left: Porthcawl seawall and tarmacadam pavement today – compare with Figure 1.4; tarmacadam smooth pavement junction with cobbles and tar (Photographs by Rob Dickson); Remains of the Brogden (1887) seawall on the lower promenade (Photograph by Viv Griffiths); Seawall constructions through time*

■ **Table 10.2** *Some beach nourishment schemes in Wales (after McCue et al., 2010)*

Date	Location	Circa amount ($10^3 m^3$)
1990s	Llanelli, Llandudno	100–159
2000	Sker near Porthcawl	14
2003	Talacre near Prestatyn	180
2004	Aberavon	5
2007	Aberavon	115
2008	Crymlin Burrows near Neath	220

■ **Figure 10.9** *Towyn: Clockwise from top left: Old groins; construction in progress; breakwater; final new protection structure (Source: Atkins Consulting Engineers)*

reach a height of some 3 m against the current seawall, and 12 m seaward, again abandoned in August 2004. ABPmer, unveiled a new, favourably received, 'Offshore Headlands Scheme', amended in 2009 by Atkins Consulting Engineers, and construction commenced in January 2010. The scheme involved rock structures, 62 per cent coming from San Malo, France, the rest from north Wales; concrete-stepped revetment; and timber groins to replace 34 old dilapidated ones, whose lengths were *circa* 124 m. followed by two short ones of *circa* 70 m, with a spacing of the order of 50 m (Figure 10.9). Rock armour block sizes of 1–3 tonnes (5,000 tonnes), 3–6 tonnes (15,000 tonnes) and 6-10 tonnes (32,000 tonnes) were needed.

Soft engineering approaches

Beach nourishment Beach nourishment schemes in Wales have been on a very minor scale. For example, dredge spoil material from harbour excavation was used to nourish nearby beaches at Pwllheli, Aberystwyth and Aberaeron. In the 1990s, locations such as Llanelli and Llandudno received nourishment as part of coastal protection schemes (Table 10.2.). In 2010, a study was undertaken to assess the feasibility of undertaking nourishment at ten mixed sand-

shingle beaches: Talacre, Abergele-Pensarn, Traeth Crugan, Morfa Duffryn, Broadwater-Towyn-Aberdovey, Tenby North Beach, Port Eynon, North Swansea Bay, Aberavon and Porthcawl (McCue *et al.*, 2010; Winnard *et al.*, 2011).

Tenby North beach is a small (1 km long) bay head beach and the town is a very popular tourist destination (Figure 10.10). The beach is backed by steep cliff slopes, which have a seawall at its foot. To the south, the low tide beach is wide (250 m) with a gentle gradient and due to its easterly orientation, the site is sheltered from the direct force of south-westerly waves. Sediment movement is essentially onshore/offshore with little historical shoreline change occurring since 1887. In time, sea level rise will reduce the width of the available dry beach and consequent tourist amenity. The preferred policy is to 'hold the line' and no record of previous beach nourishment exists.

Beach nourishment would increase the high tide dry beach area, as well as the beach capacity to dissipate storm wave energy – which would reduce seawall maintenance, but excessive nourishment could cause harbour sedimentation problems. The beach width requiring nourishment would vary from 63 to 80 m along the beach, quadrupling the area available for recreation at high tide. The nourishment amount required to achieve an initial profile would be 46×10^3 m^3. The additional volume needed to maintain this profile under future sea level rise conditions could vary from 2×10^3 m^3 over 20 years to 45×10^3 m^3 over 100 years. To maintain the average

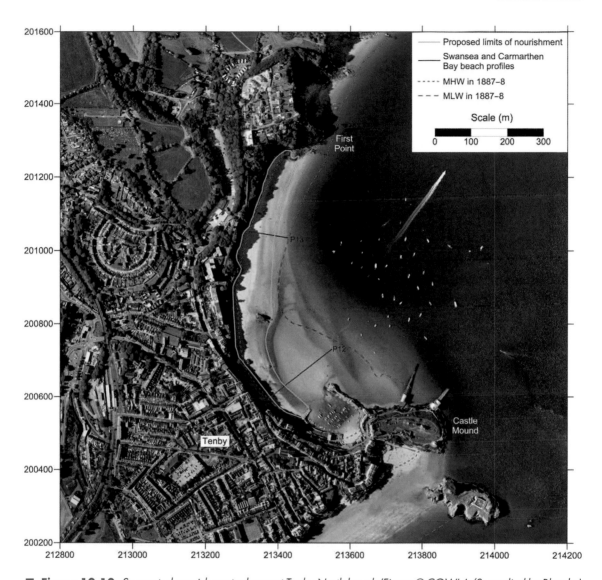

■ **Figure 10.10** *Suggested nourishment scheme at Tenby North beach (Figure © COWI A/S supplied by Bluesky)*

beach level with respect to expected sea level rise in the next 20 years and allowing for 50 per cent loss and subsequent five yearly re-nourishment, would need a sediment volume between 140 and 146×10^3 m^3 depending upon the sea level rise rate.

Dunes Sand dunes systems in Wales cover more than 8,150 ha (Dargie, 1995) and 11 systems have

Special Area of Conservation (SAC) status (Table 10.3). In historical times many systems have been destroyed or damaged through afforestation policies, intensive recreational usage or urban and industrial development e.g. at Rhyl and Prestatyn (Williams and Davies, 2001). In the past 50 years Welsh dune systems have become increasingly stabilized by vegetation through a combination of natural processes,

■ **Table 10.3** *Sea defences and stabilisation measures in some selected Welsh dune systems > 50 ha. (after: Williams and Davies, 2001; Pye et al, 2007)*

SITE	Area (ha)	Stability	Defense	NNR	SAC	GCR	HER	NP	SSSI
Merthyr Mawr	342	6	b,f	x	x		x		x
Kenfig Burrows	602			x	x				x
Baglan Bay	78	1	e,g						
Crymlyn Burrows	118	1	e,g						x
Pennard, Penmaen & Nicholaston ***	87						x		x
Oxwich Burrows***	93		a	x		x	x		x
Hillend to Hills Tor Burrows	224		g						
Whiteford Burrows***	142	6		x	x	x	x		x
Pembrey Burrows	591	1,6	a,e,g			x			x
Laugharne Burrows	431	1	e,g		x	x			x
Pendine Burrows	173	1			x	x			x
Tenby Burrows	92		a,g				x	x	x
Stackpole Warren	179	5,6		x	x		x	x	x
Brownslade & Linney Burrows	253	5,6			x				x
Broomhill Burrows	183				x		x	x	x
Ynyslas*	68		f	x		x			
Tywyn to Aberdovey	111	1,4	a,b,d,g					x	x
Morfa Dyffryn	313	1,3		x	x	x		x	x
Morfa Harlech	341	1,3		x	x	x		x	x
Morfa Bychan**	169	1							x
Morfa Dinlle	67				x	x			x
Newborough Warren***	529	6	g	x	x	x			x
Aberffraw***	248				x	x			x
Valley Airfield	192								
Conwy & Deganwy Dunes	75	3	d,g						
Rhyl to Prestatyn	53	3							
Gronant to Talacre	190		d,g			x			x

Abbreviations

SSSI	Site of Special Scientific Interest	a	Sea wall	1	Fencing	*a Biosphere Reserve
NNR	National Nature Reserve	b	Revetment	2	Xmas trees	and Ramsar Site
SAC	Special Area of Conservation	c	Gabions	3	Marram	** is the sole LNR
LNR	Local Nature Reserve	d	Groins	4	Contouring	*** an AONB
AONB	Area of Outstanding Natural Beauty	e	Breakwater	5	Thatching/binding	
HER	Heritage Coast	f	Fencing	6	Afforestation	
NP	National Park	g	Rock Armour			
GCR	Geological Review Sites					

changes in grazing regime and over-management. Fixed dunes are replacing mobile dune forms and there has been extensive scrub encroachment. This is an issue of major concern to the Countryside Council for Wales and a number of trial schemes are now being implemented in an attempt to reverse the trend (Pye and Blott, 2012).

The Countryside Council for Wales, Local Authorities, National Trust, National Parks, Ministry of Defence and private owners all make contributions to dune management. Many systems also cross county boundaries so proactive measures are sometimes difficult to implement, especially as many systems are in an erosion phase (Davies, 2001). Dargie (1995) indicated that at least 27 out of 45 systems studied are subject to frontal dune erosion. Pye and Saye (2005) using a sea level rise of 0.41 m by 2100 (median IPCC prediction) suggest that areas, such as, Morfa Duffryn, Newborough, Whiteford, and Kenfig are likely to experience significant dune loss, whilst areas where sediment supply rates are probably going to remain high are likely to prograde, e.g. Pendine, Morfa Harlech, and Ynyslas (Table 10.3). Rock armouring appears to be the main protection utilized (Table 10.3). Christmas trees are used at, for example, the Talacre-Gronant system and seawalls tend to protect a host of small systems along the northern coast. Today, most systems are heavily vegetated which restrict landward migration and many have golf courses built within them, e.g. Kenfig.

There is some uncertainty regarding dune evolution with future climate change, especially a temperature rise, which can affect vegetation e.g. *Leymus arenarius*. Most grasses are thermophilic and more stabilization is likely to result from any future rise in temperature. To maintain early successional phases some 20 per cent of bare dune sand is needed but this runs counter to flood protection schemes. Counter-stabilization measures include grazing – sheep, cattle, e.g. Merthyr Mawr; encouraging rabbits, mowing, reactivation of blow-outs, mechanical ploughing, e.g. Talacre; dune slack re-invigoration, e.g. Kenfig, Pendine. However, grey dunes have Priority Habitat status, given by the EC Habitat and Species Directive and are therefore protected on some Special Areas of Conservation sites, so that any disruption could breach article 6 of EC law.

Scotland

Introduction

Just as Italy is the 'most coastal country' in the Mediterranean Sea, Scotland can readily lay claim to that title for the component countries of the United Kingdom and arguably for the whole of the European continent. For a relatively small country in terms of surface area, Scotland has a very long, complex and diverse coastline. The total length has apparently increased over the years, an artefact of the continually improving accuracy by which it can be measured and the scale of maps used as the basis for measurement. An early estimate derived digitally from 1:63,360 scale Ordnance Survey maps, which omitted the vast numbers of islands, located principally off the western and northern coasts, indicated the length of mainland Scotland to be 5,340 +/- 20 km (Baugh and Boreham, 1976). Subsequently, inclusion of the island coastlines has resulted in length estimates ranging from 'over 11,000 km' (Scottish Executive, 2005), c.11,800 km, and as such representing over 8 per cent of the coastline of Europe (Doody, 1999), to 16,490 km. The most recent measurement by Scottish Natural Heritage, based on analysis of the Ordnance Survey's 1:10,000 coastal outline using ArcGIS, is, however, 18,670 km. Thus, Scotland's coastline is equivalent to approximately one-eighth of the European total according to the most recent estimate.

In general, the pressures leading to coastal defence in Scotland have historically been far less than in England. Although around one-fifth of Scotland's population live within 1 km of the coast and about 70 per cent within 10 km (Scottish Executive, 2005), the low total population (*circa* 5 million people) relative to the area of the landmass results in many areas of essentially undeveloped coastline, especially in the north and west. Of the country's estimated 790 islands (Scottish Executive, 2003), only 130 are inhabited (Scottish Executive, 2005). As a consequence of the generally much lower population density than south of the border, combined with typically more durable rock types characterizing the coast, only about 6 per cent of Scotland's coastlines are defended compared with some 44 per cent of those in England and Wales (DEFRA, 2001). Nevertheless there are several important 'hotspots' of erosion around Scotland's fringes, many of which have been aggravated by human intervention, and overall it has been estimated recently that some 12 per cent of the country's coastline is subject to erosion (Baxter *et al.*, 2008). The latter value is a substantial increase compared with that of 8 per cent estimated two decades previously by Ritchie and McLean (1988). Similarly, at that time, Carr (1988) noted that some form of artificial structure protected approximately 20 per cent of the English coast.

Coastal geology and geomorphology

Three major SW–NE trending crustal fractures, the largest faults in the UK, sub-divide Scotland into four geological provinces: from north to south; the Northern Highlands, the Grampian Highlands, the Midland Valley and the Southern Uplands, the first of which is bisected by the Moine Thrust (Figure 10.11). The oldest rocks in the country, comprising the Precambrian age Lewisian Gneisses and the overlying Torridonian Sandstones, crop out to the north and west of the Moine Thrust, forming much of the mainland coastal fringe along with the Outer Hebridean islands. In very general terms, the rocks become younger in age to the south and east of the country. The bulk of the Northern Highlands and Grampian Highlands comprise mainly Precambrian metamorphic rocks, with Moine and Dalradian Schists dominating south of the Great Glen Fault. Exceptions are the Orkney Islands and Caithness, in the north eastern extremity of the mainland. Here, sedimentary rocks of Devonian age, sandstones and siltstones, form spectacular vertical cliffs. South of Caithness, the coast is formed of younger Jurassic formations followed by sandstones, primarily of Devonian age, around the shores of the inner Moray Firth. The southern shore of the outer part of this major firth is fringed with a variety of Precambrian metamorphic formations – schists and quartzites – that continue to crop out as far south as the eastern end of Highland Boundary Fault. To the south of this fracture, the northern part of the Midland Valley is dominated by sedimentary formations of Devonian age, which comprise mainly conglomerates, and sandstones with contemporaneous lavas. The southern section of the Midland Valley is dominated by sedimentary rocks of Carboniferous age including sandstones, mudstones,

shales, limestones and coal measures, along with local lavas and igneous intrusions. The southern part of Scotland, the Southern Uplands, to the south of the Southern Upland Fault, consists of tightly folded, layered rocks of the Ordovician and Silurian, comprising mainly shales and greywackes. Permian and Triassic sandstones and mudstones underlie the inner part of the Solway Firth. By contrast with the Outer Hebrides, the Inner Hebrides are formed largely from a wide range of igneous rock varieties, volcanic and intrusive, of Tertiary age – in sharp contrast to the unconsolidated sedimentary rocks of the same age that occur in south-east England.

This geological variation, overprinted by the impact of repeated glacial activity during the Pleistocene, has a profound influence on the nature and morphodynamics of Scotland's coasts. In the north and west the characteristically rocky coast is deeply embayed and indented with numerous structurally controlled NW–SE trending sea lochs and islands. The lochs and the narrow straits between the islands and islets are the legacy of glacial erosion, exploiting the structural grain to produce over-deepened troughs which became flooded as relative sea levels rose when deglaciation commenced some 18,000 years ago. By contrast, the generally softer rock types that characterize the eastern fringes have given rise to a much simpler coastal planform, with numerous depositional forms, and three major estuaries: the Moray Firth, the Tay Estuary and the Firth of Forth (Figure 10.11). The highly complex, indented west coast results in a low potential for longshore sediment exchange between adjacent coastal sections. In the east, however, the degree of sediment interconnectivity is much higher owing to the much simpler coastal morphology that results in far longer, continuous sections of soft materials that facilitate sediment exchange (Hansom *et al.*, 2004). Pressures on the Scottish coast and the need to install defences are far from uniform and are most acute where major conurbations coincide with the coasts of the Midland Valley as in the Firths of Clyde in the west and the Forth in the east.

Global sea level rise by 2090 (under the IPCC SRES A1B Scenario) is projected at *circa* 4 mm/yr.

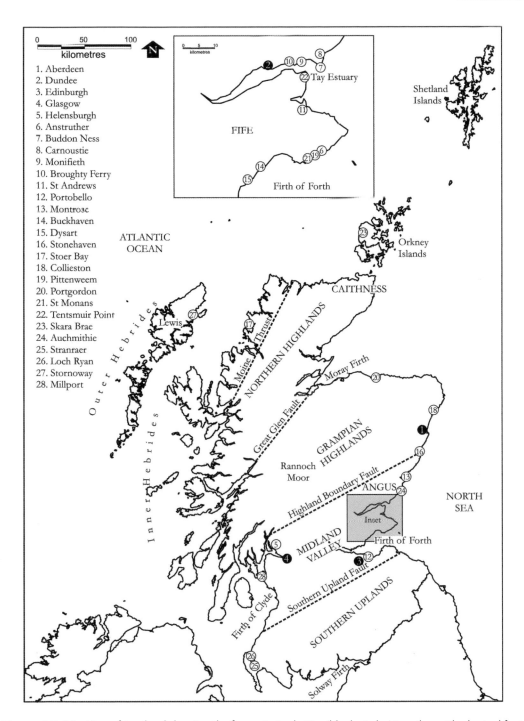

1. Aberdeen
2. Dundee
3. Edinburgh
4. Glasgow
5. Helensburgh
6. Anstruther
7. Buddon Ness
8. Carnoustie
9. Monifieth
10. Broughty Ferry
11. St Andrews
12. Portobello
13. Montrose
14. Buckhaven
15. Dysart
16. Stonehaven
17. Stoer Bay
18. Collieston
19. Pittenweem
20. Portgordon
21. St Monans
22. Tentsmuir Point
23. Skara Brae
24. Auchmithie
25. Stranraer
26. Loch Ryan
27. Stornoway
28. Millport

■ **Figure 10.11** *Map of Scotland showing the four principal cities (black circles) together with physical features and locations referred to in the text (Figure courtesy of the School of the Environment, University of Dundee)*

However, in Scotland eustatic sea level rise is, in some parts, moderated by isostatic uplift of the land following the melting of the last Scottish ice sheet around 10,000 years ago. Such uplift is greatest where the ice attained its maximum thickness over the Rannoch Moor area of the western Grampian Highlands (Figure 10.11). Away from this principal centre of rebound, especially in the Outer Hebrides, Caithness, Orkney and Shetland, relative sea level is rising (Shennan and Horton, 2002; Ball *et al.*, 2008) and erosion is particularly prevalent. However, it is by no means restricted to these regions (Lees *et al.*, 1998). Future sea level rise (by the 2080s), incorporating best estimates of glacio–isostatic adjustment coupled with a 4 mm/yr eustatic sea level rise, yield, for example, relative sea levels 0.2 m higher in the Firth of Clyde, 0.24 m higher in the Solway Firth, 0.28 m higher in the Moray Firth and over 0.32 m higher in the Orkney and Shetland Islands (Ball *et al.*, 2008). These estimates of increased flood risk attributed to relative sea level rise are lower than that caused by surges and wave action. However, no significant evidence of a trend towards increased storminess has been detected in analyses of mean annual surges at Aberdeen, Stornoway and Millport (Ball *et al.*, 2008).

Series of raised shorelines, also referred to as raised beaches, are commonly developed on both the western and eastern isostatically uplifted coasts of the Scottish mainland and around the islands of the Inner Hebrides. In the Outer Hebrides, the far north of the country and in the Orkney and Shetland Islands such features are absent, reflecting relative sea level rise at locations most distant from the zone of maximal ice thickness. Many raised shorelines in remote western locations are completely undeveloped and unprotected, typically serving as near-horizontal surfaces on which roads and tracks have long been established (Figure 10.12). Where coastal settlements have been developed in such terrain, the raised terraces provide the horizontal foundations for dwellings at the different levels above the high water mark and the lines of houses themselves often serve to highlight the form of the sequentially raised, step and stair, coastline above (Figure 10.13). Typically, wall or rock armour defends the lower shoreline. Where raised shorelines are sufficiently extensive they form the ideal topography, especially when capped with undulating dunes, for links golf courses, as at the world-famous St Andrews links, for caravan sites, and for sheep and cattle grazing on machair grasslands.

■ **Figure 10.12** *Raised shoreline and associated lagoon developed in Precambrian terrain at Stoer Bay on the NW coast of Scotland*

■ **Figure 10.13** *The village of Collieston on the Aberdeenshire coast developed on a series of raised shoreline terraces, the lowermost of which is defended by a masonry wall*

Hard engineering

Virtually all of Scotland's coastal towns and villages have some form of hard-engineered coastal protection that typically fronts onto a promenade, a road or a railway formation. In consequence, basal scour at the toe has led in many instances to beach draw-down as, for example, at Helensburgh on the Firth of Clyde coast. Many walls were typically constructed from locally sourced materials thus exacerbating the problem of beach draw-down and basal scour. All types of sea walls are represented in the country, from vertical masonry structures to inclined, stepped and curving designs constructed of concrete (e.g. Figure 10.14) or as gabions. Energy absorbing rock armour revetments have seen increasing usage since the mid-1970s. Vertical masonry walls dating back to the 1800s typically front old-established fishing settlements (Figure 10.15) often showing a patchwork of repairs made over the decades. Large portions of a late eighteenth-century sandstone sea wall at Anstruther in Fife and two corner walls of a house above were severely damaged in a violent storm on 30 March 2010 (Figure 10.16). In many such situations flanking erosion, along with the starvation of longshore sediment transport, is a localized problem that has typically been tackled by progressively extending walls, adding sections of rock armour or concrete rubble in an attempt to prevent the impacts of wave attack (Figure 10.17).

■ **Figure 10.14** *Gravel beach at Stonehaven backed by a stepped concrete sea wall with upper, curved, wave reflecting section that protects a promenade and car park*

■ **Figure 10.15** *The fishing village of Pittenweem, Fife, the frontage of which is protected by a vertical masonry wall dating back to the eighteenth century*

■ **Figure 10.16** *Erosion damage at Anstruther, Fife, caused by the storm of 30 March 2010. Repairs to the sea wall have begun*

■ **Figure 10.17** *Flanking erosion and attempts to arrest it at the downdrift (western) end of the low masonry sea wall that protects the frontage of the small, former fishing village of Portgordon on the southern shore of the Moray Firth*

In many areas, the piecemeal approach to coastal erosion management, which for decades was so typical in Scotland, has led to the installation of unsightly defences (e.g. Figure 10.18). At Skara Brae in the Orkney Islands, Europe's most complete Neolithic village, which has been accorded UNESCO World Heritage Site status, is under threat from flanking erosion at the ends of the sea wall built initially in the 1920s to protect it. Increased erosion along the shores of Loch Ryan in south-west Scotland, at the head of which is located the port of Stranraer, has been attributed to the wash created by the new breed of fast ferries that are now deployed on the route to Northern Ireland. Starvation of longshore sediment transport has occurred in several places owing to the installation of groins and many such structures today lie semi-derelict, as at Broughty Ferry and Monifieth. Harbour walls and jetties have unintentionally interrupted longshore sediment transport rates at numerous localities around the country (e.g. Figure 10.19).

■ **Figure 10.18**
Fourteenth-century church at St. Monans, Fife, said to be the closest to the sea in Scotland, protected by a high, vertical masonry wall. The nearby coastal footpath has been defended over recent years by a series of unsightly, ad hoc defences; gabions, sleepers, rock armour and concrete

■ **Figure 10.19**
The old harbour wall at Auchmithie, dating back to the 1890s, on the Angus coast is acting like a large groin, trapping gravel on the updrift side. As a result, an erosional bight has been formed on the opposite side of the wall and this has now scoured around the landward end of the structure

All of the major firths and estuaries have been subject to land claim, particularly during the nineteenth century. In the Firth of Forth, for instance, over 50 per cent of the intertidal mudflats and salt marshes have been progressively isolated from the tidal prism to create agricultural and industrial sites bounded by earth embankments (dykes) with cores of impermeable clay. These were faced on the water side with cut stone blocks to prevent wave erosion. The practice of land claim for industry continued into the 1920s, often using nearby colliery and oil shale mine waste to raise the land level behind the retaining embankments. Land claimed for industry included large areas of the foreshores on the southern bank, which now host not only the largest container port in Scotland but also the adjacent oil refinery that is Scotland's main fuel supplier. Virtually the whole of the inner part of the Firth of Clyde upstream to Glasgow, famous for its ship building industry, has been hard engineered and canalized to improve navigation.

Unlike England, Scotland has relatively few very extensive reaches of hard engineered defences along its open coasts. One of the largest and most influential was a 3 km long rock armour revetment built at a cost of Sterling £3 million by the Ministry of Defence (Hansom, 1999) in 1992–93 to protect the east-facing side of the promontory of Buddon Ness, on which the Barry Links military training area has long been located. This massive, visually intrusive structure (Figure 10.20), a continuation southward of an earlier defence near Carnoustie, has served to protect the promontory from wave attack, but in so doing has cut off what is a natural source of beach

■ **Figure 10.20** *Part of the largest extent of rock armour revetment in Scotland that protects the eastern side of the promontory of Buddon Ness, looking towards the town of Carnoustie, Angus. Interchange of sand between the dunes and the beach is now inhibited and the dunes in the middle distance are degrading. This defence has caused depletion in the amount of sediment entering the Tay Estuary on the north side, starving beaches further to the west of their natural sand supply*

sand for the Tay Estuary. Longshore currents on the north side naturally transport sand from east to west around Buddon Ness and into the Tay feeding the beaches downdrift at Monifieth and Broughty Ferry. These sites are now starved of their natural sediment supply, which has contributed to the necessity to install further defences since these sites have become more vulnerable to marine erosion.

Soft engineering

It was not until the 1980s that more enlightened approaches to erosion control and coastal zone management in general were accepted and became adopted in Scotland. The planting of eroding dunes with marram grass (*Ammophila arenaria*) and lyme grass (*Leymus arenarius*) together with the placement of geotextile or jute membranes and with prevention of public access by the erection of fencing are now common occurrences. A recent example (2010) of such coastal stabilization is provided by a section of restored dunes of the West Sands of St Andrews, which are not only subject to wave attack from the North Sea but also, as a Blue Flag Beach in a popular tourist destination, to very considerable recreational pressures during the summer months. Beach recharge schemes, however, are comparatively rare in Scotland with one of the earliest and most successful being at Portobello, on the eastern side of Edinburgh, which dates back to the early 1970s. This utilizes sand dredged from offshore which is sourced from a site 3 km to the east, ferried to Portobello by barges and then pumped onshore. In common with many other British schemes, Portobello beach needs to have its sand replenished roughly every ten years. However, its original deterioration was due ironically to direct extraction of sand from the beach to feed the local glass bottle industry between 1824 and the mid-1930s. By 1970, the promenade at the beach crest was damaged by wave erosion and the buildings to landward periodically flooded (DEFRA, 2005). This periodic recharge has successfully restored an important amenity and recreational asset for the country's capital city.

Brief case studies

Buckhaven to Dysart

For nearly 800 years, coal mining was an important industry in Fife, in eastern Scotland. Its decline demonstrates an interesting perspective on the long-term dumping of collicry waste along the shore. Along the NE to SW trending stretch of coast between Buckhaven and Dysart, the custom for centuries was to tip large amounts of mine waste directly onto the beaches and rocky foreshore, locally building headlands of loose fragments of coal, shale and sandstone that were then reworked along the coastline by prevailing longshore currents. Such was the extent of dumping that, along this stretch, the coastline moved seaward by 100 to 170 m (Saiu *et al.*, 1994). Colliery waste migrated north-eastwards and gradually infilled the harbour at Buckhaven; thus by 1946 this fishing harbour had to be abandoned owing to siltation. This practice of dumping continued until the closure of the last of the three deep mines in the area; the Wellesley in 1967, the Michael in 1968 and the Frances in 1984. As a result, this progressive reduction of mine waste discharge to the coast led to a greatly improved beach amenity in terms of aesthetics and ecology. However, the dumping of waste had another and more positive effect, neither realised nor appreciated until the practice ceased. A very important consequence was the reduction in material available for transport by the longshore current system of the area, a source that had been depleted when the dumping stopped. This, in turn, led to greatly increased erosion of the underlying natural beach and Carboniferous age rock materials, which was exacerbated by coastal subsidence due to the long history of coal mining in the area (Saiu *et al.*, 1994). The response to the enhanced erosion was the installation of substantial sections of rock armour to provide coastal protection. This is a prime example of an environmentally unfriendly practice imperceptibly protecting a section of coast from erosion for centuries.

■ **Figure 10.21** *Oblique aerial photograph over Tentsmuir Point across the outer Tay Estuary to Dundee. The line immediately to seaward of the forest edge marks the line of World War II anti-tank blocks placed on a high water mark in 1940 to the east of which accretion has taken place. By contrast, the area in the foreground (downdrift) is eroding and the line of blocks is now below High Water Mark (Photograph © Aerial Photography Solutions)*

Tentsmuir Point

Relative to a line of concrete anti-tank blocks placed at the High Water Mark in 1940 in an attempt to prevent an enemy invasion, Tentsmuir Point in Fife has advanced, according to location, from 7 to as much as 14 m per annum (Figure 10.21). Since the first accurate mapping of the area in 1812, what is now a National Nature Reserve has been accreting in a north-eastward direction at a long-term average rate of *circa* 5 m per year and is thus one of the most rapidly accreting parts of the British coast. In this area, the dominant direction of longshore sediment transport is from south to north towards the mouth of the Tay Estuary. To the south, as the natural supply of sediment becomes depleted, the undeveloped coast is eroding and the concrete blocks lie detached, some displaced by wave erosion, now located several tens of metres seaward from the high water mark. Overall, from 1812 to 1990, the land at Tentsmuir Point advanced around 870 m in a north-eastward direction and eroded approximately the same distance inland in the south (Whittington, 1996).

Discussion

Wide varieties of rock types and structures, differing exposure to wave activity in a range of relative sea level change scenarios and variations in sediment supply have resulted in a broad range of coastal landforms and coastal processes in Scotland. Although coastal erosion is not as extensive as in England, many of Scotland's beach-dune systems are actively receding and there are growing pressures along many soft coasts in the country. It was noted in the late 1990s that "No strategic approaches to coastal defence in Scotland, nationally or regionally, have been adopted or promoted to date and consequently protective work has proliferated in an *ad hoc* and unregulated fashion . . . ." (Lees *et al.*, 1998). This has been the catalyst for many of the erosion problems of today. Since that time, however, management of the Scottish coast has evolved through the formation of seven voluntary Local Coastal Partnerships (LCPs), such as the Tay Estuary Forum (Burningham *et al.*, 2000; Booth and Duck, 2010), and the implementation of Shoreline Management Plans (SMPs) on the basis of sediment cells.

HR Wallingford (1997) introduced the concept of coastal cells and sub-cells to Scotland after their initial development in England. Each of the 11 cells and 40 sub-cells was analysed by Hansom *et al.* (2004) with regard to their internal coarse sediment interconnectivity and thus their suitability for the SMP approach and it was found that only four were suitable. The deeply indented west coast, in contrast to the much simpler planform in the east, is largely unsuited to sub-division into sediment cells. Non-statutory SMPs were first conceived and introduced in England and Wales in 1993. Their aim is to provide a strategic framework for decision making along the coast, especially with respect to defence, taking account of the natural coastal processes, human and other environmental influences and needs (Environment Agency). Today the whole length of the English and Welsh coast is covered by such plans, some in their second generation. Only eight exist for Scotland, however, of which the most fully developed are those for the coasts of Angus, Fife and the northern (*i.e.* Scottish) shore of the Solway Firth. Each plan evaluates the natural processes that are acting on a length of shoreline and predicts, as far as possible, the way in which it will evolve into the future. The principal issues of concern relating to coastal erosion and flooding are determined, along with the ways in which the natural processes are managed and identification of coastal assets that may be affected by erosion or the current management practices. As such, each plan must take account of the potential impact of present and future coastal defence schemes, hard or soft engineered, on the natural environment and the likely environmental, financial and social costs involved. The Marine (Scotland) Act was approved in 2010. It differs from the UK Marine and Coastal Access Act in that it provides for the delineation of marine regions and the ability to delegate marine planning to a local level. At the time of writing, the alternative proposals for the boundaries of the 'Scottish Marine Regions' are under consultation (Scottish Government, 2010). This new Scottish legislation, however, will bring many challenges; not least how to join marine spatial planning in whatever regions that emerge from the consultation with management of the country's highly complex coastal zone for which there is an incomplete set of both LCPs and SMPs. Initially, at least, it is suggested that this aspect of the Marine (Scotland) Act could lead to far more problems than it does solutions.

England

Introduction

The length of coastline, including the Isle of Wight, and the Scilly Isles, is approximately 10,000 km (as defined on Ordnance Survey 1:10,000 scale maps up to the normal tidal limit within estuaries). Some official documents have suggested that approximately 44 per cent of the coast is protected by some form of structure (e.g. DEFRA, 2010), including flood embankments along tidal rivers and estuaries and 'offshore' structures such as training walls. The proportion of the 'open' coast which is protected by structures is much lower, closer to 15 per cent overall, although up-to-date definitive figures are not available. The highest proportion of defended coast is found in southern and eastern England, which is relatively low-lying and heavily populated. The lowest proportions of defended coast are found in north-west and north-east England, where there is more high ground and a lower population density.

The primary purpose of many structures is to provide protection against coastal erosion, others are present mainly to provide defence against sea flooding, and some perform a dual role. Under the 1949 Coast Protection Act, the responsibility for coast protection and sea defence was placed in the hands of a number of Operating Authorities, many of which were Maritime District Councils. Responsibility for some sections of the open coast and estuarine tidal areas at risk of flooding was initially given to internal drainage boards and subsequently transferred to the Environment Agency, which, in 2008, was also given overall responsibility for Sea Flooding and Coastal Erosion Risk Management. Overall guidance on coastal defence policy at the regional and sub-regional scale is provided by a series of Shoreline Management Plans. Coastal Strategy Studies prepared by the Operating Authorities provide more specific frameworks for shoreline management at the sub-regional scale. All applications for new coastal defence structures are now judged against national government priorities and a series of 'outcomes' measures. Over the course of the past decade, there has been a move away from the construction of new 'hard' defences except in areas where there is a requirement to protect land or infrastructure of high asset value, principally, but not exclusively, in urban and industrial areas. Even in such areas, there has been increasing use of 'soft' engineering methods in beach and dune management.

Patterns and rates of erosion

The main areas at risk of erosion are those where the coast is formed by soft cliffs or unconsolidated sediments. Soft 'rock' cliffs occur in many places in eastern and southern England, notably in east Yorkshire and Humberside, East Anglia, Dorset and east Devon. Some of the cliffs are composed geologically of weakly consolidated late Mesozoic and Cenozoic rocks, while others are composed of unconsolidated Quaternary sediments, including glacial till and outwash deposits. A classic example is provided by the Holderness coast in the East Riding of Yorkshire and North Humberside. In this area, glacial till and outwash deposits form cliffs up to 30 m high. Analysis of historical maps and aerial photographs has demonstrated that parts of the coast have been eroded by more than 300 m over the last 150 years, at an average rate of *circa* 2 m per year (Valentin, 1971). At any given location, erosion tends to be episodic, with periods of stability lasting up to ten years interspersed with periods of rapid cliff recession. This episodic behaviour is partly due to the movement of oblique longshore bar and trough features, locally known as 'ords' (Pringle, 1985). Where the ends of the troughs (runnels) join the coastline, beach levels are low and waves are frequently able to break against the cliff toe. On the other hand, where the ridges join the coastline the upper beach levels are high and waves break well before they reach the cliff foot. The trough and bar features show a long-term tendency to move southwards along the coast at an average rate of 400 to 500 m/yr. The Holderness cliffs are mostly composed of fine sand and mud grade material, with only about 1 per cent gravel, although there are localized occurrences of fluvio-glacial gravelly sands and windblown sands. Much of the fine sand and mud released by cliff erosion is dispersed into the waters of the southern North Sea, but the medium to coarse sand and gravel fractions are retained within the beach and nearshore zones. Longshore drift occurs towards the south, leading to net accumulation in the relatively lower energy nearshore area between Kilnsea and Spurn Head. Man-made defences, mainly around the towns of Bridlington, Hornsea, Mappleton, Withernsea and Easington (Table 10.4), protect approximately 13 per cent of the coastline between Flamborough Head and Spurn Point.

■ **Table 10.4** *Summary of defended and undefended lengths of frontage along the coast of the East Riding of Yorkshire (modified after East Riding of Yorkshire Council (ERYC); source: www.eastriding.gov.uk/coastalexplorer)*

Location	Undefended cliff line	Defended frontage (ERYC)	Defended frontage (private)	Frontage type
Bempton to Bridlington	18.29 km			High chalk cliffs
Bridlington		3.60 km		Masonry and concrete sea walls with groins
Bridlington to Barmston	5.62 km			Low clay cliffs
Barmston private defences			0.13 km	Rock and concrete armour revetment
Cliffs south to Barmston drain	0.62 km			Low clay cliffs
Barmston drain defences			0.20 km	Rock and concrete armour revetment
Barmston to Ulrome	1.47 km			Clay cliffs
Ulrome north defences			0.35 km	Concrete sea walls
Ulrome cliffs	0.20 km			Clay cliffs
Ulrome south defences			0.09 km	Concrete sea walls
Ulrome to Hornsea	8.26 km			Variable height clay cliffs
Hornsea		1.86 km		Concrete sea walls, groins, rock armour
Hornsea to Mappleton	3.10 km			High clay cliffs
Mappleton		0.45 km		Rock armour revetment with rock groins
Mappleton to Tunstall	15.32 km			High clay cliffs
Tunstall north defences			0.18 km	Rock and concrete armouring
Tunstall cliffs	0.12 km			Low clay cliffs
Tunstall south defences			0.14 km	Rock armour revetment
Tunstall to Withernsea	2.81 km			Variable height clay cliffs
Withernsea		2.26 km		Concrete sea walls, groins, rock armour
Withernsea to Easington	8.38 km			High clay cliffs
Easington defences		1.03 km		Rock armour revetment
Easington to Spurn	5.67 km			Variable height clay cliffs
Spurn defences (derelict)			1.06 km	Concrete sea walls, groins, rock armour
Spurn dunes to Spurn Point	3.25 km			Low clay cliffs and sand dunes
Totals:	73.11 km	9.20 km	2.15 km	
Total length of coastline	85 km			

Other unconsolidated sediments of late Quaternary to Holocene age which are potentially susceptible to erosion include beach ridge plains composed of sand and gravel, sand dune systems, and saltmarshes. Risks of erosion are greatest where such 'soft' sediments are located on the open coast and are exposed to large wave fetch distances. At Formby Point, located just north of Liverpool, the coastal dunes have experienced up to 800 m of erosion since 1906, following a period of rapid seaward progradation during the second half of the nineteenth century (Pye and Neal, 1994). Erosion occurs mainly during severe winter storms, when relatively large waves from the west and northwest coincide with high tides. Storm surges in the eastern Irish Sea frequently raise tidal levels by 1 to 2 m, and under such conditions, waves may break directly against the dune cliff. Recession of up to 14 m during a single storm event has been recorded. Long-term average erosion rates in this area lie in the range of 2 to 3 m per year (Pye and Blott, 2008). However, such relatively high rates are uncommon in England. At most other dune sites, including the Suffolk coast near Sizewell) average rates of frontal dune erosion of 1 m are more typical (Pye *et al.*, 2007).

Types of coastal structures

Most coastal defence structures in England are constructed of wood, rock, concrete or a combination of these. Steel sheet piling, old tyres and other materials have also been used on a localised basis. Different types of defences have both advantages and disadvantages in terms of cost, effectiveness and visual attractiveness (Table 10.5). Wooden structures are relatively cheap to build but tend to have limited durability (<20 years). Many different types of wooden structure have been used, including shore-parallel wooden revetments, wooden fences and wooden groins.

The shore-parallel revetments are normally erected on the mid foreshore and are designed to reduce, though not to entirely prevent, wave erosion of the soft cliffs behind. Some sediment input from the cliffs, and longshore movement along the upper beach, is maintained. Examples are found between Cromer and Happisburgh, in north Norfolk, and near Corton, in north Suffolk (Figure 10.22). Wooden fences have been widely used to trap sand and mud, and occasionally shingle, on the upper beach or foreshore. Examples are found near Lepe, on the Solent, and along the Dengie Peninsula, in Essex (Figure 10.22b). Wooden fences have also been widely used to trap windblown sand on the backshore in order to build a wider frontal dune system. Numerous examples are found in Cornwall, Devon, and Northwest England (Pye *et al.*, 2007; Figure 10.22c). Shore-normal wooden groins, designed to restrict the rate of longshore sediment transport, were once common around the English coast but in recent years have tended to be replaced by rock groins, which have greater durability. However, many examples of wooden groins are still to be found, as near Aldeburgh, on the Suffolk coast (Figure 10.22d).

Rock has been used in many different ways. Massive rock armour revetments have been built in a large number of locations, including Hurst Castle Spit in Dorset (Figure 10.23a). Rock has also been used to construct shore-normal groins, fish-tail groins, and shore-parallel offshore breakwaters (Figure 10.23b). The size and type of rock chosen is based on wave energy conditions, local availability and transport costs. Large blocks of hard, durable igneous or metamorphic rock weighing 40 tonnes or more are used in the most exposed settings (Figure 10.24a). In lower energy settings, cobble- or small-boulder-sized rock is often used to form artificial cobble beaches (Figure 10.24b) and to fill wire basket gabions (Figure 10.24c). The rock material may derive from quarry sources or be waste demolition material including bricks and concrete. Finer-grained gravel and sand materials are often placed inside geotextile bags (Figure 10.24d) or geotextile-line gabions such as Hesco concertainer™ gabions. These have the advantage that they can be used to 'build' protective walls and other 'temporary' structures that can be readily removed when the requirement ends.

The Happisburgh to Winterton scheme in Northeast Norfolk (Fleming and Hamer, 2000) provides an example of the use of rock offshore

■ *Table 10.5* *Advantages and disadvantages associated with different types of coastal defence structure (modified after East Riding of Yorkshire Council; www.eastriding.gov.uk/ coastalexplorer)*

Structure Type	Advantages	Disadvantages
Seawalls Vertical or near vertical masonry or concrete wall. Can incorporate a wave return profile to improve overtopping performance and a stopped apron toe to reduce scour.	1 Effective prevention of erosion 2 Effective protection against overtopping 3 Strong enough to resist severe exposure sites 4 Many different types 5 Can incorporate promenade amenity features 6 Generally safe for public use	1 Poor energy absorption and high wave reflection rates 2 Wave reflection and scour can destabilize beach 3 Often requires additional energy-absorbing apron 4 Tends to be an expensive option
Revetments Sloping structures of either solid or open construction.	**Rock armour** 1 Good hydraulic performance and energy dissipation 2 Can be used in exposed sites 3 Construction costs generally cheaper than solid structures 4 Requires little ongoing maintenance 5 Relatively easy and quick to construct 6 Often used in conjunction with seawalls to reduce toe scour	1 Difficult to provide amenity value if used as primary defence 2 Often needs to be massive wide structures 3 Can be visually less appealing 4 Tends to be less safe for public use
	Solid reinforced concrete construction 1 Better hydraulic performance than vertical seawalls 2 Can incorporate promenade amenity features 3 Generally safe for public use	1 Disadvantages similar to vertical seawalls 2 Often requires toe scour protection 3 Tend to require more ongoing maintenance than seawalls
Sand dunes Created and maintained through the deposition of sand, dunes can be artificially or naturally created.	1 Provide a valuable store of sand helping to regulate beach levels 2 In maintaining beach levels they aid dissipation of wave energy 3 Provide an important amenity and wildlife value	1 Highly susceptible to erosion
Splash walls Used as secondary defences to control the effects of overtopping or flooding. Splash walls are usually of reinforced concrete design.	1 Allowing some overtopping greatly reduces the scale of the primary defence with associated cost savings 2 Can incorporate promenade amenity features	1 Requires space and promenade width to provide a floodable area 2 Promenade may require increased cleaning and maintenance cost

■ **Table 10.5** *continued*

Structure Type	Advantages	Disadvantages
Flood banks Flood banks tend to be of simple soil/ clay or gabion construction.	1 Used in sheltered locations the control of flooding through the use of a flood bank can relieve the need for a primary defence 2 Set back from the main defence line they provide a cheap solution to control flooding	1 Can only be used as a primary defence at sheltered locations 2 Often requires additional toe protection
Beaches Beaches are effective in harmlessly dissipating wave energy and constitute an excellent form of natural defence.	1 A healthy beach provides effective control of erosion and overtopping 2 Beaches provide a valuable amenity feature 3 Provision of a beach reduces the exposure of the main backstop defence 4 Generally safe for public use	1 A constant source of sand is required 2 To be effective beach levels need to be maintained, this may require costly beach control and/or regular nourishment 3 Maintenance of a beach using natural supplies can starve down-drift areas 4 As a defence beaches are highly sensitive to draw-down during storms
Groins Groins help to build and maintain beach levels by intercepting the long-shore movement of sand.	1 Can be effective in beach building 2 Provision of a beach provides a valuable amenity feature 3 Can be constructed relatively easily from a wide range of materials 4 Maintenance of a beach reduces the exposure of the main backstop defence 5 Can be relatively quick to construct	1 Can produce local scour and increased down-drift erosion 2 Require sand supplies of either natural long-shore drift or artificial nourishment 3 Less effective in controlling cross shore sand movements 4 When constructed of materials other than rock they can have a high maintenance cost 5 Rock groins tend to be less safe for public use
Offshore structures Forcing waves to break offshore reduces wave activity in their lee. A reduction in wave energy at the shoreline encourages the deposition of sand and reduces erosion potential.	1 Promotes the natural build up of beach levels 2 Maintenance of a beach reduces the exposure of the main backstop defence 3 Require little ongoing maintenance	1 Offshore constructions tend to be more massive and therefore more costly 2 Can create a navigation hazard and cause public safety issues 3 Can produce increased downdrift erosion 4 Difficult to construct in deep water

■ **Figure 10.22** *Examples of wooden coastal defence structures: (a) displaced, shore-parallel revetment, Corton, Suffolk; (b) 'polder' structure designed to trap muddy sediment introduced by tidal action, Dengie, Essex; (c) chestnut paling fencing placed to trap windblown sand and slow dune erosion, Formby Point, Merseyside; (d) wooden groins and artificially recharged groin bays, Slaughden, Suffolk*

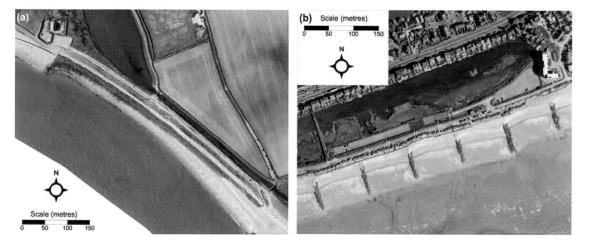

■ **Figure 10.23** *Aerial views of two contrasting rock structures: (a) shore-parallel rock armour revetment, Hurst Castle Spit, Dorset; (b) shore-normal rock groins, Lancing Beach, West Sussex*

■ **Figure 10.24** *Examples of rock and sediment-filled structures: (a) rock armour revetment around the Royal West Norfolk Golf Club, Brancaster, Norfolk; (b) demolition rubble from the city of Liverpool used to form an artificial 'shingle' bank at Crosby, Merseyside; (c) rock-filled gabion baskets, Thorpeness, Suffolk; (d) sand and shingle-filled geotextile bags, Thorpeness, Suffolk*

breakwaters. In Phase 1 of the construction programme four offshore reefs, each approximately 230 m long, were built at a distance of approximately 200 m from the high water mark. The beach behind the reefs was nourished with sand from an offshore source with the intention of developing a stable crenulated shoreline landward of the reefs, which would slow but not entirely stop alongshore transport. However, a series of tombolos formed, linking the reefs to the shore, and acting as a significant impediment to alongshore transport. Consequently, in Phase 2 of the scheme an additional five, shorter reefs, each approximately 160 m long, were built at the same distance from the shoreline. These were successful in preventing the development of tombolos linking the reefs with the shore.

The earliest sea walls in England were mainly constructed of stone, usually from local quarry sources, but concrete has been used for at least 150 years (French, 2001; Reeve *et al.*, 2004; Dupray *et al.*, 2010). During later Victorian times, when the initial development of many seaside resorts took place, extensive promenades were constructed along many beaches, including the Seebees revetment and sea wall at Skegness, Lincolnshire (Figure 10.25a). Following the 1949 Coast Protection Act, and especially after the disastrous North Sea floods of 1953, concrete revetments and facing walls were also built to protect earth flood embankments, sand dunes and soft cliffs (Lee and Clark, 2002; Figure 10.25b, c, d). Unfortunately, in many instances the effect of sea walls has been to cause wave reflection, toe scour

■ **Figure 10.25** *Examples of concrete sea defence structures: (a) Seebees revetment and sea wall at Skegness, Lincolnshire; (b) soft cliff toe protection at Corton, Suffolk; (c) stepped lower wall with curved wave return upper wall, Sea Palling, Norfolk; (d) vertical concrete wall protecting erodible shale cliffs, Robin Hood's Bay, Yorkshire*

and falling levels on the beaches to seawards, resulting in extreme cases in the exposure of steel sheet piling at the toe of the wall, and even collapse. In some instances added protection is provided by building asphalt or concrete aprons at the toe of the wall, or by placing rock armour, rip-rap, tetrapods or 'seebees' in front of the sea wall (Pallett and Young, 1989; Reeve *et al.*, 2004).

Massive sea walls, in some cases more than 10 m high, have also often had a negative effect from the beach user point of view, both in terms of visual impact and by cutting off the beach from facilities on the landward side. In recent coastal engineering schemes, including that at Cleveleys, north of Blackpool, attempts have been made to reduce these adverse impacts by careful design that incorporates a stepped seaward profile and an indented outline to the sea wall in plan (Figure 10.26). The aim has been to connect shops and cafes to the beach and encourage use of both the promenade and beach for a diverse range of leisure events. Such designs also reduce toe scour and longshore sediment drift rates. The Cleveleys project as a whole has been given the progressive nickname 'The New Wave' (Box 10.1).

Alternatives to 'hard' defences

Over the past 20 years in England there has been a move away from the construction of 'hard defences' towards a more balanced approach to beach management, which aims to reduce coastal flood and erosion risk using a combination of detailed beach

BOX 10.1 CLEVELEYS PROMENADE CASE EXAMPLE

■ **Figure 10.26** *Cleveleys Promenade (With permission from Wyre Council)*

Cleveleys is a small coastal town located just north of Blackpool of Lancashire's Fylde coast in Northwest England. Following a strategic review of coastal defence requirements undertaken for Wyre Borough Council by the consultants Halcrow in 2002, a six-year programme of sea defence improvements was started in 2004. The scheme received Grant Aid funding from the Department of Environment, Food and Rural Affairs and The Environment Agency, with subsidiary financial contributions from the European Regional Development Fund and Wyre Borough Council. The total cost of the works undertaken between 2004 and 2010 was in excess of £26 million.

The surface geology of the Fylde Peninsula consists principally of unconsolidated Pleistocene glacial deposits, with areas of Holocene marine, estuarine and riverine alluvium. Until the first artificial defences were constructed in the mid nineteenth century most of the exposed west-facing coast experienced erosion at rates of up to 1.5 m per year. Between 1850 and 1920, the area developed rapidly as a tourist and retirement destination, leading to construction of hard defences and promenades along almost the entire frontage between Lytham in the South and Fleetwood in the north.

By the late 1990s the existing defences at Cleveleys were approaching the end of their design lives and recommendations were made for a new defences scheme to protect approximately 8,000 properties. A

Phase 1 Masterplan was conceived and designed by landscape architects Ferguson and MacIlveen (now part of URS Scott Wilson). The works were planned to be carried out in three subsequent phases, starting in the north of the area and moving southwards in the direction of Blackpool, under a partnership between Wyre Borough Council and Birse Coastal Ltd. The combined cost of Phase 2 and 3, undertaken between 2005 and 2008, was approximately £20 million. Phase 4 was completed, in July 2010 at an additional cost of more than £6 million.

The design sought to combine engineering and visual aspirations by incorporating a stepped revetment on the seaward side providing access to the beach, a wave form in plan, and a wide promenade behind the seawall for multifunctional use. Construction involved a combination of sheet piling, *in-situ* cast foundation concrete, pre-cast revetment slabs made of bright white concrete, and exposed aggregate, sand-coloured concrete for the surface of the promenade. The project has achieved several awards, including the British Construction Industry's Environmental award and the North West Construction Industry 'project of the year' award.

monitoring and maintenance (Rogers *et al.*, 2010). This change has been driven partly on grounds of cost and partly on environmental grounds. Beach management seeks to identify problems associated with falling beach levels at an early stage, where possible to remedy the problem in the short term by beach re-profiling and sediment re-cycling, and over the longer term by beach nourishment. Large-scale beach nourishment projects, involving 400–1,000 10^3 m^3 of sediment, have been undertaken in more than 50 locations, including Bournemouth on the South coast, on the Lincolnshire coast between Mablethorpe and Skegness (Thomas, 1992; Blott and Pye, 2004), and on the Norfolk coast between Horsey and Winterton. Beach nourishment is not a cheap option; some £45 million has been spent on the Lincshore scheme alone since the late 1990s, with further expenditure of about £30 million planned over the five-year period 2010–15. In some cases, beach nourishment has been undertaken as a stand-alone method of beach management, but in the majority of cases, it has been used in conjunction with the construction of rock groins and offshore breakwaters. Sediment nourishment has also been undertaken within estuaries to raise the level of tidal flats, to create artificial sand and shingle bars and to create saline or brackish lagoons for nature conservation reasons.

Concerns about the potential impacts of global climate change and sea level rise have led to a widespread recognition that in the medium to longer term (50 to 100 years) maintenance and/or replacement of some coastal defences will not be sustainable. Consequently, the most recent Shoreline Management Plans recommend a policy of long-term 'Hold the Line' only in areas of highest asset value. Elsewhere, policies of 'No Active Intervention' or 'Managed Realignment' have been recommended. Detailed design guidelines for managed realignment schemes have been published (e.g. Leggett *et al.*, 2004), several schemes have already been completed and others are in progress, principally in the estuaries and on the open coast of southeast and eastern England, including the Blackwater and Roach-Crouch estuaries in Essex. By allowing partial breaching of existing sea defences, it is intended that ground levels will be raised as new sediment accumulates, thereby allowing natural sedimentation processes to build up a new line of defence against sea flooding.

Acknowledgement

Grateful thanks are extended to Tracey Dixon for expertly producing Figure 10.11.

References

Balas, C. E., Ergin, A. and Williams, A. T., 2000. Reliability-based risk assessment in project management of coastal structures. In: *Proceedings of the International Conference on Offshore Mechanics and Arctic Engineering*, Grundy, P., Koo, J., Langen, I. and Ueada, Y. (Eds), International Society of Offshore and Polar Engineers, Cupertino, CA, 369–374.

Ball, T., Werritty, A., Duck, R. W., Edwards, A., Booth, L. and Black, A. R., 2008. *Coastal flooding in Scotland: A scoping study.* Report for Scotland and Northern Ireland Forum for Environmental Research (SNIFFER), 86 pp.

Baugh, I. D. H. and Boreham, J. R., 1976. Measuring the coastline from maps: A study of the Scottish mainland. *The Cartographic Journal*, 13, 167–171.

Baxter, J. M., Boyd, I. L., Cox, M., Cunningham, L., Holmes, P. and Moffat, C. F., 2008. *Scotland's Seas: Towards understanding their state.* Fisheries Research Services, Aberdeen, 174 pp.

Blott, S. J. and Pye, K., 2004. Morphological and sedimentological changes on an artificially nourished beach, Lincolnshire. *Journal of Coastal Research*, 20, 214–233.

Booth, L. M. and Duck, R. W., 2010. A decade of delivering sustainable coastal zone management: The Tay Estuary Forum, a voluntary local coastal partnership in Scotland. *Proceedings of the Littoral 2010 Conference, London.* Available at: http://www.dundee.ac.uk/crsem/TEF/TEF%20Littoral%202010%20Paper%20BoothDuck.pdf (accessed 31 January 2011).

Burgess, K., Jay, H. and Hosking, A., 2004. Futurecoast: Predicting the future evolution of the coast of England and Wales. *Journal of Coastal Conservation*, 20, 65–71.

Burningham, H., Duck, R. W. and Watt, A. M., 2000. Perspectives from a newly formed ICZM voluntary partnership: The Tay Estuary Forum (Scotland). *Periodicum Biologorum*, 102, Suppl. 1, 101–105.

Carr, A. P., 1988. UK – England. In: *Artificial structures and shorelines*, Walker, H. J. (Ed.), Kluwer Academic Publishers, London, 137–144.

Clayton, K. and Shamoon, N., 1998. New approach to the relief of Great Britain II: A classification of rocks based on relative resistance to denudation. *Geomorphology*, 25, 155–171.

Dargie, T. C. D., 1995. *Sand dune vegetation survey of Great Britain: A national inventory*. Part 3: Wales. Joint Nature Conservation Committee, Peterborough, UK.

Davies, R. H., 2001. The management of the sand dune areas in Wales: The findings of a management inventory. In: *Coastal dune management: Shared experience of European conservation practice*, Houston, J.A., Edmondson S.E. and Rooney, P. J. (Eds.), Liverpool, Liverpool University Press, 233–242.

DEFRA, 2001. *Charting progress: Section 3.2 Coastal Defence*. Available at: http://chartingprogress.defra.gov.uk/feeder/Section_3.2_Coastal%20Defence.pdf (last accessed 18 August 2012).

DEFRA, 2005. *Beach lowering in front of coastal structures, Appendices 1–3.* R&D Project Record FD1916/PR, 44 pp.

DEFRA, 2010. *Charting progress 2 - Feeder report: Productive seas.* DEFRA, London, vi + 462 pp.Doody, J. P., 1999. The Scottish coast in a European perspective. In: *Scotland's living coastline*, Baxter, J.M., Duncan, K., Atkins, S. M. and Lees, G. (Eds), The Stationery Office, London, 15–29.

Dupray, S., Knights, S., Robertshaw, G., Wimpenny, D. and Woods Ballard, B., 2010. *The use of concrete in maritime engineering: A guide to good practice.* CIRIA Technical Manual No. C674, CIRIA, London, 450 pp.

Environment Agency (EA), 1991. Environment Agency, *Survey of sea defences in Wales owned by local authorities and private or public organizations*, Final Report Phases of II and III (JC148 W2SD), Environment Agency, Cardiff, F105.1–F105.17.

Environment Agency (EA), 2010. *Future flooding in Wales: Flood defences.* Environment Agency, Cardiff, Wales, 32 pp.

Environment Agency, n.d. *Shoreline Management Plans (SMPs).* Available at: http://www.environment-agency.gov.uk/research/planning/104939.aspx (last accessed 18 August 2012).

Evans, E. P., Ashley, R. M., Hall, J., Penning-Rowsell, E., Sayers, P., Thorne, C. and Watkinson, A., 2004. *Foresight, Volume 1*, Chapter 6, 151 pp. Available at: http://www.bis.gov.uk/assets/foresight/docs/flood-and-coastal-defence/vol1chapter6.pdf (last accessed 31 August 2012).

Fleming, C. A. and Hamer, B., 2000. Successful implementation of an offshore reef scheme. *Proceedings of the 27th International Conference on Coastal Engineering*, American Society of Civil Engineers, Sydney, 1813–1820.

French, P. W., 2001. *Coastal defences: Processes, problems and solutions*. Routledge, London and New York, 366 pp.

Hydrographer of the Navy (HN), 1999. *Admiralty tide tables*, Vol. 1, European Waters, Taunton, UK.

Halcrow, 2002. *Futurecoast*. CD produced by Halcrow, Swindon, on behalf of DEFRA, London.

Halcrow, 2010. *Shoreline Management Plan SMP2: Lavernock Point to St Anne's Head, identification of preferred policy,* Report on behalf of the Swansea and Carmarthen Bay Coastal Engineering Group, Halcrow, Swindon.

Hansom, J. D., 1999. The coastal geomorphology of Scotland: Understanding sediment budgets for effective coastal management. In: *Scotland's Living Coastline*. Baxter, J. M., Duncan, K., Atkins, S. M. and Lees, G. (Eds), The Stationery Office, London, 34–44.

Hansom, J. D., Lees, G., McGlashan, D. J. and John, S., 2004. Shoreline Management Plans and coastal cells in Scotland. *Coastal Management*, 32, 227–242.

H. N, 1999, Hydrographer of the Navy. *Admiralty Tide Tables*, Vol.1, European Waters, UK.

HR Wallingford (HRW), 1995. *Dredging application on the Nash Bank, Bristol Channel, dispersion of dredged material*, HR Wallingford Report No: EX 3312, UK.

HR Wallingford (HRW), 1997. *Coastal cells in Scotland*. Scottish Natural Heritage Research, Survey and Monitoring Report, 56, Edinburgh.

Institute of Oceanographic Sciences (IOS), 1980. *Wave data and computed wave climates*, Topic Report 5.

Lamb, H. H., 1991. *Historic storms of the North Sea, British Isles and Northwest Europe*. Cambridge University Press, Cambridge, UK.

Lee, M. and Clark, A., 2002. *Investigation and management of soft rock cliffs*. Thomas Telford, London, 382 pp.

Lees, R. G., Gordon, J. E. and McKirdy, A. P., 1998. Coastal erosion, coastal defences and Earth Heritage in Scotland. In: *Coastal defence and earth science conservation,* Hooke, J. (Ed), The Geological Society, London, 133–150.

Leggett, D. J., Cooper, N. and Harvey, R., 2004. *Coastal and estuarine managed realignment: Design issues*. CIRIA Report C628, CIRIA, London, 215 pp.

Masselink, G. and Russell, P., 2008. *Marine climate change impact partnership: Annual report card 2007–2008, Scientific Review*, University of Plymouth, UK, 15 pp.

McCue, J., Pye, K. and Wareing, A., 2010. *Beach nourishment operations in Wales and likely future requirements for beach nourishment in an era of sea level rise and climate change: A pilot study*. CCW Science Report 928, Countryside Council for Wales, Bangor.

National Trust (NT), 2007. *Shifting shores: Living with a changing coastline*, National Trust, London, 23 pp.

Pallett, N. and Young, S. W., 1989. Coastal structures. In: *Coastal Management*, Institution of Civil Engineers (Ed.), Thomas Telford, London, 211–226.

Pringle, A., 1985. Holderness coast erosion and the significance of ords. *Earth Surface Processes and Landforms*, 10, 107–124.

Pye, K. and Blott, S. J., 2008. Decadal scale variation in dune erosion and accretion rates: An investigation of the significance of changing storm tide frequency and magnitude on the Sefton coast, UK. *Geomorphology*, 102, 652–666.

Pye, K. and Blott, S. J., 2012. *A geomorphological assessment of Welsh dune systems to determine best methods of dune rejuvenation*. CCW Contract Science Report No. 1002, Countryside Council for Wales, Bangor.

Pye, K. and Neal, A., 1994. Coastal dune erosion at Formby Point, north Merseyside, England: Aauses and mechanisms. *Marine Geology,* 119, 39–56.

Pye, K. and Saye, S. E., 2005. *The geomorphological response of Welsh sand dunes to sea level rise over the next 100 years and the management implications for SAC and SSSI sites.* Contract Science Report 670, Countryside Council for Wales, Bangor.

Pye, K., Saye, S. E. and Blott, S. J., 2007. *Sand dune processes and management for flood and coastal defence.* DEFRA / Environment Agency Joint R & D Programme Technical Report FD1302/TR, 5 volumes, DEFRA, London.

Reeve, D., Chadwick, A. and Fleming, C., 2004. *Coastal engineering: Processes, theory and design practice.* Spon Press, Abingdon, 461 pp.

Ritchie, W. and Mather, A. S., 1984. *The beaches of Scotland.* Countryside Commission for Scotland, Perth, 130 pp.

Ritchie, W. and McLean, L., 1988. UK: Scotland. In: *Artificial structures and shorelines,* Walker, H. J. (Ed.), Kluwer Academic Publishers, London, 127–135.

Rogers J. R., Baptiste, A. and Jeans, K., 2009. National coastal erosion risk mapping: The final furlong. *Proc. 44th Flood and Coastal Risk Management Conference,* Telford, UK.

Rogers, J., Hamer, B., Brampton, A., Challinor, S., Glennerster, M., Brenton, P. and Bradbury, A., 2010. *Beach management manual (second edition).* CIRIA Report C685, CIRIA, London, 915 pp.

Saiu, E., McManus, J. and Duck, R.W., 1994. Impact of industrial growth and decline on coastal equilibrium, eastern Scotland. *Coastal Zone Canada '94,* Coastal Zone Canada Association, 2205–2219.

Scottish Executive, 2003. Scotland in Short: Factsheets. Available at: http://www.scribd.com/doc/334117/Scotland-in-Short (accessed 31 January 2011).

Scottish Executive, 2005. *Seas the opportunity: A strategy for the long term sustainability of Scotland's coasts and seas.* Scottish Executive, Edinburgh, 40 pp.

Scottish Government, 2010. Marine Scotland. In: *Scottish Marine Regions: Defining their Boundaries – A consultation.* Scottish Government, Edinburgh, p. 47.

Shennan, I. and Horton, B., 2002. Holocene land- and sea-level changes in Great Britain. *Journal of Quaternary Science,* 17, 11–526.

Shoreline Management Partnership (SMP), 1999. *Shoreline Management Plan for sub cell 11a (Great Ormes Head to Formby Point)* Plan document and context report, carried out for Liverpool Bay by Shoreline Management Partnership, Rossett, UK.

Shoreline Management Partnership and HR Wallingford (SMP and HRW), 1999. *Swansea Bay Shoreline Management Plan, Worms Head to Lavernock Point,* Stage 1, Vol. 3, 74 pp.

Thomas, R. S., 1992. The defences for Lincolnshire. In: *Coastal zone planning and management,* Barrett, M. G. (Ed.), Thomas Telford, London, 269–281.

Valentin, H., 1971. Land loss at Holderness. In: *Applied coastal geomorphology,* Steers, J.A. (ed.), Macmillan, London, 116–137.

Welsh Assembly Government (WAG), 2008. *Coastal tourism strategy: Visit Wales.* Report by the Tourism and Marketing Division to the Welsh Assembly Government, Cardiff.

Welsh Audit Office (WAO), 2009. *Coastal erosion and tidal flooding risks in Wales,* Welsh Assembly, Cardiff, Wales, 61 pp.

Whittington, G., 1996. *Fragile environments: The use and management of Tentsmuir National Nature Reserve.* Scottish Cultural Press, Dalkeith, 120 pp.

Williams, A. T. and Davies, P., 2001. Coastal dunes of Wales: Vulnerability and protection. *Journal of Coastal Conservation,* 7, 145–154.

Williams, A. T., Ergin, A., Micallef, A. and Phillips, M. R., 2005. The perception of coastal structures: groined beaches. *Zeitschrift fur Geomorphologie,* 141, 111–122.

Winnard, K., McCue, J., Pye, K., Blott, S. J. and Dearnaley, M., 2011. *Rebuilding Welsh beaches to deliver multiple benefits: Overview report.* CCW Contract Science Report No. 974, Countryside Council for Wales, Bangor.

11 Ireland

Andrew Cooper

Introduction

Ireland, comprising Northern Ireland and the Republic of Ireland, has a long and geologically diverse coastline in relation to the size of the island (Devoy, 2008). Of the approximately 6,000 km of coastline, about 50 per cent is soft coast (modern sediments or unconsolidated Quaternary material) and is subject to periodic erosion (National Coastal Erosion Committee, 1992). The rock coast does undergo progressive retreat but at much lower rates. Most of the coast is exposed to direct high-energy influences from the North Atlantic, while the more sheltered eastern coast is fronted by the shallow Irish Sea. From a coastal processes perspective the island can therefore be broadly divided into (1) a high-energy, highly indented rocky coast (in the south, west and north); and (ii) a lower energy, more linear east coast (Wexford to Antrim) where outcrops of unconsolidated Quaternary sediment are common but the underlying bedrock crops out intermittently (Figure 11.1). The coastlines of the sheltered large marine embayments or loughs constitute a third morphodynamic domain in which the coast is subject to much reduced levels of wave energy. The processes operating in these three coastal domains at decadal to century timescales are quite distinctive and they set the context for understanding contemporary coastal erosion and concerns around it (Carter *et al.*, 1987; Devoy, 2008).

There is also a marked north–south gradient in relative sea level (RSL) history as a result of the influence of the last Ice Sheet. The northern part of Ireland has been dominated by land uplift (isostatic rebound following ice removal) and sea levels have been static or falling for the past few thousand years. The southern part of Ireland, which was deglaciated much earlier, has experienced continuous rise in sea levels over the same period. These RSL trends have important implications for coastal processes and medium term shoreline evolution.

Coastal erosion is, of course, just one element in a natural coastal sedimentary system that involves sediment erosion, transport, storage and deposition. Viewed in this light, erosion is essential to the maintenance of a healthy coastal ecosystem and the ecosystem services that it provides both directly and indirectly (e.g. recreation, natural coastal defence, food, etc.) (Defeo *et al.*, 2009). However, as elsewhere in the world, coastal erosion is more commonly viewed in Ireland simply as a problem. In some cases, this is because it affects human infrastructure, but in others, it is simply perceived to be undesirable. Cooper and McKenna (2008a) contend that this deeply ingrained view originates in a strong attachment to land as property and pre-dates the growth of environmental consciousness that has contributed to public understanding of coastal ecosystems. This attitude is important in shaping the human response to coastal erosion in Ireland and resulted in a situation whereby property owners resist coastal erosion by hard defences if they have the technical and economic resources to do so. In this regard, the demographic and economic context is of some importance in understanding the impact of, and response to coastal erosion in Ireland.

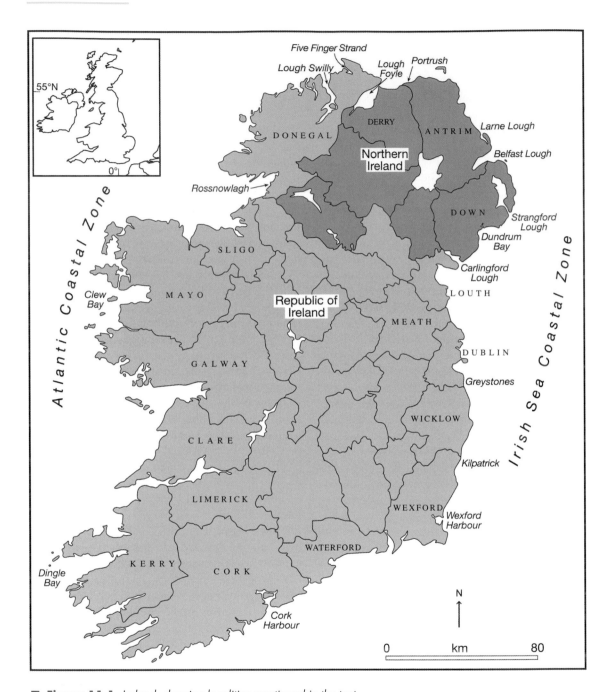

■ **Figure 11.1** *Ireland, showing localities mentioned in the text*

Natural shoreline evolution

Atlantic coast

The Atlantic coast is dominated by bedrock outcrop, although sediment of glacial origin deposited on the shelf has been driven into embayments to form depositional complexes (beaches, barriers and dunes; Carter et al., 1987; Carter and Wilson, 1991). In some areas, there are carbonate beaches composed of calcareous algae, molluscs, barnacles and foraminifera while others derive their sediment from contemporary erosion of soft cliffs (Guilcher and King, 1961). In broad terms, coastal processes on the western coast of Ireland are driven by long-period swell waves that begin to interact with the seabed far from shore and consequently arrive at the coast fully refracted. On the highly indented coastline, sediment is confined within individual sediment cells and longshore drift is very much subordinate to cross-shore sediment exchange. Most of the sedimentary shorelines are swash-aligned and show historically low rates of net change. High dunes prevent overwash in most places, except in the few gravel barriers (Delaney and Devoy, 1995).

In this setting, progressive shoreline erosion occurs at slow rates on the hard cliff coasts and somewhat faster on the relatively rare glacial sediment outcrops. In a few localities, e.g. Clew Bay, Galway Bay and the south coast, contemporary erosion of onshore outcrops of Quaternary sediment supplies material for beach and barrier formation. These barriers respond to fluctuations in sediment supply and energy levels via periodic erosion and accretion (see Orford and Carter, 1982).

Sand beaches on the Atlantic coast are subject to modally high wave energy and they typically assume a highly dissipative state. Consequently, only extreme storms can overcome the high morphological threshold and cause shoreline erosion. These rare storms, however, can cause dramatic changes from which the coast can take decades to recover (Orford et al., 1999; Cooper et al., 2004). At Inch Strand in Dingle Bay, for example, Orford et al. (1999) attributed dune front recession of over 100 m to a single storm (Hurricane Debbie) in 1961. Nonetheless, the highly compartmentalized shoreline means that eroded sediment is moved offshore from whence it is eventually returned to the beach and dunes during fair-weather conditions. At Inch Strand, recovery after the 1961 storm continued over three decades.

On embayed beach complexes associated with estuary mouths dramatic shoreline erosion has also been noted in association with changes in tidal inlet orientation (Burningham and Cooper, 2004; Cooper et al., 2007, O'Connor et al., 2011). These changes can amount to hundreds of metres in a few months to years (Figure 11.2), but also tend to be cyclic in nature since sediment is not lost from the system, but redistributed within it between dune, beach, shoreface and tidal deltas. At Five Finger Strand, County Donegal, the periodicity of the inlet movement is two to three decades and frontal dune erosion (Figure 11.3) persists for decades only to be succeeded by decades of accretion when the inlet reverts to its former position.

The scale of changes driven by storms and inlet movement is much greater than any attributed to sea level rise. On the Irish Sea coast, longshore drift is more important and erosion of Quaternary sediments from coastal bluffs is essential to the maintenance of contemporary beaches. The general pattern here is of progressive shoreline erosion, but with periodic cliff collapse delivering sediment to the coast and then its progressive redistribution before the next cliff collapse occurs.

In summary, therefore, erosion in embayments along the Atlantic coast is dominated by cyclic coastal behaviour. There are exchanges between beach, shoreface and dunes at decadal timescales but overall retention of the sediment volume. Smaller scale cross-shore exchange of sediment during winter storms can involve trimming of dunes, but such erosional events tend to be followed by recovery in subsequent years or even decades. Extreme storms can cause dramatic recession but can be followed by recovery at decadal timescales. In those sections of the coast where unconsolidated Quaternary outcrops, long-term recession and the sand and gravel component of the eroded material supplies adjacent beach systems.

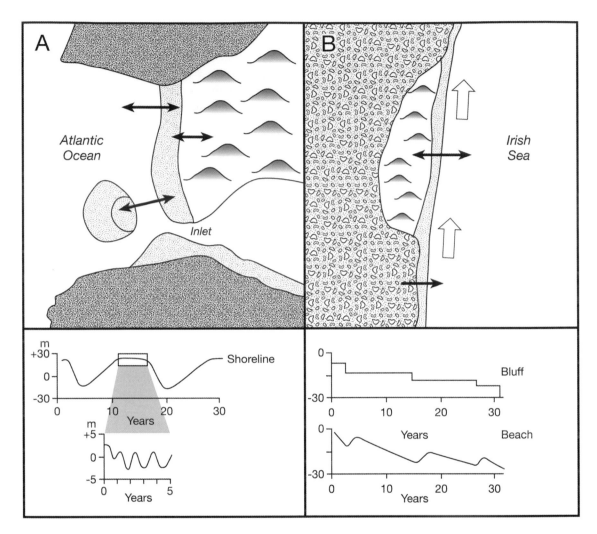

Figure 11.2 *Schematic representation of sediment exchanges in a. the compartmentalized Atlantic coast and b. the drift-dominated Irish Sea coast. Graphs illustrate the temporal trends in shoreline behaviour. Sediment exchanges in embayments of the Atlantic coast typically involve retention of the total sediment volume but cycling between sub-environments. This leads to cycles of erosion and accretion at different timescales but little long-term retreat of the shoreline*

Figure 11.3 *Dramatic erosion of the sand dune at Five Finger Strand (quad motorcycle for scale). Although dramatic, this is part of a natural decadal cycle of erosion and accretion, driven by periodic fluctuations in the position of a tidal inlet*

Irish Sea coast

On the east coast, short-period sea waves generated in the Irish Sea are the dominant driver of coastal processes (Carter *et al.*, 1987). This creates a strong south–north longshore drift while the high incidence of Quaternary sediments onshore provides the major source of sediment for beach and barrier formation and maintenance. The coast generally comprises narrow beaches of reworked Quaternary sediments whose planform is influenced by occasional outcrops of bedrock. Coastal dunes have accumulated at the rear of several beaches and often undergo seasonal erosion during winter storms (Figure 11.4). Long-term changes in sediment flux control temporal trends in shoreline erosion and recovery (Carter *et al.*, 1987), but much of the coastline has experienced long-term recession as the sediment on beaches has been transported alongshore and offshore, exposing cliffs to renewed wave attack and introducing fresh sediment into the coastal system (Figure 11.2). In the north-east, relative sea level fall in the late Holocene has left onshore sediment sources (raised beaches and Quaternary outcrops) stranded above the reach of current waves and has consequently reduced the contemporary sediment supply.

Long-term erosion rates in eastern Ireland are variable both in time and space depending on past sediment supply and dispersal to a large extent. Typical recession rates at the century timescale are about 1 m/yr, but long term averages as high as 2 m/yr have been reported in some locations (National Coastal Erosion Committee, 1992).

There are localized exceptions to this pattern, particularly where large offshore sediment sources exist. In Dundrum Bay, County Down, for example, relict sands on the shelf have yielded abundant sand to the nearshore zone, causing coastal progradation from the late Holocene into the historical period and building extensive coastal dunes (Cooper and Navas, 2004; Orford *et al.*, 2003).

In summary, most of the Irish Sea coast is experiencing long-term recession as Quaternary sediment is eroded to build modern beaches. However, the net loss of sediment from the coast appears to exceed inputs and beaches and dunes are also retreating. The rate of recession is highly variable in time and space.

■ **Figure 11.4** *Erosion of the dune front at Brittas Bay, County Wicklow. This is a seasonal phenomenon driven by winter storms, but a slow long-term coastal retreat is evident on most Irish Sea beaches*

Large marine embayments

Within many of Ireland's large marine embayments (except for west-facing examples that receive the full impact of the Atlantic), there is a degree of protection from direct wave influences. These areas are subject to tidal currents and waves formed under fetch-limited conditions, the largest include the Shannon estuary, Cork Harbour, Wexford Harbour, Carlingford, Strangford, Belfast and Larne Loughs and Loughs Foyle and Swilly. Natural shoreline evolution within these bays is highly site-specific and influenced strongly by local sediment availability, pre-existing topography and exposure to wave and tide processes. Marsh, beach, tidal flat, barrier, bluff and rock shorelines are found in these diverse coastal environments and almost all of Ireland's embayments have experienced a high degree of human modification by port development, shoreline armouring and reclamation.

Human infrastructure and the Irish coast

Ireland has experienced a turbulent population history over the past two centuries. Its population grew from 2.5 million in 1700 to 4.5 million in 1800. It peaked in the 1840s at 8 million (the population of Great Britain at the same time was only 18 million) and fell to six million in a single decade during the famine of the mid nineteenth century. It continued to fall to 4 million by 1920, where it remained until 1970 and it has since increased to about 4.5 million. Throughout most of this period, the island had a poorly developed economy and was largely agricultural. The relative wealth of the island increased dramatically during the 1980s and 1990s, a period known as the 'Celtic Tiger' economic boom. The financial crisis of 2007–present, has, however, revealed underlying economic weaknesses.

The relevance of demography and economics for understanding the relationship between people and the coast is quite profound. Throughout most of the nineteenth and twentieth centuries, Ireland was relatively impoverished and sparsely populated, with a largely agricultural economy outside its major towns and cities. This meant that there was limited pressure on the coast and limited financial resources for intervention in coastal processes. As a consequence, most of the coast was free to fluctuate in response to natural processes. Infrastructure or property that was threatened by coastal erosion was generally not protected (Figure 11.5) and the coast remained free to operate naturally, except in the vicinity of harbours and several seaside resorts where promenades were built.

In the nineteenth century, major land claim works were undertaken in most of Ireland's loughs and estuaries, to the extent that much intertidal flat and salt marsh was transformed to agricultural land (Healy and Hickey, 2002; McErlean, 2011). Much of this is still maintained today by active drainage schemes, although a few have fallen into disrepair, but the reclamation impacts are still evident in the landscape. In Wexford harbour (Orford, 1988) and in estuaries north of Dublin (Mulrennan, 1993) reclamation of back-barrier intertidal areas has resulted in significant changes at adjacent tidal inlets.

The advent of widespread mechanized transport had two major influences. First, the road and rail infrastructure was improved and much of this followed the coast. Second, people became more

■ **Figure 11.5** *House collapsing through erosion of cliffs of soft Quaternary sediments, Blackwater Head, Wexford*

mobile and began to visit previously inaccessible coastal areas. This still confined human infrastructure at the coast to a relatively few coastal resorts and port cities and it was only with the 'Celtic Tiger' economy of the 1990s that enhanced wealth began to lead to an intensive development of the coast for housing and in particular holiday homes. This process had begun somewhat earlier in Northern Ireland but it also took off there during the property boom of the 1990s and early 2000s (Cooper and McKenna, 2009). The European Environment Agency's assessment of the state of Europe's coasts (EEA, 2006) highlighted the massive increase in the area of developed land between 1990 and 2000 in Ireland. According to the 1981 census, around 1.16 million people live near the Irish coast (McWilliams, 1994) which is a 5.2 per cent increase on the preceding decade. Most of these increases were in low-lying areas adjoining the cities of Dublin, Cork and Galway. In the case of Dublin, these developments have also carried transport routes on to areas that could be endangered by increased coastal erosion.

Legislative and administrative background

A series of storms in the 1980s focused attention on the lack of a strategy for coping with coastal erosion and a National Coastal Erosion Committee was constituted to make recommendations on erosion management. This body made several recommendations in its 1992 report, but neither Ireland (O'Connor et al., 2009) nor Northern Ireland (Dodds et al., 2010) has yet adopted a strategic approach to shoreline management. Consequently, the response to coastal erosion occurs in a policy vacuum (O'Connor et al., 2010) and depends largely on a combination of politics, economics and environmental considerations in descending order of importance. O'Hagan and Cooper (2002) found that at a county level in Ireland, no two local authorities operated the same procedures for dealing with coastal erosion.

A national coastal protection strategy study has been underway in Ireland since 2002 (it is scheduled for completion in 2013) and an ongoing review of flood and erosion protection in Northern Ireland is being undertaken in response to the European Floods Directive (Dodds et al., 2010). These are delivering useful tools such as flood and erosion risk maps, but responsibility for coastal erosion management remains fragmented and practice-led. The emphasis in both cases still appears to remain on protection schemes but with an anticipated increased reliance on beach nourishment in Ireland (Casey, 2009), as suggested previously by the National Coastal Erosion Committee (1992). The cost of maintenance of coastal protection works and emplacement of new works in Ireland in 2008 was €2.97 million.

Recent initiatives in response to the EU Recommendation in ICZM (McKenna et al., 2008) have failed to embrace its principle of 'working with natural processes' (Cooper and McKenna, 2008b) and substantial barriers exist to the attainment of such a goal or implementation of a strategic approach to shoreline management. In this policy vacuum, a number of guidelines have been developed to aid coastal management practitioners. The Government of Ireland (1996) published a manual on 'Environmentally Friendly Coastal Protection' that was issued to every local authority senior engineer. McKenna et al. (2000) presented practical guidance for managing rural beaches including issues related to erosion. This approach has been termed 'practice-led shoreline management' (O'Connor et al., 2009).

The prevailing view of coastal erosion as a problem is evident in the National Coastal Erosion Committee's 1992 report. Despite acknowledging the importance of the coast (p. 9) for its 'environmental, ecological and recreational' values, the report is focuses on the costs of protection of assets and envisaged beach nourishment becoming a more prevalent activity.

Coastal erosion management in Ireland

The historically limited development of the Irish coastline, coupled with financial constraints, created

■ **Figure 11.6** Collage of typical shore protection structures in Ireland: (a) informal rubble protecting private house; (b) pre-cast concrete structure protecting footpath; (c) gabion baskets protecting footpath; (d) urban seafront with several generations of formal walls; (e) informal rubble protecting caravan site; (f) formal rock armour protecting coastal road a situation whereby in 1988 Carter and Orford were able to state that the extent of coastal defence works was small in relation to the length of the coast. Since then, the 'Celtic Tiger' economy has changed the situation; the extent of coastal development has increased dramatically and so too has the extent of coastal defence works.

No thorough study of the nature and extent of coastal defence structures has yet been undertaken, although Carter and Orford (1988) estimated that 238 km, or 3.8 per cent, of the Republic of Ireland's coast is protected by artificial structures. A subsequent survey by the National Coastal Erosion Committee (1992) estimated that seawalls were present on 274 km of the 3,017 km of soft coastline in the Republic of Ireland, with around half of that length being County Cork. The same survey contained an estimate by county engineers that a further 493 km of sea defences was required because of the risk from erosion.

The extent of shoreline armouring in Northern Ireland was recently estimated at 100 km out of a total shoreline of 650 km (Gibson, 2011). Like the estimates for the Irish republic, this is almost certainly an underestimate, as it does not take account of many small stretches of armour fronting individual packages of land (Figure 11.6). Of the 100 km of defences identified, 36 km is 'statutory coastal defence' built and maintained by government for the express purpose of maintaining public safety. Examples include sea defences on Strangford Lough, that protect low-lying ground containing a small airfield and housing development from occasional marine flooding, and defences on Lough Foyle, that protect agricultural land that was artificially reclaimed in the nineteenth century. The non-statutory defences are a mix of many short stretches of sea defence built by government departments, local councils, private individuals, sporting clubs and farmers. In many cases they were built in response to episodic storms (from which natural recovery was ensured in any case) and are intended to protect both built infrastructure and undeveloped land (including sand dunes). This serves to illustrate both the lack of appreciation of coastal processes and the underlying view that erosion is undesirable.

Case studies: Hard stabilization

A number of case studies illustrate contemporary approaches to coastal erosion in Ireland, with installation of hard defences being the dominant approach. Commonly these involve seawalls or rock armour, but occasionally they are accompanied by groins, particularly on the Irish Sea coast where longshore drift is important. The case studies illustrate the issues at each site and the decision-making process behind hard defence construction.

Kilpatrick, County Wexford

At Kilpatrick a 2 km long mixed sand and gravel beach flanked by rocky headlands is backed by a wide and vegetated sand dune system that is designated as a Special Area of Conservation (SAC) under the EU Habitats directive. The site is described as "a good example of a mature sand dune system which shows the developmental stages of dunes from foredunes to mature grey dunes" (NPWS, 1997, p. 1). Map evidence shows that the coastline has been undergoing net erosion over the past century (Figure 11.7) and contemporary exposure of evidence of mediaeval cultivation in the eroding dune face indicates this is a long-term progressive trend.

A succession of minor storms in December 2006 led to retreat of the dune face by more than 10 m within a single month. This undermined an access road to three houses, making it impossible to pass in a vehicle. Management options are constrained by the SAC designation, which, amongst other things, precluded simple construction of a new access road through the SAC and required that the habitat integrity be maintained. The local authority (Wexford County Council) consequently commissioned a study into possible management options for the site.

The options considered were (1) emplacement of rock armouring, (2) construction of offshore breakwaters and (3) no intervention. Rock armouring was recognised as problematic in that it would fundamentally halt the operation of natural processes necessary to link the dune and beach and thus affect the SAC designation. In contrast, the no-intervention option would permit processes to operate but would

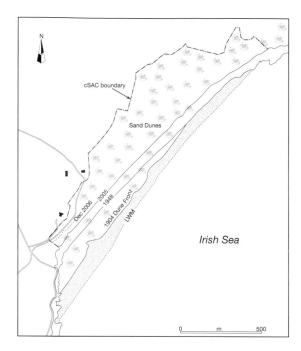

N

cSAC boundary

Sand Dunes

Dec 2006
2005
1948
1904 Dune Front
LWM

Irish Sea

0   m   500

■ **Figure 11.7** *The coastal site at Kilpatrick, County Wexford (after Cooper and McKenna, 2008b), showing the progressive retreat of the shoreline*

likely lead to loss of much of the site through natural erosion, while still leaving the problem of access to the houses. Offshore breakwaters, intended to slow the rate of sediment loss by creating wave-shelter, would be many times more costly than the value of the properties being defended and at best might simply slow the rate of recession. At worst, they could focus wave action and cause rip currents to denude the beach more rapidly. A further option (the creation of a different access route to the houses, landward of the dune field) was not considered because the lands across which such a route would run are in private ownership and no system existed within the local authority's remit to take forward such a proposal. At the time of writing (2012), no work has yet been undertaken, although there is considerable public and political pressure for a seawall to be constructed.

In County Wexford, the planning authority considers coastal erosion in planning decisions and

will prohibit development in coastal areas . . . where the erosion is likely to threaten the viability of such developments. . . Within all coastal areas the council will prohibit any new buildings or development including caravans and temporary dwellings within 50 metres of soft shore lines.

(An Bord Pleanála, 2001, p. 5)

In contrast, where existing development is threatened the policy states

Coastal erosion poses a serious threat to public infrastructure, private property, etc. Coastal protection works have been carried out at a number of locations in recent years. The council will cooperate . . . in undertaking coastal defence works at . . . locations where need is greatest.

(Wexford County Council, 2007)

This difference in policy with regard to new and existing dwellings appears to emphasise that protection of property is a higher priority than preservation of the coastal ecosystem. Nonetheless it does seek to ensure that future problems are not exacerbated by badly sited construction.

Greystones, County Wicklow

At Greystones, a prolonged planning application and resulting appeal process ended in a 2007 decision to permit the construction of a large marina and associated housing development. It was acknowledged that this would interrupt the longshore drift, reducing supply to the beach on the northern (downdrift) side of the marina. It was also recognised that this would lead to accelerated erosion of the bluff of Quaternary sediments at the rear of the beach (Figure 11.8) which in turn threatened a landfill site with erosion. However, the planning application was, however, passed on condition of mitigation of this impact with a 20-year beach nourishment scheme involving the emplacement by truck, of sediment from inland sources.

With the subsequent collapse in the Irish economy, work on the partially completed marina has halted.

The situation regarding beach nourishment both now and in the future is unclear, but is unlikely to be high on the agenda, given the bleak economic outlook. In any case, the commitment to nourishment was time-limited in terms of the developer's obligations and no provision was made for its operation beyond that time.

Rossnowlagh, County Donegal

The 2 km long sand beach at Rossnowlagh is backed by a narrow gravel ridge together with vegetated sand dunes. Several caravan sites, a hotel and numerous holiday homes are located adjacent to the beach. Between 1970 and 2000 over 25 per cent of the c. 20 ha dune area was covered with built structures (roads, car park, houses, caravan parks, a holiday home village, hotel and shops). Much of the new development is close to a dune scarp that is progressively eroding at average long-term rates of 0.6 m/yr. In 1972, property owners began to emplace rock armour to defend buildings located behind the dune scarp. This has proceeded to the point where by 2010 over 50 per cent of the dune front is armoured. Aesthetically the effect has been negative as some armour sections are *ad hoc* in nature and are poorly designed and constructed. Erosion has continued (and due to armour 'end-effects' has probably accelerated) on the remaining sections of natural dune. The result is an unsightly and fragmented dune front consisting of linear armour sections alternating with eroding natural dune sections, the latter cut by deep gaps and hollows (Figure 11.9).

The flat beach profile and absence of sand accumulation structures suggest that sand supply may be quite limited. Storm-wave sand release from the dunes to augment the beach and improve its wave-damping capabilities is now prevented by the armouring. It is likely that erosion will intensify as ever lower beach levels allow accelerated backshore erosion. The outcome is likely to be an increase in extent of defences.

The proliferation of coastal defences at Rossnowlagh was driven by private owners rushing to realize the value of coastal sites, aided by an

■ **Figure 11.8** *The beach at Greystones County Wicklow. (a) The natural beach is backed by a bluff of unconsolidated Quaternary sediments that provides sand and gravel for the modern beach. (b) The developers of Greystones Marina (seen here under construction in 2009) acknowledge that it will reduce longshore sediment supply and increase erosion rates on the adjacent beach. The proposed solution was to nourish the beach*

uninformed and pro-development planning ethos (Cooper and McKenna, 2009). The presence of World War II pill boxes on the beach, 40 m seaward of the dune front, provided clear evidence of long-term erosion that was either ignored or not deemed important when permission was granted for construction of the dwellings.

■ **Figure 11.9** *The shoreline and beach at Rossnowlagh has been defended by individual rock armour emplacements in front of badly sited newly constructed houses and caravan parks*

Alternative approaches

Although shoreline armouring is the dominant response to coastal erosion, there are a few examples of other approaches. In response to coastal erosion of adjacent soft cliffs, the east coast railway line south of Dublin has been realigned several times since its construction in 1856, although sections of the cliff have recently been armoured. Rather than any deliberate decision to retreat in the face of coastal erosion, there are a few instances where property has been lost through erosion (Figure 11.5). Most of these situations arose several decades ago and reflect a lack of resources rather than formal policy. In Lough Swilly, erosion has breached nineteenth century embankments that protected reclaimed agricultural land. These have not been repaired and the land is reverting to intertidal flat and salt marsh. This situation reflects the comparatively low value of agricultural land compared to the cost of maintaining

it. An important policy development in erosion management is that the National Trust (the single biggest coastal landowner in Northern Ireland) has recently adopted a policy of non-interference in coastal processes (National Trust, 2006). Sea defences will not be built or maintained and some of its beach infrastructure has been constructed in such a way as to render it demountable and moveable should it be threatened by future coastal erosion.

Soft engineering in the form of beach nourishment is uncommon in Ireland. A nourishment scheme at Rosslare, immediately south of Wexford harbour was carried out in the winter of 1994–95. This involved the emplacement of 160,000 m^3 of sediment dredged from an offshore borrow area in association with installation of rock groins and protection of the cliff toe with rock armour. Following an initial beach accretion in the summer after emplacement, the nourished sand had been lost and the shoreline retreated to approximately its pre-nourishment position within a year. In a similar scheme at Bray, south of Dublin, following rock groin and breakwater emplacement, 240,000 m^3 of shingle was pumped onshore to restore the beach and protect the promenade at its rear. Nourishment was also implemented at Malahide, north of Dublin, to dispose of dredge spoil from an adjacent marina.

At Five Finger Strand, County Donegal, O'Connor et al. (2010) reported an instance where unnecessary installation of rock armouring was avoided. Five Finger Strand is a 350 m wide, 1,700 m long sand beach on the rural northwest coast of Donegal. It is backed by a wide, well-vegetated sand dune system and has a tidal inlet at its southern margin.

Erosion is very evident at this site and a dune scarp over 20 m high extends along much of the beach. The retreat of the dune is part of a cyclic pattern of erosion and accretion linked to switches in the position of the tidal inlet and its associated ebb delta (Cooper et al., 2007). This causes the dune front position to fluctuate by over 100 m. The current phase of erosion began in 1995 and continues to the present, but there is abundant evidence that this is a reversing pattern that has occurred several times over the past two centuries. Only sand dunes, used as pasture, are within the erosion zone and no infrastructure is at risk. Despite this, there is a public perception of an erosion "problem" and consequently intense pressure on the local authority to 'protect' the coast.

The perception that erosion at Five Finger Strand constitutes a 'problem' is related to human value judgements that beach and dune erosion is fundamentally undesirable, and a vegetated frontal dune is more desirable than a steep dune scarp. While it is true that the physical conditions of the erosive phase are somewhat less suitable for recreation, in themselves they do not indicate that the physical functioning or integrity of the beach/dune system has been compromised.

If public clamour for rock armour was to be satisfied, the defences would damage the scenic quality of the site (Figure 11.10), and by encouraging wave reflection and scouring may put the beach at risk, as it would not be economically feasible to armour the entire dune scarp. Even if a limited stretch of dune front were armoured, erosion would continue elsewhere and would accelerate at the armour end points. If the dune scarp position were fixed by armour, an essential mechanism for beach survival is threatened, because the sand stored in the dunes permits the beach system to adjust to and absorb the impacts of storms.

Given that all that is at risk for the foreseeable future is low intensity dune pasture and the seaward edge of a parking area, there is no argument for stabilization on cost-effectiveness grounds. Alternative soft engineering approaches, such as, sand fences and marram grass planting, do not have the negative side effects of armour but require an input of blown sand, which the presently sand-starved beach could not supply. Although it was recognized that the system does not require intervention, the local authority deemed it necessary to do something tangible in order to assuage public opinion. To address the political realities, but not create long-term damage to the coastal system, a low-cost and temporary approach was adopted. A line of straw bales was emplaced at the dune scarp toe in the expectation that it might eventually be likely to be washed away, but in the short term might assist in the accumulation of wind-blown sand. As it turned out, the bales were largely eroded in the following six months, but their presence did not affect the ability of the coast to operate dynamically (Figure 11.11).

The decision against hard defences was probably only overcome because of a close working relationship between university researchers and local officials established during the EU-funded Corepoint project (O'Connor et al., 2010). Under routine operational procedures it would have been easier (but more costly and damaging) to emplace armouring than resist public pressure.

Discussion

For most of its history, Ireland saw comparatively little human intervention in coastal processes, due to a

■ **Figure 11.10** *Like most Irish beaches, the scenic view of Five Finger Strand as an element in the landscape is one of its greatest attributes. Armouring the shoreline would not only destroy the natural beach system by cutting off the beach-dune link which is an essential component of the system, but would damage its visual appeal*

combination of scattered rural populations and limited financial resources. Consequently, even late in the twentieth century, various studies (Carter and Orford, 1988; National Coastal Erosion Committee, 1992) reported a relatively small proportion of coastal defences in comparison with other European countries. The result was that around most of the coast, coastal erosion was unimpeded and the natural coastal sedimentary system was intact. Throughout the nineteenth and much of the twentieth centuries, human intervention in response to coastal erosion was limited to a few locations where coastal roads were threatened, and then usually over small distances (Carter *et al.*, 1993). There were some local instances where erosion

was caused by human interventions that unwittingly altered coastal processes through pier construction (e.g. at Bray and Courtown), estuary reclamation (e.g. at Wexford Harbour), and construction of promenades at seaside resorts (e.g. Portrush, Bray).

With increasing prosperity in the late twentieth century, however, came an increased ability (both financial and technical) to undertake coastal protection works and a consequent increase in the extent of shoreline armouring ensued.

The National Coastal Erosion Committee (1992) focused attention on how much of the coast was 'at risk' from coastal erosion and prompted a national assessment of the 'coastal protection' needs. It is note-

■ **Figure 11.11** *A sacrificial line of straw bales was deployed to satisfy public demands for something to be done, but in the knowledge that if unsuccessful, it was of minimal cost and would not cause long-term damage*

worthy that the focus was solely on coastal protection and the potential loss of land, rather than preserving the natural attributes of the coast. The zenith of this phase was reached in the 1990s and early 2000s when a dramatic increase in prosperity, fuelled by a construction boom, led to unprecedented levels of coastal development. Scattered individual homes and entire housing developments were built in attractive coastal locations with little regard for coastal erosion, reflecting a historic lack of experience in the land planning system in considering erosion (County Wexford is a notable exception, although as noted above, there are distinctly different approaches for new and existing developments). The construction of houses in areas at risk from erosion was probably underpinned by the belief that there were adequate resources for building shore protection works should erosion threaten the developments. Indeed many such developments are now fronted by rock armouring at public or private expense.

Throughout its history, and to the present day, coastal defence is 'the only show in town' when it comes to coastal erosion in Ireland (Figure 11.12). There is a very poor public understanding of coastal erosion and it is viewed only as a threat to human activities and property and not as an integral part of a functioning coastal ecosystem. The completion of a few beach nourishment schemes and recognition of the economic value of sand beaches indicate that policy-makers are beginning to recognise the damaging impacts of hard defences and therefore foresee an increase in beach nourishment as a coastal protection measure. However, the widespread adoption of this response to infrastructure being threatened will likely impede progress toward other management options (such as retreat and proactive discouragement of ill-sited development) that will enable coasts to fluctuate freely. In spite of past situations where buildings have collapsed under coastal erosion (e.g. Figure 11.3), it is difficult to foresee this being allowed to happen in the present circumstances (outside areas in the ownership of the National Trust), despite the need to do so in order to maintain the coastal ecosystem.

The sectoral nature of coastal management (nature conservation, recreation and tourism, transport, housing, etc. are under different government departments), coupled with the lack of a strategic approach to shoreline management, mean that coastal erosion is not considered from a sustainability perspective.

■ **Figure 11.12** *Hard coastal defence is currently the 'only show in town' as far as the Irish coast is concerned. Hard protection structures are used to protect any property at the coast including (a) a single mobile home; (b) scattered holiday homes built in sand dunes; and (c) to facilitate beach access at coastal resorts*

Instead, the fundamental principle underlying erosion management is property protection. The minor advances in implementing 'soft engineering' in practice at a few locations are at the mercy of local politics and individual personalities, as illustrated by the Five Finger Strand example. They have not contributed significantly to changing general practice. Aside from coastal restoration projects that reinstate damaged ecosystems, and despite the title of an Irish government manual (Government of Ireland, 1996), there is no public understanding that there is no such thing as 'environmentally friendly coastal protection'.

The future is likely to see an increase in the rate of sea level rise in the south and a change from sea level fall to sea level rise in the north. This will inevitably lead to natural coastline responses, some of which will be manifest as erosion as sediment is redistributed. Coasts with existing sea defences will be subject to coastal squeeze, and this is particularly likely to affect recreational beaches at tourist resorts (Cooper and Boyd, 2010). The nature conservation responsibilities associated with European designated sites do create a driver for considering coastal ecosystems, but this is often viewed only as a secondary consideration by bodies making decisions on coastal 'protection'. In an attempt to stimulate government action, the National Trust in Northern Ireland commissioned a study (Orford *et al.*, 2008) into three of its coastal sites to illustrate the multifarious issues that arise from climate change and sea level rise. This illustrated the conflicts around coastal squeeze for example, whereby maintaining sea defences would lead to a 50 per cent reduction in the area of adjacent tidal flats. The report highlighted the fact that these issues are currently viewed quite independently.

Although the benefits of a policy of prohibition of shoreline armouring in conserving dynamic coastal environments has been demonstrated elsewhere (Kittinger and Ayers, 2010), Ireland is far from this position. There is a pressing need for an approach to shoreline management that rather than only seeking to defend infrastructure, rather seeks to maintain coastal ecosystems.

References

An Bord Pleanála, 2001. Inspector's Report on development in County Wexford. *Report PL 26.123495*, Dublin, Ireland.

Burningham, H. and Cooper, J. A. G., 2004. Morphology and historical evolution of north-east Atlantic coastal deposits: The west Donegal estuaries. *Journal of Coastal Research, Special Issue 41*, 148–159.

Carter, R. W. G., Eastwood, D. A. and Pollard, H. J., 1993. Man's impact on the coast of Ireland, in P.P. Wong (ed.) *Tourism vs environment: The case for coastal areas*. Kluwer, Dordrecht, The Netherlands, 211–220.

Carter, R. W. G. and Orford, J. D., 1988. Ireland, in H. J. Walker (ed.) *Artificial structures on shorelines*. Kluwer, Amsterdam, 155–164.

Carter, R. W. G., Johnston, T. W., McKenna, J. and Orford, J. D., 1987. Sea-level, sediment supply and coastal changes: Examples from the coast of Ireland. *Progress in Oceanography*, 18, 79–101.

Carter, R. W. G. and Wilson, P., 1991. Chronology and geomorphology of the Irish dunes, in: M. B. Quigley (ed.) *A guide to the sand dunes of Ireland*. European Union for Dune Conservation and Coastal Management, Dublin, 18–41.

Casey, J., 2009. Development of Ireland's coastal protection strategy. Flood defences and coastal structures seminar, University College Cork, January 2009. Available at: http://www.ucc.ie/en/hmrc/research/coastalres/ (accessed 24 August 2012).

Cooper, J. A. G. and Boyd, S., 2010. Climate change and coastal tourism in Ireland, In: A. Jones and M. R. Philips (eds) *Disappearing destinations: Climate change and future challenges for coastal tourism*. CAB International, Wallingford, 125–143.

Cooper, J. A. G., Jackson, D. W. T., Navas, F., McKenna, J. and Malvarez, G., 2004. Identifying storm impacts on an embayed, high energy coastline: Western Ireland. *Marine Geology*, 210, 261–280.

Cooper, J. A. G., McKenna, J., Jackson, D. W. T. and O'Connor, M., 2007. Mesoscale coastal behavior related to morphological self-adjustment. *Geology*, 35, 187–190.

Cooper, J. A. G. and McKenna, J., 2008a. Social justice and coastal erosion management: The temporal and spatial dimensions. *Geoforum*, 39, 294–306.

Cooper, J. A. G. and McKenna, J., 2008b. Working with natural processes: The challenge for Coastal Protection Strategies. *Geographical Journal*, 174, 315–331.

Cooper, J. A. G. and McKenna, J., 2009. Boom and bust: The influence of macroscale economics on the world's coast. *Journal of Coastal Research*, 25, 533–538.

Cooper, J. A. G. and Navas, F., 2004. Natural bathymetric change as a control on century-scale shoreline behaviour. *Geology*, 32, 513–516.

Defeo, O., McLachlan, A., Schoeman, D. S., Schlacher, T. A., Dugan, J., Jones, A., Lastra, M. and Scapini, F., 2009. Threats to sandy beach ecosystems: A review. *Estuarine Coastal and Shelf Science*, 81, 1–12.

Delaney, C. and Devoy, R., 1995. Evidence from sites in western Ireland of late Holocene changes in coastal environments. *Marine Geology*, 124, 273–287.

Devoy, R. J. N., 2008. Coastal vulnerability and the implications of sea-level rise for Ireland. *Journal of Coastal Research*, 24, 325–341.

Dodds, W., Cooper, J. A. G. and McKenna, J., 2010. Flood and coastal erosion risk management policy evolution in Northern Ireland: Incremental or Leapfrogging? *Ocean and Coastal Management*, 53, 779–786.

EEA (European Environment Agency), 2006. *The changing face of Europe's coastal areas*. EEA Report, June, Copenhagen.

Gibson, C. E. (Editor), 2011. *Northern Ireland State of the Seas Report*. AFBI/NIEA, Belfast, 112 pp.

Government of Ireland, 1996. *Environmentally friendly coastal protection: ECOPRO Code of Practice*. Dublin: The Stationery Office.

Guilcher, A. and King, C. A. M., 1961. Spits, tombolos and tidal marshes in Connemara and west Kerry, Ireland. *Proc. Royal Irish Academy*, 61, 283–336.

Healy, M. G. and Hickey, K. R., 2002. Historic land reclamation in the intertidal wetlands of the Shannon Estuary, western Ireland. *Journal of Coastal Research, Special Issue* 36, 365–373.

Kittinger, J. N. and Ayers, A. L., 2010. Shoreline armoring, risk management, and coastal resilience under rising seas. *Coastal Management*, 38, 634–653.

McErlean, T. C., 2011. The Maritime Heritage of Lough Swilly in J.A.G. Cooper, (editor) *Lough Swilly: A living landscape*. Four Courts Press, Dublin, Ireland.

McKenna, J., Power, J., Macleod, M. and Cooper, J. A. G., 2000. *Rural beach management: A good practice guide*. Donegal County Council, Lifford, Ireland, 109 pp.

McKenna, J., Cooper, J. A. G. and O'Hagan, A. M., 2008. Managing by principle: A critical assessment of the EU principles of ICZM. *Marine Policy*, 32, 941–955.

McWilliams, B. E., 1994. Executive summary in B. E. M., McWilliams (ed.), *Climate change: The implications for Ireland*. Department of the Environment, Government of Ireland, Dublin.

Mulrennan, M. 1993. Changes since the nineteenth century to the estuary-barrier complexes of north County Dublin. *Irish Geography*, 26, 1–13.

National Coastal Erosion Committee, 1992. *Coastal management: A case for action*. EOLAS, Dublin, Ireland (2 volumes).

NPWS (National Parks and Wildlife Service), 1997. Site Synopsis: Kilpatrick Sandhills, Site Code 001742, National Parks and Wildlife Service, Dublin. Available at: http://www.npws.ie/protectedsites/specialareasofconservationsac/kilpatricksandhillssac/ (accessed 24 August 2012).

National Trust, 2006. *The National Trust Coastal Policy*. Available at: http://www.nationaltrust.org.uk/ (accessed 24 August 2012).

O'Connor, M., Lymbery, G., Cooper, J. A. G., Gault, J. and McKenna, J., 2009. Practice versus policy-led coastal defence management. *Marine Policy*, 33, 923–929.

O'Connor, M., Cooper, J. A. G., McKenna, J. and Jackson, D. W. T., 2010. Shoreline management in a policy vacuum: A local authority perspective. *Ocean and Coastal Management*, 53, 769–778.

O'Connor, M., Cooper, J. A. G. and Jackson, D. W. T., 2011. Decadal behaviour of tidal inlet-associated beach systems, northwest Ireland, in relation to climate forcing. *Journal of Sedimentary Research*, 81, 38–51.

O'Hagan, A. M. and Cooper, J. A. G., 2002. Spatial variability in approaches to coastal protection in Ireland. *Journal of Coastal Research Special Issue*, 36, 544–551.

Orford, J. D., 1988. Alternative interpretation of man-induced shoreline changes in Rosslare Bay, south-east Ireland. *Transactions Institute of British Geographers*, 13, 65–78.

Orford, J.D., Betts, N., Cooper, J. A. G. and Smith, B. J., 2008. *Future coastal scenarios for Northern Ireland*. Unpublished Report to National Trust, Northern Ireland.

Orford, J. D. and Carter, R. W. G., 1982. Geomorphological changes in the barrier coasts of south Wexford. *Irish Geography*, 15, 70–84.

Orford, J. D., Cooper, J. A. G. and McKenna, J., 1999. Mesoscale temporal changes to foredunes at Inch Spit, south-west Ireland. *Zeitschrift fur Geomorphologie*, 43, 439–461.

Orford, J. D., Murdy, J. M. and Wintle, A. G., 2003. Prograded Holocene beach ridges with super-imposed dunes in north-east Ireland: Mechanisms and timescales of fine and coarse beach sediment decoupling and deposition. *Marine Geology*, 194, 47–64.

Wexford County Council, 2007. *Wexford County development plan 2007–2013*. Available at: http://www.wexford.ie/wex/Departments/Planning/DevelopmentPlans/ (accessed 24 August 2012).

12 France

Edward J. Anthony and François Sabatier

Introduction

France has a unique position in Europe in terms of its 5,500 km-long seaboard, as it is the only country with three maritime fronts: the southern North Sea and eastern English Channel, the Atlantic and the Mediterranean. The compact shape of the country, nationally referred to as 'the Hexagon', implies that large areas inland are several hundreds of kilometres distant from the coast. The presence of an important seaboard, the high degree of economic development, and attractiveness of the country, in terms of both international investments (consistently in the top five countries in attracting investments over the last decade) and tourism (the world's leading tourist destination over the last decade, according to statistics from the World Tourism Organization of the United Nations – http://unwto.org/), are a number of decisive factors that have, directly and indirectly, brought pressure to bear on the coasts of France.

Communes with a seaboard represent only 4 per cent of the total surface area of communes in France, but account for 10 per cent (6.5 million) of the permanently resident population, a figure that has been increasing strongly over the years. The population density of these coastal communes is 272 inhabitants per km², which is over 2.5 times the national density. Urbanized zones make up for 9.8 per cent of the territory of coastal communes compared to only 4 per cent for the national territory. Summer tourist population levels can increase up to a hundred-fold in certain coastal communes, as millions of French and foreign holidaymakers,

notably from northern Europe, flock to the coast, especially the Mediterranean. Beach recreation has, by far, the lion's share of this seasonal increase in coastal population. Given the limited beach-holding capacity, this has brought a number of problems to bear on the stability of these beaches, exacerbated in the past, and still to a certain degree, by a lack of concerted beach protection and management effort, with every commune fending for itself in the defence of its own beach space.

The coast of France

The geological heritage

The coast of France is a heritage of the complex geological history of Western Europe and the Mediterranean, dominated by the last major events associated with the Alpine orogeny. The western seaboard, comprising the coasts of the southern North Sea, the English Channel and the Atlantic, forms the maritime fringes of tectonically mildly uplifted marine sediments that form plateaux and plains (Figure 12.1). These are limited inland by the Alpine massifs that extend south to the Mediterranean, and by the Pyrenees between France and Spain. The Alpine massifs serve as watersheds for the Loire and Seine, two of the largest rivers in France, the estuaries of which debouch respectively on the Atlantic coast and the eastern English Channel (Figure 12.1). The basins between these mountains form the relatively low-lying western seaboard, comprising, respectively, cliffs cut in variably consolidated strata where uplift has been more significant, and coastal plains and river valleys in less uplifted or synclinal areas, especially along the Aquitaine coast. This pattern is interrupted along the rugged coast of Britanny, which consists of a predominantly granitic mass, inherited from the earlier Hercynian orogeny. The Garonne river debouches from the Pyrenees into the Atlantic (Figure 12.1). The Pyrenees, and especially the Southern Alps in the south-eastern corner of the country, serve as catchments for numerous small steep Mediterranean rivers, many of which

formed deltas during the Holocene. By far the most important delta in France, and in the western Mediterranean is that of the Rhône river (Figure 12.1) and is associated with a graben structure between the Massif Central and the Alps, therefore it has been abundantly sourced in sediments in the past by these two mountain masses.

Coastal morphology

The geological heritage resulting from the spatial distribution of the Alpine and older eroded

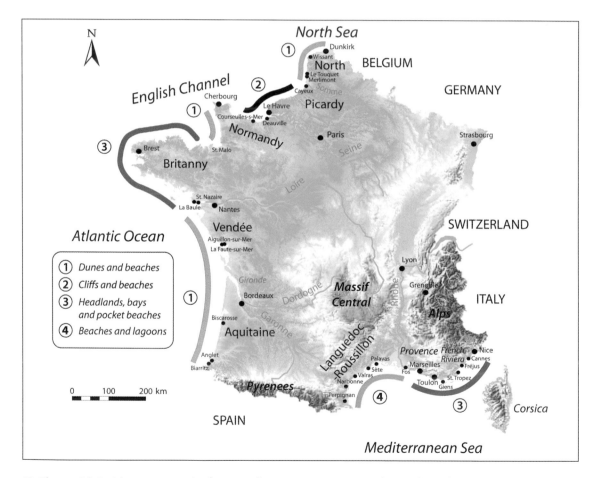

■ **Figure 12.1** *Major topographic features of France, rivers, towns and coastal morphotypes. Large areas of the west and north of the country consist of fertile plains and plateaux. The map also shows locations of the numerous coastal towns and resorts cited in the text*

Hercynian massif in Britanny has imprinted a predominantly rock headland and pocket beach shoreline in France (2. in figure 10.1), which form 2,270 km of the total length (41 per cent, of which 13 per cent are cliffs). Open beaches account for 1,950 km (35 per cent) of the total length, while sand flats, mudflats and salt marshes, largely empoldered today, form 1,320 km (24 per cent). The proportion of these percentages taken up by artificial shorelines is unknown. Artificial shorelines are mainly associated with ports, marinas, and reclamation fill, and include the open-coast commercial ports of Marseilles and Dunkirk (Dunkerque), the military port of Toulon, and the French Riviera airport of Nice, built on the reclaimed delta of the Var river enlarged by engineering works (Anthony, 1994).

An important aspect of the coastal geomorphology of the North Sea, English Channel and Atlantic sectors is an absence of deltas. With the exception of the Rhône, all major rivers debouching on the coast of France are characterized by still infilling estuaries that trap sediments from both inland and the sea. As a result, these coastal sectors have mainly been sourced in sand and gravel by shoreface sediments reworked by storms, tidal currents, and, on shores facing the Atlantic, by long swell, and cliff erosion where active cliffs are exposed. Active cliffs cut into various types of sedimentary rocks are common along the English Channel and parts of the Britanny coast. Chalk and clay cliffs tend to exhibit rapid retreat rates (> 1 m/yr), while hard sandstone and limestone sectors exhibit relatively low retreat rates (< 0.08 m/yr) except in a few critical eroding segments where natural changes in the platform sedimentary budget, facilitated by extraction of pebbles for more than half a century, have occurred (Pierre, 2006). Mass movements on cliff faces are essentially shallow-seated translational slides along with small debris falls and mudflows.

Cliffs occur in places as hard-rock headland projections but are most commonly associated with sectors where low sand stocks mantle the shallow shoreface, especially between the Somme and Seine estuaries. The occurrence of cliff outcrops in this sector is jointly hinged on the inherited coastal morphology following the Post-Glacial Marine Transgression and on the abundance of sand stocks on the shallow shoreface. Differences in sand abundance on the shoreface in the Channel/North Sea sector are related to large-scale bedload sorting processes controlled by a combination of tides, dominant westerly winds and Coriolis forcing (Anthony, 2002). Patterns of regional dune development in relation to the wind are well illustrated by the Cherbourg Peninsula, in Normandy. The eastern and northern flanks of this Peninsula, which also lies astride the Channel and the Atlantic, comprise cliffs and numerous sand pocket beaches associated with limited dune development due to the dominantly offshore westerly winds. The north–south orientation of the western flank of the peninsula, the only major headland relative to the axial hydrodynamic and wind-forced Channel circulation, favours large-scale tidal gyres that have been important in creating nearshore sand pools that fed coastal dunes and the infill of numerous small estuaries. These are associated with inlets deflected southwards by longshore drift (Robin et al., 2009). The middle part of the Atlantic coast forms a transitional zone between the rugged coast of Britanny and the long, straight sand coast south of the Gironde estuary.

Dunes are best developed along west-facing coastal sectors: from the North Sea to the Somme estuary in Picardy, along the western flank of the Cherbourg Peninsula, and along the Atlantic coast. Massive dune accumulations have especially developed during the late Holocene, forming barriers bounding now largely reclaimed coastal-plain marshlands. The dunes formed by sand transport from the inner shoreface on the North Sea/English Channel coast have fossilized former cliffs in places. On this sector, cliffs commonly alternate with aeolian dunes. Dunes have been obliterated in many places along the coast by urban and port development (Figure 12.2). Along parts of the Atlantic coast, which exhibits the lowest coastal population densities in France, dunes were fixed in the past using pine forest plantations. Some of these areas have become neglected due to the rural exodus and dunes are becoming more mobile.

Polders have been established along various parts of the low-lying French coast (Figure 12.3) where

■ **Figure 12.2**
Urban extension to the detriment of coastal dunes, Merlimont, northern France

■ **Figure 12.3**
Reclaimed wetlands protected by the gravel barrier of Cayeux in Picardy

pressures for land have come from relatively high local population densities and where a long maritime tradition has existed. The empoldering of parts of the North Sea, English Channel and Atlantic coasts of France dates back to the Middle Ages. In the southern North Sea, polders are associated with Flemish settlements, while in other parts of the French coast, the practice of empoldering largely benefited from Dutch engineering expertise as early as the Middle Ages.

Beaches are sand, gravel or a combination of both clast types. Pure gravel beaches are relatively rare; mixed sand and gravel beaches are commoner. In the English Channel, these coarse-clastic beaches are usually limited to sectors with eroding chalk cliffs rich in flint clasts, as in the case of the Cayeux barrier (Figure 12.3), the most important gravel barrier in France. Coarse sediments are also typical of some of the pocket beaches in Britanny. The French Riviera beaches, from near Nice to the Italian border, also comprise gravel derived from puddingstone formations rich in rounded clasts.

Sand beaches are the dominant beach types and are well developed in association with the coastal dune systems of the southern North Sea and English Channel down to the Somme estuary, on the west-facing coast of the Cherbourg Peninsula, and on the Atlantic and Mediterranean coast. Beaches vary in setting from open, along the afore-mentioned coasts, to embayed, and occur as both barriers and fringing beaches. The Britanny coast is characterized by granite headlands, comprising numerous more or less highly embayed sand and gravel pocket beaches, some of which are associated with dune development. Embayed and pocket beaches are also very common along the rocky Atlantic Pyrenees coast between Biarritz and the border with Spain and in the eastern (Pyrenees) and western (Alps) extremities of the Mediterranean coast.

The beaches of the North Sea and eastern English Channel are characterized by multiple intertidal bars (Reichmüth and Anthony, 2007). Beaches on the more wave-exposed western shores of the Cherbourg Peninsula exhibit a flat barless dissipative low-tide zone and a relatively steep and much narrower sand to gravel high-tide zone (Levoy et al., 2000). The long, straight beaches of the Atlantic exhibit dominantly intermediate crescentric-bar morphology (Castelle et al., 2007). Sand beaches are also the dominant type in the Mediterranean, forming long stretches along the central coastal sector between the mountainous extremities. Where the Pyrenees and Southern Alps meet the sea, rocky shores comprising bold plunging cliffs, pocket beaches and a few larger embayed beaches abound. These sectors contrast with the low central sector where beaches are capped by low to poorly developed dunes due to strong offshore winds. This central sector forms a barrier-lagoon system associated with sand supply from several small rivers. The barriers are generally narrow, stationary forms perched atop a substrate of lagoonal mud and show little net progradation since sea level stabilized in the middle Holocene (Barusseau et al., 1996). Unlike classical barrier island-lagoon systems associated with well-developed tidal inlets, these barriers lack inlets in this low-tide range setting, but are subject to overwash in places during storms.

East of the headland of Sète, the beach system is largely dominated by the Rhône delta. Build-up of the Rhône delta plain began at about 7,000 years BP and took place in several stages which are still visible in the present morphology, including palaeochannels, sandy beach ridges, dunes and spits (Vella et al., 2005). The present delta shoreline assumed its shape at the beginning of the eighteenth century, characterized by a series of well-defined littoral drift cells (Sabatier et al., 2006). Between the Rhône delta and Italy, the rocky shores of Provence and the French Riviera comprise numerous pocket beaches and several longer embayed beaches, the most important of which are those on either side of the remarkable tombolo of Giens, and those of St. Tropez, Fréjus, Cannes and Nice. These beaches are generally associated with small wave-dominated river deltas that have sourced sediment supply in the past. Similar headland-bay beaches characterize the island of Corsica.

Wave climate and tides

The morphological diversity of the coast also goes with a wide variety of hydrodynamic contexts (Figure 12.4). Tidal ranges are low microtidal (< 0.7 m at spring tides) in the Mediterranean; micro- to macrotidal (2–5 m at spring tides) along the Atlantic coast; macrotidal (5–8 m at spring tides) in the eastern English Channel and southern North Sea; and peak to megatidal (> 8 m at spring tides) near St. Malo. These variations are hinged on continental shelf width, as along the Mediterranean and Atlantic seaboards, and on the prevalence of a shallow bathymetry bordering the epicontinental English Channel and North Sea, with tidal range peaks occurring in shallow embayments and estuaries.

Wave contexts range from relatively short fetch conditions in the southern North Sea and Mediterranean, to large fetch swells along the Atlantic coast (Figure 12.4). The dominant winds affecting the southern North Sea and eastern English Channel coast are from south to west-southwest, north to northeast, and southeast. The wave regime progressively evolves from a storm-dominated short-fetch sea context in the southern North Sea and eastern English Channel to a mixed sea and swell regime, as the coast opens up to the Atlantic on the western façade of the Cherbourg Peninsula. Between

the eastern façade of the Cherbourg Peninsula and Belgium, offshore modal significant wave heights are less than 1.5 m, but may attain up to 3 m during storms. Considerable refraction and dissipation of waves occur over the shallow nearshore zone, resulting in modal inshore wave heights less than 0.5 m high more than 80 per cent of the time. Wave periods are in the range of 4 to 6 s. Currents recorded during the course of numerous field experiments (e.g., Levoy et al., 2000; Sedrati and Anthony, 2007; Héquette et al., 2008) show a clear hydrodynamic pattern modulated by tidal range and by wind and wave activity. In calm weather, current directions are closely conditioned by the tide. During storms, tidal control of the mean currents becomes subordinate to that of wind forcing and the current velocities

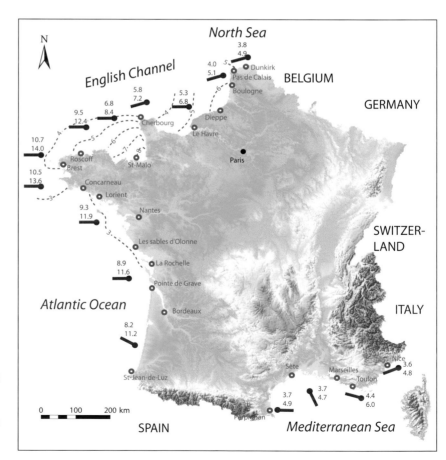

■ **Figure 12.4**
Significant wave heights with 1-yr and 10-yr return periods and wave approach directions, and mean equinox tidal range lines

increase significantly. Waves impinging on the Atlantic coast are generated by W–E depressions in the North Atlantic Ocean and the climate is therefore, strongly seasonal. Winters are characterized by strong northwest to southwest winds that generate dominantly west–northwest waves with heights ≥ 10 m and periods of up to 18 s during large storms. Summers are characterized by relatively low energy conditions. Intermediate conditions between high-energy winter waves and low-energy summer waves prevail in spring and autumn. The central part of the Atlantic coast of France forms a transitional zone between the rugged coast of Britanny and the long, straight sand coast south of the Gironde estuary. Island masses formed by Oléron and Yeu Islands induce variable coastal penetration of high-energy swells. Wave heights range from 1 to 2 m. Longshore sand drift in this sector is generally directed southward and ranges from 50,000 ± 20,000 m^3/yr to 140,000 ± 30,000 m^3/yr (Bertin *et al.*, 2008). The long straight dune-bound sandy Atlantic coast is interrupted by the 5 km wide Arcachon lagoon tidal inlet, which contains a complex sand bar and inlet system. Tidal currents are intense close to the Arcachon lagoon inlet, but are not significant on the rest of this open coast in comparison to wave-induced currents. Castelle *et al.* (2007) reported, from observations by fishermen, lifeguards and surfers, strong tidally modulated rip currents over the intertidal morphology, causing deaths by drowning each summer. The high waves impinging on this coast also generate strong longshore currents that induce a southerly net longshore drift of approximately 700,000 m^3 of sand per year (Castelle *et al.*, 2006).

The French Mediterranean coast is exposed to fetch-limited, low-energy wind waves (Figure 12.4). The annual and 100-year significant wave height values offshore are respectively 3.4 m and 6 m. The mean deepwater wave height is 0.77 m and the mean peak period 4.7 s. Waves approach from the northeast to southwest, with the easterly window being dominant. Surges associated with local depressions during storms may add up to 1 m of vertical wave excursion.

Coastal development

Although open to the sea along three seaboards, France cannot be considered as a truly 'maritime' country, such as could be said for the Netherlands, Britain and Italy, for instance, with only 10 per cent of the population living on the coast, although this percentage has increased significantly over the last five decades. There are few major cities on the open coast, the most important being Marseilles (more than 1 million inhabitants).

Compared to many other developed countries where competition for coastal space has been high, especially between industry and tourism, as in Italy (Pranzini, this book), there has been little competition of this type in France. The development of major space-consuming industrial estates has been hinged on the urbanization pattern in France, with, at most, significant coastal infrastructures being limited to major ports with an affirmed industrial tradition, such as Fos near Marseilles, and Dunkirk, while several other estates are located in ports built within estuaries, notably Le Havre on the Seine river estuary and St. Nazaire facing Nantes on the Loire river estuary, while the once bustling port of Bordeaux on the Gironde estuary has become moribund. A few industrial estates have developed on the Mediterranean coast, such as near Sète but much development of estates on the French coast over the last fifty years reflects the pressure from mass tourism.

Apart from the Flanders, Normandy and Britanny coasts, where there has been a long maritime tradition, much of the development of urban centres fronting the sea has been associated with the exponential rise in tourism and associated leisure port development since the early 1960s, especially in the Mediterranean. Prior to this, large parts of the coast were devoid of human settlements for various reasons. This situation prevailed, for instance, up to the early 1960s along the low-lying beach-bound Languedoc-Roussillon coast in the western Mediterranean, and still prevails along large stretches of the Atlantic coast.

The geographical context of France, characterized by large, attractive fertile plains, plateaux, river valleys, and rich mountain hinterlands, and by the development, since Roman times, of several favourably situated inland urban entities located at strategic crossroads that rose to prominence (Paris, Lyon, Toulouse, Lille, Grenoble and Strasbourg), are factors that did not particularly favour large-scale development of coastal settlements. In many low-relief areas, the coast was repulsive due to exposure to storms and flooding, the presence of wetlands, mosquito infestation and military invasions. Under these environmental conditions, coastal attractiveness weighed little compared to the rich and agriculturally attractive hinterland, especially given the relatively low population relative to the country size. Large areas of the low-lying barrier-lagoon Mediterranean coast were mosquito-infested until major eradication projects were undertaken in order to foster tourism in the 1960s. The Rhône delta is still largely mosquito-infested, thus keeping the area free from strong tourist pressures! High grounds, as in the eastern and western extremities of the Mediterranean, and protected embayed or estuarine settings, e.g. parts of the Britanny coast and the Channel and Atlantic estuaries, provided advantages for coastal settlements. Population densities and development pressures on the coast were not high enough as to significantly affect beaches and dunes. Elsewhere, as on the Flanders coast, a long maritime and empoldering tradition enabled the creation of the port of Dunkirk in the Middle Ages, this being one of the rare examples of a French port built on the beach.

Winter tourism as practised by rich European aristocracy and rising bourgeois families led to the early development, in the late 1880s, of a few resorts, alongside small fishing ports, such as Malo-les-Bains near Dunkirk on the North Sea, Deauville in Normandy, Biarritz on the Atlantic coast, and Nice and Cannes on the French Riviera. The flourishing of these resorts was favoured by development of railway lines. Nice and the French Riviera became connected to the national network in 1864, and La Baule on the Atlantic coast in 1879. Two major sources of impetus to the development of coastal settlements in France were the commencement of paid holidays in 1945 and the economic boom of the 1960s. Paid holidays paved the way for mass summer tourism that had to be accommodated by development of fast motorways and airports linking the major cities of France (and beyond, connections with European cities) to the sea, and coastal camp-sites, hotels, marinas and leisure ports. Many of the more inland communes were duplicated with new beach-sited equivalents carrying the name 'Plage' (beach) (Biscarosse and Biscarosse Plage, Narbonne and Narbonne Plage, etc.), intended as an 'attractor' of mass, beach-oriented summer tourism.

Development generated by mass tourism and the economic boom of the 1960s left their imprint on all three seaboards of France, although pressures have been very variable, attaining their maximum in the Mediterranean, the only sector where large-scale planned development involving joint state and private capital ventures was implemented. The large-scale changes brought about to the Mediterranean coast west of the Rhône delta in the 1960s reflected state efforts to:

1. diversify the economy of the coastal communes of the Languedoc-Roussillon region, which largely depended on agriculture, and especially on table wine production;
2. capture mass tourism from northern Europe and France that was routed directly down to the booming sunny resorts of Spain;
3. provide mass summer tourism to alleviate pressure on the French Riviera where limited tourist accommodation space went hand in hand with the prevalence of 'high-class' tourism based largely on leisure boating (Anthony, 1994).

State involvement in what was a large-scale planned venture implemented by several French ministries following recommendations of the Racine Report, on the Languedoc-Roussillon coast, was also a way of controlling tourism development and the ensuing increasingly artificial shorelines generated by such development on the French Riviera. On this corner of the French coast, the situation was entirely out of state control, and investments involving major leisure port and marina development projects were benefiting from funds of dubious origin that were laundered in casinos on the Riviera and in Italy (Anthony, 1997). Indeed, much of the coast east of the Rhône delta has been under strong pressure associated with one of the earliest forms of tourism development, especially on the afore-mentioned French Riviera. The Riviera started catering in the late nineteenth century for European aristocracy and the international gentry seeking sun and mild winters. In this area, the few beaches and predominantly rock coasts form the narrow shoreline rim of a steep continental margin, where the southern French Alps abruptly meet the Mediterranean Sea. The high mountainous relief has resulted in a concentration of over 90 per cent of the French Riviera population of 2 million on the narrow coastal fringe, thereby bringing pressure to bear on the Riviera beaches. These have been at the forefront of urban and socio-economic development pressures on the French Mediterranean coast.

Short-haul tourism from Paris, favoured by the short distances to the southern North Sea and the English Channel, promoted the prosperity of many villages in northern France and Normandy hitherto dedicated to fishing, such as Le Touquet and Deauville, while fostering creation of new coastal resorts. Similar developments have been much more limited on the more wave-exposed Atlantic coast where winter and summer resorts are mainly located on rock coast and embayed beaches adjacent to the foot of the Pyrenees, where the climate has less rain and winter temperatures are milder than on the coast further north. Large sectors of the coast south of the Gironde estuary are still devoid of settlements, and plans to develop this coast by joint state and private ventures along lines, such as those used for the Languedoc-Roussillon coast in the 1960s and 1970s, were never implemented to the scale initially planned. This was purportedly due to a combination of a rather wet climate, the presence of significant portions of wetlands that required massive drainage, together with relatively low population densities.

Coastal erosion and management

The present status of coastal erosion

Appropriation of the shores that were often avoided in ancient times and in the Middle Ages for safety, health and other reasons, has resulted in progressive destabilization of the beaches and dunes. Over the twentieth century, this situation has been characterized, especially along the Mediterranean coast, by misguided patterns of economic development that have exacerbated coastal erosion while endangering coastal ecosystems. It must be stated at the outset that reliable statistics are lacking on the actual state of the French coast in terms of stretches affected by erosion, stability or accretion. The situation appears to be relatively well restrained for parts of the Mediterranean, together with parts of the coast of Normandy and the North where rates of natural and/or human-induced erosion have been fairly well documented (Box. 12.1). Few of the numerous case studies in France purportedly highlighting unceasing erosion are, however, actually based on long-term and rigorous datasets from which such generalized chronic erosion can be distinguished from local or temporary retreat. This is especially the case for the sandy, dynamic high-wave energy Atlantic coast. There is, therefore, a recognized need for a rigorous characterization of the status of the coast in terms of stability, erosion or accretion.

Causes of coastal erosion

In order to understand the mechanisms that have led to unremitting erosion of several sectors of the French coast over the last two centuries, and notably of the beaches, it is necessary to recall the sediment sources that have sustained these beaches. Low coasts are associated with various sources and modes of sediment inputs, from the shoreface, from commonly heavily managed rivers, from retreating cliffs, and via variably developed littoral drift cells.

Beach erosion patterns differ considerably, depending on the geomorphic context and sediment supply modulation by both natural and anthropogenic factors

BOX 12.1 WISSANT BAY, NORTHERN FRANCE, A MAJOR EROSION HOTSPOT ON THE FRENCH COAST

Wissant Bay is a mildly embayed coast that forms a longshore sediment transport cell between the headland of Cape Gris Nez and the bold chalk cliffs of Cape Blanc Nez in the Dover Strait (Figure 12.5a). Parts of Wissant Bay show some of the highest rates of historical shoreline retreat in France. The southwestern sector of the bay exhibited a retreat of the shoreline of up to 250 m between 1949 and 1997, following an early period of stability (Aernouts and Héquette, 2006). Although gross rates of annual shoreline retreat by time slices of several decades have been identified from careful interpretation of long series of ortho-rectified

■ **Figure 12.5** *Wissant bay, northern France, a major erosion hotspot on the French coast. (a) and (b): map and aerial photograph of bay; (c) and (d): ground photograph showing dune erosion and remnants of World War II wall on the beach, and beach lowering and erosion at the base of the inclined seawall. Note the World War II blockhouse in the background in d.*

aerial photographs by these authors, which may attain up to 4 m a year, little is known of the mechanisms of such retreat. The nearshore zone exhibits numerous tidal ridges and relict and presently active sand banks, collectively named the Flemish banks. The most important of these banks to the long-term evolution of Wissant Bay is the Line Bank, which lies very close to the bay shoreline (Figure 12.5a). Aernouts and Héquette (2006) showed that shoreline retreat in the southwestern part of the bay has been matched by net sand loss by the Line Bank offshore with attendant bathymetric lowering. A SWAN wave simulation by these authors further showed that incident wave energy in this eroding sector had increased in 2002 relative to conditions in 1977 due to the lower bathymetry of the Line Bank. It is possible that the dynamics of the Line Bank itself are embedded in a clearly identified process of larger-scale storm and tide-controlled sand migration from the eastern English Channel towards the Dover Strait and the southern North Sea (Anthony, 2002). Lowering of the Line Bank may also have been exacerbated by seabed aggregate extractions in the past. Erosion of Wissant Bay over the last few decades has led to the exposure of peat outcrops on the beach representing former backbarrier vegetation, and in areas close to Cape Gris Nez, an upper beach frame of gravel has accumulated as the dune front has retreated (Anthony and Dolique, 2001). This chronic erosion poses a threat to the beachfront commune of Wissant (Figure 12.5b), as there is a likelihood of storm breaching of the narrow dune barrier. Figure 12.5c shows remnants of the Atlantic Wall destroyed by this persistant erosion. As a result of this erosion, the profile of the beach in Wissant (Figure 12.5d) shows significant lowering compared to sectors of beach farther away from this beachfront resort (Sedrati and Anthony, 2008). Note the partially collapsed World War II blockhouse in the background on this photo. Sand released from unceasing shoreline retreat is progressively evacuated by strong longshore currents towards the northeastern sector of the bay (Sedrati and Anthony, 2007), which is presently undergoing accretion, thus highlighting a pattern of progressive beach rotation that may be embedded in larger-scale changes associated with the tidal sand ridges and banks that characterize the adjacent storm- and tide-dominated shoreface (Sedrati and Anthony, 2008).

(Box 12.1). Most beaches in France were stable or accreted during historical times because sediment supplies were not perturbed by urban growth, seabed extractions, cliff stabilization and river damming. Leaving the Mediterranean aside, the prevalence of still infilling estuaries on much of the coast has implied that sediments for beaches, especially those having a significant gravel component, have been derived mainly from cliff erosion. Cliff stabilization has essentially been aimed at protecting rich farmland and a few settlements that progressive natural cliff retreat had brought to the brink of collapse over time, such as on the Picardy coast (Figure 12.5). Such cliff stabilization has been an important cause of gravel beach erosion on the coasts of Normandy and Picardy, leading coastal settlements into a spiral of constructing beach protection structures. The relatively high density of coastal settlements on this

part of the coast has, of course, had a snowball effect on the implantation of structures on the shoreline, each commune bent on assuring its share of a dwindling sediment supply.

In the Mediterranean, where beaches have been sustained by sediment supply from river deltas, erosion is set within an overall background of lesser river sediment input related to a decrease in the frequency of major floods (end of the Little Ice Age), catchment reforestation, dam construction and dredging activities since the 1950s. This erosion is likely to continue in the future because of dwindling fluvial sediment supply from river catchments. In the case of the Rhône delta, following shifts in river channels and mouths, relict prodeltaic lobes are being reworked by waves, and their sediments are contributing partially to spit growth. These spits exhibit strong progradation and form the main sand drift

■ **Figure 12.6** *Cliff-top urbanization and stabilization has led to a reduction in beach sediment supply along the coasts of Normandy and Picardy*

termini associated with the afore-mentioned sediment cells, which, in addition to sediment supply by the prodeltaic lobes, are also sourced by chronic erosion of the delta shoreline (Figure 12.7) updrift of the spits (Sabatier *et al.*, 2009a).

The most direct and most important development effect has been a drastic reduction of beach width due to the growth of urban fronts. This situation is best illustrated along the steep, narrow French Riviera margin, where the railway to Italy, the coastal route and urbanization were all squeezed within the narrow coastal zone. In Nice, the capital of the French Riviera, the famous sea-front *Promenade des Anglais*, which is protected by a seawall, was built directly on the gravel beach, resulting in beach narrowing and further beachface destabilization. These changes included the reclamation and extension of the nearby Var delta plain through infill, and armouring of the shoreline for construction of the Nice-Riviera airport, thus completely cutting off the former natural gravel supply from this river (Anthony, 1994; Anthony and Julian, 1999). The present beach sediment budget is one of zero natural sediment inputs, resulting in chronic beach erosion (Anthony *et al.*, 2011).

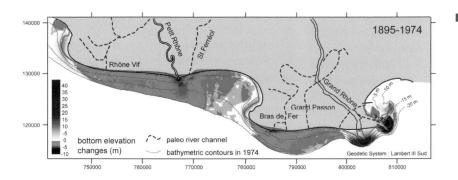

■ **Figure 12.7**
Changes in the shoreface bathymetry of the Rhône delta from 1895 to 1974 (from Sabatier et al., 2009a)

Coastal management practice

Coastal protection in France has a long history dating back to the Middle Ages, especially with regards to protection of reclaimed coastal plains. Prior to the Middle Ages, a few rare ports in the Mediterranean constructed by the Romans were protected by breakwaters, a fine example being that of the ancient port of Fréjus. Developments in the Middle Ages essentially concerned empoldering operations, especially in the southern North Sea and eastern English Channel, where coastal and estuarine plains were embanked throughout the Middle Ages and up to the twentieth century in order to gain both agricultural land and protect low-lying areas from the sea. These embankment techniques progressively gained the central sector of the Atlantic coast, where many marshland areas were reclaimed in the Middle Ages using techniques derived from the Netherlands.

There are, however, marked disparities in the extent to which the coast has been protected using artificial structures or harnessing existing dunes. The present distribution of coastal protection works reflects both the history of human occupation of the coast in the nineteenth century and the exponential development of coastal tourism and leisure activities in the twentieth century. Coastal management over the last two centuries has been regulated by a number of laws, the application of which has not been really efficient, as far as coastal protection has been concerned. The Law of 1807 conferred on the Ministry of Public Works the role of identifying the need for coastal defence works. This law concerned the coast, lakes and river banks and stipulated the setting up of landowner associations that had to bear, proportionately to the interests of each landowner, all costs of defence or maintenance works carried out, except where government interests were concerned, therefore providing a source of state subsidy for such works. In reality, subsidies for coastal defence were negligible, as this was not considered a priority area in France. Enforcement of this law was never really assured and in fact large-scale coastal development pressures rapidly outpaced the problem of private landowner responsibility in coastal defence.

Significant development of engineering structures associated with the coast essentially dates from the 1880s, when numerous coastal resorts, initially geared to winter tourism, were built. Since 1970, the local government representative of the Public Works Department, a member of the Corps of Engineers, takes decisions concerning the defence of urbanized coasts and this decision determines whether funding is appropriate. Funding is exceptional, and it may be argued that funds have been granted more on the basis of political lobbying or affiliation than on other considerations. Under the continuing pressure of coastal urbanization, generally spurred on by large-scale lucrative estate acquisition and development, the state decided, in the Law of 1973, to allow local communities to undertake coastal defence works where this was deemed necessary to preserve the common interest. This is presently the situation in France, where the municipality or community bears the costs of local defence operations, with the possibility of additional funding by the Regional Council.

With respect to beach nourishment projects, the Law of 1977, geared to environmental conservation, stipulated the necessity of an environmental impact assessment study when the project budget exceeded €1 million euros. This law was improved by a whole set of bills regarding the administrative procedures, bidding offers, accreditation, and public-interest inquiries, when the surface area of the projected works exceeded 2,000 m². Its efficiency has been impaired by these amendments, which are generally not attained or difficult to define. As a result beach nourishment projects have, by and large, circumvented this law (Hanson *et al.*, 2002). Further bills promulgated in 1979 stipulated the necessity of an administrative approval by the Prefect (the State representative), via the Public Works engineer, for hard defence structures built on public coastal land, which leaves the door open for nourishment operations to circumvent this regulation. The 'Littoral' Law of 1985, relative to management, protection and valorization of the coast, was promulgated with the aim of controlling the rampant and disorganized coastal urbanization. The law instituted, in particular, a development setback line of 100 m. A cursory examination of the coast, especially in the Mediterranean, shows that this law has not been respected and despite a few spectacular cases of demolition of illegal beachfront restaurants, the state has not been exemplary in enforcing the law.

Coastal management in France is now essentially under the control of local communal authorities, except on coasts adjacent to the three major ports of Marseilles, Le Havre and Dunkirk, which are managed by the port authority. An overall view of management practice in terms, for instance, of coastal sediment cells, has generally been lacking, although this situation has been changing in recent years. As a result, the spread of beach erosion has commonly been aggravated by individual communal efforts lacking a common view of the effects on downdrift sectors of engineering structures implanted in updrift sectors.

Coastal conservation

The relative inefficiency of the law and increasing environmental awareness led to the creation, in 1975, of the *Conservatoire du Littoral et des Rivages Lacustres* (Coastal and Lake Shores Heritage Trust), a government-funded institution, under the authority of the Ministry of the Environment, aimed at acquiring and protecting outstanding natural areas on the coast and lake shores. Apart from acquisitions by donations or legacy, the Conservatoire acquires coastal land by private agreement, by pre-emption, and sometimes by expropriation. The Conservatoire is empowered to carry out coastal restoration works, most commonly dune restoration (see earlier). The Conservatoire entrusts management of its lands to local authorities or to recognized local conservation groups. The Conservatoire decides on the land use, and is advised by a wide panel of specialists ranging from botanists to geomorphologists. By 2006, the Conservatoire was in charge of 732 km² of land, and owned 660 km, or 12 per cent, of the French coast. This included 21 per cent of the coast of Corsica, 11 per cent of the Mediterranean coast, 13 per cent of the English Channel and southern North Sea coast, and 8 per cent of the Atlantic coast. The annual budget of the Conservatoire is about €30 million, €25 million of which is set for acquisitions and development. The national government provides most of this budget, but contributions also come from local authorities, the European Union, corporations and private donors.

Coastal protection

France has a long tradition of civil and coastal engineering dating back to the Napoleonic era, when specialized higher education schools of mining and engineering were created. These schools have continued to form the engineering elite in France employed by both public authorities and private consultancies involved in coastal defence. This situation has also nurtured a form of consanguinity, in terms of coastal defence strategy, between engineers employed by communes or the state, and engineers employed by private consultancies and firms responsible for implementing defence works. Since defence projects involving construction of hard engineering structures are highly lucrative, the trend has been one of essentially hard engineering, especially during the phase of rapid coastal economic development from the 1960s to the 1980s.

There has been a clear opening in the 1980s towards softer coastal and beach protection methods such as beach nourishment. This was nurtured by several factors: the lack of structured coastal management in France, the relative inefficiency of the legislative framework related to coastal management, increasing pressure from development, and consequent beach erosion, larger environmental awareness, the failure or poor performance of many projects based on 'hard' and costly engineering structures, and the desire to test new softer methods that could provide viable coastal protection solutions. This opening has gone hand in hand with a diversification of the actors involved in coastal management and planning leading to greater checks and balances, especially under the impetus of pressure from environmental groups.

Design methods in coastal protection

Coastal protection projects have generally involved more or less detailed design studies, for instance, determination of the design water levels and wave climate, field surveys on beach topography, on shoreface bathymetry, and on sediment composition and distribution. This preliminary phase commonly includes a collation of all available useful environmental data that could enhance project implementation. Such data are generally culled from the abundant grey literature on beaches generated either by academic studies (from undergraduate to Ph.D. theses) or by studies commissioned by local and regional authorities. A detailed beach morphodynamic analysis is commonly carried out and a mobile-bed scale model has been the traditional tool in morphodynamic impact investigations since the 1950s (Hanson *et al.*, 2002). Shoreline numerical modelling is now almost systematically resorted to in design determination.

'Hard' coastal defence structures: rock armouring and seawalls

The earliest forms of coastal protection in France, as in many other countries in Europe, are rock armouring and seawalls. These essentially concerned initially empoldered areas reclaimed from the sea. Rock armouring is still widely used in France, but in many areas it has been replaced by masoned seawalls. Some rock armouring used in the late nineteenth to early twentieth centuries as winter resorts started developing, consisted of rip-rap chaotically emplaced on the upper beach or on the dune front. These generally turned out to be inefficient, progressively being

■ **Figure 12.8** *Rock armour protection of a cliff base in Normandy*

replaced by arranged blocks comprising various embedded levels of size from small fractions at the core to an outer casing of massive blocks. Rock armouring is still prevalent along areas where urban seafronts are absent, especially for cliff base protection (Figure 12.8).

The oldest beachfront seawalls were vertical wooden structures, hardly any of which exist today. These wooden walls were rapidly replaced by stone ones and later by walls constructed of concrete. The earliest vertical stone seawalls were constructed in the eighteenth century, initially mainly for coastal defence purposes (Pinot, 1998). This defence function was in fact amplified during World War II when the Atlantic Wall was built along large parts of the French coast by the German army. Much of this wall has been eroded (see Box 12.1) or removed but some coastal resorts have adapted or modified it for use as a coastal defence against storm flooding. Masoned or stone seawalls are presently the most ubiquitous form of coastal protection, as most coastal resorts and urban centres comprise a seafront promenade and a coastal route. Vertical seawalls still exist, sometimes alongside

inclined walls of various slope configurations. Seawalls have become much more popular than rock armouring because they are seen to be more aesthetic and, once enlarged, are generally backed by a seafront promenade (Figure 12.9), but some seawalls are still fronted by rock armouring, especially in northern France and Normandy. In many coastal resorts, the seafront promenade on the seawall serves as a hallmark of quality, important in attracting tourists, a fine example being the *Promenade des Anglais* in Nice (Figure 12.10). This advantage complements the defensive role of many seawalls in France, but there are examples, such as Wissant, in northern France, where a deficient sand supply has resulted in beach lowering in front of the seawall (Box. 12.1).

Another example of seawall failure is that of Véran, on the Rhône delta beachfront. The seawall was built in the 1970s 300 m behind the beach in order to protect a salt-making estate. Continuous beach retreat finally directly exposed the seawall to waves and the wall was destroyed by a storm with a 100-year return interval in 1997, thus exposing the industrial estate. The wall was hastily reconstructed, based on the same

■ **Figure 12.9** *An example of an inclined seawall near Dunkirk in northern France. Seawalls are the most common form of urban coastal defence in France*

■ **Figure 12.10** *The role of seawalls both as a defence structure and as a landscape element integrated in certain prestigious urban fronts such as that of Nice*

design and dimensions as the earlier wall, but the chronic beach erosion (Sabatier *et al.*, 2009a) has been destabilizing it since the early 2000s. The salt-extraction estate in this area of the delta no longer exists and the Conservatoire du Littoral has purchased this coastal sector. After 40 years of vain efforts to contain beach erosion and maintain an economic activity in this area, the hard-defence protection culture has been replaced by one of pragmatic adaptation to retreat.

Coastal dune rehabilitation

Dune rehabilitation is widely practised in France, and has gained ground as environmental protection groups have gained impetus. Various sand-trapping designs have been implemented, e.g. fences, plantations (Figure 12.11), and geotextiles, the efficiency

of which has been demonstrated in many case studies (e.g., Anthony *et al.*, 2007). In certain cases, these operations are combined with significant sand nourishment and beach and dune profile modifications aimed at optimizing sand immobilization. A fine example of a recent project of successful large-scale dune rehabilitation is that of the Gracieuse Spit, at the eastern extremity of the Rhône delta. This spit protects the entrance to the port of Marseilles from storm waves. Following a test phase in 1988–89, construction of a 3.5 km artificial dune ridge, reinforced by fencing, was completed in 1993. The fences favoured rapid accretion on either side by onshore and offshore winds.

■ **Figure 12.11** *Dune rehabilitation in northern France using a combination of fences and grass plantations*

'Hard' beach protection structures: Groins

Groins are the most ubiquitous form of beach protection (Figure 12.12), although there is no indication as to how far back they have been in use in France. Groins vary from wooden structures, still largely used along parts of the embayed central coast of France between Britanny and Vendée, to rock and masonry structures. Wooden groins are probably not solely used for beach stabilization, as they are commonly efficient in trapping kelp collected by farmers. These groins can be easily removed to facilitate the movement of bulldozers used to collect the kelp. Over 95 per cent of the groins in France are the usual long or short orthogonal groin. T-groins are much less common, and are generally associated with artificial beaches. Groins are particularly abundant as beach protection structures on cliff shores that have been stabilized in Normandy and Picardy (Box. 12.2) and may be found associated with a seawall. Along many seafront beach sectors, especially on the coast of Normandy, the groin field in several beachfront communes generally ends downdrift with a jetty, which acts as a long terminal groin (Figure 12.13), an efficient way of depriving the neighbouring commune downdrift of sand (!) but many terminal groins are used to canalize river outlets. Jetties are also used to canalize urban terrestrial runoff to the sea, and to delimit small river mouths that are very commonly used as harbours.

■ **Figure 12.12** *A set of rock groins on the Vendée coast*

■ **Figure 12.13** *A jetty on the Normandy coast. Note the wider beach downdrift*

BOX 12.2 THE MASSIVE RECOURSE TO GROINS ON THE COASTS OF NORMANDY AND PICARDY

■ **Figure 12.14** *The massive recourse to groins on the coasts of Normandy and Picardy: a. the dense groin field on the gravel barrier of Cayeux; b. rock groins and an armoured seawall in Berck-sur-Mer just north of the Somme estuary; c. rock groins and rock seawall in Luc-sur-Mer, Normandy; d. rock groins and a masoned vertical seawall at Courseulles-sur-Mer, Normandy*

The pressure on beachfronts in France has been matched by a significant increase in the number of implanted groins, which is close to 1,000, although their distribution on the three coastal façades of France is quite variable, depending on the density of coastal resorts. They are generally constructed in sets of two or more groins, and the groin field tends to extend downdrift, as erosion ensues where updrift sediment trapping has been efficient. The largest number of groins (about 548 in 2011) is found along the English Channel coast and the western coast of the Cherbourg Peninsula, where a high concentration of coastal resorts exists. A particularly high concentration exists along the proximal and central sectors of the Cayeux gravel spit. As the cliffs downdrift were stabilized, the drop in gravel supply to the spit, which protects polders on the Picardy coast (Figure 12.14a) decreased dramatically, leading to a succession of groins, which now number 85, on a 9 km stretch of the gravel spit. High densities of groins are also found on the dune coast of Picardy (Figure 12.14b), along the cliff coast of Normandy, between the Somme and Seine estuaries, and especially along the north-facing sandy coast of Normandy (Figure 12.14c, d), where the number of groins attains a maximum of 97 on the beach fronting the resort of Courseulles-sur-Mer. The density decreases dramatically in Brittany where embayed beaches prevail. Groins become numerous once again along the central Atlantic coast, between La Baule and the Gironde estuary, and in the Mediterranean.

Interestingly, the number of groins has stagnated over the last decade, as stricter controls are imposed by the DREAL (a state agency responsible for overseeing, among several of its missions, coastal protection in each region). This situation also reflects the fact that groins on many coastal sectors have not solved the problem of beach erosion and where dense groin fields have been built, erosion has been displaced downdrift, creating problems for other communes. Groin dimensions have also been a problem in places, especially in the Mediterranean, where certain short groins constructed in the early 1970s turned out to be totally inadequate in stopping sand from drifting alongshore over inshore bars, or were destabilized and became detached from the beach as a result of never-ending erosion. There are grounds for believing that the massive construction of groins in France from the 1970s to the early 1990s was in part motivated by the bonus received by engineers for every groin. This bonus system was terminated in the mid 1990s, leading to a drastic drop in groin construction! Detractors have it that a similar bonus system instituted for the construction of roundabouts in France, to the benefit of the powerful lobby of civil engineers in the country, is deemed as being responsible for the sharp increase in these roundabouts on French roads since the mid-1990s!

'Hard' beach protection structures: Breakwaters

Breakwaters are much less commonly used in France for beach protection, no doubt due to their much more expensive implementation costs, especially along the southern North Sea/English Channel and Atlantic coasts, where larger tidal ranges require higher breakwater configurations. A few breakwaters have been constructed on the high-energy mesotidal beaches of the Atlantic coast. Not surprisingly, breakwaters are commoner in the Mediterranean (Figure 12.15). A rare example of breakwaters in a macrotidal setting is that of Malo-les-Bains on the Dunkirk seafront in the southern North Sea, near the border with Belgium (Figure 12.16). Following the extension of the port of Dunkirk, three breakwaters were built, two in 1978 and one in 1984, in order to stabilize the macrotidal beach fronting the residential and resort centre located downdrift of the port. The breakwaters were accompanied by two nourishment operations, in 1978 (200,000 m^3 of sand) and 1988 (160,000 m^3). Oblinger and Anthony (2008) have shown that the breakwaters are extremely efficient in completely attenuating impinging waves during a large part of the ebb and flow of

■ **Figure 12.15**
Breakwaters built in the early 1990s near Fréjus to stabilize a severely eroding portion of the beach. On the right of the photograph, the smallest of the four breakwaters is in fact a T-groin. Note the rock armouring in the left corner of the photograph, emplaced to protect the road

■ **Figure 12.16** *Breakwaters on a macrotidal beach, Dunkirk, northern France*

the tide. This has, however, generated a wave shadow zone associated with mud sedimentation in the lee of the breakwaters. The breakwaters generate long-shore flows, due essentially to wind and tidal forcing, rather than to waves, that lead to long-term beach stabilization. The structure of these currents behind the breakwater and the wave energy dissipation engendered by the breakwaters has led to a long-term balance that generates only mild morphological change and maintains sand *in situ*. In this regard, the breakwaters have been efficient. However, this efficiency is also favoured by both tidal translation rates and wave energy conditions. The overall fetch limited low to moderate energy context (due also largely to significant refraction and dissipation by the nearshore sand banks) and the relatively high tidal translation rates of wave fronts imposed by the large tidal range, especially in the mid-beach zone, inevitably lead to low morphological and sediment budget changes.

'Soft' beach protection structures: Submerged geotextile tubes

Although submerged geotextile tubes aimed at wave attenuation have been implemented on beaches for over a decade, the method is still relatively novel in France and has been experimented with only on a couple of beaches in the Mediterranean (Isebe *et al.*, 2008). The most recent project on a Marseillan beach, near Sète, implemented in 2010, concerns a 1 km stretch of shore on which a 2.5 m high and 6 m wide geotextile tube has been implanted at a distance of 350 m from the beach and a depth of 4.5 m below mean sea level. The beach has also been recharged to the tune of 600,000 m³. A study to assess the efficiency of the operation, initially scheduled for autumn 2011, has now been deferred to 2013.

Beach nourishment

There is a long history of beach nourishment in France and it has been inadvertently practised on many beaches which served as waste sites for materials excavated during building operations, especially hotels, such as in Cannes and Nice. Such waste

■ **Figure 12.17** *Oblique aerial photograph of Cannes showing the beach which was significantly widened from 1962 to 1964 in what was one of the world's first large-scale engineered beach nourishment projects. Also note the three leisure ports and an artificial beach with a central T-groin*

disposal on beaches was banned in 1926. One of the earliest and most well planned beach nourishment projects in the world concerned the famous 1 km long Croisette beach in Cannes (Figure 12.17). In their review of the situation of beach nourishment in Europe, Hanson *et al.* (2002) provided a good summary of practice in France, as well as a list of operations carried out up to about 2000. Beach nourishment in France has traditionally been coupled with hard structures, groins and breakwaters, used as supporting measures to minimize sand losses and maintenance. Since the 1990s, the protection of the Cayeux barrier (Box 12.2) by an impressive groin field has been matched by large-scale periodic recharge of gravel in the erosional proximal sector of the barrier, transported by trucks from the downdrift accretion sector.

Hanson *et al.* (2002) showed that in the most important nourishment projects, the nourishment option was chosen as a means of getting rid of available sand dredged to maintain navigable depths in a nearby harbour. Apart from the disposal of dredged sand from harbour extensions or maintenance, other project motivations have included the creation of artificial beaches, coastal defence and dune restoration. Over 30 artificial beaches, generally no more than a few hundred metres long, have been constructed on the Mediterranean coast between Marseilles and the Italian border, generally to create extra beach space for recreation in this area, where rock shores constrain available recreational space. Such beaches commonly require periodic nourishment and are generally protected by breakwaters, groins and T-groins (Figure 12.18).

Since one motivation is getting rid of dredged material, grain size has varied from shingle to fine sand depending on the location providing sediments that may be more or less suitable for beach nourishment, the use of fine sand in several projects having been associated with major losses (Hanson *et al.*, 2002). Nourishment material is generally dumped on the beach by trucks in the case of shingle and pebbles, and hydraulically in the case of dredged sand. Hanson *et al.* (2002) have also shown that in several cases, *in situ* tests were performed to check the design, but that

■ **Figure 12.18** *Part of the artificial beach of Larvotto in Monaco. The breakwater protecting the beach is reinforced by rock armouring on its seaward face and is connected to the beach during the tourist season*

monitoring after nourishment was, in most cases, not systematic.

The most important nourishment project in France has been that of Anglet, on the Atlantic coast near Spain. The shoreface nourishment involved a total volume of over 8 million m³ between 1974 and 1992. Another important but little known beach nourishment scheme concerned the urban front of Nice (Figure 12.10). The solution to the Nice beach narrowing has resided in artificial nourishment operations over the last 35 years in order to restore the sediment budget. The 4.5 km long beach was recharged from 1976 to 2005 to the tune of 558,000 m³, making this long-term operation one of the most important in France, and certainly one of the most significant for gravel beaches in the world. Nourishment ranged from nil in certain years (1979, 1980, 1983–85, 2001–2) to a peak of over 97,000 m³ in 2000. Curiously, the Nice beach nourishment scheme was not inventoried by Hanson *et al.* (2002) despite its importance and operation over the years. The total volume of this nourishment scheme at the time of publication of the inventory of Hanson *et al.* (2002), compared to total nourishment volumes for France, was about 5 per cent, and about 50 per cent when only gravel/shingle nourishment volumes are considered. Analyses of 50 transects covering the beach highlight no significant change in net beach width over this 30-year period of massive gravel nourishment (Anthony *et al.*, 2011). Since there is no alongshore gravel leakage away from the strongly embayed Nice beach, the relative stability in beach width clearly implies loss of recharged gravel to the steep inner shoreface inherited from the geological context of Nice beach at the flanks of the southern Alps. This heavy loss is probably exacerbated by the practice of artificial beach widening through the flattening of a narrow (5–15 m wide) mobile zone of the profile in order to enhance the 'carrying capacity' of this highly touristic beach. Beach widening and flattening following nourishment bring close to the very steep inner shoreface zone, several cubic metres of gravel for each metre of beach that may be permanently lost downslope during autumn and winter storms. Recharged gravel is redistributed alongshore and offshore leakage is probably enhanced where small narrow submarine canyon heads impinge on the beach, resulting in a very narrow shoreface. Mean beach width shows an oscillating alongshore pattern that may be due to the influence of these canyons as pathways of gravel loss offshore.

Beach nourishment is now tending to supersede other forms of beach protection and many communes or communities are resorting to this practice, especially in Britanny and the Mediterranean. Over the last decade, beach nourishment has emerged as the most commonly employed form of beach protection and the three sectors of the French coast are replete with examples conducted under the auspices of numerous beachfront communes that have tended to move away from hard structures such as groins and breakwaters. The case of the Mediterranean beach of Petit Travers is a good example of the recent massive scale of beach nourishment in the Mediterranean. The beach is situated downdrift of a field of breakwaters and groins and unending beach erosion has directly threatened the coastal road behind the beach (Sabatier *et al.*, 2009b), a now classic situation on the Languedoc-Roussillon coast where leisure port breakwaters have been responsible for erosion of beaches in a whole host of tourist resorts such as Palavas and Valras. More than 1 million m³ of sand was recharged on these beaches from January to April 2008. The monitoring carried out by the DREAL in the year following this operation has highlighted progressive loss of the recharged material. Although beach nourishment is considered as a 'soft' approach to management, there are strong ecological concerns in the Mediterranean regarding nourishment, especially on the shoreface, as such operations can damage *Posidonia oceanica* seagrass colonies, which play a passive role in natural beach protection.

■ **Figure 12.19**
Beach drainage and pump system (inset) using the Ecoplage® *procedure on the Atlantic beach of Sables d'Olonnes, where the operation has turned out to be successful*

Beach drainage

Beach drainage has been gaining ground in France over the last decades, via a unique operator and a procedure called *Ecoplage®* (*Ecobeach*). Drainage comprises perforated drainage pipes buried below the upper beach surface, and connected to a pump and discharge system (Figure 12.17). The concept is based on the principle that sand beach accretion is favoured by enhanced permeability due to an artificially lowered water table. One advantage of this system is that the drainage devices are largely buried and therefore the technique has no visual impact. Results have been rather mixed on the trial sites in France, partly as a result of poor site selection, such as in macrotidal areas, inadequate design, and lack of management. While installation costs are relatively low, both maintenance and management costs are high, and communes have tended to overlook this point in several of the cases where the system has been installed.

Other forms of coastal protection

Increasing attention is being paid to the preservation of seagrass colonies that play a role in dissipating wave energy on Mediterranean beaches. This is notably the case of *Posidonia oceanica*, largely destroyed in places by large-scale beach protection, beach nourishment, and leisure port structures and by invasions by the seaweed *Caulerpa taxifolia*. Studies are being carried out on both the wave-energy dissipation capacities of *Posidonia* and on possibilities of seabed recolonization. These studies are promoted by a Scientific Interest Group, the *GIS-Posidonies*.

There are several other types of new or innovative coastal protection being experimented with. One example is the Elastocoast® procedure which uses highly porous revetments meant to absorb wave energy. The principle consists of a mixture of resin components and rocks, creating a very robust structure. This structure is slightly elastic and highly porous, and contains numerous cavities that absorb wave energy while serving as refuges colonized by plants and animals.

Concluding remarks: Emerging paradigms on coastal protection

Following a rather piecemeal approach to coastal protection for decades, during which France lacked an integrated view of coastal management practice and a vigorous view of coastal protection, the country has been moving over the last decade into what may be considered as a truly 'maritime' state, aware of the richness and diversity of its coast and of the necessity of protecting it. While enforcement of the legal framework of coastal protection has not proven to be as efficient as it should have, there has been progress, but a lot still needs to be done, and the state itself needs to have an exemplary approach in protecting fragile coastal areas.

Two points that need to be mentioned regarding recent developments in coastal management in France are the French *Grenelle* on the Environment, and the effects of Storm Xynthia in February 2010. The word Grenelle, which comes from a town in France, is now used in the international literature to characterize debates of wide-ranging international or national scope. The French Grenelle on the Environment was a nation-wide, multi-party debate that brought together, in 2007, representatives of national and local government and organizations aimed at identifying the key points of public policy on ecological and sustainable development over the next five years. Among key developments from the Grenelle was the creation in 2009 of a new directorate on sustainable development (the DREAL) that included coastal protection. The DREAL officer in each administrative region represents the state, oversees projects related to coastal protection, and may or may not authorize coastal protection works that a commune may intend to commission. The new logic prevailing in this regard is one of wide concert agreement, commonly encouraged by the DREAL officer, on the basis that a more proactive and upfront approach to coastal management will be beneficial, instead of the logic of a one-shot immediate response to storm erosion problems that tended to prevail in the past. In this regard, it is interesting to note that regions are individually setting up a 'Sea and Coast

Forum' aimed at prospective multi-party management of the coast and of marine resources. This approach is a forerunner to the setting up of a National Coastal Observatory, probably within the next two years.

Storm Xynthia crossed Western Europe between 27 February and 1 March 2010. In France, the storm led to the deaths of at least 51 people, with 12 more said to be missing. Most of these casualties occurred as a result of the coincidence of a powerful storm surge, high tide, and waves up to 7.5 m high. The storm overtopped and destroyed in places the 200-year-old seawall, built at the time of Napoleon to protect the towns of Aiguillon-sur-Mer and Faute-sur-Mer in Vendée. It also destroyed several homes, including a mobile home park. Widespread flooding with water levels up to 1.5 m high affected several parts of the coast. The dramatic effects of Storm Xynthia brought considerable attention to bear on the problems of coastal management and coastal defence in France, including the necessity of better management practice involving storm warning systems, seawall dimensions, building regulations in the coastal zone and set-back lines. The implementation of setback lines in Faute-sur-Mer will involve the destruction of several hundreds of homes. The damage and trauma caused by Storm Xynthia also heralds the problems of protection faced by low-lying coastal areas in the face of sea-level rise and climate change.

References

Aernouts, D., Héquette, A., 2006. L'évolution du rivage et des petits-fonds en Baie de Wissant pendant le XXème siècle, Pas-de-Calais, France. *Géomorphologie: Relief, Processus et Environnement.* 2006, 1, 49–64.

Anthony, E. J., 1994. Natural and artificial shores of the French Riviera: An analysis of their inter-relationship. *Journal of Coastal Research*, 10, 48–58.

Anthony, E. J., 1997. The status of beaches and shoreline development options on the French Riviera: A perspective and a prognosis. *Journal of Coastal Conservation*, 3, 169–178.

Anthony, E. J., 2002. Long-term marine bedload segregation and sandy versus gravelly Holocene shorelines in the eastern English Channel. *Marine Geology*, 187, 221–234.

Anthony, E. J., Cohen, O., Sabatier, F., 2011. Chronic offshore loss of nourishment on Nice Beach, French Riviera: A case of over-nourishment of a steep beach? *Coastal Engineering*, 58, 374–383.

Anthony, E. J., Dolique, F., 2001. Natural and human influences on the contemporary evolution of gravel shorelines between the Seine estuary and Belgium. In: Packham, J. R., Randall, R. E., Barnes, R. S. K., Neal, A. (eds), *The ecology and geomorphology of coastal shingle.* Westbury Academic and Scientific Publishers, Otley, UK, pp. 132–148.

Anthony, E .J., Julian, M., 1999. Source-to-sink sediment transfers, environmental engineering and hazard mitigation in the steep Var river catchment, French Riviera, southeastern France. *Geomorphology*, 31, 337–354.

Anthony, E. J., Vanhée, S., Ruz, M. H., 2007. An assessment of the impact of experimental brush-wood fences on foredune sand accumulation based on digital elevation models. *Ecological Engineering*, 31, 41–46.

Barusseau, J. P., Akouango, E., Bâ, M., Descamps, C., Golf, A., 1996. Evidence for short term retreat of the barrier shorelines. *Quaternary Science Reviews*, 15, 763–771.

Bertin, X., Castelle, B., Chaumillon, E., Butel, R., Quique, R., 2008. Longshore transport estimation and inter-annual variability at a high-energy dissipative beach: St. Trojan beach, SW Oléron Island, France. *Continental Shelf Research*, 28, 1316–1332.

Castelle, B., Bonneton, P., Dupuis, H., Sénéchal, N., 2007. Double bar beach dynamics on the high-energy meso-macrotidal French Aquitanian Coast: A review. *Marine Geology*, 245, 141–159.

Castelle, B., Bonneton, P., Sénéchal, N., Dupuis, H., Butel, R., Michel, D., 2006. Dynamics of wave-induced currents over an alongshore non-uniform multiple-barred sandy beach on the Aquitanian Coast, France. *Continental Shelf Research*, 26, 113–131.

Héquette, A., Hemdane, Y., Anthony, E. J., 2008. Sediment transport under wave and current combined flows on a tide-dominated shoreface, northern coast of France. *Marine Geology*, 249, 226–242.

Hanson, H., Brampton, A., Capobianco, M., Dette, H. H., Hamm, L., Laustrup, A., Lechuga, A., Spanhoff, R., 2002. Beach nourishment projects, practices and objectives: A European overview. *Coastal Engineering*, 47, 81–111.

Isebe, D., Azerad, P., Bouchette, F., Ivorra, B., Mohammadi, B., 2008. Shape optimization of geotextile tubes for sandy beach protection. *International Journal for Numerical Methods in Engineering*, 74, 1262–1267.

Levoy, F., Anthony, E. J., Monfort, O., Larsonneur, C., 2000. The morphodynamics of megatidal beaches in Normandy, France. *Marine Geology*, 171, 39–59.

Oblinger, A., Anthony, E. J., 2008. Wave attenuation and intertidal morphology of a macrotidal barred (ridge-and-runnel) beach behind a breakwater: Dunkerque, northern France. *Zeitschrift für Geomorphologie*, 52, Suppl. 3, 167–177.

Pierre, G., 2006. The processes and rate of retreat of the clay and sandstone sea cliffs of the northern Boulonnais (France). *Geomorphology*, 73, 64–77.

Pinot, J. P., 1998. La Gestion du Littoral. 2 vols, Institut Océanographique de Paris, 759 pp.

Reichmüth, B., Anthony, E. J., 2007. Tidal influence on the intertidal bar morphology of two contrasting macrotidal beaches. *Geomorphology*, 90, 101–114.

Robin, N., Levoy, F., Monfort, O., Anthony, E. J., 2009. Onshore bar migration and shoreline changes in the vicinity of an ebb delta in a megatidal environment: A short-term to multi-decadal scale approach. *Journal of Geophysical Research*, 114, F04024.

Sabatier, F., Maillet, G., Fleury, J., Provansal, M., Antonelli, C., Suanez, S., Vella, C., 2006. Sediment budget of the Rhône delta shoreface since the middle of the nineteenth century. *Marine Geology*, 234, 143–157.

Sabatier, F., Samat, O., Ullmann, A., Suanez, S., 2009a. Connecting large-scale coastal behaviour with coastal management of the Rhône delta. *Geomorphology*, 107, 79–89.

Sabatier, F., Samat, O., Brunel, C., Heurtefeux, H., Delanghe-Sabatier, D., 2009b. Determination of setback lines on eroding coasts: Example of the Gulf of Lions (French Mediterranean Coast). *Journal of Coastal Conservation*, 13, 57–64.

Sedrati, M., Anthony, E. J., 2007. Storm-generated morphological change and longshore sand transport in the intertidal zone of a multi-barred macrotidal beach. *Marine Geology*, 244, 201–229.

Sedrati, M., Anthony, E. J., 2008. Sediment dynamics and morphological change on the upper beach of a multi-barred macrotidal foreshore, and implications for mesoscale shoreline retreat: Wissant Bay, northern France. *Zeitschrift für Geomorphologie*, 52, Suppl. 3, 91–106.

Vella, C., Fleury, T. J., Raccasi, G., Provansal, M., Sabatier, F., Bourcier, M., 2005. Evolution of the Rhône delta plain in the Holocene. *Marine Geology*, 222–223, 235–265.

13 Spain

Vicente Gracia, Agustín Sánchez-Arcilla and Giorgio Anfuso

Introduction

Coastal morphology

The Spanish coastline exhibits a straight and rectilinear trend when compared to the coastal fringe of other European countries, such as Denmark or Greece. The main factor determining the abrupt topography and morphology derives from its tectonic character: the collision of different sub-plates favouring formation of a long mountain chain which generally runs parallel to the coastline.

Of the 7,870 km of coastline (excluding estuaries), about 50 per cent consists of hard and soft cliffs, which are particularly abundant in the Atlantic area, e.g. Galicia and Asturias (Figure 13.1). Cliffs are also well developed in the Spanish archipelagos, e.g. the Canary and Balearic Islands. At many places, cliffs form prominent headlands delimiting pocket beaches of different dimensions and characteristics. A further 25 per cent of the coast is occupied by beaches and some 17 per cent consists of low-lying areas, particularly abundant in the Mediterranean and essentially linked to the existence of sedimentary basins where rivers have formed coastal plains with deltas, such as, the Ebro, Llobregat, Turia, Jucar, etc., and albuferas (coastal lagoons such as Valencia, Adra, etc.). Near Gibraltar the strong persistent winds have formed important dune fields. Finally, the remaining 8 per cent consists of human transformed areas, which are mainly located along the Mediterranean coastline.

The continental shelf is generally narrow, with the 150 m isobath running close to the coastline. The

■ **Figure 13.1** *Spain, showing the various coastal Autonomous Communities (see Table 13.1 for details)*

main exceptions are south of the Ebro Delta, between Valencia and Catalonia, it is some 60 km from the coast and on the Atlantic side of Andalusia, where it is about 35 km offshore.

Oceanographic conditions

The variety of coastal and shelf shapes and dimensions (from narrow to moderately wide) together with the water bodies that encircle the Iberian Peninsula has resulted in a wide range of oceanographic settings. The most relevant driving factors that control coastal processes are described below, *i.e.* waves, currents and tides.

Table 13.1 Coastal uses (%) along the coast of the Spanish Autonomous Communities (from DGPC, 1991)

Region	Tourist/ Recreational	Industrial	Fishing	Port/ Commercial	Natural	Not defined
Basque C.	13.2	1.3	3.6	8.9	2.0	71.0
Cantabria	23.4	4.2	2.8	1.0	4.8	63.8
Asturias	10.7	7.0	12.3	3.4	0.6	66.0
Galicia	15.3	3.4	9.5	1.5	27.5	42.8
Andalusia	41.4	2.3	0.9	2.5	4.7	48.2
Murcia	28.2	3.6	1.3	2.2	–	64.7
Valencia	52.3	1.4	1.4	2.1		42.7
Catalonia	60.5	15.0	5.1	2.6	0.1	16.7
Balearic I.	24.3	0.2	0.8	2.5	25.0	47.2
Canary I.	13.1	0.8	1.2	1.2	0.7	83.0
Total	25.5	3.2	4.1	2.2	11.3	53.7

The significant wave height characteristics have been analyzed by *Puerto del Estado* (Spanish National Port Authority) by dividing the littoral into ten zones (ROM, 1992; Figure 13.2). The significant offshore wave height threshold (H_{S0}) to define a storm event is set at 3.0 m in the Northern Cantabrian coast, 1.5 m in the Southern Atlantic coast and presents a value of only 1.0 m in the Southern Mediterranean coast. Finally, the H_{S0} values for the Catalan coast (Northern Mediterranean) and the Canary Islands range between 1.5 m and 2.0 m. These differences reflect the length of the available fetch and intensity of prevailing winds for each area.

The average significant wave height value recorded during a typical storm event is 5.0 m on the North and North-Western Atlantic coast and 2.0 m and 3.5 m in the Mediterranean coast and the Canary Islands, respectively. Swell peak periods, Tp, show values greater than 10 s. for the Atlantic and the Cantabric areas and about 7 s. for Mediterranean waters.

Generally, the more relevant currents for coastal processes are those associated with oblique breaking waves. Their magnitude ranges between 0.50 m/s and 2.00 m/s. Wind and tide-induced currents can locally achieve some relevance, with values higher than 1.0 m/s. The general circulation pattern plays only a secondary role, since it only exceeds 0.2 m/s at few locations.

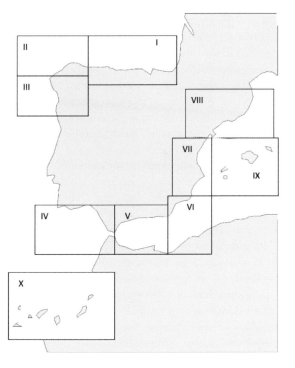

Figure 13.2 Sectors in which the Spanish littoral is divided according to the meteo-oceanographic prevailing conditions (ROM, 1992)

The mean sea level, defined as the combined contribution of the astronomical and meteorological tides, varies depending on the analyzed region. Along the Mediterranean coast (regions V to IX in Figure 13.2), the meteorological tide is the main factor accounting for most of the water-level variations, whereas the astronomical component is more relevant in the Atlantic and Cantabric coasts (regions I to IV and X in Figure 13.2).

The astronomical tidal range varies from 5.40 m at Santander (on the Cantabric Northern coast) to 1.6 m at Tarifa, near the Straits of Gibraltar. Tidal range is about 2.5 m at the Canary Islands and it goes from 0.80 m at Malaga, in the Southern Mediterranean coast, to about 0.20 m in the Catalan (Northern Mediterranean) coast.

Along most of the Spanish coast, storm surges associated with a return period of 20 years, range from 0.40 m to 0.90 m. Surges in the Canary Islands associated with the aforementioned return period, show values from 0.25 to 0.30 m. The meteorological tide may exceed 1.0 m in Mediterranean regions which are prone to resonance events (with an important port activity impact) and it is therefore the dominant mean water level factor for these microtidal environments.

All the above-mentioned oceanographic conditions are monitored by different observational networks which depend upon the central government (e.g. Puertos del Estado) or local governments (e.g. the XIOM network from the Generalitat de Catalunya). Buoys are deployed at depths which go from *circa* 1 km, to characterize offshore conditions, to about 50 m for corresponding coastal or shelf conditions. The time series available spans less than 30 years, which implies wide uncertainty intervals for extreme distributions. An illustration of the discrepancies in the extreme probability distribution functions for the significant wave height off the Catalan Coast appears in Figure 13.3. In this figure the comparison is between the extreme distributions derived from observed wave values and hind-cast ones, without any imposed correction on the numerical fields.

Early settlements

The earlier morphology presented numerous islands, hosting earlier historic settlements. This fact forced the necessity for establishing secure connections between islands and the mainland, by means of port and harbour construction. These activities continued and even increased with time because of the increasing importance of maritime commerce with Mediterranean and Northern European regions and later with America. Hence, many harbours and coastal cities developed along the Spanish coast, good examples being harbours built inside natural bays (Cadiz, Algeciras, Majorca, Santander, etc.), rivers (Seville) or river mouths and estuarine areas (Bilbao, Vigo, Ferrol, etc.). Lastly, the need for port construction in the Mediterranean coast prompted emplacement of open coast structures, such as the ports of Valencia and Barcelona.

The main settlements during the Phoenician, Greek and Roman ages were established along the Mediterranean and Andalusian Atlantic coasts. There was a direct relationship between development of coastal towns and harbour location; all built in natural sheltered areas with almost no initial artificial protection structures (Rodríguez-Villasante, 1994).

In the first millennium BC, the Phoenicians started to colonize the Spanish Mediterranean coast, establishing important settlements at Malaga, Almuñécar, Adra and Ibiza (Figure 13.1). These settlements enhanced marine trade, fishing activities and salt production. As an illustration, the city of Cartagena and its associated harbour were established behind a small peninsula sheltering a natural bay. On the Atlantic side, the Phoenicians founded an earlier settlement at Cadiz around 1100 BC, the oldest well-established coastal town in the Iberian Peninsula.

During the Greek age there were many other examples of coastal towns and associated harbours. The present city of Empúries (Emporion) featured a harbour in a natural sheltered area, protected by two small islands. The settlement of Rhode (the present Roses) at the Muga river mouth in Catalonia was another example of a dynamic coastal town, showing important commercial activities with other Mediterranean towns, especially Marseilles.

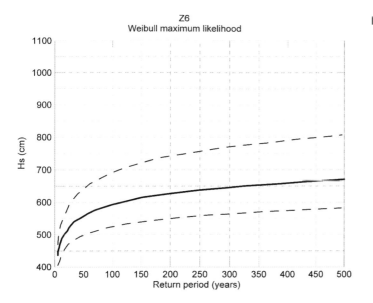

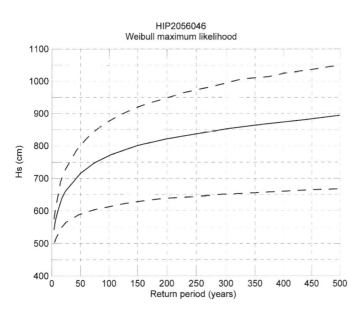

■ **Figure 13.3** *Extreme probability distribution function for the deep water significant wave height, Hs, corresponding to offshore wave data in front of the Ebro delta at the border between Catalunya and Valencia in the Spanish Mediterranean coast. The values come from the XIOM network (top) and numerically hind-cast fields (bottom) from Puertos del Estado*

In the Roman period, a number of new settlements appeared in the Mediterranean and Atlantic coasts of Spain. The city and harbour of Tarragona, established behind a hill close to a natural harbour, the city of Brigantium (an early location of La Coruña) and its famous lighthouse, constituted important examples. Other existing coastal towns were occupied and re-established by Romans, e.g. Cartagena and Cadiz (Bonet, 1994). Development of the Roman settlements along the Atlantic coast of Andalusia was strongly associated with the existence of bays (e.g. Cadiz, Algeciras, etc.) or took place on open coasts (at Bael Claudia, Barbate or Sancti Petri, this later location including probably the Hercules temple). The special location and characteristics of Cadiz Bay favoured establishment of a natural port devoted to maritime commerce in the Atlantic Ocean. In the Straits of Gibraltar, navigation was quite complex and some manmade floating structures were needed in the Bay of Gibraltar to protect vessels from waves.

The Atlantic routes, promoted by Julius Caesar and the increasing commercial activity between the Iberian Peninsula and Northern Africa (especially with Tangiers), favoured development of several coastal cities. However, many of the human settlements founded during the Roman Age along the Atlantic littoral of Spain and Morocco were destroyed over the following centuries by tsunamis generated off the Portuguese coast (Luque *et al.*, 2000).

Historical settlements from the Middle Ages

These were influenced by the Muslim domination and subsequent Christian reconquest. The kingdom of Aragon, covering roughly the eastern part of Spain, favoured development of Valencia and Barcelona harbours. The Muslim territories also developed important harbours such as Malaga (close to Granada, the capital of Al-Andalus) and Algeciras, the natural link with Northern Africa (Anonymous, 2009). The kingdom of Castile (roughly the Western part of Spain) promoted harbour construction in the Northern and Southern Spanish littoral.

All coastal cities had the same structure: an elevated mountain at the back, a castle protecting the settlement from land attack and a harbour in a bay or sheltered area. Because of the economic recession which took place in Spain in the seventeenth and eighteenth centuries, many coastal structures were abandoned or destroyed in the nineteenth century by the French or English navies.

The development of recent coastal cities (examples are Barcelona, Tarragona, Tortosa, Valencia, Alicante and Malaga) was based on a fortified town and a natural harbour. City and trade development gave rise to the emplacement of new port facilities in areas close to existing cities (Rodriguez-Villasante, 1994). After the discovery of America, harbour structures had to be adapted to larger ships, with larger mooring areas and deeper waters (Konvitz, 1978). At that time a change was also recorded in the engineering practices for harbour structure construction. The new typology of port that emerged was formed by an outer harbour. In the nineteenth century dredging works commenced which were carried out at many ports without a general management plan. New breakwaters were built to stop sediment impoundment at port structures at Barcelona, Tarragona, Valencia (Mediterranean coast) and Vigo (Atlantic coast).

Present land uses

Since the nineteenth century, human pressure along most of the Spanish coast has steadily increased. The previous concept of the coastal zone, considered as a risky and unhealthy area, changed because of the increase in security and of land reclamation processes, essentially the drying of coastal lagoons and wetlands. Land reclamation became a wide-spread activity along the Spanish coast and, as an example, the lagoon area of the Valencian Albufera was reduced by half between the seventeenth and nineteenth centuries.

Regarding human occupation, a coastal fringe of 5 km (7 per cent of the Iberian Peninsula area) hosted 12 per cent of the total population in 1990 (DGPC, 1999). Today more than 35 per cent of the Spanish population lives within a coastal fringe of 50 km width (DGPC, 1999). Coastal municipalities present the largest densities of inhabitants (350 inhab./km², compared to the average value for Spain which is 77 inhab./km²). Moreover, during the summer months, the tourism pressure produces a density increase of 1,000 inhab./km².

Andalusia and Catalonia are among the most populated areas and the land cover has experienced important changes, mostly in Andalusia, with a great increase of urban uses and a decrease of forest cover (CORINE, Land Cover EU project: Heymann *et al.*, 1994). A summary of coastal developments and land uses along the Spanish coastal autonomous communities is presented in Table 13.1.

Natural uses are low along the Mediterranean coast and, in general, most of the Spanish coastal fringe. The main exceptions are observed in the northwest of Spain (Galicia) and in some sectors of the Balearic Islands that focus predominantly on fishing or other natural functions/uses. It is apparent that tourism and recreational activities are, by far, the main activities along most Spanish regions, particularly those in the Mediterranean.

Eight per cent of the total world's tourists holiday in Spain, Italy, France, Greece and Turkey and these countries account for "the most significant flow of tourists . . . a sun, sea and sand (3S) market" (Dodds and Kelman, 2008, p. 58). Specifically, tourist activities in Spain, which represent 11 per cent of its GDP, take place mainly in coastal areas (Yepes, 1999), with an average number of visitors exceeding 40 million per year and a rate of increase between 6–8 per cent per year (Barragán, 2004).

Industrial fishing and aquaculture activities have also experienced an important growth since the 1950s. Many ports were constructed in recent decades for fishing, commercial and tourist purposes. They are managed by the central or regional administrations and represent an important percentage of coastal length at many municipalities. For many Mediterranean sectors, located on open coasts, ports produce an important barrier effect for sediment transport, with accretion in the northern side and erosion in the southern one (Rosselló, 1988). Ancient coastal protection works (mainly done during the middle of the last century) have also played an important role in this coastal reshaping.

Marinas have also experienced an important development; they were constructed following the US model, which consists of an interior basin, often a coastal lagoon, and very few or no external protection structures. Concerning their distribution, 27 per cent are built in the northern regions of Spain, while 48 per cent are located along the Mediterranean and southwest Atlantic coasts and the remaining 25 per cent in the islands (Barragán, 2004). For instance, in recent decades, 22 marinas were built in

the Costa del Sol (Mediterranean coast of Andalusia), with an average distance between structures of 12 km.

Coastal issues

Coastal protection evolution

Urban development of the Spanish coastal zone started in the 1970s under low regulatory conditions (Malvárez and Domínguez, 2000) and has affected coastal dynamics and dune systems, resulting in enhanced erosion and flooding. This was followed by construction of coastal promenades and protection structures, the latter designed to protect fast-growing urban communities and tourist activities. The lack of a general management strategy to fight coastal erosion and the short-term urgency to protect specific stretches has resulted in a reactive approach based initially on hard engineering structures, in the best of cases, which solved the problem only locally. The net longer-term effect was degradation of many coastal areas and sediment starvation, also enhanced by the progressive damming of rivers and associated river regulation policies.

In the 1970s, coastal protection was mainly based on construction of a groin or groin field. To stop sediment losses towards the offshore, groins were soon built curved, with 'L' or 'T' shapes. Examples are the structures at Estepona, Málaga, Benalmádena or Marbella in the Southern Mediterranean coast. This was done in parallel to the development of harbours and marinas.

In the 1990s, some groins were removed because the general sediment starvation along Spanish coasts hampered their functionality as sediment barriers or trapping structures. Benalmádena in the Mediterranean coast of Andalusia is a good example, with groins built at 200 m intervals during the late 1980s and early 1990s. In spite of that there was little sediment deposit and a seawall had to be built to protect the coastal promenade (the seawall was later also damaged by wave action).

Over the period between the second half of the

1980s and early 1990s, beach nourishment was adopted as a 'cure-all' solution for most coastal erosion problems. It was largely based on offshore sediment sources, quite often composed of too fine sands, which were not stable at beaches located along open coasts such as those in the Mediterranean Sea. About 7 million m^3 of sediments were artificially provided to the 27 km of beaches along the Costa del Sol in 1992.

The tourist development experienced in the second half of the twentieth century has resulted in more than 80 per cent of tourists being concentrated in the coastal fringe. This implies a programme to maintain some minimum beach widths for about 6,000 to 7,000 tourists per km of coast. However, the decreasing capacity of rivers in supplying sediment and coastal armouring has produced a scarcity of sediments which is not compatible with present and planned uses. This situation, also pointed out by the European project EUROSION and its continuation, the CONSCIENCE project, affects many of the tourist shores along the Mediterranean coast. The same conclusion is supported by more detailed analyses such as, for instance, the ones presented by CIIRC (2010) and commissioned by the *Generalitat de Catalunya*, where it was calculated that about 71 per cent of its 260 km of sand beaches has experienced a shoreline retreat at an average rate of 2.1 m/yr during the past 10 years.

Shore protection

The main problem is the lack of suitable sediments, essentially appropriate sand fractions stable under present wave conditions. The origin of this problem, which has heavily increased in the past decades, is linked to river regulation and the progressive occupation of the coastal fringe. River regulation involves damming, the management of river catchments (including the fight against land erosion) and extraction of sediment from river beds (e.g. for construction), plus the armouring of river margins. Progressive occupation of the coastal fringe has consisted of construction of promenades, buildings, etc. together with the associated withdrawal of the back beach and dune sediment volumes from present coastal dynamics. Coastal occupation has included also the emplacement of groins and detached breakwaters, which reduce longshore sediment transport and increase offshore sediment losses.

A number of special monitoring plans intended to determine the final equilibrium position of the beach and present evolution trends have been designed for several specific and problematic beaches, including particularly sensitive urban coasts or beaches with valuable assets or complex dynamics. The monitoring programme normally aims at reconstruction of seasonal and medium-term evolution trends. As an illustration, Figure 13.4 shows the shoreline trend obtained by means of sequences of beach profiles at the Ebro delta coast: a clear seasonal oscillation co-exists with a general, long-term erosive trend.

In order to reduce erosion problems, two main types of coastal engineering solutions have been applied. The 'soft' solution consists in beach nourishment works, *i.e.* the artificial supply of sand at specific coastal sectors. The 'hard' engineering solution consists in construction of seawalls, groins or detached breakwaters.

Beach nourishment has been adopted for maintenance of eroding beaches, for the creation of artificial beaches or to protect particularly valuable hinterlands. Hard coastal engineering solutions have been applied in cases where further shoreline erosion was not acceptable.

A common problem, also found in the Spanish Mediterranean, comes associated with coastal armouring. It is called the 'coastal squeeze' (Doody, 2004), and takes place when a coastline is prevented from its landward migration by seawalls or other man-made structures. Coastal erosive processes due to sediment imbalances or associated with sea level rise and increasing storminess can cause the complete disappearance of the beach or salt marsh and a deepening of nearshore areas fronting coastal structures (Pilkey and Dixon, 1996; Doody, 2004). The beach 'anthropisation' experienced by the Spanish coast shows good examples of these processes and has

■ Figure 13.4
Shoreline evolution trend off the Ebro delta, obtained from periodic profile surveys

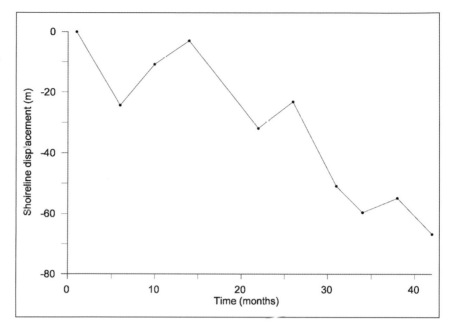

produced a loss of ecological value, decrease in biodiversity and diminution of landscape value and tourist attractiveness. This also has been reported for other coastal stretches (Williams and Micallef, 2009).

Both types of solutions are now being re-assessed in terms of their sustainability at medium- and long-term scales, involving central, regional and municipal administrations.

Legal and administrative settings

The Spanish coast is regulated by the 'Spanish Coastal Law' (Law 22/1988 for the Coast) which declares this piece of territory public domain and establishes the main activities that can be carried out within it. The coastal law does not include the public domain in the lower course of rivers, nor does it include the public domain fringe of wetlands. When all these lengths are added, the resulting Spanish coastal zone significantly increases its length from 7,800 km to about 10,000 km. Specifically, the Coastal Law establishes the main activities allowed in the seaside public domain, and includes also a 'protection' and

an 'influence' zone. Within the former, which extends 100 m landward, any kind of construction is prohibited, and within the latter, extending 500 m landward, only buildings with a low density of construction are allowed.

The coast is managed by the Ministry of Environment (General Direction for Coasts) in Madrid, for all aspects considered of general (public) interest. However, most issues related to land planning and environmental protection are managed by the governments of the Autonomous Communities (at regional level). Coastal studies and related projects are normally analyzed in terms of the beach functions and main driving agents (waves, mean water level oscillations and currents). Regarding functions, the Ministry of Environment recognizes three main beach functionalities: to protect the hinterland, to provide a natural and geologic support to the coastal environment and to support social and economic activities.

The high concentration of population and variety of existing uses and interests in the coastal zone generate frequent conflicts. Hence, a zoning is required in order to obtain an efficient management plan of

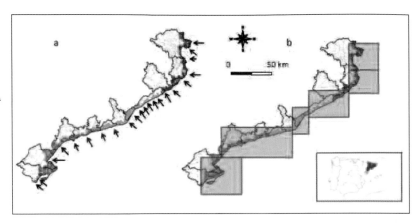

the available sand volume, which is much less than the one demanded by the present level of uses and activities. Figure 13.5a shows an example of coastal segmentation according to the sediment transport characteristics in the Catalan coast, and Figure 13.5b reflects the segmentation of the same coastal area according to wave features. Along the Spanish coast, waves are considered the main driving factor producing coastal changes at engineering time scales (from a few days to a few decades).

From a practical point of view, the Spanish coastal framework directive is based on the definition of a threshold beach width taking into account:

* the expected erosion in the considered time interval;
* the seasonal variation of beach width;
* the shoreline retreat due to relative sea level rise on a long-term scale;
* an additional beach width acting as a buffer area to account for uncertainties in driving terms, coastal responses or the inertia of the corresponding management solutions.

The diagnosis of a given coastal stretch in Spain is normally based on the following elements:

* the sedimentary budget of the analysed beach;
* the beach evolution trends (mainly obtained from shoreline mapping in aerial photographs);
* the relation between the considered coastal stretch and the nearby coast, *i.e.* investigation of its boundary conditions.

The functional design of coastal engineering structures is normally carried out for a design life between 10 and 25 years, with the exception of equilibrium bays, which are taken to be stable at longer time scales, e.g. about 50 years. Longer-term trends, such as those associated with climatic change, are considered only in terms of sea level rise (SLR). Usually, an increase in SLR of about 50 cm for the year 2100 is taken into account following the IPCC reports (IPCC, 2001; 2007).

The project of small-craft harbours is normally required to consider the barrier effect they exert upon long shore transport. The resulting bypass, to compensate for that barrier effect, is normally required as one of the conditions for approving the environmental impact assessment of such a harbour. It is to be carried out with a yearly frequency, or as required by the filling rate of up-stream deposition. The follow-up of these operations and, in general, of coastal engineering projects, is only implemented for the most sensitive or problematic projects, although it is not specifically regulated.

Coastal protection

Hard structures

The most conventional response to counteract erosion, and the one employed for a longer period of time, has been the construction of groins or groin fields, with the aim of reducing long shore transport. Table 13.2 shows an inventory of shore perpendicular groins along the Spanish coast, classified by coastal Autonomous Communities. In the 657 groins which form this inventory, there is a large variety of forms and dimensions. Most are mound structures, with a crest width between 5 to 7 m (due to the land-based construction procedure) and normally oriented perpendicular to the main shoreline direction. However, it is only in the last two decades that the true morphodynamic behaviour of these structures

■ *Table 13.2* Inventory of shore-perpendicular coastal works along the Spanish coastline, indicating Province and Autonomous Community

Province	Km of coast	Autonomous Community	Number of groins	Km of coast	Km of coast per groin
Guipúzcoa	92	Euskadi	12	246	20.5
Vizcaya	154				
Cantabria	284	Cantabria	2	284	142.0
Asturias	401	Asturias	11	401	36.5
Lugo	144	Galicia	6	1498	249.7
Coruña (A)	956				
Pontevedra	398				
Huelva	122	Andalusia	250	945	3.8
Cadiz	285				
Malaga	208				
Granada	81				
Almería	249				
Murcia	274	Murcia	15	274	18.3
Alicante/Alacant	244	Valencia	204	518	2.5
Valencia/València	135				
Castellón/Castelló	139				
Tarragona	278	Catalonia	75	699	9.3
Barcelona	161				
Girona	260				
Balearic Islands	1,428	Balearic Islands	21	1428	68.0
Palmas (Las)	815	Canary Islands	52	1583	30.4
Santa Cruz de Tenerife	768				
Ceuta	21	Ceuta	9	21	2.3
Melilla	9	Melilla	0	9	0.0
TOTAL	7,906		657		

has been incorporated into the functional design. In other words, it is only in recent years that the coastal engineering community and coastal users have realized that these structures only produce an upstream deposit if there is enough sediment transport, and that longshore transport calculations only estimate the potential transport capacity.

The distribution of these groins along the coast reflects the meteo-oceanographic and geomorphological features of the Spanish coast. The north and north-west coastal sectors, called Cantabric and Galician coasts respectively, show a smaller number of groins per kilometre. This is because this part of the Spanish coast features a large number of cliffs and pocket beaches; most of the sedimentary deposits appear in river mouths or estuaries, where there is a partial natural shelter. Because of that the more common type of groin is an L-shaped curved jetty which fixes the inlet and serves as support for the neighbouring beaches.

The Mediterranean coast is more exposed to open sea conditions, which in this case are milder both in terms of waves and tidal range, and thus presents beaches which were originally much longer. In these coastal stretches there was a higher longshore transport, although in many regions the steep slopes and presence of submarine canyons have also favoured important cross-shore transport rates. The fact that the weather in this part of the Spanish coast is milder and therefore there are higher tourist numbers has resulted in a larger density of groins per kilometre of coast, between 2.5 to 9.3 groins per kilometre.

The groins which are normally built in Spain can be classified into four main categories, although many of the actually built structures also share more than one function.

1. Barriers to longshore transport;
2. Barriers to wave action, producing a sheltered area which only receives diffracted waves;
3. Headland control structures, which act as a rigid point for shore line evolution;
4. Jetties, whose main function is to guarantee a stable inlet in plan and in depth (profile).

Most groins built in Spain fall into the first category, since their main objective is to provide a barrier for longshore transport. Some are intended to provide a fixed support for a long beach, and in this case their length is between 50 to 200 m. If more than one groin is used, then they are widely spaced, since their objective is, in that case, to provide two support points for the beach lying between them. The Spanish coast also presents a number of groin fields, which are designed so that structures decrease in length as the number of groins grows; an extreme example is the La Habana beach in Almería (part of the Andalusian coast), where a sand and gravel beach about 4 km in length has been stabilized with a field of about 100 groins, with lengths shorter than 20 m (Figure 13.6). These headland control structures generate coastal cells of about 20 m length and 10 m width.

Another example of a large groin field is the 6 km coast between the industrial harbour of Sagunto, near Valencia City, acting as a total barrier for longshore sediment transport, and the small craft harbour of Pobla de Farnals, with breakwaters ending at smaller depths and therefore acting only as a partial barrier.

■ **Figure 13.6** *Headland control for shoreline stabilisation, obtained by the emplacement of a groin field at the La Habana beach in the Almeria coast of Andalusia (Source: Ministerio de Medio Ambiente Rural y Marino)*

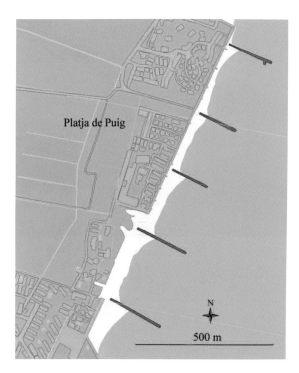

■ **Figure 13.7** *Groins at a small coastal sector close to Sagunto harbour in the Valencia Autonomous Community (Source: E. Pranzini)*

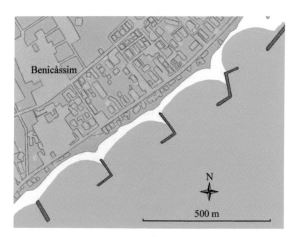

■ **Figure 13.8** *The Benicàssim beaches in the Valencia Autonomous Community, showing a number of L-shaped groins (Source: E. Pranzini)*

This coastal stretch features 18 groins; of about 200 m length each and which act as a partial barrier for long shore transport (Figure 13.7).

Other groin types to prevent sediment losses from the considered coastal cells have also been employed. As an illustration, there are a number of beaches where L-shaped groins have been built to prevent the offshore sediment transport due to incident storms and the currents which tend to occur near shore perpendicular barriers (Figure 13.8).

Some other coastal stretches combine shore perpendicular and shore parallel structures, with the aim of reducing as much as possible the mobility of the shore in these areas.

Coastal engineering practice in Spain also includes groins with partially submerged sections and fully submerged groins and detached breakwaters. These structures, depending on its location with respect to the shoreline and the width of the breaker zone, may act as partial or total barriers (Sánchez-Arcilla *et al.*, 2005).

The groins of the second category, *i.e.* those which favour diffraction and therefore create a sheltered beach subject 'only' to diffracted wave trains, are illustrated in Figures 13.8 and 13.9. In these cases the L-shape (e.g. groins at the Benicàssim beach or the Santapola beach on the Valencia coast) or the T-shape (on the Garrucha beach, Andalusian coast) illustrates the resulting morphodynamic behaviour. The same effect has been obtained for some beaches in the Canary Islands, using oblique groins with respect to the main shoreline orientation.

Finally, there are also a large number of examples along the Spanish coast with jetties which stabilize an inlet and also serve as support for the neighbouring beach. They are usually long, curved structures and they tend to be built near harbour structures, which induce local erosion processes and which also require a stabilization of the corresponding inlet (Figure 13.9).

Detached breakwaters have been traditionally considered by the Spanish coastal engineering community as a class of structures difficult to design from a functional stand point (see e.g. Cáceres *et al.*, 2005; Peña and Sánchez,, 2006; Sánchez-Arcilla *et al.*, 2005; Alsina *et al.*, 2007). The coastal response has not

■ **Figure 13.9** *Jetty at the mouth of the Llobregat river, south of Barcelona harbour and city on the northeastern Spanish Mediterranean coast (Source: Institute Cartogràfic de Catalunya, www.icc.cat)*

■ *Table 13.3* Inventory of detached breakwaters along the Spanish coast, indicating the Autonomous Community where they are located, the number of breakwaters and the corresponding length of coast

Province	Km of coast	Autonomous Community	Number of detached breakwaters
Asturias	401	Asturias	1
Huelva	122	Andalucía	22
Caádiz	285		
Maálaga	208		
Murcia	274	Murcia	6
Alicante/Alacant	244	Valencia	10
Valencia/València	135		
Castellón/Castelló	139		
Tarragona	278	Catalonia	34
Barcelona	161		
Girona	260		
Balearic Islands (Illes)	1,428	I. Balearic Islands	4
Palmas (Las)	815	I. Canary Islands	9
Santa Cruz de Tenerife	768		
Melilla	9		1
TOTAL	5,527		87

always been as calculated, showing in some cases the expected evolution (see e.g. Peña and Sánchez, 2008) while in some other cases there were larger discrepancies (see e.g. Bricio, 2009). In any case, the most common initial design criteria employed are those of Ming and Chiew (2000) and Hsu and Silvester (1990). However it is nowadays apparent that those criteria only incorporate part of the complex hydro-morphodynamic fluxes that happen around and even through a low crested detached breakwater, which condition the final resulting coastal evolution. Issues such as the overtopping volume have only been considered during the last years (see e.g. Cáceres *et al.*, 2008).

■ **Figure 13.10** *(a) Single detached breakwater in the Barcelona urban beaches, just north of Barcelona harbour, also shown (partly) in the image; (b) tombolos formed because of a field of shore parallel detached breakwaters in the coast of Tarragona (Source: Institut Cartogràfic de Catalunya, www.icc.cat)*

In spite of that, there are about 87 detached break-waters along the Spanish coast, as described in Table 13.3. About 40 per cent of that number (34) are built on the Catalonian coast, followed closely by the coast of Andalusia with 22 detached breakwaters. That clearly illustrates the suitability of this type of structure for seas in which there is only a microtidal range. The reason is that, for a significant tidal range, the position of the structure (or set of detached structures) varies across the surf zone, with the phase of the tide and that complicates even further the functional design.

Regarding structural configuration, these struc-tures have been built in Spain either as an isolated breakwater (Figure 13.10, top) or as a set or field of detached breakwaters (Figure 13.10, bottom).

Nourishment

Artificial nourishment along the Spanish coast started in the 1980s, after the relative functional failure of the more conventional 'hard' coastal engineering works. During the past decades (1980–2002) the Spanish coastline has been artificially nourished in a manner similar to other European Union coun-tries. However, the number of beaches artificially maintained (about 400) and the volume of sand (10 million m^3 of sand per year) make the Spanish coastline the foremost in Europe from this perspec-tive, being equivalent to the combined volumes of Italy, Germany, France and the Netherlands together.

In the period 2004–2008 the volume artificially supplied to Spanish beaches has decreased, reaching a value between 3 and 5 million m³ per year, although this is difficult to quantify, since in many cases the works are executed under the generic name of 'maintenance and conservation of the coast'. During this period, there have been nine large nourishment operations, with volumes exceeding 120,000 m³ while the rest of maintenance activities had volumes of between 10,000 and 70,000 m³. The reported origin of the nourished sediment for this period is mainly marine, including shelf deposits and down drift accumulations in groins, harbours, or natural obstacles. Only for smaller volumes did the source have an inland origin, including the bed of the lower course of rivers and creeks. These point out the difficulty of incorporating 'new' sand into the coastal system, due to high costs and/or impacts.

Most of the large volume operations (exceeding 100,000 m³) have been done in the Mediterranean coast (Girona, Barcelona, Castellón and Valencia provinces). As an alternative, in Castellón, a gravel beach was created together with some stabilizing groins. In general terms, the main aim was to regenerate previously existing beaches or restore the longshore transport interrupted by obstacles. On the Mediterranean coast of Andalusia (the provinces of Almería, Granada, Malaga and Cadiz) there have been a number of large-volume operations (for instance more than 300,000 m³ of sand in Almeria from marine deposits) and also from accretion areas, generated mainly by harbour breakwaters (e.g. Malaga). There have also been nourishment operations in Ceuta and Melilla (north coast of Africa), in this latter case using material from land origin. In the province of Cadiz, as described with more detail elsewhere, there have been important nourishment operations from a variety of origins.

In the Atlantic coast of Galicia and the Cantabric coast of Asturias, Cantabria and the Basque country there have also been a number of artificial nourishments from a variety of sources, including the natural deposits of estuaries and access channels to harbours (e.g. Santander) and a variety of objectives (beaches of adequate quality, maintenance of the public domain, etc.). There have also been artificial nourishment operations in the Balearic and Canary Islands, in the latter case modifying the typology/sedimentology of the beach (e.g. from sand mixed with cobbles to only sand). In the island of Las Palmas (Canary Islands) some sand beaches have been substituted by gravel deposits to prevent aeolian transport which resulted in disturbances for nearby roads and villages.

The main aims behind the artificial nourishment policy are to: i) improve beach functionality; ii) create a new beach; and iii) restore a previously existing beach. According to these objectives, artificial nourishment operations can be carried out to counteract background erosion, using either marine sediments or land sediments. It can also be addressed to counteract a temporary, undesirable shape, in plan or profile, of the beach.

These 'scrapping' operations always use the same native sediments. Whenever non-native sediment is used, due consideration must be paid to the sediment durability (*i.e.* loss rates) and to the newly generated deposit functionality. This can be illustrated by considering that coarser sediments last longer but produce a steeper slope, which may not be desired by most beach users.

Any artificial nourishment operation must be compatible with the driving terms (basically wave conditions), with the social preferences (sediment texture and colour, beach slope, etc.) and with the natural ecosystem in that given beach. In general, artificial nourishment operations are normally accepted in Spain only for two types of coastal environment:

1. Urban beaches, which are in fact artificial and require, therefore, nearly permanent nourishment;
2. In front of well-defined tourist areas, which depend critically on beach resorts.

In all cases the Ministry of the Environment considers artificial nourishment as a long-term solution, assuming the responsibility to re-nourish periodically, without any limitation in time or volume. The required volume is calculated as a function of the sediment deficit, based on the long shore sediment

BOX 13.1 CADIZ CASE STUDY

On the Atlantic side of Andalusia, the Spanish Coastal Authority started coastal protective plans mainly consisting of beach nourishment, sometimes accompanied by construction of small groins, seawalls and revetments. Following Muñoz et al. (2001), the characteristics and location of nourishment works carried out on the southwestern Andalusian littoral coast between 1989 and 1999 have been presented in Table 13.4 and in Figure 13.11.

According to Table 13.4, nourished volumes ranged greatly, from a few thousands to two million m³, injected at La Victoria beach in 1991, with a total amount of 12 million m³ nourished during the decade investigated. The unitary cost of nourished sand ranged from 1–2 to 6–7 US$/m³, the aforementioned values being quite cheap when compared with figures observed in other European countries. Mining and transport methods presented in Table 13.4 depended on the origin of borrow sediments: i) sediments obtained in land on near accreting beaches (Cortadura, Rompidillo, etc.) or migrating dunes (Trafalgar and Valdevaqueros dunes) or from shallow water deposits in harbours, river mouths, etc. (Figure 13.11), are dredged according to different modalities and then transported by truck to areas to be nourished; ii) sediments dredged from marine placers (i.e., San Jacinto, Meca, etc.), in deeper areas of harbours (i.e., at Huelva, Rota, Barbate) or harbour entrances and channels

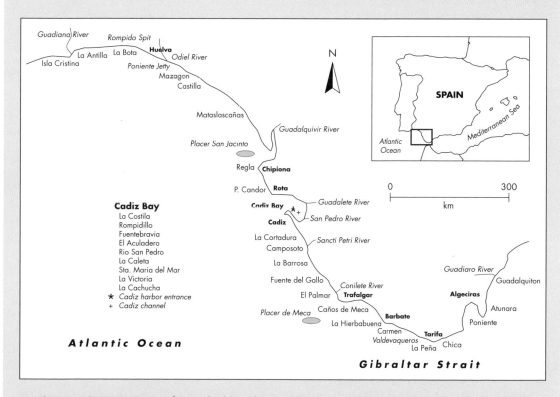

■ **Figure 13.11** Location of nourished beaches and borrow areas (grey stipple) along the southwestern coast of Andalusia 1989–99

■ **Table 13.4** *Characteristics of nourishment works carried out on the southwestern littoral of Andalusia, 1989–1999 (Muñoz et al., 2001)*

Year	Beach Name	Nourished volume (m3)	Cost (US$/m3)	Mining and transport methods(1)	Borrow area location	Distance from nourished beach (km)
1989	Castilla	1,690,000	2.26	M	Poniente Jetty	6.3
1990	La Antilla	1,300,000	4.10	M	Rompido Spit	6.1
1990	La Cachucha	82,000	6.16	L	San Pedro River mouth	3.8
1991	La Caleta	41,000	1.96	L	La Cortadura beach	1.4
1991	La Victoria	2,000,000	2.76	M	Cadiz channel	5.2
1991	Sta Maria del Mar	306,000	1.82	M	Cadiz harbor entrance	6.0
1991	La Peña	67,000	3.23	L	Valdevaqueros dune	11.2
1991	Caños de Meca	16,000	2.48	L	Trafalgar dune	2.4
1991	Rio San Pedro	23,160	2.58	M	San Pedro River mouth	0.5
1992	Regla	502,000	5.28	M	Placer San Jacinto	4.2
1992	Caños de Meca	124,230	2.12	L	Trafalgar dune	2.4
1992	Fuentebravia	75,000	2.84	L	Rota Harbor	0.9
1993	Carmen	23,150	1.88	L	Barbate harbor	0.6
1993	Poniente	131,820	1.42	L	Atunara beach	6.7
1993	El Palmar	18,790	1.79	L	Conilete mouth	2.0
1993	La Peña	170,000	2.43	L	Valdevaqueros dune	11.2
1994	La Barrosa	463,000	2.68	M	Placer de Meca	25.1
1994	Fuentebravia	275,000	1.97	M	Cadiz channel	8.1
1993	El Aculadero	172,450	1.55	L-M	Cadiz channel	4.5
1994	La Hierbabuena	16,210	2.50	L	Barbate harbor	0.4
1994	Canos de Meca	496,000	2.06	M	Placer de Meca	4.0
1994	Fuente del Gallo	400,000	2.60	M	Placer de Meca	13.0
1994	Isla Cristina	330,000	2.69	M	Isla Cristina harbor	4.1
1995	La Bota	930,000	3.43	M	Poniente Jetty	13.0
1995	Chica	12,780	3.17	L	Valdevaqueros dune	11.2
1995	El Rompidillo	7,735	1.50	L	Rota harbor	0.6
1996	Rota	197,000	4.82	M	Cadiz channel	11.0
1996	El Aculadero	75,625	3.19	M	Cadiz channel	4.5
1996	La Costilla	94,566	3.09	M	Cadiz channel	11.0
1996	Fuentebravia	134,800	2.61	M	Cadiz channel	8.1
1996	Mazagon	425,000	3.31	M	Huelva harbor	7.2
1996	Mataslascañas	125,000	3.31	M	Huelva harbor	18.3
1997	La Antilla	300,000	6.67	M	Huelva harbor	35.1
1997	Sta Maria del Mar	60,000	3.79	M	Cadiz channel	6.0
1997	La Barrosa	30,000	4.75	M	Placer de Meca	25.0
1998	Camposoto	737,000	3.50	M	Placer de Meca	31.5
1998	P. Candor	19,210	5.22	M	Cadiz channel	14.2
1998	Guadalquiton	175,000	3.08	M	Guadiaro river mouth	0.8

(1): M: marine origin; L: land or shallow water origin.

and updrift of coastal structures (Poniente Jetty, Figure 13.11), are stored in a dredging boat and pumped onto the beach using the trail suction method and redistributed on the beach surface by trucks. The first modality was performed to nourish 13 beaches accounting for the 6.5 per cent of the total nourished volume with an average price of 2.7 US$/m³, and distances from borrow areas to nourished beaches ranging from 0.4 to 11.2 km with an average value of 4.2 km. Sediment dredged and transported according to the second modality presented an average cost of 3.1 US$/m³ and a distance ranging from 0.5 to 35 km, with an average value of 10.9 km.

Special attention was devoted to the characteristics and evolution of a few beaches, e.g., Rota, Fuentebravia, Aculadero, La Barrosa and Santa Maria del Mar, where different monitoring programs were carried out with a monthly periodicity (Anfuso et al., 2001). In all cases except La Barrosa beach, nourished sand was dredged from Cadiz harbour entrance channel (Table 13.4), at water depths between 8 and 14 m. Short groins were emplaced at Rota, Fuentebravia and Aculadero to stop longshore transport, generally made of armoured stone blocks without any core of quarry-run material, the surface being covered with slab-shaped calcareous sandstone in order to have a walking surface used as a promenade. Groin profiles reproduced natural winter profiles of adjacent beaches with an elevation of 0.5 to 1 m over the beach surface, therefore reducing visual impact. At Santa Maria del Mar, several protection works and artificial fills have been built since the seventeenth century when construction of a seawall seemed imperative to protect this part of the city.

Concerning beach morphology, a general trend was observed in all nourished beaches which showed an initial narrow dry beach area and a smooth foreshore slope (Table 13.5) close to the dissipative state. Aculadero beach was a totally artificial beach created in an area with a rock shore platform and a very narrow dry beach (Table 13.5). Nourished sand was essentially redistributed along the dry beach and upper foreshore by bulldozers in order to create an artificial berm and a steep and narrow foreshore, transforming the initial dissipative profile in a reflective one (Table 13.5).

Nourished works usually finish in September and filled beaches experienced important and rapid erosion processes in the following months because of storm impacts. Artificial reflective states were not able to dissipate

■ **Table 13.5** *Main characteristics of investigated beaches before and after nourishment works*

Beach characteristics	Rota	Fuentebravia	Aculadero	La Barrosa
Initial dry beach width (m)	5	52	0	60
Foreshore width (m)	85	30–40	5	120–130
Foreshore slope (%)	1.5	3.13	1	2
Beach profile	Dissip.	Dissip.	Dissip.	Dissip.
Medium grain size (mm)	0.35	0.20	0.47	0.20
Date of nourishment	Sept. 1996	Sept. 1996	Sept. 1996	Sept. 1994
Total split sand volume (m³)	95,000	135,000	160,000	460,000
Split sand volume per profile (m³/m)	113	129	111	652
Nourished length (m)	500	700	750	800
Increase of dry beach width (m)	55	22	47	100
Design of intertidal gradient (%)	5	4.7	7.8	5
Design of beach profile	Reflect.	Moderate Reflect.	High Reflect.	Reflect.
Borrow sand grain size (mm)	0.33	0.22	0.31	0.26

the incident wave energy as natural dissipative beaches do, especially when the first storms hit the coast during autumn and a great disequilibrium between beach slope and energy exists. The change in morphodynamic state also involves a change in breaking wave type: spilling breakers, which do not significantly affect bottom sediments, were replaced by plunging breakers associated with reflective states and removed large quantities of bottom sediments so favouring beach erosion. Further, the steeper intertidal slope favoured an off-shore transport due to the gravitational component of breaking waves. When considering the erosive trends in studied beaches, a similar behaviour was observed with an exponential trend of sediment loss, in fact almost 100 per cent of nourished sand was lost during the two months following the fill works. After a few months beaches tended to achieve a minimum equilibrium volume, the time to acquire it essentially depending on initial beach volume and particular energy conditions. La Barrosa beach presented a different behaviour during the first year (1994–1995) when no morphological changes were observed. This was due to the absence of storms, in this sense low energy conditions allowed beach sand accommodation under natural conditions. Shell fragments (abundant in nourished sediments) presented great susceptibility to erosive processes due to their low density and form (that give raise to a low shear stress and a small setting velocity). As a result, sediment coarsening, linked to the presence of shell fragments, produced erosive beach processes.

transport gradient and the desired or target beach profile. The along shore sediment transport is normally calculated using the CERC formula, although Kamphuis and Van Rijn expressions are also employed.

Conclusions

The analysis presented shows the value of the Spanish coastal fringe for the country economy. It also shows the excess of pressure for the limited territorial extent. This results in a threatened sustainability of present-day uses and functions that can be illustrated by the relative functional failure of many hard and soft coastal engineering solutions. The limited durability of many artificial nourishment operations and the limited ability of groins to promote sediment deposits, where there is no longer sand available, have reduced social confidence in these engineering interventions.

It is, therefore, necessary to revisit the employed engineering approaches, considering not only the coastal fringe but also river catchment basis. The same applies to a proper evaluation of sedimentary balances on a carefully selected time scale, particularly now that climate variability and future (different) scenarios are beginning to be established at regional scales. The overall situation is one of generalized erosion, particularly for areas near river mouths and urban (artificial) beaches.

The simultaneous consideration of all these aspects, at the core of any integrated coastal zone management plan, should result in an easier to maintain coastal fringe, where engineering interventions are designed structurally, functionally and also with an explicit impact evaluation and maintenance plan.

Acknowledgements

This work is a contribution to the Andalusia PAI Research Group RNM-328. Thanks go to employers of the Centro de Arqueología Subacuática del Instituto Andaluz del Patrimonio Histórico who helped in the collection of information regarding ancient human settlements.

References

Alsina, J. M., Sierra, J.P., González-Marco, D., Cáceres, I., Sánchez-Arcilla, A., 2007. Estudio comparativo del comportamiento hidro-morfodinámico de estructuras emergidas y sumergidas. *Ingeniería hidráulica en Mexico*, 22, 2, 21–3.

Anfuso, G., Benavente, J. and Gracia, F. J., 2001. Morphodynamic response of nourished beaches in SW Spain. *Journal of Coastal Conservation*, 7, 71–80.

Anonymous, 2009. *Atlas de la historia del patrimonio de Andalucía*, Junta de Andalucía, Spain, 252 pp.

Barragán, J. M., 2004. *Las Áreas litorales de España*, Editorial Ariel, Barcelona, 214 pp.

Bonet, A., 1994. Las ciudades marítimas españolas y su iconografía. In, *Puertos españoles en la Historia,* MOPTMA (ed.), Madrid, MOPTMA, Centro de Publicaciones, 389 pp.

Bricio, L., 2009. *Comportamiento funcional y ambiental de los diques exentos de baja cota de coronación y su importancia en la ingeniería de costas.* Tesis Doctoral, E.T.S.I. Caminos, Canales y Puertos (UPM).

Cáceres, I., Sánchez-Arcilla, A., Zanuttigh, B., Lamberti, A., Franco, L., 2005. Wave overtopping and induced currents at emergent low crested structures. *Coastal Engineering*, 52, 931–947

Cáceres, I., Stive, M. J. F., Sánchez-Arcilla, A., Trung, L. H., 2008. Quantification of changes in current intensities induced by wave overtopping around low-crested structures. *Coastal Engineering*, 55, 113–124.

CIIRC, 2010. *Estat de la zona costanera a Catalunya.* Generalitat de Catalunya.

Heymann, Y., Steenmans, Ch., Croissille, G. and Bossard, M., 1994. *CORINE land cover: Technical guide.*Office for Official Publications of European Communities, Luxembourg, 137 pp.

DGPC, Dirección General de Puertos y Costas, 1991. *Actuaciones en la costa 1988–1990,* MOPU, Spain, 307 pp.

Dodds, R. and Kelman, I., 2008. How climate change is considered in sustainable tourism policies: A case of the Mediterranean islands of Malta and Mallorca. *Tourism Review International*, 12, 57–70.

Doody, J. P., 2004. Coastal squeeze: An historical perspective. *Journal of Coastal Conservation*, 10, 1–2, 129–138.

Hsu, J. R. and Silvester, R., 1990. Accretion behind single offshore breakwater. *Journal of Waterway, Port, Coastal, and Ocean Engineering, ASCE*, 116, 3, 362–380.

IPCC, 2001. *Climate Change 2001: Synthesis report,* R. T. Watson and Core Writing Team (eds). Cambridge University Press, Cambridge, UK and New York, 145 pp.

IPCC, 2007. *Climate Change 2007: Synthesis report,* Contribution of Working Groups I, II and III to the Fourth Assessment Report of the Intergovernmental Panel on Climate Change, Core Writing Team, R. K. Pachauri and A. Reisinger (eds). IPCC, Geneva, Switzerland, 104 pp.

Konvitz, J., 1978. *Cities and the Sea: Port city planning in early modern Europe.* Johns Hopkins University Press, Baltimore, MD, Part I, 3–69.

Luque, L., Lario, J., Zazo, C., Goy, J .L., Dabrio, C. J., Silva, P. G. and Bardají, T., 2000. Sedimentary record and tsunami hazard in the Gulf of Cádiz (Spain). In *3° Simp. sobre el Margen Ibérico Atlántico*, Univ. do Algarve, Faro, 371–372.

Malvárez, G. and Domínguez, R., 2000. Origins, management and measurement of stress on the coast of southern Spain. *Coastal Management*, 28, 215–234.

Ming, D. and Chiew, M., 2000. Shoreline changes behind detached breakwaters. *Journal of Waterways, Port, Coastal and Ocean Engineering, ASCE*, 126, 2, 63–70.

Muñoz, J. J., López, B., Gutierrez, J. M., Moreno, L. and Cuena, G., 2001. Cost of beach maintenance in the Gulf of Cadiz (SW Spain). *Coastal Engineering*, 42, 143–153.

Peña J. M. and Sánchez, F. J., 2006. *Inventario y efectos morfológicos de los diques exentos en las costas españolas.* CEDEX, Technical report n° 22-404-5-11.

Peña, J. M. and Sánchez, F .J.. 2008. *Directrices para el diseño de diques exentos en las costas españolas.* Monografía, Madrid, 153 pp.

Rodríguez-Villasante, J. A., 1994. La evolución de los puertos españoles en la edad moderna. In

Puertos españoles en la Historia, MOPTMA, Spain, 389 pp.

Pilkey, O. and Dixon, K., 1996. *The corps and the shore*. Island Press, Washington DC, 272 pp.

ROM (0.3-91), 1992. *Recomendaciones para Obras Marítimas. Oleaje. Anejo I clima marítimo en el litoral español*. Ministerio de Obras Públicas y Transporte (M.O.P.T.), Dirección General de Puertos, Madrid, 76 pp.

Rosselló, V. M., 1988. La defensa del litoral. *Boletín Asociación de Geógrafos Españoles*, 7, 1, 3–28.

Sánchez-Arcilla, A., Sierra, J. P., Cáceres, I., González-Marco, D., Alsina, J. M., Montoya, F. and Galofré, J., 2005. Beach dynamics in the presence of a low crested structure: The Altafulla case. *Journal of Coastal Research*, SI no. 39, 759–764.

Yepes, V., 1999. Las playas en la gestión sostenible del litoral. *Cuadernos de Turismo*, 4, 89–110.

Williams, A. T. and Micallef, A., 2009. *Beach management: Principles and practices*. Earthscan, London, 445 pp.

14 Portugal

Óscar Ferreira and Ana Matias

Introduction

Coastal erosion in Portugal is a complex issue that relates not only to its main causes, but also to a number of other factors, namely the inherited physiographic coastal characteristics, the type of human development and its evolution over time, and the options of past coastal zone protection. Erosion in Portugal is a public issue discussed by the population, a topic for legislation, a cause of loss of human lives and therefore an important problem for political and social debates. Simultaneously, erosion is often neglected in times of urban development but also when restrictive measurements are implemented. Some major flaws can be identified to explain how erosion became an important problem:

- lack of continuous and systematic monitoring of the coastline;
- unarticulated management strategy by local, regional and national authorities; and
- underestimation of erosion complexity since coastal management has been frequently dealt with by non-experts in coastal processes.

However, this chapter will not address in detail these socio-economic aspects related to erosion; rather it will focus on the most important physical processes identified so far. Additionally, it is devoted to erosion and shore protection of mainland Portugal, and consequently excludes the Portuguese archipelagos of the Azores and Madeira. Accordingly, throughout the text all references to the Portuguese coast exclude the insular coastline.

Coastal zone characteristics

Morphology

The mainland Portuguese coastline is more than 900 km long and morphologically diverse (Figure 14.1). Geomorphological features include amongst others, extensive sand shores backed by dunes, rock coasts with low and high cliffs, pocket beaches, bays, estuaries, lagoons and barrier islands. Along the west and south coasts, a variety of physiographic alternations can be observed, particularly between rock coasts and sand shores. Ferreira *et al.* (2008) defined three main morphological types for the Portuguese coast (Figure 14.1): sand shores, cliffed coasts (the dominant type) and low-lying rock shores (the least frequent). Estuaries and coastal lagoons also have an important morphological role and the main estuaries are associated with the Tagus and Sado Rivers, located on the central part of the western coast. The most important coastal lagoons are the Aveiro Lagoon and the Ria Formosa Lagoon (Figure 14.1). According to its physiographic properties, Andrade *et al.* (2002) divided the Portuguese coastline into eight stretches (Figure 14.1):

1. Minho River to Douro Estuary: a low irregular rocky coast with sand and gravel beaches, rock

shoals and sea stacks. The larger sand beaches develop southward of estuaries associated with sand spit formation.

2. Douro Estuary to Nazaré: dominated by linear sand beaches backed by dunes, which in the northern part protects the Aveiro coastal lagoon.
3. Nazaré to Tagus Estuary: a rocky coastal stretch with actively eroding cliffs, several headlands, bays, pocket beaches and small coastal lagoons. It ends at the large Tagus Estuary that includes a wide variety of pristine and developed estuarine sand beaches, tidal flats and salt marshes.
4. Tagus Estuary to Sado Estuary: includes an arcuate bay with sand beaches backed by dunes or cliffs at the north, whilst cliffs rising directly from the sea or with small pocket beaches occur to the south.
5. Sado Estuary to Sines Cape: Sado Estuary is a large estuary with tidal flats and salt marshes. South of the estuary a large bay includes sand beaches backed by dunes (north) and cliffs (south).
6. Sines Cape to Cape St. Vincent: dominated by cliffs occasionally connected to shore platforms. Pocket beaches occur on sheltered areas or near rivers and stream mouths.
7. Cape St. Vincent to Ancão: rock and cliffed coast with alternating headlands and bays where pocket beaches are frequent. Sea stacks, shoals and shore platforms often occur in front of cliffs. Two large sand embayments (Lagos-Alvor and Armação-Salgados) with large dune fields and lagoons contrast with the surrounding rock coast.
8. Ancão to Guadiana Estuary: composed of a multi-inlet barrier island system that encloses the Ria Formosa Lagoon. The lagoon includes tidal channels, backbarrier beaches, tidal flats and salt marshes. The Guadiana Estuary has a beach-dune system at its mouth and marshes and tidal flats at the lower estuary.

Wave regime and sea level

The wave regime on the west and south coasts of Portugal is different because of the different exposure to the North Atlantic Ocean where the most

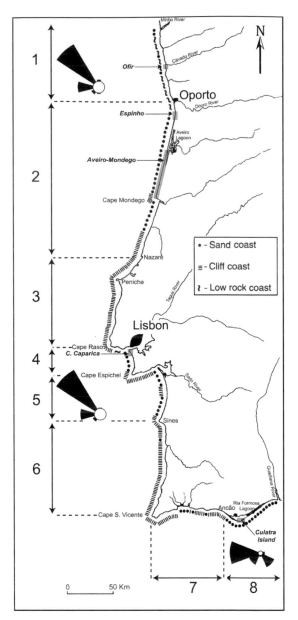

■ **Figure 14.1** *Portuguese shoreline, including the coastal stretches (1 to 8) defined by Andrade et al. (2002), wave roses for Leixões, Sines and Faro (data from Costa et al., 2001 and Costa and Esteves, 2010), main coastal types, rivers and coastal lagoons (adapted from Ferreira et al., 2008)*

energetic waves are generated. Consequently the west coast wave energy level is much higher than on the more sheltered south coast. The west coast wave regime is dominated by high energy swells, which occur 73 per cent of the year, with annual mean offshore significant wave heights (Hs_0) ranging from 1.7 m (at Sines, Figure 14.1) to 2.0 m (at Leixões, 5 km north of Oporto) (Costa, 1994; Costa et al., 2001; Costa and Esteves, 2010). Annual average peak period (Tp) is about 11 s at both locations. About 50 per cent of the waves have a combined occurrence of northwest directions (Figure 14.1), Hs_0 = 1–3 m and Tp= 9–13 s (Costa, 1994). High energy waves are relatively frequent, with $Hs_0 > 3$ m during 17 per cent of the time at Leixões and 9 per cent at Sines (Costa and Esteves, 2010). Storms along the western Portuguese coast have $Hs_0 > 5$ m (Pita and Santos, 1989), which represent 1–2 per cent of the wave occurrences (Costa and Esteves, 2010). These waves are mainly generated by high to mid-latitude depressions crossing the North Atlantic towards Western Europe (Costa et al., 2001). Maximum observed Hs_0 are 8.9 m and 8.1 m respectively for Leixões and Sines (Costa and Esteves, 2010). Storm groups occur frequently on the west coast of Portugal, on average once a year for groups of two storms and once every four years for a group of three storms (Ferreira, 2005).

The wave regime at the south coast of Portugal is less severe, with $Hs_0 < 1$ m during 68 per cent of the year (Costa et al., 2001). Annual average Hs_0 and Tp are 1 m and 8.2 s, respectively (Costa et al., 2001). The dominant incident wave direction is W-SW (71 per cent of the year, Figure 14.1). E-SE waves correspond to waves generated by local winds, termed *Levante,* and occur during 27 per cent of the year. The most frequent south coast storms are from the W-SW with Hs_0 generally smaller than 5 m but with maximum values above 6 m (Costa, 1994; Costa et al., 2001).

Storminess changes are a hot topic worldwide. Dodet et al. (2010) analysed the wave climate variability in the North-East Atlantic Ocean over the last six decades and showed spatially variable long-term trends, with a fairly constant trend for Hs_0 and

a clockwise shift of wave direction (up to + 0.15°/yr) at southern European latitudes (including Portugal). Almeida et al. (2011) analysed the variation and trends in storminess along the Portuguese south coast, using data from wave buoy measurements and from modelling, for the period 1952 to 2009, but no statistically significant linear trend was found for any of the analysed storm parameters. The main pattern of storm characteristics and extreme wave heights is an oscillatory variability with intensity peaks every seven to eight years and the magnitude of recent variations is comparable with that of variations observed in the earlier parts of the record.

Tides along the Portuguese coast are semidiurnal and mesotidal. The average neap tidal range is 1 m and average spring tidal range is 2.8 m. However, the equinoctial spring tidal range can reach about 3.8 m, with maximum tidal elevations approximately 2 m above mean sea level (MSL). Significant tide variations are noticed along particular segments of the coast, especially at estuaries and lagoons.

Storm surge works in Portugal are still scarce, but occurrences of storm surges between 0.5 m and 1 m have been reported by several authors (e.g. Taborda and Dias, 1992; Gama et al., 1994a, b). Between 1986 and 1988 a maximum storm surge of 1.10 m was recorded in the northwest coast and a maximum of 0.75 m in the south coast (Gama et al., 1994a).

Dias and Taborda (1992) analyzed the record of the Cascais and Lagos tide gauges (data since 1880 for Cascais and early twentieth century for Lagos) and indicate a significant rise of relative sea level in Portugal during the twentieth century, within the range of most global estimates. The temporal structure of the Cascais tide-gauge data suggests that relative sea level has not risen consistently. From the end of the nineteenth century until 1920, mean sea level dropped at a mean rate of 0.5 mm/yr. Subsequently, relative sea level has shown a net upward trend of 1.7 mm/yr (Dias and Taborda, 1992). Recently Antunes and Taborda (2009) stated that between 1977 and 2000 relative sea level rose at Cascais at a rate of 2.1 mm/yr and that there is a trend towards acceleration. For that site, the same authors predict a mean sea level 47 cm higher in 2100 than

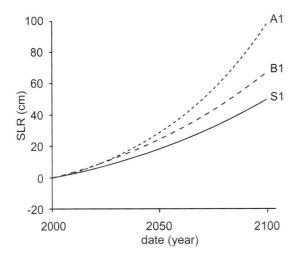

in 1990 (Figure 14.2, line S1) with a 95 per cent confidence interval between 19 and 75 cm.

Littoral drift

The west coast of Portugal is exposed to waves with important transport capabilities. These waves, predominantly from W–NW, induce a net littoral southward drift, with a potential transport capability of 2×10^6 m³/yr (e.g. Oliveira *et al.*, 1982; Taborda, 2000). This potential transport is variable along the west coast according to the coastline orientation and gradually decreases southwards because wave energy decreases. In linear sections slightly rotated clockwise along equilibrium arcuate embayments and sections sheltered from prevailing NW waves, the net littoral drift can be null or even reversed in direction (Andrade *et al.*, 2002).

At the south coast of Portugal (South Algarve) dominant waves drive an easterly directed littoral drift of about 10^4 to 10^5 m³/yr, depending on coastal orientation. At the south coast, the littoral drift is only continuous at the coastal stretch Ancão to Guadiana River, being higher at the west flank of the Ria

Formosa barrier island system because it is the most exposed sandy coast stretch (Figure 14.1).

Human development and tourism

Human occupation of Portugal is very varied: the most densely populated areas are found in the west coastal zone (from the northern border until the Sado River) and in the Algarve south coast (Pimentel, 2005). Nearly 40 per cent of the Portuguese population live in two coastal metropolitan areas (Lisbon and Oporto) and its demographic importance has been increasing in the last 40 years (Pimentel, 2005). All international airports (Lisbon, Oporto and Faro) were built near coastal cities, and obviously, ports and harbours. Ports and harbours demand expensive coastal engineering structures and maintenance operations and used to be crucial for the Portuguese economy. Historically (until the 1960s) maritime activities were important for the economy and employment structures, and included fisheries, salt extraction, shipbuilding and ship repair industry, food processing industry and ship transport activities (Rocha, 1984). Nowadays, fisheries represent only less than 1 per cent of the gross domestic product (GDP), whereas 68 per cent of the GDP comes from the service sector (Souto, 2005). In turn, the service sector (e.g. education, healthcare, transports, banking) is associated with densely populated areas, mostly located in the coastal zone. Therefore, *circa* 85 per cent of the GDP is generated within 60 km of the coast (Andrade *et al.*, 2002). An important part is derived from coastal tourism, one of the main activities at the coast. About 90 per cent of Portuguese hotel accommodation capacity is located in the coastal zone, with the Algarve concentrating 35 per cent of this capacity (INAG, 2006; using data from Census 2001). Approximately 30 per cent of the Portuguese coast was occupied by housing, harbours, industrial facilities and tourism infrastructures (Andrade *et al.*, 2002; referring to data from 1994). This value has most probably increased in the past years, mainly by an increase in housing and tourism infrastructures. Despite the existing occupation, the coastal zone of Portugal holds natural and landscape values that underlie the designation of protected areas for *circa* 50 per cent of its extension (INAG, 2006).

Coastal erosion

Causes

At present, the coastal zone is strongly impacted by erosion. Shoreline retreat rates of more than 1 m/yr are recorded in more than 50 per cent of the sand shores, with maximum rates of 5 to 10 m/yr south of Espinho (Ferreira *et al.*, 2008; Pedrosa and Freitas, 2008). Three main causes contribute to current coastal erosion: interventions at river basins, coastal engineering structures and sea level rise.

Since the 1940s, hundreds of dams and water reservoirs have been constructed, and by the end of the 1980s, only 15 per cent of Portuguese river basins were directly unaffected by dams (Figure 14.3; Dias, 1990). Dam construction was driven by the increasing demand of water and energy by the Portuguese and Spanish. Dams reduced river flow and controlled water discharge (decrease of flood frequency and magnitude) leading to a significant decrease of fluvial sediment transport to the coastal area (Oliveira *et al.*, 1982; Dias, 1990). For example, the Tagus river supplies less than a third of its potential solid load to the inner estuary (Ramos and Reis, 2001) while the estimated Douro river sediment supply to the ocean (250,000 m³/yr) is one order of magnitude smaller than its potential transport (Oliveira *et al.*, 1982). This overall decrease of sediment supply to the littoral and shelf systems enhances a sediment deficit and promotes coastal erosion. Sediment starvation at coastal areas is amplified by

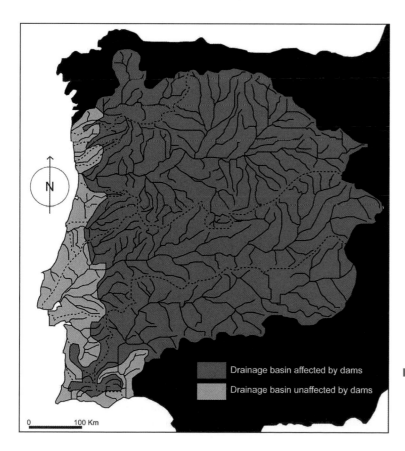

Drainage basin affected by dams

Drainage basin unaffected by dams

0   100 Km

■ **Figure 14.3** *Comparison between Portuguese rivers drainage basins affected and unaffected by dams (adapted from Dias, 1990)*

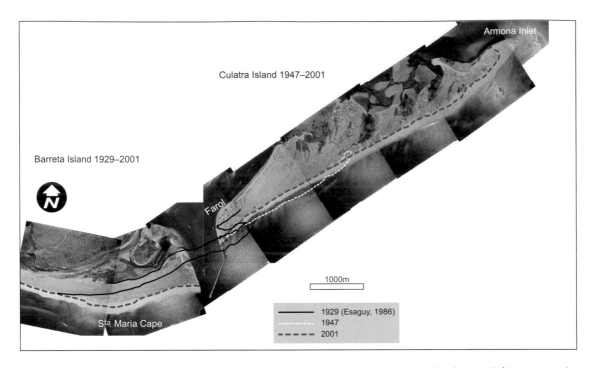

■ **Figure 14.4** *Faro-Olhão Inlet breakwaters (built between 1927 and 1955), with clear updrift (western side) accumulation and downdrift (eastern side) shoreline retreat. The shorelines of 1929, 1947 and 2001 indicate relative shoreline displacements on the vicinity of the breakwaters (adapted from Pacheco et al., 2008)*

mining activities at the river basins, estuaries and harbours, for navigational and construction purposes (Andrade *et al.*, 2002; Veloso Gomes *et al.*, 2006; Ferreira *et al.*, 2008).

Coastal erosion problems are often exacerbated downdrift of harbour jetties, breakwaters, and hard protection structures. These structures act as barriers to longshore transport and induce a reduction on sediment supply to downdrift areas (Figure 14.4). Hard protection structures were extensively used during the 1970s and 1980s, leading to the armouring of significant parts of the Portuguese coastline, e.g. Espinho, south of Aveiro, Caparica and Quarteira.

Shoreline retreat rates related to sea level rise were computed for sand coasts by application of Bruun's Rule (Andrade, 1990; Teixeira, 1990; Ferreira, 1993). Retreat rates are of the order of tens of centimetres per year, which only account for 10 per cent

to 15 per cent of the actual shoreline retreat (Andrade *et al.*, 2002; Ferreira *et al.*, 2008). Hence, sea level rise is apparently a minor cause of shoreline retreat at sand shores. However, on coasts where there is no space to accommodate beach landward translation (such as beaches backed by cliffs or seawalls), sea level rise can become an important factor for beach width reduction, as observed in a number of beaches in the Cascais municipality (Taborda *et al.*, 2010).

A fourth, although less known, possible cause is a change in storminess and wave climate, with consequent shoreline adaptation. Storminess and wave climate changes at the Portuguese coast are not evident (Dodet *et al.*, 2010; Almeida *et al.*, 2011) although a clockwise shift of the mean winter wave direction was observed at southern European latitudes (Dodet *et al.*, 2010). That shift would be responsible for higher longshore transport and consequent coastal readjustment and erosion (Andrade *et al.*, 1996). This subject deserves future attention and further investigation.

Consequences

Acknowledgement of coastal erosion is not new. At the close of the nineteenth century, Espinho already faced retreat values of about 1 m/yr (Pedrosa and Freitas, 2008). Despite recognition of erosion problems, Portuguese institutions did not devote a significant effort in the systematic prosecution of in-depth coastal evolution studies. Studies of the coast were particularly related to engineering purposes (channel navigation and harbour protection). Therefore, reliable information about coastline evolution prior to the 1960s is very limited. Shoreline evolution observed during the 1940s and 1950s do not consistently indicate erosion or accretion but point towards a dominant erosive behaviour, with shoreline retreat rates generally smaller than 0.5 m/yr (Ferreira *et al.*, 2008).

The increase of sediment starvation during the second half of the twentieth century, by causes mentioned earlier in this text, led to increased retreat rates where erosion was ongoing and expansion of areas undergoing erosion. Coastal stretches dominated by cliffs or low-lying rocky shores (e.g. Minho River to Douro Estuary, Nazaré to Tagus Estuary, Sines Cape to Cape St Vincent and Cape St. Vincent to Ancão) were already under sediment starvation, which increased during the second half of the twentieth century. Between the Minho and Douro rivers, sediment starvation led to the disappearance of wide sand beaches, that in some places had been replaced by steeper and narrow gravel beaches (Loureiro, 2006). The most dramatic changes occurred on west coast sand shores (e.g. Douro Estuary to Nazaré, and the northern part of the Tagus Estuary to Sado Estuary). Retreat rates reached more than 8 m/yr south of Espinho (Ferreira and Dias, 1991; Ângelo, 1991; Pedrosa and Freitas, 2008), 4 m/yr south of Aveiro (Ferreira and Dias, 1992, Ferreira, 1993, Teixeira, 1994), or 7 m/yr south of the Tagus Estuary (Ferreira *et al.*, 2005). The recent evolution of five of the most relevant coastal areas under erosion that represent different coastal settings and hydrodynamic conditions is described below.

Ofir

Ofir sand spit (Figure 14.1) constitutes the south margin of the Cávado river mouth. Despite human development, the sand spit is a fragile feature, a result of ongoing erosion processes during the past decades, which are a direct consequence of dredging activities on the Cávado river and dam construction in its basin (Veloso Gomes *et al.*, 2006). Ofir sand spit has recently been subject to breaching, namely in 1992 and 2005 (ICN, 2007). As a consequence, nourishment with estuarine dredged sand was carried out to prevent further inlet opening. The sand coast south of the spit is also eroding, which compelled construction of several coastal protection structures (Carvalho *et al.*, 1986; Granja and Carvalho, 1995). A set of four groins was built between Ofir sand spit and Apúlia village, and seawalls and revetments placed in front of the main buildings, e.g. houses and hotels. One of the groins (Pedrinhas) was cut to half its original length to allow some littoral drift to bypass southwards and therefore reduce downdrift erosion. However, the coastal trend was not reversed and the area maintains a marked negative sediment budget (Loureiro, 2006) and dune retreat is a constant threat to dwellings and people.

South of Espinho

The coastal area south of Espinho, located on the northwestern coast of Portugal (Figure 14.1), is a heavily protected coastline, with a high density of groins and seawalls (Dias, 1990; Ferreira and Dias, 1991). Permanent human occupation dates from the mid- nineteenth century, with the settlement of fishermen families (Pedrosa and Freitas, 2008), which developed over the years in association with both fisheries and tourism. The Espinho shorefront showed the first signs of coastal retreat in the late nineteenth century (Dias *et al.*, 1994; Pedrosa and Freitas, 2008). At that time, significant parts of Espinho were destroyed by the sea, including the main street and tens of houses. Between 1900 and 1933, the southern part of Espinho had an average retreat rate of *circa* 1 m/yr, with neighbouring coastal areas remaining stable but at present accreting, up to 3.5 m/yr (Pedrosa and Freitas, 2008). Coastal destruction in Espinho required management intervention to protect the town from further damage. The first hard protection structures in Portugal were built in Espinho, at the beginning of the twentieth century (Dias *et al.*, 1994), including a seawall in 1909 (destroyed by the sea in 1911), two test groins (in 1910) and two groins (between 1911 and 1912; Pedrosa and Freitas, 2008). Despite these efforts, the retreat rate over the entire oceanfront of Espinho increased between 1933 and 1970, reaching 3.2 m/yr in the northern sector of the city and over 6 m/yr in the remaining oceanfront, with a maximum of more than 8 m/yr (Pedrosa and Freitas, 2008). Coastal protection structures interrupted the littoral drift and led to increased retreat rates south of the town (Ferreira and Dias, 1991), in association with sand reduction to the coastal system contributed by the Douro river (Oliveira *et al.*, 1982).

South of Espinho, shoreline retreat became more accentuated, reaching up to 8.8 m/yr between 1933 and 1970 (Pedrosa and Freitas, 2008). During the second half of the twentieth century, groins and seawalls were successively built, changed and increased in length and number both in Espinho and southwards. Shoreline retreat values south of Cortegaça (8 km south of Espinho; Figure 14.5) reached more than 10 m/yr during the 1980s (Ferreira and Dias, 1991). Recently (1998–2003) coastline retreat has been ongoing at Espinho and to the south, regardless

■ **Figure 14.5** *Example of fast coastline retreat (> 5 m/yr) at Maceda (10 km south of Espinho) at a non-occupied area (photograph by Carlos Loureiro)*

of coastal engineering efforts, with average retreat rates ranging between 3 and 5 m/yr, with a maximum of up to 10 m/yr (Portela and Freitas, 2008). The conclusion is that coastal protection schemes adopted during the twentieth century were effective in protecting urban development in the shorefront, but ineffective in preventing coastline retreat. Moreover, engineering structures contributed to amplification and propagation of the phenomenon downdrift (southward).

Aveiro–Cape Mondego

This coastal sector is located at the central part of the west coast (Figure 14.1), immediately south of the Aveiro Lagoon inlet, and extends over 50 km southward. Aveiro Inlet is stabilized by two long jetties which were 800 m long in 1959 (Abecasis *et al.*, 1970)

and have been further extended. The Aveiro jetties interrupt littoral drift, induce sediment deviation towards deeper parts of the shelf and cause sediment starvation and severe erosion on downdrift areas (see for example Abecasis *et al.*, 1970; Dias, 1990; Ferreira, 1993; 1998). Erosion was increased by sediment mining updrift and at the navigational channels inside Aveiro harbour (Dias, 1990; Veloso Gomes *et al.*, 2006). Several authors (Ferreira and Dias, 1991; 1992; Bettencourt and Ângelo, 1992; Ferreira, 1993; Teixeira, 1994; Veloso Gomes *et al.*, 2002; Baptista, 2006) quantified the coastal erosion process in this area, proving that there is a direct relationship between Aveiro harbour structures and subsequence maintenance and shoreline retreat. Shoreline retreat rates were initially (1940s until 1970s) higher immediately southward of the jetties (Barra–Costa Nova; Figure 14.6) leading to construction of the first groins

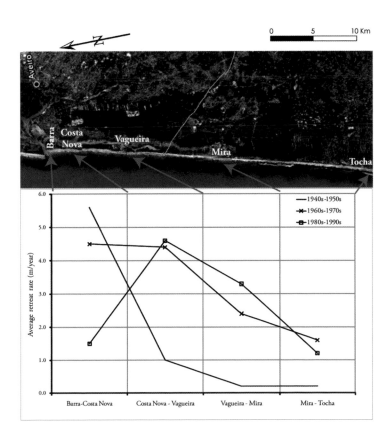

■ **Figure 14.6** *Average retreat rates between Aveiro and Tocha by decades (1940s–1950s; 1960s–1970s; 1980s–1990s) based on data integration from Ângelo (1991), Ferreira and Dias (1992), Ferreira (1993), Bettencourt and Ângelo (1992), Teixeira (1994), Veloso Gomes et al. (2002), and Baptista (2006)*

at Costa Nova in 1972. Extension of the northern jetty of Aveiro harbour sheltered the Barra–Costa Nova sector so during the 1980s and 1990s there was a reduction in the retreat rate of just under than 25 per cent from the previous decades (Figure 14.6). The erosion trend migrated downdrift affecting more than 20 km of coastline south of the Aveiro harbour. The erosion peak moved south (Figure 14.6) being located between Costa Nova and Vagueira during the 1980s and 1990s. Average shoreline retreat rates were at that time higher than 4 m/yr (Ferreira and Dias, 1991; 1992; Ferreira, 1993; Figure 14.6).

For the coastal area south of Costa Nova, retreat rates of 6–7 m/yr (Veloso Gomes et al., 2002) and 5–6 m/yr (Baptista, 2006) were recorded for the periods 1996–2001 and 2000–2003, respectively. Based on retreat rates obtained since the 1940s, Ferreira and Dias (1992) predicted that continuity of the erosion process would lead to extensive overwash and inlet breaching south of Vagueira. Overwash was extensively observed but breaching did not occur because of emergency interventions (including dune nourishment). If breaching was allowed to occur, low-lying inland areas protected by the vulnerable dune system would be easily flooded, leading to the loss of important agricultural areas and hundreds of houses (Veloso Gomes et al., 2006).

Costa da Caparica

Costa da Caparica is located immediately south of the Tagus River mouth (Figure 14.1) and its coastal evolution depends heavily on Tagus sediment exportation. This area is densely urbanized and is the main recreational beach for people living in Lisbon and surrounding towns. Severe shoreline retreat began during the first half of the twentieth century (see Castanho et al., 1974), after the 1930s and particularly after 1947. Erosion was directly related with evolution of the northern sand spit that connected the Tagus Estuary with Costa da Caparica (loss of sand, inland migration and finally disappearance). Between 1929 and 1957, the spit head retreated about 1.5 km and subsequently attached itself to the coast.

This evolution has been attributed to large amounts of dredged sediment at this area and also to associated activities (dredging and dam construction) at the Tagus river basin (Castanho et al., 1974; Veloso Gomes et al., 2006). Between 1957 and 1964, shoreline retreat at the shorefront of Costa da Caparica was about 100 to 150 m, and dune height lowered 6 m (Barceló, 1971). The first coastal structures (three groins) were built north of Caparica in 1959 and the early 1960s. Further structures (dikes, seawalls and groins) have been built until the beginning of the 1970s (Castanho et al., 1974), so that most of the Costa da Caparica shorefront has become artificial. Between 1972 and 2000 the coastline was relatively stable (Veloso Gomes et al., 2006). Since 2003, new engineering interventions have been performed including reshaping of the existing groins and seawalls and intensive beach/dune nourishment with about 2.5 million m^3 of sand placed between 2007 and 2009 (Rodrigues and Veloso Gomes, 2011). However, during the first years of the twenty-first century, severe winter storms promoted shoreline retreat, damaged coastal protection structures and exposed some beach leisure facilities.

Culatra Island

Culatra Island is part of the Ria Formosa barrier system located in the Algarve, southern Portugal (Figure 14.1). Two tidal inlets limit the island: the Faro-Olhão Inlet at the west and the Armona Inlet at the east. The updrift Faro-Olhão Inlet was artificially built between 1927 and 1955, being stabilized with jetties that disrupted the longshore drift (Figure 14.4). Culatra Island is 7 km long and its west end is artificial due to the jetties and a seawall/groin system built to protect the lighthouse and settlement in the early 1980s. Sediment starvation at the western part of the island caused by the jetties and groin led to important shoreline retreat that reached rates of more than 3 m/yr between 1958 and 1976, with a total shoreline displacement of more than 100 m between 1947 and 2001 (Garcia et al., 2002; Figure 14.4). In contrast, the pristine eastern half of Culatra Island experienced significant accretion, with development of longshore curved spits and island growth eastward (*circa* 3200 m; Figure 14.4) together with a reduction in the width of the Armona Inlet. The island growth rate was higher until 1976, and at the same time extensive overwash occurred in recently formed low-lying areas. Since 1976, the island grew at slower rates and overwash occurrence has been limited. The major sediment source for the fast growth of Culatra Island is the inland attachment of the former Armona Inlet ebb deltas at the island's eastern tip. Abandonment of the formerly active ebb deltas is a direct consequence of the Armona Inlet tidal prism loss due to the Faro-Olhão Inlet opening (Salles, 2001; Pacheco *et al.*, 2008).

Coastal protection

Portugal has been a pioneer in the development of laws for protection of the coastal zone, with the establishment of the Public Maritime Domain (DPM, Domínio Público Marítimo) at the end of the nineteenth century. This law was imposed along the entire Portuguese coast, stating that the entire area was public and no permanent settlement could be constructed without governmental agreement. Although in some (exceptional) cases the DPM was reverted into private (or municipal) property, this law turned most coastal areas public and was a bulwark against urbanization for a long time. However, pressure from the building industry and tourism agency lobbies has been powerful, particularly after the tourism boom of the 1970s. Until the late 1980s, there was no integrated coastal management perspective in Portugal (Veloso Gomes *et al.*, 2006). Local authorities often defined management of the coastal area as primarily based on the perspective of short-term profits from construction incomes. As no consistent integrated management approach was set for the coastal area, each local management plan had its own method and strategy. The Portuguese legislative framework was (and still is) complex and several national and regional plans overlap for the same coastal area (e.g. Regional Land Use Management Plans, Natural Parks Land Use Management Plans, Municipalities Director Plans, and National Ecological Reserve).

Nowadays, the most important management tools are the POOCs (Coastal Zone Management Plans) which define the spatial planning and land-use from 30 m depth until a maximum distance of 500 m from the coastline. POOCs also include the establishment of coastal hazards and the implementation of protection measures. All nine POOCs for Continental Portugal are in force and their implementation can be regarded as a first step towards reduction of coastal hazards, restriction of urban development and definition of solid rules for coastal management. The European Union defined a set of principles, strategies and procedures for integrated coastal zone management, with a recommendation to be implemented in all Member States. Accordingly, in 2009 Portugal approved the National Strategy towards an Integrated Coastal Zone Management (*Estratégia Nacional para a Gestão Integrada da Zona Costeira*; INAG, 2009). From 2009 onwards, all management plans and actions dealing with coastal subjects have to follow this National Strategy. All existing plans define natural approaches, the reduction of coastal hazards, ecosystem restoration and natural function of systems, as major objectives. In addition, these goals should not be reached at the expenses of loss of social and economic development. Coastal protection is a paramount factor in achieving the intended goals and one of the most challenging and debatable topics in coastal management programs. As stated earlier, coastal protection in Portugal has been carried out both through hard protection (e.g. groins and seawalls) and soft protection (e.g. beach nourishment and dune construction).

Hard protection

Since the beginning of the twentieth century hard protection structures, mostly groins (Figure 14.7a) and seawalls, were the main solution to the frequent coastal hazard problems. When properly designed, these coastal structures are an effective solution for risk reduction of areas immediately under its protection, avoiding damage to people and properties.

However, hard protection does not address the causes of erosion and increases the hazard to downdrift areas, which often compel construction of more protection structures. That was the case in Ofir, Espinho, Aveiro, Caparica and many other areas. As a result there was a proliferation of hard protection structures in Portugal during the second half of the twentieth century. From 1958 until 1989 an exponential growth in the number of groins and seawalls was observed between Oporto and Costa Nova (south of Aveiro; Dias, 1990). At this coastal area, the construction of new coastal defence structures and maintenance of old ones was intense until the end of the 1990s.

Parts of the Portuguese coast can be considered 'artificial' (armoured) with the contact between land and sea entirely achieved by hard structures (Figure 14.7b). In the late 1980s, the coast between Espinho and Cortegaça had a hard protection density of some two groins/km and 320 m/km of seawalls and revetments (Dias, 1990). Today, several other coastal areas, such as Caparica and Quarteira, also have a few kilometres of armoured shoreline. Current coastal protection policy in Portugal is no longer based on hard protection structures; most interventions are now dedicated to the maintenance or redesign of existing structures. Moreover, new structures are occasionally built to provide support to beach nourishment projects, as for example in Três Castelos (Praia da Rocha) at the rock coast of Algarve (south Portugal).

■ **Figure 14.7** *(a) Example of a groin built to protect Vagueira village with visible sand starvation at the downdrift side; (b) example of coastal armouring by revetments and groin at Cortegaça (south of Espinho; photograph by Carlos Loureiro)*

Soft protection

Soft protection intends to return the coastal zone to a more natural state using beach nourishment, dune nourishment, sand fencing, access restriction, elevated footpaths, social awareness and education, and housing removal. The first nourishment in Portugal was performed in 1950 at Estoril, near Lisbon, with the placement of 15,000 m³ of sand. Due to the high tourist value of this beach four nourishments were undertaken between 1950 and 1954 with the placement of 45,000 m³ in total (Martins, 1977). The first large-scale intervention was carried out at Praia da Rocha (Algarve) involving some 1×10^6 m³ of sand (Psuty and Moreira, 1992), and it is a success as the sand is still in place.

Between 1950 and 1997, about 30 fills were recorded at 12 different sites, adding up a total fill volume of 6.6×10^6 m³ (Hanson et al., 2002). About two-thirds of this sediment was placed along the Portuguese south coast (low to moderate wave energy) with only about 170 km, as compared to the more than 800 km of the west coast (high wave energy). The first major Portuguese intervention that had dune preservation as a primary concern rather than property protection was on the Cacela peninsula

(easternmost Ria Formosa barrier; Figure 14.1), part of the Ria Formosa Natural Park. Between October 1996 and February 1997, 325,000 m³ of sand was placed on the Cacela peninsula dune crest to avoid overwash and breaching (Matias et al., 1999; 2004). Protection was further carried out with sand fencing, grass planting, dune access restriction and elevated footpaths, placed to allow tourist activities in the area. The biggest soft protection intervention so far is the nourishment of more than 13 km of beaches and dunes along the Ria Formosa barrier islands (Figure 14.1). Sediments were from the lagoon and its channels and some 2,650,000 m³ of sediments were deposited between April 1999 and July 2000 (Ramos and Dias, 2000). This nourishment was followed by the placement of circa 13 km of fences to enhance dune recovery. In this case, the areas to be protected were mostly pristine and the main goal of intervention was to protect and reconstruct the dune ecosystem while improving lagoon water quality.

Extensive dune nourishment has also been performed on the west coast, south of Aveiro (Figures 14.1 and 14.6), to avoid overwash and inlet breaching (Veloso Gomes et al., 2006; Taveira Pinto et al., 2009). Since the 1990s, beach nourishment has become a common practice, and recent examples

■ **Figure 14.8** *Beach nourishment in front of the cliffs on the coastal stretch Quarteira–Vale do Lobo in summer 2010 (see box 14.1 for further details)*

prove that it is one of the most popular coastal management tools in Portugal. On the west coast, Costa da Caparica (Figure 14.1) is subject to recent coastal management plans which include beach nourishment with 3×10^6 m^3 of sand, of which 2.5×10^6 m^3 have already been placed, between 2007 and 2009. On the south coast, in an extension of 4.5 km between Quarteira and Vale do Lobo (8 km west of Ancão, Figure 14.1), the beach was nourished with 1×10^6 m^3 of sand (Figure 14.8). This nourishment was undertaken in the summer of 2010 and is expected to last for about ten years. Albufeira (on the south coast) is the most recent case of beach nourishment in Portugal (April–May 2011) with about 600,000 m^3 of sand deposited along 2 km of an embayed beach (oral communication by Antonio Rodrigues, Water Institute).

BOX 14.1   FORTE NOVO–GARRÃO CLIFFS EVOLUTION AND COASTAL PROTECTION

The coastal stretch of Forte Novo–Garrão (Figure 14.9) is a 4 km long open coast located southeast of the village of Quarteira (South Portugal). The cliffs have a total length of about 1,800 m, with clifftop elevations that range from 7 m to 20 m. Cliffs are cut into poorly or non-consolidated red sand formations from the Plio-

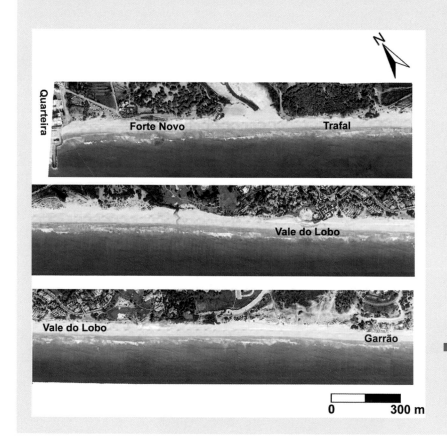

■ **Figure 14.9**
Forte Novo–Garrão Coastal Stretch (aerial photograph from November 2009)

■ **Figure 14.10** *Example of cliff erosion at Forte Novo (Photograph by Mara Nunes)*

Pleistocene. These formations are extremely fractured, with variable states of consolidation and poor internal cohesion. The coastal stretch has been affected by coastal erosion, in particular since the 1970s due to construction of the updrift Vilamoura Marina jetties, riprap seawalls and an extensive groin field in front of Quarteira. These engineering structures constrain longshore littoral drift to Forte Novo–Garrão beaches (to the east) and are the main cause of beach erosion and cliff retreat. Nevertheless, other important factors contribute to rapid coastal recession, including cliff lithological properties (soft cohesion); a narrow protective beach; wave storm events; vibration caused by building works or traffic near the cliff edge; golf courses extending until the cliff edge, which provide cliff watering; occasional ruptures in the pluvial drainage system and sewer pipes; and heavy rainfall episodes in very short-time spans, that cause formation and enlargement of gullies. All these factors combine in various ways to produce a rapid landward migration of cliffs in this area, in some cases measurable on a daily scale (during storm events; Figure 14.10).

Cliff retreat endangered houses and other infrastructures at the Vale do Lobo tourist resort which forced local authorities (in partnership with the national authorities) to undertake emergency actions to mitigate coastal erosion. With the main objective of 'holding the line', a beach nourishment was carried out between October 1998 and January 1999. About 700,000 m³ of sediment was placed to form a 1.4 km beach platform, 80 m wide and 4 m in elevation (above mean sea level). Sediment was dredged 4 km offshore Vale do Lobo beach at a depth of about 20 m. Retreat rates before nourishment (1991–99) ranged from 2.27 m/yr at Forte Novo to a minimum of 0.14 m/yr at Garrão. After beach nourishment (1999–2001), almost no cliff erosion took place in the vicinity of the nourished area. In 2006, the beach was not wide enough to protect the Vale do Lobo cliffs and therefore a new beach nourishment was needed. This intervention used about 400,000 m³ of sand, which remained for some four years. Although Vale do Lobo cliffs were relatively protected by the interventions, nearby cliffs were still facing undergoing retreat. As an example, Forte Novo beach lowered and narrowed between November 2009 and March 2010, which allowed the cliff base to be reached by wave swash and therefore enhanced cliff retreat. Mean cliff retreat during this period was 2 m, with a maximum of 7 m, being mainly associated with a four-storm cluster that struck the coast in just 23 days.

Following the storm cluster of 2010, one of the largest nourishment operations ever made in Portugal was undertaken. More than 1×10^6 m³ of sand dredged offshore (circa 25 m depth) was deposited between Forte Novo and Garrão, and this protected the cliffs and existing settlement. This operation is expected to have a life-span of about ten years.

(References: Nunes et al., 2011; Oliveira et al., 2008)

Conclusions

The coastal zone of Portugal faces several threats and hazards because of a combination of intensive human development in coastal areas under severe erosion. The erosion process has been caused by sediment starvation mainly because of interventions at river basins. It is also increased by updrift coastal engineering structures (e.g. harbours, jetties or protection groins) and it is expected to further increase, because of sea level rise. The rise in the littoral population during the twentieth century expanded the urban coastal areas and the tourism boom at the end of the 1970s contributed to an occupation of the oceanfront's most hazardous zones. Consequently, the frequency and magnitude of coastal problems increased, which in turn exacerbated the need to pursue integrated coastal management actions. Examples of areas under threat and facing serious problems are: Ofir, South of Espinho, South of Aveiro and Costa da Caparica on the west coast; Quarteira–Vale do Lobo (see box) and the occupied areas of Ria Formosa Natural Park on the south coast. These examples were chosen amongst many others,

as more than half of the Portuguese sand coast is facing important shoreline retreat.

A product of the lack of clear rules and an integrated coastal management strategy, coastal protection was until the 1990s primarily driven by the institutional obligation to respond to population and the need to protect property. Local problems were addressed with hard protection measures because of its immediate effectiveness, regardless of erosion problem transference to downdrift areas. As a result, parts of the Portuguese coast became armoured whilst neighbouring downdrift areas experienced ongoing severe shoreline retreat. Soft protection in Portugal started in the 1950s, but it was only during the 1990s that it became a major frequent management approach. Despite the rather restrictive Portuguese legal framework, an integrated coastal management approach is still uncommon, and interventions are rarely designed on the scale of the entire coastal cell. Positive examples, such as the Ria Formosa Natural Park beach/dune recovery (1999/2000) and the Quarteira–Vale do Lobo beach nourishment (2010) will probably increase in the near future, within the framework of the National Strategy towards an Integrated Coastal Zone Management.

References

Abecasis, F., Castanho, J., Matias, M. F., 1970. *Coastal regime. Carriage of material by swell and currents. Model studies and in situ observations. Influence of port structures. Coastal defence works. Breakwaters.* Memória no 362, LNEC, Lisbon, Portugal, 30 pp.

Almeida, L. P., Ferreira, Ó., Vousdoukas, M., Dodet, G., 2011. Historical Variation and Trends in Storminess along the Portuguese South Coast, *Natural Hazards and Earth Systems Science*, 11, 2407–2417.

Andrade, C., 1990. *O ambiente de barreira da Ria Formosa. Algarve – Portugal*, Ph.D. thesis, Universidade de Lisboa, Portugal, 626 pp.

Andrade, C., Teixeira, S., Reis, R., Freitas, C., 1996. The record of storminess of the Portuguese NW coast in newspaper sources, in J. Taussik and J. Mitchell (eds) *Partnership in Coastal Zone Management,* Samara Publishing, Cardigan, 159–166.

Andrade, C., Freitas, M. C., Cachado, C., Cardoso, A. C., Monteiro, J. H., Brito, P., Rebelo, L., 2002. 'Coastal zones', in F. D. Santos, K. Forbes and R. Moita (eds) *Climate change in Portugal: Scenarios, impacts and adaptation measures. SIAM Project,* Gradiva, Lisbon, 175 219.

Ângelo, C., 1991. Taxas de variação do litoral Oeste: Uma avaliação temporal e espacial, *Comunicações do Seminário "A zona costeira e os problemas ambientais",* Eurocoast, Portugal, 109–120.

Antunes, C., Taborda, R., 2009. Sea level at Cascais tide gauge: Data, analysis and results, *Journal of Coastal Research*, SI 56, 218–222.

Baptista, P. R., 2006. *O sistema de posicionamento global aplicado ao estudo de litorais arenosos*, PhD thesis, Universidade de Aveiro, Portugal, 278 pp.

Barceló, J. P., 1971. *Experimental study of the hydraulic behaviour of inclined groin systems.* Memória no 394, LNEC, Lisbon, Portugal, 20 pp.

Bettencourt, P., Ângelo, C., 1992. Faixa costeira centro oeste (Espinho–Nazaré): Enquadramento geomorfológico e evolução recente, *Geonovas*, nº Especial 1, A Geologia e o Ambiente, 7–30.

Carvalho, G. S., Alves, A. C., Granja, H., 1986. A evolução e o ordenamento do litoral do Minho, *II Congresso Nacional de Geologia*, Lisbon, 5–33.

Castanho, J. P., Gomes, N. A., Carvalho, J. R, Vera-Cruz, D., Araújo, O., Teixeira, A. A., Weinholtz, M. B., 1974. *Means of controlling littoral drift to protect beaches, dunes, estuaries and harbour entrances: Establishment of artificial beaches.* Memória nº 448, LNEC, Lisbon, Portugal, 26 pp.

Costa, M., 1994. Agitação marítima na costa portuguesa, *Anais do Instituto Hidrográfico*, 8, 23–28.

Costa, M., Esteves, R., 2010. Clima de agitação marítima na costa oeste de Portugal Continental, in C. Guedes Soares and C. Costa Monteiro (eds.) *Proceedings 11ª Jornadas de Engenharia Naval: O Sector Marítimo Português*, Edições Salamandra, Lisbon, 413–426.

Costa, M., Silva, R., Vitorino, J., 2001. Contribuição para o estudo do clima de agitação marítima na costa portuguesa, *2nd Jornadas Engenharia Costeira Portuguesas*, Sines, 20 pp (CD-ROM).

Dias, J. M. A., 1990. A evolução actual do litoral português, *Geonovas*, 11, 15–28.

Dias, J. M. A., Taborda, R., 1992. Tidal gauge in deducing secular trends of relative sea level and crustal movements in Portugal, *Journal of Coastal Research*, 8, 3, 655–659.

Dias, J. A. M., Ferreira, Ó., Pereira, A. R., 1994. *Estudo Sintético de Diagnóstico da Geomorfologia e da Dinâmica Sedimentar dos Troços Costeiros entre Espinho e Nazaré*, http://w3.ualg.pt/~jdias/JAD/eb_EspinhoNazare.html (accessed 25 August 2012).

Dias, J.A.M., Ferreira, Ó., Matias, A., Vila-Concejo, A., Sá-Pires, C., 2003. Evaluation of soft protection techniques in barrier islands by monitoring programs: case studies from Ria Formosa (Algarve-Portugal), *Journal of Coastal Research*, SI 35, 117–131.

Dodet, G., Bertin, X., Taborda, R., 2010. Wave climate variability in the North-East Atlantic Ocean over the last six decades, *Ocean Modelling*, 31, 120–131,

Ferreira, J. C., Rocha, J., Tenedório, J. A., 2005. Land use evolution patterns on developed coastal areas: Fuzzy data integration through neural network and cellular automata modelling, in F. Veloso Gomes, F. Taveira Pinto, L. das Neves, A. Sena and Ó. Ferreira (eds.) *Proceedings of the First Conference on Coastal Conservation and Management in the Atlantic and Mediterranean*, IHRH, Oporto, 165–174.

Ferreira, Ó., 1993. *Caracterização dos principais factores condicionantes do balanço sedimentar e da evolução da linha de costa entre Aveiro e o Cabo Mondego*, MSc thesis, Universidade de Lisboa, Portugal, 168 pp.

Ferreira, Ó., 1998. *Morfodinâmica de praias expostas: aplicação ao sector costeiro Aveiro–Cabo Mondego*, PhD thesis, Universidade do Algarve, Portugal, 337 pp.

Ferreira, Ó., 2005 Storm groups versus extreme single storms: Predicted erosion and management consequences, *Journal of Coastal Research*, 42, 221–227.

Ferreira, Ó., Dias, J. M. A., 1991. Evolução recente de alguns troços do litoral entre Espinho e o Cabo Mondego, *Actas do 2º Simpósio sobre a Protecção e Valorização da Faixa Costeira do Minho ao Liz*, IHRH, Oporto, 85–95.

Ferreira, Ó., Dias, J. M. A., 1992. Dune erosion and shoreline retreat between Aveiro and Cape Mondego (Portugal): Prediction of future evolution, in H. Sterr, J. Hofstede and H. P. Plag (eds,) *Interdisciplinary discussions of coastal research and coastal management issues and problems: Proceedings of the International Coastal Congress*, Peter Lang, Frankfurt am Main, 187–200.

Ferreira, Ó., Dias, J. M. A., Taborda, R., 2008. Implications of sea-level rise for Continental Portugal, *Journal of Coastal Research*, 24, 2, 317–324.

Gama, C., Dias J. M. A., Ferreira, Ó., Taborda, R., 1994a. Analysis of storm surge in Portugal between June 1986 and May 1988, *Proceedings of Littoral'94*, Eurocoast, IHRH/ICN, Lisbon, 381–187.

Gama, C., Taborda, R., Dias, J. M. A., Ferreira, Ó., 1994b. Return periods of extreme sea levels in Portugal, *Gaia*, 8, 59–61.

Garcia, T., Ferreira, Ó., Matias, A., Dias, J. M. A., 2002. Recent evolution of Culatra Island (Algarve-Portugal), in F. Veloso Gomes, F. Taveira Pinto, L. das Neves (eds.) *The changing coast*, IHRH, Oporto, 289–294.

Granja, H. M., Carvalho, G. S., 1995. Is the coastline 'protection' of Portugal by hard engineering structures effective?, *Journal of Coastal Research*, 11, 4, 1229–1241.

ICN., 2007. *Plano de ordenamento e gestão do parque natural do litoral norte: Fase I – Caracterização, Parte I – Descrição, Volume II – Caracterização física*, ICN, Lisbon, 84 pp.

INAG., 2006. *Execução da recomendação sobre gestão integrada da zona costeira em Portugal: Relatório de Progresso*, INAG, Lisbon, 98 pp.

INAG., 2009 *Estratégia nacional para a gestão integrada da zona costeira*, INAG, Lisbon, 83 pp.

Loureiro, E., 2006. *Indicadores geomorfológicos e sedimentológicos na avaliação da tendência evolutiva da zona costeira (aplicação ao concelho de Esposende)*, PhD thesis, Universidade do Minho, Portugal, 621 pp.

Martins, M., 1977. Alimentação artificial de praias: Casos portugueses, *Seminário sobre obras de protecção costeira*, LNEC, Lisbon, 1–6.

Matias, A., Ferreira, Ó., Dias, J. M. A., 1999. Preliminary results on the Cacela Peninsula dune replenishment, *Boletin del Instituto Español de Oceanografía*, 15, 1–4, 283–288.

Matias, A., Ferreira, Ó., Dias, J. M. A., Vila-Concejo, A., 2004. Development of indices for the evaluation of dune recovery techniques, *Coastal Engineering*, 51, 261–276.

Nunes, M., Ferreira, Ó., Loureiro, C., Baily, B., 2011. Beach and cliff retreat induced by storm groups at Forte Novo, Algarve (Portugal), *Journal of Coastal Research*, SI 64, 795–799.

Oliveira, I. M., Valle, A. J., Miranda, F. C., 1982. Littoral problems in the Portuguese west coast, *International Conference on Coastal Engineering '88*, 1857–1867.

Oliveira, S., Catalão, J., Ferreira, Ó., Dias, J. M. A., 2008. Evaluation of cliff retreat and beach nourishment in southern Portugal using photogrammetric techniques, *Journal of Coastal Research*, 24, 4C, 184–193.

Pacheco, A., Vila-Concejo, A., Ferreira, Ó., Dias, J. M. A., 2008. Assessment of tidal inlet evolution and stability using sediment budget computations and hydraulic parameter analysis, *Marine Geology*, 247, 104–127.

Pedrosa, A., Freitas, C., 2008. The human impact on the Espinho-Paramos coast in the 20th century, *Journal of Iberian Geology*, 34, 2, 253–270.

Pimentel, D., 2005. A população, *Atlas de Portugal*, Instituto Geográfico de Portugal, 86–97.

Pita, C., Santos, J., 1989. *Análise de temporais da costa oeste de Portugal Continental*, Relatório PO-WAVES 1/89-A, IH/LNEC, Lisbon, Portugal, 29 pp.

Psuty, N. P., Moreira, M. E., 1992. Characteristics and longevity of beach nourishment at Praia da Rocha, Portugal, *Journal of Coastal Research*, 8, 3, 660–676.

Ramos, C., Reis, E., 2001. As cheias no sul de Portugal em diferentes tipos de bacias hidrográficas, *Finisterra*, 36, 71, 61–82.

Ramos, L., Dias, J. M. A., 2000. Atenuação da vulnerabilidade a galgamentos oceânicos no sistema da Ria Formosa mediante intervenções suaves, *Proceedings of the 3rd Symposium on the Iberian Margin*, 361–362.

Rocha, E., 1984. Crescimento económico em Portugal nos anos 1960–73: Alteração estrutural e ajustamento da oferta à procura de trabalho, *Análise Social*, 20, 84, 621–644.

Rodrigues, A., Veloso Gomes, F., 2011. Alimentações artificiais na Costa da Caparica, *VI Congresso sobre Planeamento e Gestão das Zonas Costeiras dos Países de Expressão Portuguesa*, Boavista, Cape Verde, 16 pp. (pen-drive)

Salles, P., 2001. *Hydrodynamic controls on multiple tidal inlet persistence*, PhD thesis, Massachusetts Institute of Technology and Woods Hole Oceanographic Institution, 272 pp.

Souto, H., 2005. Recursos vivos marinhos, *Atlas de Portugal*, Instituto Geográfico de Portugal, 168–175.

Taborda, R., 2000. *Modelação da dinâmica sedimentar na plataforma continental Portuguesa*, PhD thesis, Universidade de Lisboa, Portugal, 366 pp.

Taborda, R., Dias, J. M. A., 1992. Análise da sobreelevação do nível do mar de origem meteorological durante os temporais de Fev/Mar de 1978 e Dez de 1981. *Geonovas*, nº Especial 1, A Geologia e o Ambiente, 89–98.

Taborda, R., Andrade, C., Marques, F., Freitas, M. C., Rodrigues, R., Antunes, C., Pólvora, C., 2010. *Plano estratégico de Cascais face às alterações climáticas: Sector zonas costeiras*, Cascais, Portugal, 48 pp.

Taveira Pinto, F., Pais Barbosa, J., Veloso Gomes, F., 2009. Coastline evolution at Esmoriz-Furadouro stretch (Portugal), *Journal of Coastal Research*, SI 56, 673–677.

Teixeira, S., 1990. *Dinâmica das praias da Península de Setúbal (Portugal)*, MSc thesis, Universidade de Lisboa, Portugal, 189 pp.

Teixeira, S., 1994. *Dinâmica morfosedimentar da Ria de Aveiro (Portugal)*, PhD thesis, Universidade de Lisboa, Portugal, 396 pp.

Veloso Gomes, F., Taveira Pinto, F., das Neves, L., Pais Barbosa, J., Coelho, C., 2002. High risk in the NW Portuguese coast: Douro River – Cape Mondego, in F. Veloso Gomes, F. Taveira Pinto, L. das Neves (eds) *The changing coast*, IHRH, Oporto, 411–421.

Veloso Gomes, F., Taveira Pinto, F., das Neves, L., Pais Barbosa, J., 2006. *Pilot studies of river Douro–Cape Mondego and case studies of Estela, Aveiro, Caparica, Vale do Lobo and Azores*, IHRH, Oporto, Portugal, 317 pp.

15 Italy

Enzo Pranzini

Introduction

Italy, with a surface of 301,338 km², protrudes into the Mediterranean Sea with approximately 7,900 km of coast and a very few small areas wedged in Swiss territory located more than 200 km away from the sea. Facing the Ligurian, Tyrrhenian, Ionian and Adriatic Seas and with the two largest islands in the Mediterranean (Sicily and Sardinia) it is certainly the most coastal country in the Mediterranean. This 'maritime' characteristic has favoured trade and cultural exchanges since prehistory and is now the basis for a blooming tourist industry, but coastal erosion has begun to affect Italian beaches and a long coastline presents large problems.

Coastal morphology

Sand or gravel strands stretch for approximately 3,240 km, whereas the remaining 4,660 km are constituted by cliffs plunging into the sea and enclosing small pocket beaches (Figure 15.1). The latter prevails in Liguria, where beaches are present only at the outflow of a few rivers or in very deep bays. Tuscany, Latium, Campania and Calabria have coastal plains built up by large rivers, such as the Arno, Ombrone, Tevere, Volturno and Sele, whilst several embayed beaches are present where the Apennines (or their marginal mountains) approach the coast. The Basilicata region on the Ionian Sea has consequent rivers which flow parallel to each other resulting in a coalescent coastal plain bordered by sand beaches.

A limestone rock coast characterizes Puglia, both on the Ionian and South Adriatic seas. Northwards, after a long but narrow coastal plain, interrupted by several promontories in Molise, Abruzzo and Marche, the long Emilia-Romagna beach area merges into the River Po delta. Several spits and barrier islands close the Northern Adriatic lagoons of Venice, Marano and Grado, before reaching the rock coast at the Slovenian border.

Sicily and Sardinia have 1,623 and 1,897 km of coastline respectively. Sand or gravel beaches are prevalent on the former (1,117 km) and rock coasts on the latter (1,438 km) and one of the best examples of a ria coast in the Mediterranean can be found in the northern part of Sardinia.

Volcanic coasts occur from central to southern Italy, both on the continent (Campi Flegrei and Vesuvius) and the islands [Isole Ponziane (Pontine Islands), Ischia, and the Isole Eolie (Aeolian Islands)], where very unstable cliffs plunge deep into the sea forming a few narrow beaches. Etna, the highest volcano in Europe, has for centuries been sending tongues of lava into the sea at the Strait of Messina. Beach rock, exposed by coastal erosion, frequently fringes sand beaches, with offshore platforms effectively protecting the coast. Exiting currents, capable of taking sediment offshore, can flow through some gaps. Where beach rock is below the backshore, it inhibits infiltration, thereby increasing water saturation and sediment instability (Vousdoukas et al., 2007).

Artificial coasts (harbours and shore protection structures) represent 7.5 per cent of the entire

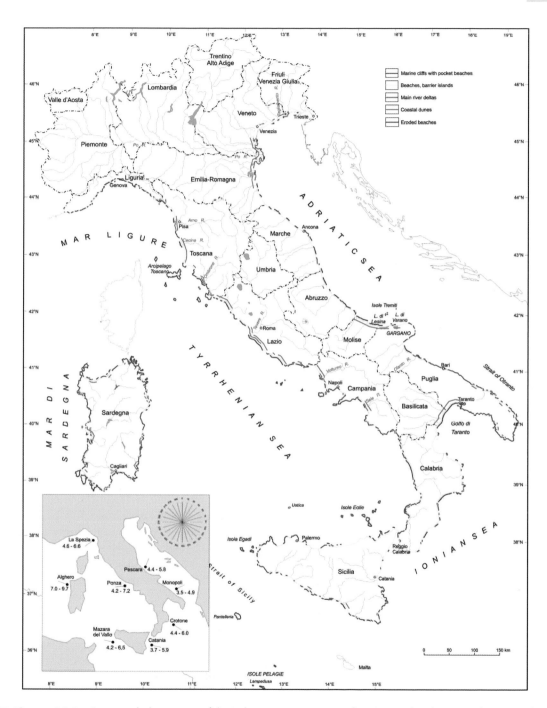

■ **Figure 15.1** *Geomorphologic map of the Italian coast. Inset: Significant wave height (Hs) with one- and ten-year return period at the National Sea Wave Measurement Network (RON) from 1989 to 2001. Severest storm provenance direction for each gauge is also shown (Data from Corsini et al., 2006)*

coastline, totalling 592 km (Valloni et al., 2003). Large dune systems exist only on the western Sardinian coast, where strong north-west winds blow over dry sand beaches; here wide dune fields (Is Arenas, Porto Ferro, Pistis, San Niccolò) are present along with some cliff top dunes reaching 60 m a.m.s.l. elevation at Piscinas. Minor dune systems are present on the Tyrrhenian peninsular coast, where river deltas show converging foredunes and beach ridges. However, at most sites dunes have been bulldozed in order to allow for development of coastal settlements and construction of industries, streets and railways (Macchia et al., 2006).

During the last 2,500 years, all the largest Italian rivers have built up wave-dominated deltas (Marinelli, 1926), with the exception of the Po. This river began to build up such a morphology 4,000 years ago, prior to shifting to a river-dominated delta during the 1500s (Gabbianelli et al., 2000) and has recently returned to its original category due to reduction in river sediment input.

Most lagoons and wetlands present in historical times were reclaimed during the last two centuries, after several failed attempts by the rulers of fragmented Italy during the Renaissance (Barsanti and Rombai, 1986). At the beginning of the twentieth century 700,000 ha of wetlands were present along the Italian coast; this had been reduced to 192,000 in 1972 and to less than 100,000 in 1994 (Benedetto et al., 2000). Those remaining increase the morphological richness of Italian coasts, such as Laguna di Orbetello, delimited by the two tombolos connecting Monte Argentario to the southern Tuscan coast. In Agro Pontino (Pontine Marshes, Lazio), after intense land reclamation by the Papacy and, later, by the government during Fascist times, several lakes remain: some are shore parallel for bar accretion (Fogliano, Monaci and Caprolace Lakes) but preserve an indented inland coast cut by creeks in red aeolian sand deposited during the last glacial age. Lago di Fondi, on the contrary, has several wide branches (ancient river courses submerged during Holocene transgression) and a sand bar closing this ria, forming an irregular lake. Lesina and Varano Lakes, near the Gargano promontory, are gulfs enclosed by two bay barriers. The shape and origin of Venice lagoon, in the northern Adriatic sea, is far more complex: closed by a barrier island, whose origin can be related both to onshore bar displacement induced by Holocene sea level rise and sediment moving longshore coming from the Po, Adige, Brenta and Piave rivers. The Republic of Venice had to develop innovative hydraulic systems to maintain depth in the lagoon and entrance through the inlets, since these rivers represent very conspicuous sediment transport systems. Near the Slovenian border, the Marano lagoon is located between the Tagliamento and Isonzo rivers, and contributes to its lateral closure, but has less of a problem with siltation because rivers discharging into it are short and drain Karst areas, thus having limited bedload.

Wave climate

In spite of being classified as sheltered (Davies, 1980), some Mediterranean coasts are exposed to severe storms. Most Italian shores enter this category, with extremes on the western Sardinia coast, where annual significant wave high (H_s1yr) is 7.0 m, and the 50-year return period wave (H_s50yr) reaches 9.7 m (Figure 15.1). The Ligurian and Tyrrhenian Seas are more sheltered, but average H_s1yr is >4.2 m in all sites, and H_s50yr = 7.2 m at Ponza. The Ionian Sea climate is milder, with H_s1yr lower than 4.5 m and H_s50yr = 6.5 at Mazara del Vallo. On the long and narrow Adriatic sea East and North-East gales can generate waves higher than expected, with H_s1yr = 4.4 m and H_s50yr = 6.6 m at Pescara. The Western basin has a unimodal wave provenance distribution, generally from NW or SW, whereas distribution in the Eastern seas is bimodal, with waves coming from N and from S or SE (Corsini et al., 2006). Tidal range is very limited, approximately 0.30 m along most of the Italian coast, with higher values on the northern Adriatic, where resonance increases a 0.1 m tide at the basin entrance (Strait of Otranto) to a 1 m tide at Trieste.

Evolution of coastal settlements

The Italian peninsula morphology is characterized by long, high mountain ranges, which favour urban and industrial development (including construction of roads and railways) along the narrow part of the territory that separates mountains from the sea. However, this land was exposed to piracy and was especially devastated by malaria, so for many centuries people had to retreat to the first hill line bordering the coast. Some of these settlements, although located several kilometres from the coast, have the epithet 'Marittimo' in their names (Rosignano Marittimo, Campiglia Marittima and Massa Marittima): land ended there!

As control over the sea increased and healthier land conditions were achieved in coastal areas, a progressive migratory flux to the coast began in the early 1800s, stimulated by the development of Italian railways. At the same time, beach erosion started to be noticed along the Italian coast. Hilltop settlements developed coastal counter-parts ('gemmations'), and added the relevant toponym to their names to distinguish themselves: 'Scalo' (meaning Station) being close to the railway, or 'Marina' being near the shore.

The modern tourism industry started at such marinas, with the first bathing establishment on pile-dwellings, moving from the Liberty cabanas of Viareggio up to the modern tourist district connecting Cesenatico, Rimini, and Riccione e Cattolica, on the Adriatic coast, which forms a unique 55 km long seaside resort. Tourism had to compete with other industries for the free land and easy connections by land and sea that were available: in the early 1970s approximately 175,000 industrial plants (18 per cent of the total) were active along the Italian coast (Giovannini, 2010).

Harbours also developed at that time, built both along rock coasts, where water was deeper and bays could be used as shelter, and on sand shores, in order to serve industrial districts. The first wooden piers, used for loading/unloading merchandise, had to be protected with breakwaters, extending offshore in order to fulfil a continuous request for deeper water.

Jetties were built at the river mouth, to prevent mouth siltation and to allow boats to enter offshore of the breaker line – on the Adriatic coast this is the most frequent harbour type to date. Except for a few historical harbour towns located on rock shores (such as Genoa, Naples, Palermo and Bari), the Italian coast gradually changed from being completely unpopulated to being a very attractive part of the territory. Today, coastal municipalities (8 per cent of Italian municipalities) host approximately 33 per cent of the population (Giovannini, 2010), indicating that in Italy coastal areas also offer better economic, social and recreational opportunities than what is available further inland (Goldberg, 1994). In Sardinia, however, 70 per cent of the coast remains completely undeveloped, representing approximately one-third of the construction-free coastal areas in the country (WWF, 2012).

Beach erosion

Historical times

In historical times most beaches accreted as a consequence of the large outflow of river sediments caused by soil erosion resulting from intensified deforestation accompanying demographic and economic development in the Italian peninsula. In fact, Etruscans had to use slope stabilization to prevent landslides on hillsides, which at that time were subject to intense forestry and agricultural uses. River deltas also formed in this period (Marinelli, 1926). The study of beach ridges and foredunes on the Arno river delta can help in reconstructing events related to the population that lived within the watershed (Figure 15.2). It is therefore possible to identify rapid delta growth during Roman times, whereas their erosion in the Early Middle Ages can be related to the social instability caused by the demise of the Roman Empire. A new period of expansion began around 1000 AD (or just prior), with intense episodes in the fourteenth, seventeenth and eighteenth centuries.

These were periods of high economic and demographic growth; widespread forest cutting was experienced particularly during the eighteenth century in order to produce coal for the industrial boom (Pranzini, 1995). Paintings of this period show landscapes with limited vegetation cover, barren soils and landslides, all depicted exactly as they are shown in current geomorphology textbooks. Erosion phases can also be identified in this period; the most evident, during the fourteenth century, can be related to the Black Death, which halved the Italian population and caused forests to expand, since most cultivated land was abandoned (Pranzini, 2001). This evolutionary pattern has been recognised in other deltas, such as the Ombrone (Innocenti and Pranzini, 1993) and the Tiber (Mikhailova *et al.*, 1998).

The nineteenth and twentieth centuries

The present phase of beach erosion in Italy started during the mid–nineteenth century as a consequence of marsh reclamation, reduction in cultivated areas, reforestation and slope stabilization. Outflow of river sediments was also reduced by riverbed quarrying and Italian dam construction. Erosion dating along the coast shows that this process started later in southern Italy than in the north (Figure 15.3); due to slower economic development in the south during the twentieth century.

River bed quarrying was intense in the Po basin in the beginning of the twentieth century (Bondesan

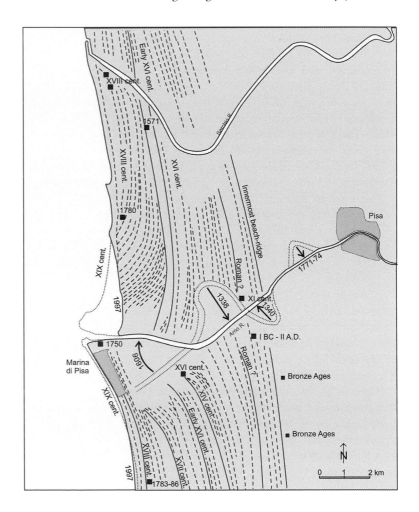

■ **Figure 15.2** *Beach ridges of the Arno River delta showing accretion and erosion phases (modified from Pranzini, 2007)*

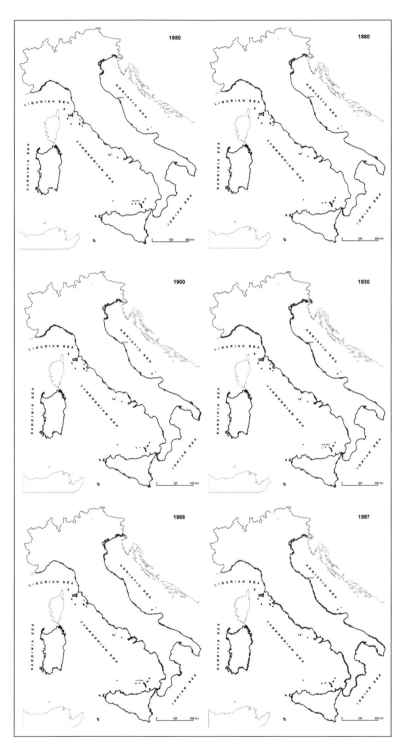

■ **Figure 15.3** *Beach erosion (red) in Italy from 1859 to 1997 (data from Albani, 1933; Commissione De Marchi, 1968; CNR, 1997)*

and Dal Cin, 1975), whereas in southern Italian rivers a similar activity can only be found in the second half (Cocco *et al.*, 1978). The same occurred with regard to abandonment of mountain and hill agriculture (due to urban primacy or to migration to the coast), which happened in the north at a time when rural societies were still fully functioning in the south and in the islands.

Dam construction commenced reducing bed load in the River Arno in the 1950s (Becchi and Paris, 1989). In Basilicata, such a problem arose only during the 1960s, nevertheless, 42 per cent of its four largest watersheds are today cut by dams (Bonora *et al.*, 2002); in Sicily similar problems arose in the 1950s (Amore and Giuffrida, 1984). In Campania, a dam on the River Sele stops sediment derived from 67 per cent of its basin and the river delta is retreating at a rate of 2.1 m/yr (Cocco and Iuliano, 1999).

An additional factor for beach erosion is land subsidence, which characterises all recently deposited sediments in river deltas; water extraction for agriculture and industrial uses increased this phenomenon by an order of magnitude (Bartolini *et al.*, 1988). Subsidence is one of the main causes for beach erosion in the River Arno delta, reaching approximately 10 mm/yr (CNR, 1986); in the River Po delta, 250 mm/yr subsidence was calculated for the 1950s (Borgia *et al.*, 1982). Along the northern Adriatic coast, effects from inshore and offshore gas extraction add to natural and water-pumping induced subsidence: Ravenna subsided at a rate of 40 mm/yr in the period 1970–1977 (Vicinanza *et al.*, 2008), when reservoir exploitation was most intense.

Erosion began when the population started to migrate from inland areas to the coast. Most new settlements had to tackle the problem of beach erosion immediately, but this did not prevent urban development from taking its course along the coast.

Easy access, large undeveloped areas to host yards, and proximity to new industrial areas favoured the location and development of harbours on low coasts, requiring long jetties. These new structures, extending offshore in order to find deeper water, interrupted sediment transport with severe effects on sediment feeding. Marina di Carrara and Viareggio harbours in

Tuscany, and Cetraro Marina in Calabria, are important cases of trapping longshore transported sediment. Although these structures prevented some short coastal segments from eroding naturally, they rapidly induced or increased erosion of downdrift sectors; a good example can be found in the new marina of Scoglitti, in Sicily, where downdrift erosion was induced before its completion. On the Adriatic coast, jetty construction, which extends offshore river courses in order to prevent mouth siltation, is the main cause of beach erosion. Porto Garibaldi, Porto Corsini, Cesenatico and Rimini (all in Emilia-Romagna), as well as Pesaro, Fano and Senigallia (in Marche), are typical examples of this phenomenon. A peculiar case is provided by the marinas of Chiavari and Lavagna (Liguria), located on the two sides of the Entella River mouth: here fluvial sediments cannot flow and feed adjacent beaches which have been undergoing severe erosion (GNRAC, 2006).

Coastal development proceeded irresponsibly alongside beach erosion. The economic boom of the 1960s and the call for holiday houses near the sea led to villa, condominium and holiday resort construction over coastal dunes (Figure 15.4); this prevented the coastal system from using a buffer, both as free space to absorb periodical shoreline retreat,

■ **Figure 15.4** *Tourist village built on the dunes of Follonica Gulf, Tuscany (1976). When shoreline retreat threatened the building stability, hard shore protection measures were applied, which triggered stronger erosion adjacent to unprotected coastal sectors*

and for sand being available for beaches during periods of crisis.

As a result, the need for shore protection emerged in these areas. At the Gulf of Follonica (Tuscany) coastal segments were protected from very limited erosion with detached breakwaters, which drained sand from unprotected sectors, increasing the problem. Had the cost of these structures been invested in beach nourishment, the 1954 shoreline position could have been maintained and landscape degradation avoided (Aminti et al., 2002).

Present status

A survey performed in 2006 (GNRAC, 2006) showed that 42 per cent of Italian beaches are eroding (Figure 15.5), not considering segments protected by coastal structures and those which are stable or

accreting due to their position updrift from harbours. Reduction in erosion is evident in some northern areas when comparing 1900 and 1997 maps (Figure 15.3); this results from shore protection (as in Liguria). However, structures are not always capable of limiting this process, and some coastal sectors, although protected by several breakwaters, are still eroding. At Montesilvano (Abruzzo), a close sequence of groins and three lines of detached breakwaters are not sufficient to stabilize the shoreline (GNRAC, 2006).

The highest erosion rates affect the major Italian river deltas, where 10 m/yr retreat can be frequently observed (the Arno, Obrone, Volturno and Po rivers), but similar rates are also seen on beaches located on smaller deltas, such as those on the deltas of the Biferno and Trigno rivers (Molise), the Metauro river (Marche) and the Amato river (Calabria) (GNRAC, 2006).

In addition, several beaches located downdrift from harbours or jetties are undergoing severe erosion: downdrift Porto Garibaldi (Emilia Romagna), 30 detached breakwaters were built to protect beaches that are intensively used for tourism, thus shifting erosion further to the north (GNRAC, 2006). In other cases, wave diffraction at the tip of harbour breakwaters is responsible for beach erosion, as occurs at Termoli (Molise) (GNRAC, 2006), at Cala Galera (Tuscany) (Bartolini et al., 1977) and on the 1 km long pocket beach of Golfo di Campo, at Elba (Cipriani et al., 1993).

Pocket beaches are also eroding, even if at a lower rate. These narrow beaches are often located in deep bays and constitute the most attractive sector of rock coasts; the economy of small islands is often linked to them: little Elba was (before conversion to the Euro) the second largest currency market in Tuscany. These beaches, delimited by bluffs or recent building, cannot shift inland for adaption to beach erosion, and become progressively narrower, with severe impacts on the local economy. According to a survey carried out in 2008 (Simeoni et al., 2012), most pocket beaches longer than 300 m are eroding in Italy.

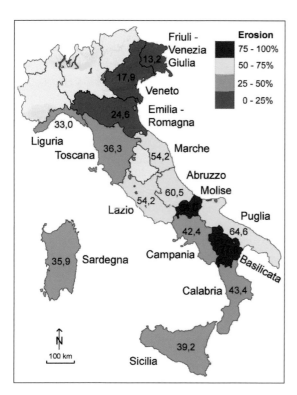

■ **Figure 15.5** *Percentage of eroding beaches in each Italian region*

Shore protection

Such an intensely developed coast, of high economic value and under severe erosion, is one of the most protected in the world, with approximately 9,650 structures (harbours included; ISPRA, 2008) over 7,900 km of coast. Protection strategies are frequently discussed amongst coastal communities, and projects widely shown in local newspapers. Private stakeholders, NGOs and public administrations are active in notifying judicial authorities of undesired shore protection works (Pranzini, 2009).

Administrative competencies on shore protection

In 1907, when erosion was affecting many coastal settlements, Italy approved a law making large financial resources available for defending threatened coastal towns. The Maritime Work Office was created under the Ministry of Public Works, and formed with technical expertise from harbour engineers who defended the coast with hard structures (mainly seawalls, detached breakwaters and groins), causing the coastal landscape to be permanently modified. An additional factor leading Italy towards the 'hard shore protection philosophy' is the large availability of rock by or near the coast; further, there was no interest in preserving beaches at that time, when their economic and landscape value was still unknown. Beach nourishment was not used, since control of the sand volumes used was considered to be impossible.

When the Regions were constituted (1970), some issued special laws (Lazio and Emilia-Romagna) allowing them to act on shore protection, experimenting with new technologies and using beach nourishment. Since 1948 Sardinia and Sicily have been regions ruled by special statutes which did not fall under the national rule on the topic, and some new projects were carried out in the 1990s. A National Law in 1989 and a Legislative Decree in 1998 shifted the responsibility for programming, planning and integrated management of the coastal zone, including coastal defence of seaside resorts and related infrastructures, from the National to Regional Administration. These laws have been gradually put into effect and today all Italian Regions are directly involved in shore protection planning – although some, such as Tuscany, transferred these competencies to the Provinces.

Ancient times

In historical times, albeit on a limited scale, Italians had to counteract the effects of the sea eroding their coasts, cutting foredunes and entering lagoons. The Romans, masters of the sea and actors in large land reclamation projects, enjoyed the luxury of building not only streets and military settlements by the sea but also those villas whose remains are still present along the coast. They also laid down the first rip-rap known in Italy: the defence of a segment (not precisely defined) of the Via Severiana, the last Imperial road, built in 198 AD by Septimius Severus to connect the harbour of Ostia with Terracina. A phrase engraved on a plaque registers this work. The excellent skills of the Romans in hydraulic construction can be demonstrated by the detached breakwater built to protect Parenza, a town in Istria (Franco, 1996).

The first important works performed to protect the coast dates from the sixth century. These were aimed at stabilizing the sand bars that delimited Venice lagoon, and consisted of poles driven into the beach which held wicker trellises containing earth and stones – similar to those used in natural engineering today. Later, these defences were flanked by groins and revetments. Beach nourishments were then used, feeding the beach with sediments that had been dredged offshore with dredgers very similar to bucket dredgers of today.

These defences were precarious and required continuous and expensive maintenance interventions. In 1738 they were replaced by *murazzi*: 14 m thick and 4.5 m high seawalls, built with Istrian stone cemented with *pozzolana* (a type of concrete, previously used by Romans, which hardens in contact with the seawater). Murazzi were then built by the Republic of Venice along approximately 20 km of coast, at Malamocco, Pellestrina and Sottomarina

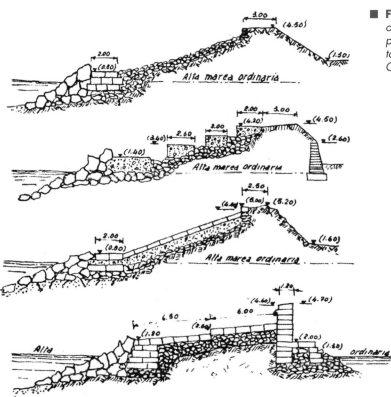

■ **Figure 15.6** *Different sections of murazzi before the reshaping performed after their collapse due to a severe storm in 1966 (from Gallareto, 1960)*

(Figure 15.6). The project had been developed by cartographer Vincenzo Coronelli in 1716, but was constructed 20 years later by the mathematician Bernardino Zendrini, overseer of waters, rivers, lagoons and bridges for the Republic of Venice.

Most Italian coastal towns lost their beaches in this context. The use of the beach, not marked at that time, became a mass phenomenon from the mid-twentieth century, when access to the sea from urban sites became desirable for providing recreation to users, independent of their social status and economic position. Such a loss affected not only large cities, like Naples and Bari, but also smaller settlements originated from the tourist use of the beach, e.g. Marina di Pisa, or whose new economic development was based on it, e.g. Pietra Ligure. The latter case is emblematic of the disaster affecting the coast: at first, this beach was eroded due to shore protection projects carried out updrift (at Ventimiglia); later, it

was occupied by the railway; and finally it was defended by revetments. Bari is a sound example of how the beach was lost by larger towns: here the beach/nearshore area had material added in order to support two lines of new housing and a promenade, all defended by a seawall; later this area was enclosed inside the harbour breakwater.

Seawalls, revetments and rip-raps

The first defences carried out in the late 1800s were rip-raps and seawalls. Coastline stabilization was the main goal of these projects, preventing shoreline retreat that threatened coastal roads and induced the collapse of buildings located by the shore. Rock (and later concrete blocks) was used at first. With polyvinyl chloride-protected iron wire, this allowed construction of gabions resistant to seawater and these elements entered several shore protection works (mostly to defend motorways and railways), reducing coastal recreational usage even more.

In all these cases, wave reflection on the structures (seawall, revetments, rip-rap) produces or increases beach erosion and deepens the profile, inducing an irreversible process as waves moving in deeper waters lose less energy. Detached breakwaters built offshore a seawall at Marina di Pisa in a 2.5 m water depth now have 7 m of water in front of them. Revetments were used not only for the protection of coastal settlements; the railway, that in many regions needs to run close to the sea (Liguria, Calabria, Marche, Abruzzo and Sicily), was defended by such structures, causing the tourist industry to lose kilometres of valuable beaches, resulting in a strong effect on the landscape. Although in some cases these 'archaeostructures' were changed into new structures that allowed beaches to be used (Guiducci *et al.*, 1993), the overall coastal usage had already been jeopardized by the railway itself. A cycle route has been recently built between Cupra Marittima and Pedaso (Marche), where a revetment was situated at a certain distance from the rail track (Figure 15.7). In Liguria the unused railway now hosts a bicycle/pedestrian route.

Detached structures

Detached breakwaters reduce wave energy onshore and allow tourists to use the beach. In Italy, it was first employed in Salerno, in 1905 (Franco, 1996). Since then, the choice for this type of protection

■ **Figure 15.7** *Railway between Cupra Marittima and Pedaso (Marche) protected by a revetment on which a cycle route was built*

became more frequent, as coastal recreation also increased. Different geometric configurations influence beach response (Pope and Dean, 1986) and several alternatives can be found, from very small structures resulting in salient formation, to larger ones that form tombolos. The first configuration, although less effective in beach expansion, allows sediment to bypass the structure and feed downcoast segments, whereas the second intercepts most of the sediments with a feeding effect. Very close structures can create two tombolos, as seen in Cirò Marina (Calabria), where a lagoon formed as a result of this process, holding stagnant water that becomes polluted in summer.

After World War II, projects based on detached breakwaters spread along the coast, frequently with the same layout, regardless of local conditions: 60–80 m long breakwaters were divided by 20–30 m gaps, all placed at 3 m water depth (Figure 15.8): a feedback process being induced, since construction of one element triggered downdrift beach erosion, requiring a new element to be built.

North of Porto Garibaldi (Emilia-Romagna), there are 74 breakwaters along 9 km of coast; between Porto San Giorgio and Casabianca (Marche), 61 breakwaters protect 5.5 km of coast.

The record belongs to Pescara littoral, where 243 detached breakwaters can be counted over 23 km of coast, from north of the town to the south. Although the Maritime Work Office was required to act only where buildings were at risk, at Gombo (Pisa) five detached breakwaters were built in the 1960s to protect a coastal segment in front of a Presidential villa that was located over 400 m inland (Box 15.1). Today these defences are built to protect artificial beach nourishments, as seen in San Felice Circeo (Lazio), and Crotone (Calabrian–Ionic coast). Although frequently named 'shore parallel', these structures can be displayed in different forms, usually facing dominant wave directions; complex dispositions are found along the Italian coast, often because of different overlapping projects, as can be seen at Lido di Riccio (Abruzzo) (Figure 15.9).

These beaches lose their natural aspect as they are no longer reshaped and cleaned by waves. They become flat and silty; fines, deposited in the very nearshore, are suspended as soon as a bather enters the sea. Along the northern Adriatic coast, where wave energy is low and fine sediments are abundantly brought by the Po River, these structures are one of the most important factors responsible for poor water quality. During storms, piling up occurs in the

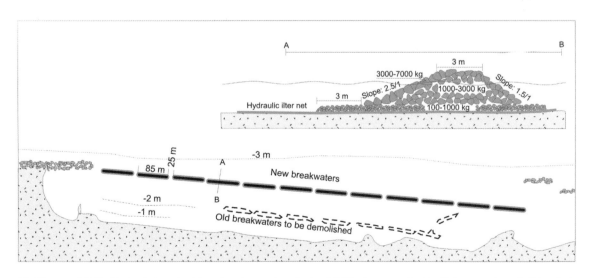

■ **Figure 15.8** *Typical detached breakwater project in the 1980s, here applied to Falconara beach (Marche)*

■ **Figure 15.9** *Detached breakwaters of uneven disposition – Lido di Riccio (Abruzzo)*

protected area and rip-currents run through the gaps, deeply eroding the sea bottom, making the structure prone to collapse (Figure 15.10). These currents are extremely dangerous for swimmers, and several drownings occur annually in these circumstances.

Submerged breakwaters

Submerged breakwaters have been adopted in the last few decades, reducing wave energy onshore; they prevent water quality from declining, but do not solve safety problems. Water exiting from the protected area carries sediments to the extremities, where two cusps can form at the expense of the beach to be protected; groins are frequently built to reduce

sediment flow, such as those constructed along the barrier island enclosing Venice lagoon (see Box 15.2).

At Marina di Massa (Tuscany), a submerged detached breakwater was built in 1980 in 2 m water depth to connect twelve groins at their tops. It was 8 m wide at the top with 1.7 m submerged, and no significant results were obtained. The height was later increased by 1.0 m and then to sea level. Today, each kilometre of coast is defended by 1.8 km of breakwaters! Similar structures have been used for beach protection in Veneto, Emilia-Romagna, Abruzzo and Lazio, frequently flanking artificial nourishments. In some cases, results were very poor; in others they strongly limited fill dispersion. At Pellestrina (Venice), the original fill volume was

■ **Figure 15.10** *3D representation of bottom morphology in a coastal segment protected by detached breakwaters (Marina di Massa, Tuscany); deepest point 14 m (not in scale; photograph from Provincia di Livorno; processed by Geocoste-UNIFI)*

■ **Figure 15.11** *Marina di Pisa (Tuscany): two out of ten segments of the detached breakwaters have been lowered to reach sea level and incoming wave energy absorbed by a gravel beach*

reduced by only 2 per cent in ten years (Box 15.2). Today, many old detached breakwaters are transformed into submerged or surfacing structures, possibly with increased width. At Follonica (Tuscany), such a reshaping did not increase shore erosion, since higher wave energy reaching the beach is compensated by reduced piling up.

A similar reshaping was performed on two of ten detached breakwaters located offshore a revetment that protects Marina di Pisa (Tuscany), where wave energy is higher. In this case a gravel beach was built at the revetment foot in order to deal with increased energy (Figure 15.11). In this breakwater reshaping, the tendency for crest enlargement is always present, to make waves break on the structure instead of on the back sheet of water. The new barrier defending Lido di Venezia will have a 60 m large berm.

Island-platform

The island-platform is a shore protection structure that can be frequently seen in Italy but is quite unusual elsewhere. Here, boulders create a round island, sometimes with a concrete ring in the middle (Figure 15.12a). It protects the coast without creating in-phase reflected waves, and disperses wave energy like a convex mirror. It was first used in Loano (Liguria) in 1967, and then in Pietra Ligure (Figure 15.12b) Ceriale and Borgo Prino (also in Liguria)

between 1969 and 1970 (Berriolo, 1985). Similar structures are present at Miliscola, near Naples, although they have a straight side facing the coast. A flat or submerged tombolo forms behind island-platforms, but sediments can overpass them during storms due to set-up, and no feeding effects are induced.

Groins

The traditional groin (made of stones, concrete and gangplanks) is widely used in Italy, due to its ease of construction and efficiency on the updrift coastal segment. This effect has been well documented since 1868, when a groin was built at the extremity of

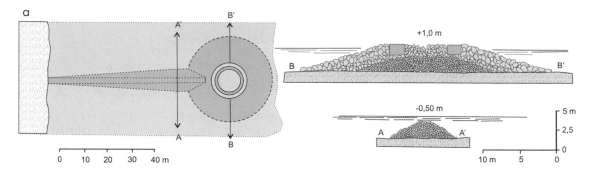

■ **Figure 15.12**
(a) island platform design (Modified from Berriolo and Sirito, 1972); (b) island platform at Pietra Ligure (Liguria): built in 1969, 400 m of coast was stabilized after artificial nourishment of 30.000 m³

Beach at Viserba, on the Adriatic coast, where groins were built in 1918 – postcard from 1950s (printed by Argia Bazzocchi)

Varazze bay (Liguria) in order to protect the newly built railway. It did limit sediment outflow from the physiographic unit and determined consistent beach expansion (Gallareto, 1960). The first groin sequence probably started at Viserba (Emilia-Romagna) in 1918 (Figure 15.13).

Due to the groins' efficiency in trapping sediments that move alongshore, downdrift erosion is a usual side-effect, forcing more groins to be built downstream (Figure 15.14). A number of groins, 126 in total, were built on the 12 km long coastal segment located between the mouths of the Piave and Livenza rivers (Veneto), whereas 206 groins can be counted today over 20 km of coast, from Lido di Rivoli to Santa Margherita di Savoia (Puglia). Where beaches are made of gravel, groins are generally shorter since coarse sediments move in a narrow belt near the swash zone.

These structures induce downdrift erosion, limit the view alongshore and are dangerous for swimmers due to topographically controlled rip-currents (Short and Masselink, 1999). They are however appreciated by fishermen and are not completely rejected by beach users, who enjoy the variability they cause (Eurosion, Tuscany). If well designed, they can be a feature of a town. Recently implemented ideas include: concrete blocks inserted in the groin axes to form a pedestrian track; a banister protecting users; and a ramp on the downcoast side allowing wheel-chair users to enter the sea in deeper water.

Carrara marble groins in Marinella di Sarzana (Liguria), built with well connected blocks of regular shape, can be walked on easily. They arouse admiration and surprise in foreign tourists, who are not aware of how much waste is produced in order to get a good marble block, capable of allowing another Michelangelo to sculpt his own David.

Submerged groins

Since most longshore transport occurs in a limited layer near the seabed, it can be intercepted by submerged groins – another Italian project (Berriolo and Sirito, 1973). These can be built with stones, concrete elements or geotextile bags and tubes filled with sand. They have the advantage that, after having trapped a certain quantity of sand (thereby inducing profile elevation and dry beach expansion), they are directly overrun by sediment, which therefore does not need to move offshore to pass in front of the groin tips. After passing the groin crest, where water velocity is at a maximum they are deposited due to a reduction in velocity. Deposition takes place both updrift and downdrift of the structure, preventing development of a saw-tooth shoreline.

■ **Figure 15.14** *Close groins at Misano (Emilia-Romagna)*

The first experimental project based on a submerged groin was carried out in Liguria in 1968, whereas the first real application occurred at Terracina (Lazio) in 1980 (Berriolo, 1993). Here polypropylene bags filled with sand were first used, and later flanked by upside-down T concrete elements with rubber wings to prevent sailing boat damage. A total of 500,000 m³ of granules was used as fill, which was immediately covered by sand reclaimed naturally from the nearshore. At Laigueglia (Liguria), Marina di Massa and Capalbio (Tuscany), submerged groins were built with geotextile sand bags (Aminti *et al.*, 2004), whereas at Capo d'Orlando (Sicily) and Capalbio (Tuscany) geotextile tubes were used.

Sometimes, submerged elements are placed as an extension of traditional groins in order to raise the beach profile without pushing the sedimentary flux too much towards offshore (or intercepting most of it). Examples can be found in Marina di Cecina and on the coast by San Rossore Regional Park (both in Tuscany), as well as Marina di Cariati (Calabria). At Marina di Minturno, the submerged extension is Y-shaped to prevent scouring the groin head and also to make incident waves diverge. In all the previous cases, submerged extensions are made of rocks, but at Marinella di Sarzana (Liguria) T-shaped concrete elements were used (Figure 15.15).

Beach response to submerged groins is similar to that induced by permeable groins, which are frequent

■ **Figure 15.15** *Marinella di Sarzana (Liguria): precast concrete elements ready to be installed as submerged extensions of the groin*

on the southern Baltic shores but almost unknown in the Mediterranean. Photographs from the 1960s portray some permeable groins at Marina di Massa (Tuscany), but their history is unknown. Their abandonment suggests that they failed in protecting the coast. At Lido di Volano (Emilia-Romagna), a permeable extension was placed at some close wooden groins fed with sand dredged at the River Po mouth.

Composite structures

A hybrid solution combining detached breakwaters and groins is to have the first connected to the shore through the second. Infinite possibilities exist, from the T groin at Marina di Torremarciano (Marche), where the parallel structure is extremely short and limits only offshore fluxes, to those in which the detached breakwater is longer, such as the ones protecting the railway at San Lucido (Calabria). Here each 320 m long cell was fed with 60,000 m^3 of sand and gravel, summed to a total of 1.1 million m^3 of sediments (Guiducci *et al.*, 1993).

Experimental and innovative projects

Several innovative and experimental shore protection projects (American Shore and Beach Preservation Association, 2007) have been applied or are close to being implemented in Italy.

- Permeable piled barriers with star-shaped elements (Ferran) were tested with good results near Ancona (Marche), but their low stability and durability, together with risks for swimmers, caused them to be abandoned (Franco, 1996).
- Beach dewatering, using BMS©, was tested at Alassio, in Liguria (Figure 15.16). Monitoring of the test site and a control beach over one year allowed morphodynamic responses of the beach to be seen, but no real stabilization effect could

be observed (Bowman *et al.*, 2007). Uncertain results were obtained in other similar projects carried out at the Lido di Ostia, Bibione, Metaponto and Procida (Ciavola *et al.*, 2008).
- Pressure Equalization Modules© (Jakobsen and Brøgger, 2007) have been installed on the coast of Pisa, without any effectiveness at least during the first year of operativity (still in progress).
- Submerged modular structures, designed to limit trawling and favour fish repopulation, have been

■ **Figure 15.16** *Schematic of the Beach Managing System installed at Alassio (Liguria) and active from 2004 to 2005*

recycled as wave energy dissipators and will be soon applied in some shore protection projects. Artificial algal carpets have also been proposed, in order to stabilize some pocket beaches on Elba.

Beach nourishment

Beach nourishment, as mentioned previously, was not part of the projects carried out by the Maritime Work Office, but some pioneering activities were performed by other institutions or by private people.

Between the two world wars, results from demolition and excavations were dumped into the sea from different headlands in Liguria; they were sorted and cleaned by waves during their movement towards bays and this material forms many beaches used for tourism today. Real nourishments were performed in the same region during the 1950s in the province of Savona (Pietra Ligure, Loano and Ceriale), where citizens rejected a project for protection of 5 km of coast, based on detached breakwaters and proposed by the Ministry of Public Works

Today it can be stated that artificial beach nourishment has been performed in all Italian Regions, ranging from small fills executed by holders of concessions for bathing establishments, to large projects carried out by the public administration. Sediment quarried inside the river bed was used at first, thus limiting future river input onto the coast. After river bed quarrying was outlawed (from the 1970s to the 1980s depending on the region), sediments from alluvial plains and crushed quarry rock were used. A rough estimate is that 15 million m^3 of this type of material has been used for beach nourishment in Italy.

Over 2 million m^3 of material, extracted during construction of an energy plant at Vado Ligure, were used in the 1970s to build a new beach at Bergeggi (Liguria) (Figure 15.17; Berriolo, 1985) and 1 million m^3 have been used since the 1980s to fill beaches in the province of Latina (Berriolo, 1993). These, together with those performed at Ostia and previously mentioned (1.2 million m^3) and at San Lucido (1.1 million m^3), are the largest nourishments carried out in Italy with the use of material from inland.

Beach nourishment proved to be effective in increasing tourist activity even when smaller projects were performed, such as the one at Cala Gonone (Sardinia). Here a new beach was created, using 23,000 m^3 of crushed white limestone at first, and 57,000 m^3 of pink granite granules later (Figure 15.18). The new beach is stabilized with artificial shoals built with rounded basalt boulders similar to those resulting from cliff collapse (Pranzini, 2009). Granules and fine pebbles were used to reconstruct beaches at Punta del Tesorino (Tuscany) and Caucana (Sicily); here the project was aimed at protecting an archaeological area attacked by the sea.

A beach nourishment using gravel was designed at Cavo (Elba) (Figure 15.19), allowing for a larger beach expansion with limited fill volume. It also aimed at providing the beach with stable material, which could cover an old beach nourishment scheme that contained a large percentage of fine sediments and created turbidity in the seawater – possibly

■ **Figure 15.17**
Bergeggi (Liguria), where a beach was created in 1971–72 using over 2 million m^3 of material, extracted during the construction of an energy plant at Vado Ligure. Pre- and post-construction pictures (Photograph by G. Berriolo)

Figure 15.18 *Artificial beach at Cala Gonone (Sardinia), completed with gravel and granules and stabilized by artificial shoals constructed with rounded boulders similar to those resulting from cliff collapse. Starting nourishment (1994) and seven years after work completion (2004)*

Figure 15.19
Gravel beach nourishment at Cavo (Elba) performed to cover a previous fill done with material containing a high percentage of fine sediments

damaging *Posidonia oceanica* prairies (Nordstrom *et al.*, 2004). Recently, gravel beaches have been created in front of seawalls (Cammelli *et al.*, 2006), where sand is not stable; this opens the possibility of giving beaches back to several towns whose waterfronts lost their beaches after construction of hard defences to protect coastal roads and inhabitants (Aminti *et al.*, 2003).

Offshore dredging is a very recent activity and was applied for the first time in 1994; nevertheless 20 million m³ have already been utilized since then. The first project was performed by Consorzio Venezia

■ **Figure 15.20** *Beach fill at Tarquinia (Lazio) performed with offshore aggregates in 2004 (Photographs by Regione Lazio)*

Nuova to restore the barrier island closing the lagoon (Box 15.2); for this, approximately 9.2 million m³ was dredged on the northern Adriatic shelf. From 2000 to 2007, approximately 1,625,000 m³ of sand were extracted 55 km offshore from the northern Emilia-Romagna coast, to be used in nine projects in this region (Preti, 2009). A peculiar case is represented by a deposit discovered by a private company in Central Adriatic: this is the only mining concession given to a private company in Italy. Only 500,000 m³ (from more than 100 million m³, according to Paltrinieri, 2007) have been dredged and sold for beach nourishment so far, after a period of storage on an inland deposit.

On the Tyrrhenian continental shelf, Regione Lazio promoted a wide search to find sediment to be used for beach nourishments (Figure 15.20). Estimated reserves amount to 580 million m³ (BEACHMED, 2003), of which approximately 9 million m³ has since been used (Lupino, pers. comm.). The first 1 million m³ was used at Ostia in 1997; the last 3 million m³ fed the beaches of S. Felice Circeo, Terracina, Fondi, Formia and Minturno in 2006 and 2007.

BOX 15.1 UPDRIFT EROSION AT A DETACHED BREAKWATERS SYSTEM AND STABILIZATION PROJECT (GOMBO, TUSCANY)

Beach erosion at the River Arno delta started at the apex in the 1880s and gradually expanded to the sides onto longer coastal sectors. On the southern lobe, where Marina di Pisa was being created at that time, groins, seawalls and detached breakwaters were constructed; the undeveloped northern lobe (now a Regional Park) was left free to retreat, and eroded by over 1,300 m (present shoreline position).

To compensate for delta asymmetry and favour river discharge, a long jetty was built on the northern side of the river mouth. When beach erosion commenced in front of the Presidential Villa at Gombo (4 km north of the river mouth), five detached breakwaters were built in a downdrift sequence from 1962 to 1965 (Figure

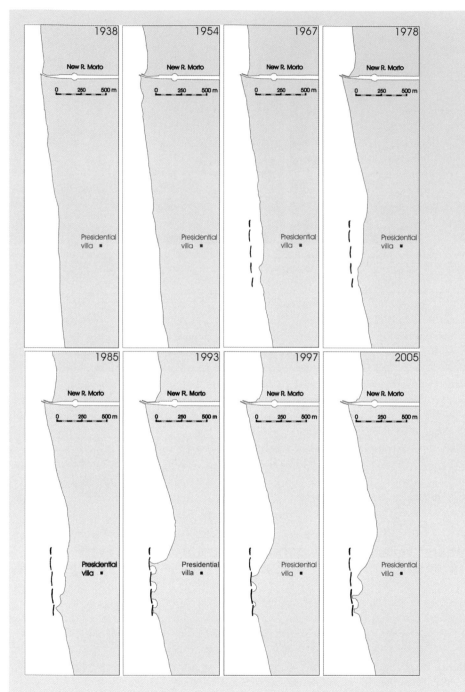

Figure 15.21 *Shoreline evolution at Gombo from 1938 to 2005 (North upward)*

15.21). Initial breakwater configuration (distance offshore/breakwater depth vs. breakwater length/gap width) was capable of inducing a salient formation behind the first structures, but most of the beach was still eroding.

In the 1980s, in order to increase efficiency of the structure, gaps were reduced between the first four elements; beach deposition began immediately and in 1993 four segments were connected to the coast by tombolos (Figure 15.22). A log-spiral planform bay, presenting a trend of progressive updrift migration of its inflection point, has been developing since then on the lee side of the disconnected detached breakwaters, connecting the jetties of an artificial channel located 1.5 km dowdrift from Gombo. Such a process was trigged by the 'groin effect' of the breakwater–tombolo system and by offshore diversion of the longshore currents induced by wave reflection in front of the structures.

This destructive process results in the gradual elimination of the entire sand storage of the tombolo and salients, i.e. it reverses the depositional trend by eroding beaches to the lee side of detached break-waters.

■ **Figure 15.22** *Breakwater-tombolo system at Gombo*

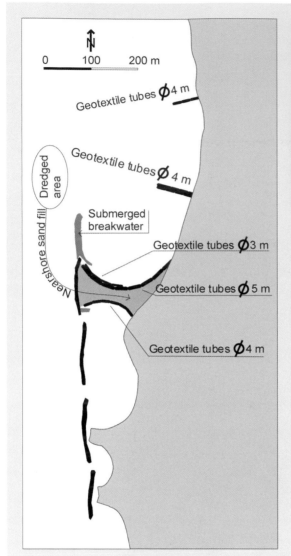

■ **Figure 15.23** *Design and construction of artificial tombolo*

Without this fixed point, a longer spiral bay should develop connecting the two jetties, with an important territory loss. To prevent this evolution, an artificial tombolo, utilising geotextile tubes filled with sand, has been built in 2009 in order to connect the fourth breakwater. Nearshore sand was dredged to fill the area inter-tubes. The ineffective fifth breakwater was lowered to sea level and two submerged groins built to stabilize the northern coast.

(Reference: Bowman and Pranzini, 2003)

BOX 15.2 SHORE PROTECTION AT VENICE LAGOON

Stabilizing the shoreline at the barrier island delimiting Venice Lagoon is one of the most important tasks for protection of the city of Venice. After the extraordinary storm that occurred on 4 November 1996, the old Murazzi, built by the Republic of Venice in the eighteenth century and continuously reinforced since then, were scoured due to profile lowering and proved to be ineffective for long-return period events.

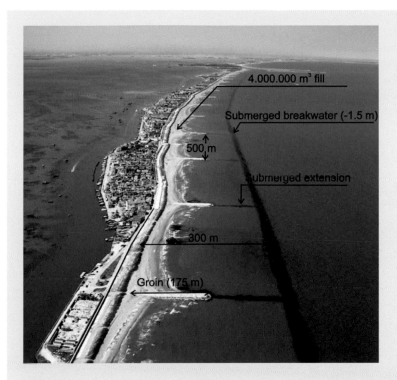

4.000.000 m³ fill

Submerged breakwater (-1.5 m)

500 m

Submerged extension

300 m

Groin (175 m)

■ **Figure 15.24** *Shore protection at Pellestrina (Photograph courtesy of Ministro delle Infrastructure e dei Trasporti – Magistrato alle Acque di Venezia – Tramite il suo concessionario Consrozio Venezia Nuova)*

The 60-km long barrier island system was narrow and fragile because of reduction in sediment input, due to river diversions (the Po River was diverted at Porto Viro in 1602), river bed quarrying and dam construction, jetty contraction at each inlet, land subsidence and sea level rise.

Magistrato delle Acqua di Venezia, a public institution founded in 1907 under the Ministry of Infrastructures and Transport, developed, through Consorzio Venezia Nuova, a project based on reconstruction of the seawall and protected nourishment of the barrier islands. The goal was to defend from a 300-year return period storm (H_s = 5.3 m; storm surge = 1.5 m), with limited groin damage (< 5 per cent), overtopping below 0.1 m³/s per km; refill volume 10 per cent in ten years.

Several tests were carried out in a 1:15 scale wave channel and in a 1:60 scale wave basin, to find the best inter-groin distance and submerged breakwater configuration. In addition, post-nourishment beach width was optimized to reduce refill time, as this was longer for extremely wide and extremely narrow beaches.

Reinforced murazzi will be faced by a 25–60 m large beach along most of the coast, protected by groins positioned at distances that will vary according to the presence/absence of a submerged breakwater (which will be placed 300 m from the seawall); the latter will be connected to the groin tips through underwater segments.

Until now, 45 km of the project have been executed, using 9.2 million m³ of fill, mostly coming from an offshore deposit in the northern Adriatic coast. Borrow sediment has approximately the same size as native grains (D_{50} B = 0.20 mm; D_{50} N = 0.17 mm).

Coastal engineering work carried out at Cavallino (11 km) and Pellestrina (9 km) are now over 10 years old, and their performance can be reliably evaluated following accurate monitoring carried out by the Consorzio Venezia Nuova.

At Pellestrina, 4 million m³ of sediments ($D_{50} = 0.2$) obtained from an offshore deposit 20 km from the coast were used. A 40 m wide beach was created in front of a rip-rap that protected the foot of the ancient Murazzi. Eighteen 500 m-spaced groins are connected to the submerged (–1.5 m) breakwater with underwater extensions. Surveys performed in 2007 show that only 2 per cent of the original fill was lost; in five out of 18 cells, the beach is narrower than 25 m (limit under which refill is necessary). This will be performed using excess sand from other cells.

At Cavallino, 31 groins of 120 m length were constructed, partially restoring some of the 65 pre-existing ones; no detached breakwater is present here. Two million m³ of the same sand was used for the fill. Approximately 450,000 m³ of sediments were lost from 1997 to 2007, but the beach is wider than the 50 m safety limit. A 50–90 m wide dune, up to 5 m high, was created along 5 km of coast over which Tamarisk and Ammophila were planted after implanting wind curtains.

The first beach re-nourishment, of 300,000 m³, was planned to be carried out at Cavallino in 2009, but only 10,000 m³ were deemed necessary; a similar volume had to be added in 2010.

(Reference: Cecconi *et al.*, 2008)

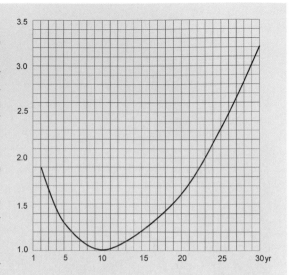

■ **Figure 15.25** *Fill volume for different renourishment periods normalized to a ten-year interval volume*

Conclusions

Due to the scarcity of inland material, as well as their higher cost, all regions are searching for marine deposits as the main source of sediments to be used in future shore protection projects. Oceanographic campaigns are today carried out by many regional administrations and new deposits are continuously being discovered. Nevertheless, the awareness that this is also a limited resource moved some administrations to look for adaptations to beach erosion.

This new position can be found in the project developed to limit coastal erosion at the Ombrone river mouth, in the Maremma Regional Park (Tuscany), where shoreline retreat is approximately 10 m/yr (Pranzini, 1994). Since this part of the coast does not have any construction and land is used only for pasture, a 'strategic' retreat has been planned: a seawall will be constructed 150 m inland from the present shoreline, as the innermost line to be defended; ten sub-soil groins will begin orthogonal to the wall, with the crest at sea level and ending at the beach.

These groins will only reduce erosion rates, but as soon as the coast is actually eroded, more of the groins will remain exposed, and this extra lengthwill increase efficiency (see Chapter 22). Shoreline retreat will proceed and the sea is expected to reach this seawall in 50–60 years. This period will be used for implementation of action plans on the Ombrone River basin, required for restoring river sediment transport up to a volume that can grant beach stability with such submerged groin.

Shoreline retreat can be managed in undeveloped areas and under a wide temporal arc, but wherever the coast is intensively developed, new defence technology, as well as new urbanism, economic and social planning is necessary. This is the challenge awaiting technicians, policy planners and all others involved in coastal management in Italy.

References

Albani D., 1933 *Ricerche sulle variazioni delle spiagge italiane: Indagine preventiva sulle recenti variazioni della linea di spiaggia della coste italiane.* Anonima Romana Editoriale, Rome, 25 pp.

Aminti P., Cammelli C., Cappietti L., Jackson N. L., Nordstrom K. F., Pranzini E., 2004. Evaluation of beach response to submerged groin construction at Marina di Ronchi, Italy, using field data and a numerical simulation model. *Journal of Coastal Research*, 33: 99–120.

Aminti P., Cipriani L.E., Pranzini E., 2002. Beach erosion control along the Golfo di Follonica (Suthern Tuscany): Actual hard protections vs. potential soft solutions. *Littoral 2002, 6th Int. Symp.*, Porto, Portugal, 355–363.

American Shore and Beach Preservation Association, ASBPA, 2007. Innovative erosion control devices. *Shore and Beach*, 75: 58–59.

Aminti P., Pranzini E., Tecchi G., 2003. Nuovi criteri di riequilibrio dei litorali intensamente protetti da opere a scogliera. *Int. Conf. CITTAM*, Napoli, 26–28 June, 487–495.

Amore C., Giuffrida E., 1984. L'influenza dell'interrimento dei bacini artificiali del F. Simeto sul litorale del Golfo di Catania. *Boll. Soc. Geol. It.*, 103: 731–753.

Barsanti D., Rombai L., 1986. *La "guerra delle acque" in Toscana.* Edizioni medicea, Florence, 169 pp.

Bartolini C., Corda L., D'Alessandro L., La Monica G. B., Regini E., 1977. Studi di geomorfologia costiera: III – Il tombolo di Feniglia *Boll. Soc. Geol. It.*, 96: 117–15 .

Bartolini C., Palla B., Pranzini E., 1988. Studi di geomorfologia costiera: X – Il ruolo della subsidenza nell'erosione litoranea della pianura del Fiume Cornia. *Boll. Soc. Geol. It.*, 108: 635–647.

BEACHMED, 2003. *Le Project BEACHMED: Récupération environnementale et entretien des littorax en erosion avec l'utilisation des dépôts sablonneux marins.* 1° Cahiere Technique, Rome, 236 pp.

Becchi I., Paris E., 1989. Il corso dell'Arno e la sua evoluzione storica. *Acqua Aria*, 6: 645–652.

Benedetto G., Fioravanti S., Martinoja D., 2000. Le coste italiane: Le norme di tutela e l'illegalità diffusa. *Rapporto dell'Area Istituzionale*, WWF Italia, Rome, 22 pp.

Berriolo G., 1985. Metodi di difesa delle spiagge. In *La gestione delle aree costiere*, edited by E. Pranzini. Edizione delle Autonomie, Rome, 145–171.

Berriolo G., 1993. Interventi di riequilibrio delle spiagge della provincia di Latina. In *La difesa dei litorali in Italia*, edited by P. Aminti and E. Pranzini. Edizioni delle Autonomie, Rome, 153–173.

Berriolo G., Sirito G., 1972. *Spiagge e porti turistici.* Hoepli, Milan, 428 pp.

Berriolo G., Sirito G., 1973. Essais sur modele de l'action des guides submerges sue le movement littoral du sable. *Ull. Ass. Int. Perm. Congres de Navigation*, 47: 91–99.

Bondesan M., Dal Cin R., 1975. Rapporti fra erosione lungo i litorali emiliano-romagnoli e del delta del Po e attività estrattiva negli alvei fluviali. In *Cave e assetto del territorio*, Italia Nostra, Regione Emilia-Romagna, 127–137.

Bonora N., Immordino F., Schiavi C., Simeoni U., Valpreda E., 2002. Interaction between catchment basin management and coastal evolution (southern Italy). *Journal of Coastal Research*, S.I. 36: 81–88.

Borgia G., Brighenti G., Vitali D., 1982. La Coltivazione dei pozzi metaniferi del bacino polesano e ferrarese. Esame critico della vicenda. *Inarco, Georisorse e Territorio*, 425: 13–23.

Bowman D., Ferri S., Pranzini E., 2007. Efficacy of beach dewatering: Alassio, Italy. *Coastal Engineering*, 54: 791–800.

Bowman, D., Pranzini, E., 2003. Reversed response within a segmented detached breakwater: The Gombo case, Tuscany coast, Italy. *Coastal Engineering*, 49: 263–274.

Cammelli C., Jackson N. L., Nordstrom K. F., Pranzini E., 2006. Assessment of a gravel-nourishment project fronting a seawall at Marina di Pisa, Italy. *Journal of Coastal Research*, S.I. 39: 770–775.

Cecconi G., Liberdo S., Stura S., 2008. La protezione del litorale della laguna di Venezia: Un caso di studio. Ass. Idrotec. It., Ass. Naz. di Ing. Sanitaria e Ambientale, Università degli Studi della Calabria, Consiglio Nazionale Ingegneri, Ordine Ingegneri Prov. Cosenza: 29° Corso di Aggiornamento in Tecniche per la difesa dall'inquinamento, Guardia Piemontese Terme (CS) 18–21 June.

Ciavola P., Vicinanza D., Fontana E., 2008. Beach drainage as a form of shoreline stabilization: Case studies in Italy. *Proc. 31th Int. Conf. on Coastal Engineering*, Hamburg, pp. 2646–2658

Cipriani L. E., Dreoni A. M., Pranzini E., 1993. Beach evolution induced by the construction of a breakwater in a small gulf and proposals for a long-term solution: A case study in the Golfo di Campo (Isola d'Elba, Italy). *Proc. Hilton Head Island Int. Coastal Symp.*, edited by P. Bruun, 1: 268–273.

CNR, 1986. Foglio 111–112: Livorno e Volterra. *Atlante delle Spiagge Italiane*. Scale 1:100,000, C.N.R.–S.El.Ca., Florence.

CNR, 1997. *Atlante delle Spiagge Italiane*. Scale 1:100,000, Selca, Florence.

Cocco E., Iuliano S., 1999. L'erosione del litorale in sinistra foce Sele (Golfo di Salerno): Dinamica evolutiva e proposte di intervento a difesa e tutela della spiaggia e della pineta litoranea di Paestum. *Il Quaternario*, 12, 125–140.

Cocco E., De Magistris M. A., De Pippo T., 1978. Studi sulle cause dell'arretramento della costa lucana ionica: I – L'estrazione degli inerti lungo le aste fluviali. *Mem. Soc. Geol. It.*, 19: 369–376.

Commissione De Marchi, 1968. Commissione Interministeriale per lo Studio della Sistemazione Idraulica e della difesa del Suolo "Rapporto De Marchi". V Sottocommissione: Difesa del mare dei territori litoranei. Carta di sintesi dei tratti di litorale in erosione.

Corsini S., Franco L., Piscopia R., Inghilesi R., 2006. *L'Atlante delle onde nei mari italiani*. APAT, Roma, 136 pp.

Davies J. L., 1980. *Geographical variations in coastal development*. Longman, New York, 2nd ed., 212 pp.

Franco L., 1996. History of coastal engineering in Italy. In: *History and Heritage of Coastal Engineering*, Am. Soc. Civ. Eng., New York, 275–335.

Gabbianelli G., Del Grande C., Simeoni U., Zamariolo A., Calderoni G., 2000. Evoluzione dell'area di Goro negli ultimi cinque secoli (Delta del Po). *Studi costieri*, 2: 45–63.

Gallareto E., 1960. *La difesa delle spiagge e delle coste basse*. Hoepli editore, Milan, 303 pp.

Giovannini C., 2010. La spiaggia come problema storico: Uso e percezione. *Riv. It. Telerilevamento*, 42: 45–54.

GNRAC, 2006. Lo stato dei litorali in Italia. *Studi costieri*, 10: 3–172.

Goldberg E. D., 1994. *Coastal Zone Space: Prelude to conflict?* Intergovernmental Oceanographic Commission Ocean Forum books series, UNESCO, Paris, 138 pp.

Guiducci F., Lo Presti F., Scalzo M., 1993. Intervento di ripascimento tra Paola e S. Lucido (CZ). In *La Difesa dei litorali in Italia*, edited by P. Aminti and E. Pranzini. Edizioni delle Autonomie, Rome, pp. 195–214.

Innocenti, L., Pranzini, E., 1993. Geomorphological evolution and sedimentology of the Ombrone River delta (Italy). *Journal of Coastal Research*, 9, 481–493.

ISPRA, 2008. *Annuario dei dati ambientali*. ISPRA, Rome.

Jakobsen P., Brøgger C., 2007. Coastal protection based on Pressure Equalization Modules (PEM). *Proc. Int. Coastal Symp.*, Gold Coast, Australia, available at: http://www.shore.dk/ICS2007%20Paper %20Poul%20Jakobsen.pdf (accessed 25 August 2012).

Macchia B., Pranzini E., Tomei P. E., 2006. *Le dune costiere in Italia: La natura e il paesaggio*. Felici Editore, Pisa, 206 pp.

Marinelli, O., 1926. Sull'età dei delta italiani. *La Geografia*, 1/2: 21–29.

Mikhailova, M. V., Bellotti, P., Valeri, P., Tortora, P., 1998. The Tiber river delta and the hydrological and morphological features of its formation. *Water Resources*, 5, 620–630.

Nordstrom K. F., Jackson N. L., Pranzini E., 2004. Beach sediment alteration by natural processes and human action: Elba Island, Italy. *Ann. Ass. Am. Geogr.*, 94: 794–806.

Paltrinieri D., 2007. Nuove fonti di approvvigionamento di sabbia, *AIOM, Bollettino*, 36: 17–21.

Pope, J., Dean, J. L., 1986. Development of design criteria for segmented breakwaters. *Proc. 20th Int. Coast. Eng. Conf. ASCE,* New York, pp. 2144–2158.

Pranzini E., 1994. The erosion of the Ombrone River delta. *EUROCOAST*, Lisbon, September, 133–147.

Pranzini, E., 1995. Cause naturali ed antropiche nelle variazioni del bilancio sedimentario del litorali. *Riv. Geogr. It.,* spec. n. 1, 47–62.

Pranzini E., 2001. Updrift river mouth migration on cuspate deltas: Two examples from the coast of Tuscany (Italy). *Geomorphology*, 1–2: 125–132.

Pranzini E., 2007. Airborne LIDAR survey applied to the analysis of the historical evolution of the Arno River delta (Italy). *Journal of Coastal Research*, SI 50 (Proceedings of the 9th International Coastal Symposium, Gold Coast, Australia), 400–409.

Pranzini E., 2009. Protection studies at two recreational beaches: Poetto and Cala Gonone beaches, Sardinia, Italy. In *Beach management*, edited by Allan Williams and Anton Micallef, Earthscan, London, 287–306.

Preti M., 2009. *Stato del litorale emiliano-romagnolo nell'anno 2007 e piano decennale di gestione*. ARPA Emilia-Romagna, Bologna, 270 pp.

Short A. D., Masselink G., 1999. Embayed and structurally controlled beaches. In *Handbook of beaches and shoreface morphodynamics*, edited by Andrew D. Short, Wiley, Chichester, 230–249.

Simeoni U., Corbau C., Pranzini E., Ginesu S., Eds., 2012. *Le pocket beach: Dinamica e gestione delle piccole spiagge*. Franco Angeli, Milan, 176 pp.

Valloni R., Ferretti O., Barsanti M., Ugolott, M., 2003. The morpho-sedimentological types of Italian coasts and their susceptibility to sea-level rise: A first approximation. *International Conference: Quaternary Coastal Morphology and Sea Level Changes*, Puglia, 22–28 September, v–vi.

Vicinanza D., Ciavola P., Biagi S., 2008. Progetto sperimentale di iniezione d'acqua in unità geologiche profonde per il controllo della subsidenza costiera: Il caso di studio di Lido Adriano (Ravenna). *Studi Costieri*, 15: 121–138.

Vousdoukas M. I., Velegrakis A. F., Plomaritis T. A., 2007. Beachrock occurrence, characteristics, formation mechanisms and impacts. *Earth-Science Reviews* 85: 23–46.

WWF, 2012. *Dossier coste. Il 'profilo' fragile dell'italia*, 76 pp. Available at: http://www.wwf.it/userfiles/file/news%20dossier%20appti/dossier/wwf_dossiercosteprofilofragile.pdf (accessed 1 September 2012).

16 Eastern Adriatic

Slovenia, Croatia and Montenegro

Kristina Pikelj, Vojislav Dragnić and Nemanja Malovrazić

Introduction

The Eastern Adriatic coast covers six countries: Italy, Slovenia, Croatia, Bosnia and Herzegovina, Montenegro and Albania; the coast discussed in this chapter lies between Italy and Albania, a direct distance of 611 km (Figure 16.1), whilst the island and mainland coastline is 6,190 km in length. Most of the coast (95 per cent) belongs to Croatia (5,835 km), with the remaining 5 per cent unevenly distributed between Montenegro (288 km; of which Boka Kotorska Bay has 105,5 km, 4 per cent), Slovenia (46 km; 0.7 per cent) and Bosnia and Herzegovina (21 km; 0.3 per cent; Riđanović, 2002). Out of 1,291 islands, 1,246 are in Croatian territorial waters (Duplančić-Leder *et al.*, 2004), while 14 islands and 31 rocks belong to Montenegro, with a total shoreline length of 25.6 km. An indentation coefficient of 10 makes it the second most indented European coast (after Norway).

Coastal geology and geomorphology

This is a function of geological composition and tectonic activity, which has resulted in high relief and hypsometric changes in relatively small areas. Compressional tectonic action has caused faulting, folding and uplift of the Dinaric and island chains (Vlahović *et al.*, 2005). On the exposed coast, mostly composed of carbonate bedrock, karst formed from the Miocene to the present. During the Holocene sea-level rise, the folded, faulted and karst relief was submerged, resulting in formation of the second

largest archipelago in the Mediterranean (Surić and Juračić, 2010), characterized by islands and channels parallel to coastal structures, aligned mostly in the same NW to SE (so-called Dinaric) direction and known in the literature as the *Dalmatian coast* type (Fairbridge, 1968). Brač, Hvar, Mljet, Lastovo and Korčula Islands (Croatia) situated in the Southern Dalmatic province (encircled with a black dashed line in Figure 16.1; Giorgetti and Mosetti, 1969), deviates from the Dinaric direction, as well as the Budva-Bar unit (Montenegro, see later). This deviation is a result of complex tectonic history associated with strike-slip movements in a late Orogenic stage along the Dinaric-Hellenic part of the Adriatic region (Picha, 2002). The northern area of the eastern Adriatic (Gulf of Trieste, Slovenia, marked as *a* on Figure 16.1) is shallow (~25 m), while the deepest (1,340 m) fronts the Montenegrin coast. Channel area depths can reach up to 100 m, even close to the mainland coast.

Most of the area is a rock transgressive coast, formed during the last sea-level rise, where most islands together with a large part of the mainland coast (over 90 per cent) were built of carbonates (limestone, dolomites and carbonate breccias; Figure 16.1). A less prevalent coastal rock assemblage (6 per cent) is the Eocene flysch (alternating marls, siltstones and sandstones). Flysch is occasionally weathered and covered by Quaternary deposits. Within the limestone-flysch ratio, considerable stretches of flysch zone face the Adriatic Sea along the Slovenian coast; approximately one third of the Montenegrin coast is flysch, whilst along the Croatian coast, it outcrops to a lesser extent (Figure 16.1) and Quaternary

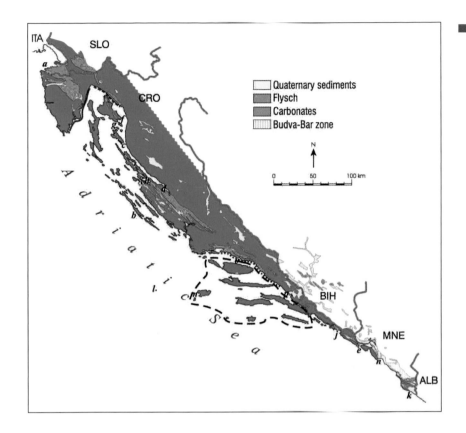

■ **Figure 16.1**
*Political and simplified geological map of the Eastern Adriatic coast. Legend: bays, beaches, islands and rivers marked with letters **a** to **p** further explained in text; blue line: Liburnian zone; blue dashed line: Vinodol zone; black dashed line: Southern Dalmatic Islands (Figure created with the support of Robert Koscal)*

alluvial deposits are also found in these areas. On a small part of the coast (inner part of the Boka Kotorska Bay and between Budva and Bar; Montenegro) pelagic basin sediments occur ranging from Middle Triassic flysch deposits, volcanic rocks, and conglomerates to Cretaceous–Palaeogene transitional sediments, termed the Budva-Bar unit (Antonijević *et al.*, 1969; Antonijević *et al.*, 1973) and for a very small area, igneous rocks and pyroclastites can also occur. Coastal areas developed under riverine influences are rare, compared to the Albanian coast. Eastern Adriatic rivers are relatively short, carrying small amounts of sediment. River mouths and part of their valleys are submerged, creating a ria coast in Slovenia and an estuary coast in Croatia (Juračić and Prohić, 1991; Vahtar, 2002; Sondi *et al.*, 2008). Small areas of lagoons and wetlands are related mostly to river mouths (Dragonja, Mirna, Raša, Neretva and Bojana).

Exact data for the artificial coast length within the region is incomplete. Some 82 per cent of Slovenian coastline is manmade (18 per cent in a natural state; Turk, 1999). Croatian data is very scarce, except for the Kvarner area on the eastern Istrian (Liburnia) coastal zone, where 14.8 per cent is artificial (marked with a blue line on Figure 16.1) and 17.6 per cent is artificial in the Vinodol zone (marked with a dashed blue line on Figure 16.1). Furthermore, 5.4 per cent of the Krk coast, 4.1 per cent of the Lošinj, 3.2 per cent of the Rab and 1.0 per cent of the Cres Islands are artificial in origin (Juračić *et al.*, 2009). Available data for Montenegro is not on a comparable scale, but all natural and artificial beaches make up 25 per cent of the total coastline. As an example, Sveti Stefan (Figure 16.2) was previously a submerged tombolo with natural beaches on both sides, on which an artificial walkway was built during the 1950s in order to connect it to the mainland (Marković *et al.*, 2003).

■ **Figure 16.2** *Sveti Stefan peninsula, Montenegro (Photograph by Š. Aščić)*

Carbonate coasts

Classic littoral processes are scarce along this carbonate coastline, as these rocks are more prone to karst processes instead of mechanical weathering. The hinterland carbonate composition influenced development of karstic relief with a poorly developed river network which had a small or restricted discharge. Amounts of terrigenous material (potential beach sediments) are negligible and most remain in river mouths and estuaries (Juračić and Prohić, 1991; Sondi *et al.*, 2008) as Adriatic Sea energy is weak.

Most mainland and island coasts are rock, steep and often inaccessible and have been tectonically formed with vertical scarps over 100 m high (both above

and below sea level). Some of these well-known and attractive 'cliffs' are located along the southwest coasts of the Northern-Dalmatian Islands (e.g. Kornati, Dugi Otok, Figure 16.3; marked as **b** on Figure 16.1). Susceptible parts of the carbonate rock coast tend to have sub-horizontal or slightly inclined upper surfaces of limestone outcrops. These structural benches can be found on many islands, as well as on the mainland coasts and are commonly used as beaches.

Most bays formed in the carbonates owe their origin to older relief features shaped by karst processes during lower sea-level stages. On the small scale, these forms were usually submerged river or torrent valleys (e.g. Zavratnica Bay, Figure 16.4; marked as

■ **Figure 16.4** *Zavratnica Bay in the Velebit Channel, Croatia; view from the Bay (Photograph by K. Fio)*

■ **Figure 16.3** *Southern coast of Dugi Otok, Croatia*

■ **Figure 16.5** *Boka Kotorska Bay, Montenegro*

c on Figure 16.1); or submerged karst forms (dolines, poljes; e.g. Novigrad Bay, marked as *d* on Figure 16.1). On the larger scale, one of the best-known deeply incised bays on the eastern Adriatic Sea, is Boka Kotorska Bay (Figure 16.5; marked as *e* on Figure 16.1). The relief is of strongly tectonized karst ridges and deeply cut flysch valleys striking in the typical Dinaric direction. In karst and non-karst areas, cirques developed from which glaciers flowed, which re-shaped the relief (Magaš, 2002; Menković *et al.*, 2004) and eventually submerged during the last sea-level rise (Magaš, 2002).

Small pocket beaches, forming limestone pebble and gravel beach systems, developed in some bays that occasionally experienced torrential surface flows,

■ **Figure 16.6** *Pocket beaches in Vrulja Bay, the Makarska riviera, Croatia*

■ **Figure 16.7** *Pocket beach Žanja on the western coast of Cres Island, Croatia*

whose lengths do not exceed a few tens of metres (Benac *et al.*, 2010). Additionally, they are partially fed with material accumulated by marine erosion and material derived from landfalls, especially where limestones are initially weaker (limestone breccias), or are significantly crushed and/or karstified (Figure 16.6). Regardless of the different material sources, pocket beaches are further shaped by waves (Figure 16.7). For the Kvarner area, Northern Adriatic Sea, Juračić *et al.* (2009) have shown that the length of gravel pocket beaches covers up to 5 per cent of the carbonate rock coast. Their geographical orientation,

shape and lengths are irregular, whether on the mainland or on islands, and they are strongly influenced by local coastal geomorphology.

Flysch coasts

As stated, coasts cut in flysch deposits are less common compared to limestone coasts, all outcrops being of Eocene age (alternate marls, siltstones and sandstones sequences, making soft, easily weathered rock systems), except in Budva-Bar, where they are of Triassic age. Flysch zones stretch along the entire Slovenian coast; whilst in Croatia there are flysch outcrops on Krk and Rab Islands, in the Zadar hinterland, in the Split-Ploče belt and south of Dubrovnik. In Montenegro, flysch coasts dominate in the Bar-Ulcinj belt (Figure 16.1). Badlands are a common erosional landscape on the flysch and cliffed coasts evolve under destructive wave impact and subaerial weathering and erosional processes.

Due to its mixed lithology, flysch does not erode at similar rates. Cliffs are usually formed in the more resistant sandstones, while rockfalls, earth flows and mass movements are commoner in the less resistant marls (Juračić *et al.*, 2009). This is especially true for Slovenian cliffs, which are particularly well developed in the Piran and Strunjan areas (Figure 16.8). These cliffs are convex-to-steep cliff types, rimmed by

■ **Figure 16.8** *Strunjan cliffs in Slovenian flysch (Photograph by I. Pepelnjak)*

almost 6.5 km of marine terraces forming pebble to cobble beach systems (Vahtar, 2002). At Rab Island (Lopar peninsula), headlands and promontories are formed of sandstone-rich flysch, while bays and beaches are formed in flysch-rich marls and siltstones (Juračić *et al.*, 2009). Over large areas, wide depressions and bays have been developed creating long (1 km) sand to fine pebble beaches. They follow the Dinaric direction (to the NW or SE; e.g. Baška on Krk Island; marked as *f* on Figure 16.1; Lopar and San Marino on the Rab Island; marked as *g* on Figure 16.1; Ljubač Bay and Nin area in the Zadar hinterland; marked as *h* and *i* on Figure 16.1; Mlini and Kupari south of Dubrovnik; marked as *j* on Figure 16.1; Igalo in the Boka Kotorska Bay, Jaz and Bečići in the Budva area; Figures 16.9 and 16.10). The longest bay formed in flysch (and recent riverborne material) is in Ulcinj (marked as *k* on Figure 16.1), where the 13 km long and on average 150 m wide sand Velika Plaža beach is located (Figure 16.11).

Significant deposits stretch in a narrow zone along the coast in the Split-Poče region. Small beaches, made of alluvial sediment, occur at the ends of submerged valleys and gullies and cuspate barriers separating them are mostly composed of considerable accumulations of the same torrential riverine material, e.g. Mali Rat, Dugi Rat and Dugi Rat-Brela.

Eastern Adriatic oceanography

The general circulation in the Adriatic Sea is driven by fresh-salt water exchanges between the Adriatic and Mediterranean additionally modified by local winds – the *Bura* and *Jugo*, in local languages, *Bora* and *Scirocco* in Italian. The Eastern Adriatic Current (EAC) enters the Adriatic Sea through the Strait of Otranto and flows in a north-western direction. This current is much weaker and wider compared to the Western Adriatic Current (WAC; Poulain and Cushman-Roisin, 2001). Due to the extremely complex geometry and relief, every bay or channel on the eastern Adriatic coast shows some specific oceanographic characteristics. The continuity of the longshore drift on the eastern side is not comparable with the western side (Orlić, 2001). Current velocities for the Eastern Adriatic are scarce and sporadic;

■ **Figure 16.9** *Baška on Krk Island, Croatia (Photograph by D. Kudrnovski)*

■ **Figure 16.10** *Jaz beach near Budva, Montenegro (Photograph courtesy of Public Enterprise for Coastal Zone Management of Montenegro)*

■ **Figure 16.11** *Velika plaža beach in the Ulcinj area, Montenegro (Photograph courtesy of Public Enterprise for Coastal Zone Management of Montenegro)*

however, surface current velocity ranges from 1 cm/s to 1 m/s (15 cm/s on average). Bottom velocities are approximately 6 cm/s. The fastest sea currents are between Boka Kotorska Bay and Vis Island (20 cm/s), while currents in the Croatian part of the Adriatic are the slowest (3–5 cm/s) (V. Dadić, *pers. comm*).

The Adriatic Sea is microtidal with a minimal measured amplitude of 22 cm in Dubrovnik (HHI, 1999). Tidal amplitudes in the Northern Adriatic along the Italian coast can reach 1.2 m during astronomical tides (Lambeck *et al.*, 2004). The most frequent waves are during winter, formed by the northeastern *Bura* and southeastern *Jugo* winds. Due

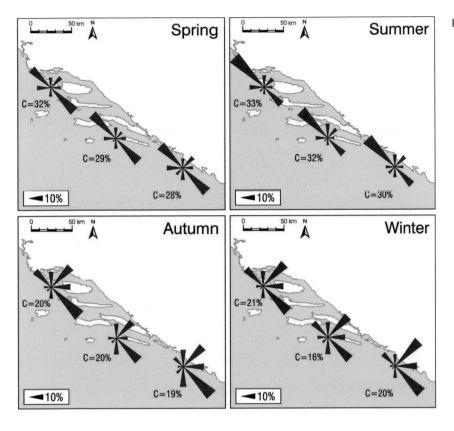

■ **Figure 16.12**
*Mean seasonal
frequencies of wave
directions and calms
in the Central and
South Dalmatia,
Croatia (after HHI,
2002)*

to the longer fetch, resulting *Jugo* waves are much higher than those generated by the *Bura* and the summer waves formed by winds with a westerly component (HHI, 1999; 2002, Figure 16.12). An absolute maximum wave height in the Adriatic Sea recorded during the long-lasting *Jugo* in the open northern Adriatic, was Hmax = 10.8 m. Maximum wave height is slightly lower in the central and south Adriatic Sea (Hmax = 8.4 m and Hmax = 8.9 m, respectively; HHI, 1999). Wave heights are usually between 0.5 and 1.5 m while waves higher than 5 m rarely occur. Wave heights and general surf energy are much lower in the inner parts of the archipelago due to protection offered by islands, bays and channels. In general, erosional processes driven by oceanographic causes are estimated to be of a low intensity.

Erosion and protection

Similar to the rest of the Mediterranean, the most pronounced problems of the coastal environment are related to anthropogenic events caused by expanding tourism, urbanization, industrial activities and rapid traffic route construction (Rupprecht Consult, 2006). These processes started during the 1960s and were temporarily interrupted during the 1990s due to political changes (civil war) on the territory of the former country of Yugoslavia.

Slovenia

The coast – 46 km long, almost entirely made of flysch and located along the Gulf of Trieste (marked as *a* on Figure 16.1) – is highly populated with a population growth between 1999 and 2002 of 1.2

per cent (Peterlin *et al.*, 2005), due to enhanced activities in transport, industry, commerce, fishery and tourism. The infrastructure for all activities (especially tourism) is markedly below the coastal carrying capacity and road network (Vahtar, 2002). Furthermore, this zone is one of sensitive landscapes, which due to its extreme narrowness, if not protected, potentially poses many problems. It is a recent ria coast, having two submerged valleys, Rižana and Dragonja, which form Koper Bay and Piran Bay respectively (Vahtar, 2002; Mavrič, 2010). The majority of the cliffs are up to 80 m in height, the highest in the eastern Adriatic Sea, and are influenced by wave erosion and sub-aerial weathering, which gives rise to gravel/pebble beaches (Figure 16.13); the rest are vertical cliffs under constant abrasive wave destruction. Erosion, induced by waves, tides and currents, is of a low intensity due to small driving forces (Vahtar, 2002), with cliff erosion rates estimated to be 1–2 cm/yr, the world average for flysch cliffs (Sunamura, 2005).

Most intensive cliff erosion occurs in uninhabited areas and natural reserves, while cliffs in inhabited areas are protected by wire mesh, concrete sills and stone walls, e.g. Piran, where part of the town is situated on the cliff top. However, in some locations beach erosion has exposed underlying layers of hard limestone, which controls cape positions and prevents further cliff erosion. Some inhabited and urbanized coastal areas have historical protection traditions which have utilized various artificial structures ranging from seawalls and submerged breakwaters to rock dikes (Vahtar, 2002). One of the most endangered anthropogenic structures is the Istrian motorway, built partially at the foot of flysch cliffs. Running parallel to the coastline and 5 m from the seafront, this road is susceptible to sea erosion (Vahtar, 2002). In the case of significant road damage, transport disruption would affect the whole Slovenian coast, as well as road transport toward Croatia. To prevent erosion, this road, where it directly fronts the sea, is mostly protected by masonry stone walls.

Coastal geomorphological characteristics are generally low and broad plains formed in alluvial, mostly fine-grained deposits, brought by the rivers Rižana, Badaševica and Dragonja. As a result, flooding occurs, with the Koper, Izola, Strunjan, Portorož and Piran areas being particularly floodprone (Kolega, 2006), especially during extremely high tide events in combination with a long lasting southeasterly *Jugo* (Pasarić and Orlić, 2001; Kolega, 2006). The plains are protected by seawalls, rock dikes and submerged breakwaters. One of the most endangered settlements on the coast is Koper port, which would be entirely under water during extreme flooding, if not protected by these reinforced concrete walls. As well as the Koper area, many remaining coastal plains are protected by seawalls and submerged rip-rap breakwaters. One of the threatened but unprotected coastal marsh wetland locations is the traditional Sečovlje saltpans on the Dragonja River plain. Protection consists only of traditional low soil barriers; erosion

■ **Figure 16.13** *Strunjan cliffs (Photograph by I. Pepelnjak)*

of these can lead to mud sediments endangering the main tourist resort of Portorož, which includes several kilometres of artificial sand beaches (Vahtar, 2002). However, erosion of the Slovenian coastline is generally well controlled and properly managed, since protection solutions have been effective for centuries (Vahtar, 2002).

Croatia

The longest portion of the Eastern Adriatic coast belongs to Croatia, and is mainly a carbonate, steep well-indented rock coast, fringed with islands and rocks (Duplančić-Leder *et al.*, 2004). Flysch occurs considerably as coastal bedrock only on Rab Island and in the Split-Ploče zone (marked with a blue dotted line on Figure 16.1), and rare outcrops in other coastal areas (Figure 16.1). Beaches in Croatia are less developed compared to those of the Slovenian and Montenegrin coast sand some 14.3 per cent is moderately urbanized, with a population density slightly below the national average (PAP/RAC Report, 2007). One of the major factors of change is tourism and tourism-related activities (traffic, urbanization, agriculture and commerce), especially during the last 10–15 years. As a product of the mass tourism boom commencing in the mid twentieth century some parts have been subjected to uncontrolled building construction. Today, the mass tourism concept is changing towards high-quality tourism, but this is not equally distributed along the coastline. It is estimated that in the next 20 years almost the same length of coast will develop as did previously (PAP/RAC Report, 2007), resulting in some parts being subjected to significant anthropogenic impact. Combinations of anthropogenic changes and natural processes may produce an irretrievably modified coast.

The coastal zone is narrow and characterized by high mountain ranges and in many areas, small agricultural-livestock settlements were originally built on the mountain slopes or in the hinterland, far from the coast, e.g. the eastern Istrian or coastal zone between Split and Ploče. A major factor for coastal development was construction of the Adriatic

Motorway (part of the Adriatic–Ionian motorway) during the 1960s. Easier access to the coast induced depopulation in mountain settlements and the emergence of new settlements, tourism facilities and other constructions on the coast (Rajčić *et al.*, 2010). The final result of coastal change is a product of the synergistic effect of natural and anthropogenic factors. Since erosion induced by waves, tides and current is of low intensity due to small driving forces, especially in the channel area (HHI, 1999), the main natural factor on the Croatian coast is coastal bedrock lithology (Juračić *et al.*, 2009).

Numerous gravel pocket beaches are found all along the coast, often serving as ports and recreational centres. Without anthropogenic intervention on sediment accumulation (drainage area changes, sealing of bedrock and soil, etc.) the only real hazards for pocket beaches are a predicted sea-level rise and increased storminess. Various man-made structures show the greatest impact affecting the beach sediment equilibrium (Figure 16.14; Juračić *et al.*, 2009; Benac *et al.*, 2010). These and similar concrete structures are usually built unplanned, mostly as moorings used by local fishermen. Increased tourist activities have led to widening and enlargement and further unplanned construction, which in turn have led to beach sediment equilibrium changes.

■ **Figure 16.14** *Concrete jetty impact on a gravel pocket beach in Medveja (Reprinted from Juračić et al., 2009; Copyright © with permission from Elsevier)*

In contrast, faster natural erosion processes occur on soft, flysch coasts, e.g. Rab Island, where currently no protection structures exist. Additional anthropogenic influence on San Marino manifests itself as enhanced erosion on one part of the beach (Figure 16.15), caused by construction of a leisure port. Lowered and flattened plains formed in flysch with overlying Quaternary alluvial sediments are often subject to intense agricultural useage. This disturbs the natural dynamics of sediment accumulation and erosion, and consequently raises the flood risk.

Coastal lithology, anthropogenic influences and marine abrasion, have created very complicated conditions with significant consequences in the flysch zone. Part of the Adriatic Motorway passing through the Split–Ploče zone is cut in flysch deposits, while in some parts the road has been rebuilt (Figure 16.16).

Erosion of natural beaches shows that the sedimentary deficit is a general problem in the Split–Ploče region.

Road construction and rapid development of the narrow coastal zone in flysch bedrock has caused sealing of natural drains, which has led to increasingly frequent rockfalls, endangering the road and beaches. Rockfall events are actively mitigated by constructing artificial drainage systems, stone walls and concrete sills, or utilizing protection by wire mesh and concrete, or combining several structures. Most of these protection structures are not planned, but randomly built.

Bedrock sealing, due to both rapid road/tourism infrastructure construction and rockfall mitigation, has led to obstruction and/or cessation of sedimentation of material transported by mountain torrents. Formerly, natural beach nourishment and longshore

■ **Figure 16.15** *Example of man-made influence on coastal erosion: Paradise Beach on the Lopar peninsula (Reprinted from Juračić et al., 2009; Copyright © with permission from Elsevier)*

■ **Figure 16.16** *The Adriatic Motorway at the exit from Omiš town before the 1960s (left), taken from the sea. This motorway is today protected by a revetment (right) (Photograph by S. Muslim)*

drift had caused the formation of pebble cuspate barriers (e.g. Dugi Rat, Mali Rat), as well as narrow pebble beaches. These accumulations were and still are very attractive for beach tourism and recreation. Many beaches are artificially covered and filled, in order to retain or improve their natural features (A. Raljević, *pers. comm*). The present lack of natural beach nourishment and continued longshore drift has resulted in increased erosion of natural beaches as well as where beaches did not naturally exist (Figure 16.17). Artificial beaches are created by sporadically hauling in truckloads of crushed stone and, in order to prevent erosion, numerous groins have been built, which have caused a classical problem of selective erosion. Groins reduce erosion rates; however, repeated annual or bi-annual nourishment and spatial sediment distribution is at present the only solution.

Construction of the Cetina River mouth jetty in Omiš (marked as *o* on Figure 16.1) is another example of sediment accumulation/erosion equilibrium disturbance. The river is torrential but carries large amounts of material, since it partly drains flysch bedrock and Quaternary deposits (Bonacci and

■ **Figure 16.17** *Accretive beach on the updrift side of the groins and erosive beach on the down-drift side of the groin on an artificial beach in the Krilo (Jesenice) area*

■ **Figure 16.18** *Mouth of Cetina River with jetty on the left-hand side*

Roje-Bonacci, 2003; Borčić *et al.*, 2007). These sands and silt-sands are partially deposited at the river mouth and further re-suspended and deposited by longshore drift moving from the south-east to the north-west (Figure 16.12). Numerous hydro-technical works (power plants, reservoirs) have been built on the river since the 1960s, which significantly altered the natural hydrological regime leading to a decrease in river sediment input (Bonacci and Roje-Bonacci, 2003). Jetty construction on the left of the rivermouth during the 1980s (Figure 16.18) and reduction of river sediment input are considered to be the main reasons for the first erosional signs along the 4 km natural sandy beach in the Duće area. However, temporary and partial stabilization of sediment has been achieved through construction of several groins.

One of the major problems of the expanding coastal tourism and traffic network construction is the significant amounts of constructional waste irresponsibly disposed of in the narrow coastal area or in the sea (Figure 16.19). Beside the negative impact on coastal communities of organisms in the waste, uncontrolled disposal of this material generates a surplus of sediment in the sea for particular beach segments (Rajčić *et al.*, 2010).

Compared to steep carbonate rock coasts, siliciclastic low coasts are more prone to significant

■ **Figure 16.19** *Construction waste disposed in the sea on Krapanj Island*

erosion, especially when decreasing amounts of river-borne material feed the coast. One such significantly endangered segment is the Neretva River delta (marked as *p* on Figure 16.1). This is the largest river on the eastern Adriatic coast (excluding Albanian rivers), flowing through Bosnia and Herzegovina and Croatia. The river's last 22 km spreads out from a canyon into an alluvial delta, before entering the Adriatic Sea. Previously, the Neretva valley was an

■ **Figure 16.20** *Neretva River mouth (Photograph by I. Kosović)*

unhealthy swampy area, but delta transformation started at the end of the nineteenth century, after which 12 reclaimed water courses were reduced to three branches (Romić *et al.*, 2008). This transformed the delta into an agricultural area that together with hydro-power plant, buildings, embankments for flood control, roads and urban centres construction and sand and gravel exploitation, significantly modified the river's regime, sediment amounts and water quality (Vranješ *et al.*, 2007). As a result, the riverbed is eroding, whilst sea water intrusion in the Neretva valley has reached Metković city, endangering agricultural production. Furthermore, lower areas of the Neretva River valley have already been significantly damaged and future adequate protection will be needed (Figure 16.20) (Vranješ *et al.*, 2007).

Bosnia and Herzegovina

Bosnia and Herzegovina occupy only 21 km of the eastern Adriatic coast in the Klek–Neum Bay and the entire coast is formed of carbonate rocks. Several beaches have formed as classical pocket beaches, but have been significantly modified due to great tourist pressure. Being a small area that receives a considerable number of tourists each year, this coast is vulnerable to land overuse by holiday home and tourist infrastructure construction. In Neum, the only town on the coast, jetties and breakwaters are common. These concrete and stone structures were originally only moorings, affecting sediment erosion and accumulation.

Montenegro

The coastal region is clearly separated from the inland area by the mountain massifs of Orjen (1,895 m), Lovćen (1,749 m), Sutorman (1,175 m) and Rumija (1,595 m). There are six municipalities: Herceg Novi, Kotor, Tivat, Budva, Bar and Ulcinj, where 23.94 per cent of the Montenegrin population lives (MONSTAT, 2011). There are 52 km of natural beaches, of which the longest are Velika plaža in Ulcinj (13 km including the sand beach at Ada island), Sutomore (1.2 km), Čanj (830 m) in Bar, Buljarica (2.3 km), Jaz (2 km), Slovenska plaža (1.9 km), Bečićka plaža (1.8 km) and Sveti Stefan (1.3 km) in Budva. In addition to natural beaches, rock shores have been adapted, as well as the creation of artificial beaches which total more than 100 beaches and bathing areas, covering a length of about 7 km, with a surface area of approximately 10 ha. With regard to composition, natural beaches are divided into sand, pebble and rock beaches, and artificial ones are either adjusted rocky shores built of concrete or artificially nourished pebble beaches.

Boka Kotorska bay is characterized by small shore platforms and many artificial concrete bathing areas, often with artificial capes, docks or piers. Due to a natural protection from higher waves, the shore and natural beaches in the bay are stable. However, due to inadequate works, some attempts to create artificial sand beaches, on a small scale, have not been successful, such as the small beaches in Bijela and Đenovići in Herceg Novi, Dobrota and Stoliv in Kotor and Lepetane in Tivat.

The middle portion is characterized by long sand or pebble pocket beaches sandwiched between rock cliffs. Most beaches are composed of sand-pebble material, while a few are covered by larger, rounded pebbles (Radojičić, 1996). Typical examples are Mala plaža in Ulcinj, Sutomore and Čanj in Bar, Buljarica, Petrovac, Bečići, Slovenska plaža, and Jaz in Budva.

Sediments on these beaches are derived partly from rivers (frequently torrents) and partly from wave coastal erosion. Velika plaža, in Ulcinj, is unique by its size, characteristics and origin. It is some 13 km in length (including Ada Island beach) and on average is 150 m wide. The beach material is composed of fine sand deposited by the Bojana River and no erosional processes have been noticed. However, there are indications of erosion on the adjacent beach on Ada Island, which could be attributed to poor regulation of the Bojana River. Today this area represents the largest potential for tourism development, as unlike other areas on the hinterland it remains protected from building activities.

Coastal erosion was not an issue in the past, but recently erosional changes have been noticed on several beaches (Petković, 2004). The beginning of the erosion processes coincided with rapid hinterland urbanization, building of settlements, their protective walls, as well as promenades and city squares immediately adjacent to the coast. These activities caused beach width reduction, which reduced the area for wave energy dissipation. On the landward side, hinterland urbanization cut off the flow of torrents that brought sediments to the beach. Beach erosion has a large influence on tourism, since the surfaces of several popular beaches have been reduced. Some beach nourishment before the tourist season has been carried out (Pržno beach – 40 m^3 of sand in 2004; the western part of Sutomore beach – 150 m^3 of sand in 2007; Petrovac – 80 m^3 of sand in 2009). However, the deposited sand amounts were not significant and usually disappeared in one or two seasons. Nowadays this activity has been reduced due to its negative impact on the marine environment (e.g. burial of sea bottom communities).

There have been no systematic measurements regarding important natural factors that influence coast and beach stability. Data on wind measurements exist, but there has been no recent measuring of wave characteristics and sea currents. Data on grain size and mineral composition of sediments on natural beaches exist only for some areas, whilst available data on isobaths and beach slopes under water are mostly outdated (Petković, 2004). After an extraordinarily strong storm during the winter period 1999/2000, several beaches suffered serious damage. The Port of Bar primary breakwater was damaged, while parts of beach walls along beaches in Petrovac, Miločer and Sutomore were completely destroyed. Beach walls were re-built in the same place, which from the standpoint of beach stability was not the best solution, since new storms of a similar strength could jeopardize those beaches.

In 2000, the Public Enterprise for Coastal Zone Management started dealing with beach erosion problems. Activities included preparation of several studies and projects and the start of monitoring coastal erosion at selected beaches. During 2001–2002, projects for the revitalization of Sutomore, Petrovac and Mogren beaches were prepared. The proposed solution for the former two beaches was a combined system of underwater reefs and artificial replenishment of beach sediment (Petković and Jovanović, 2001; Petković and Jovanović, 2002a and b). The Mogren beach revitalization project also proposed building underwater barrier reefs to reduce wave influence and widening of the beach area by pumping or dredging of sediment deposits from the sea bed from in front of the beach (Petković and Jovanović, 2001). It was suggested that the inflow of torrents be returned to the beach zone to ensure natural sediment inflow, and to return the previously natural torrent flow to the beach zone to ensure sediment deposition.

Solutions of beach stability problems must be based on results of continuous measurement and monitoring of beach characteristics, waves, winds, tides and other features (Box 16.1). It is necessary to extend field research work and create profiles for all significant beaches, including identification of the inflow of riverine sediments. Measures for shore and beach protection were proposed by the Spatial Plan for the Area of Special Purpose – the Coastal Zone of Montenegro (Skupština Crne Gore, 2007). Depending on the erosion level, economic analysis and aesthetic features, it is proposed to apply the following protective measures: (1) artificial nourishment of beaches with autochthonous sediment of suitable characteristics along the most threatened sections; (2)

BOX 16.1 BEACH EROSION MONITORING IN MONTENEGRO

Monitoring of sand movement on Petrovac, Pržno and Mogren beaches started in 2004 and included basic and simple methods for measurement of depth and sediment thickness on pre-defined beach profiles (Tatomir and Petković, 2005). This programme was extended to Sutomore beach in 2006 and to Sveti Stefan beach in 2007. In 2006, a study on the rehabilitation of Pržno beach was undertaken and also a proposal for building underwater reefs and artificial nourishment of the beach (Petković and Tatomir, 2006). However, these projects were part of a hydraulic model research and cannot be taken as a final technical solution, but as an idea that needs to be further developed, tested and revised. In order to prepare basic guidelines for this monitoring programme, studies were prepared based on experiences obtained from foreign countries, as no such experience existed in Montenegro. Monitoring is currently ongoing, however no activities have been undertaken based on these findings.

Monitoring of sand movement on these beaches includes at least four measurements at each pre-defined profile during one year, in order to include periods with the most intensive erosion, *i.e.* the fall and winter months, as well as periods of low wave intensity during summer months. In conditions of extreme storms, additional measurements should be carried out. The purpose of this field research is to establish the current beach state and characteristics of the basic natural factors in each microlocation, all of which can be used for future plans and projects. For each beach, hydraulic data, hinterland development and sea state information (waves, currents, etc.) are also collected (Petković, 2004).

At Petrovac beach, monitoring results showed that the transfer and direction of sediment transport is very intensive for the whole beach. Large amounts of sediment deposited on this beach in 2009 did not seem to have any influence on beach widening according to data comparison with 2004. The middle part of the beach was notably reduced due to inflow of sediment by torrents and due to the increase of beach slope which

■ **Figure 16.21** *Mogren beach during the 1930s (archive photograph) and today (Photograph courtesy of Public Enterprise for Coastal Zone Management of Montenegro)*

makes sediment easier to erode by waves, as they break closer to the shoreline. During winter months, the largest change (up to 3 m) in beach width was noticed in the eastern part of the beach (Tatomir and Petković, 2005; Petković *et al.*, 2008; Hydrometeorologic institute of Montenegro, 2010).

For the eastern part of Pržno beach, in 2004, at the commencement of monitoring, the beach was artificially nourished which resulted in larger initial amounts of sediment and all further measurements showed beach width reduction with a difference between measurements between 2004 and 2010 of 1.6 m. Building of the concrete walkway extension in the middle of Pržno beach in 2007 caused additional instability. However, most eroded sediment was deposited on the sea bed about 60 to 80 m from the beach (Tatomir and Petković, 2005; Petković *et al.*, 2008; Hydrometeorologic Institute of Montenegro, 2010).

Mogren beach is divided into two parts, Mogren I and Mogren II (Figure 16.21). The sediment transport at Mogren I beach was most intensive at the seaward end where wave influence was highest. Mogren II has less sand on its beach surface, but the sea floor is less steep and mostly rock, so sand is deposited between rocks on the sea bed. River torrents that naturally nourished this beach stopped during the 1970s when a road was built above the beach. Results from monitoring during the last six years show that the underwater slope in front of Mogren I significantly increased in the first 20 m from the shoreline. The most intensive change was at the easternmost part of the beach, where beach width reduced by 3 m, while the beach width of Mogren II did not significantly change (Tatomir and Petković, 2005; Petković *et al.*, 2008; Hydrometeorologic Institute of Montenegro, 2010).

Monitoring on Sutomore beach started in 2006, and results showed changes only in the middle part where the beach surface width was reduced 0.5–1m compared to measurements taken four years previously. In 2007, monitoring on the eastern edge included the Sveti Stefan beach, due to intensive hinterland development. Results from the 2007–2010 period showed that the western beach was stable, while mild erosion took place on the eastern part below the buildings (Petković *et al.*, 2008; Hydrometeorologic Institute of Montenegro, 2010).

construction of hard permanent structures on shores exposed to erosion processes, which will partly or fully prevent the effects of waves, and (3) a combined system of protection including measures (1) and (2) above.

Conclusions

Compared to other European coasts, the eastern Adriatic is less vulnerable, due to its geological characteristics. Generally, it is of high relief and rocks are of a carbonate composition, and there is a low risk of wave, currents and tidal induced erosion. In the Southern Dalmatic, rock islands occur and inlet channels are much wider and current flow is faster than those to the north, so a moderate erosion risk

exists. Siliciclastic flysch coasts are more vulnerable and in some places are already significantly endangered because of erosion. Coastal erosion in Slovenia is currently well controlled by coastal structures with most structures having been in place for centuries (Vahtar, 2002). Similar protection works are rarely found in Croatia. Although recent coastal monitoring activities of beach erosion is recognized at several in Montenegrin beaches (Pržno, Mogren, Petrovac), actual protection works have not been carried out so far.

Although it is considered that rock coasts erode mainly due to wave action (Cowell and Thom, 1997; Bush *et al.*, 1999; Masselink and Hughes, 2003), in the case of the Eastern Adriatic coast, the most pronounced threat is related to anthropogenic transformation of the coast in different ways (e.g., by

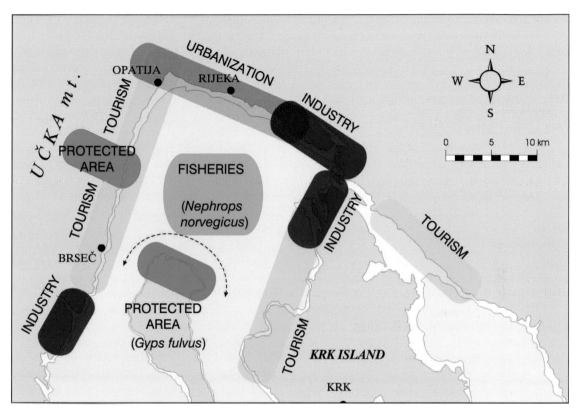

■ **Figure 16.22** *Kvarner area: an example of conflicting and overlapping interests in coastal zone usage (Reprinted from Juračić et al., 2009; Copyright © with permission from Elsevier)*

protection structures, which change sediment equilibrium or by disposal of constructional waste). Furthermore, in different areas conflicting interests of management are often present (e.g. the Kvarner area, Northern Adriatic, Figure 16.22).

Since 2002, Slovenia has implemented several highly successful regional development initiatives with a strong ICZM content (Rupprecht Consult, 2006). In Croatia laws and regulations directing coastal governance and management are not sufficient for sustainable coastal management. However, quite a number of positive and recently developed instruments are in place on which further ICZM initiatives could be founded, e.g. Spatial Strategy (1997), a Programme for Spatial Development (1999), a National Environmental Action Plan (2002), a Decree on the Protection of Coastal Area

(2004). Coastal management in Montenegro is regulated by the 1992 Coastal zone law, the 2007 Spatial Plan for the Area of Special Purpose – the Coastal Zone of Montenegro, and other regional plans, programs and strategies. The Montenegrin Government formed a special public enterprise for Coastal Zone Management to manage the use, protection and maintenance of the coast and its infrastructure. Currently numerous activities are being carried out in order to adjust legislative and institutional setup for more efficient Integrated Coastal Zone Management in accordance with the IUOP Protocol of the 1995 Barcelona Convention.

The coastal zone of the eastern Adriatic Sea makes a naturally continuous geological and geomorphological zone, with similar trends of development and problems that need solving. Therefore, cross-border cooperation mechanisms and strategies should be developed and implemented in the further conservation and development on this region.

Acknowledgements

We would like to thank K. Fio, Š. Aščić, S. Muslim, I. Kosović, D Kudrnovski, I. Pepelnjak and the Public Enterprise for Coastal Zone Management of Montenegro for providing photographs, and R. Košćal for the figure drawings.

References

Antonijević, R., Pavić, A. and Karović, J., 1969. *Osnovna geološka karta 1:100.000.* List Kotor, K 34–50 [Basic geological map 1:100.000, Kotor sheet], Federal Geological Institute, Beograd, Yugoslavia.

Antonijević, R., Pavić, A., Karović, J., Dimitrijević, M., Radoičić, R., Pejović, D., Pantić, S. and Roksandić, M., 1973. *Osnovna geološka karta 1:100.000.* Tumač za list Kotor, K 34–50 [Basic geological map 1:100.000, Geology of Kotor sheet], Federal Geological Institute, Beograd, Yugoslavia.

Benac, Č., Ružić, I. and Ilić, S., 2010. Morphodynamics of pocket beaches (Kvarner, NE Adriatic Sea), (In) *Abstract Book of 4th Croatian Geological Congress* (ed) M. Horvat, 384–385, Šibenik, Croatia.

Bonacci, O. and Roje-Bonacci, T., 2003. The influence of hydroelectrical development on the flow regime of the karstic river Cetina. *Hydrological Processes*, 17, 1–15.

Borčić, A., Bogner, D. and Juračić, M., 2007. Dynamics of foraminiferal assemblages in the sediments in front of the Cetina river mouth, Croatia, (In) *Abstract book of 38th CIESM congress* (eds.) Briand, F., Sakellariou, D., Font, J. and N. Fisher, 75–75, Istanbul, Turkey.

Bush, D. M., Neal, W. J., Young, R. S. and Pilkey, O. H., 1999. Utilization of geoindicators for rapid assessment of coastal-hazard risk and mitigation. *Ocean and Coastal Management*, 42, 647–670.

Cowell, P. J. and Thom, B. G., 1997. Morphodynamics of coastal evolution, (In) *Coastal Evolution*, (eds) R. W. G. Carter and Woodroofe, C. D., 33–86, Cambridge University Press, Cambridge.

Duplančić-Leder, T., Ujević, T. Čala, M. and Viđak, I., 2000. Categorization and number of islands in the Republic of Croatia. *Periodicum Biologorum*, 102/1, 281–284.

Fairbridge, R. W., 1968. *The encyclopedia of geomorphology*, Reinhold Book Corp., New York, 1295 pp.

Giorgetti, F. and Mosetti, F., 1969. General morphology of the Adriatic sea. *Bollettino di geofisica teorica ed applicata*, 11/41–42, 49–56.

HHI, 1999. *Peljar I. Jadransko more-istočna obala [Nautical pilot I. Adriatic Sea, eastern coast; in Croatian]*, Hydrographic Institute of the Republic of Croatia, Split, 331 pp. +D 32, Zrinski, Čakovec, Croatia.

HHI, 2002. Peljar za male brodove, II. dio. [*Pilot for small boats. II. Part; in Croatian*], (ed), Zvonko Gržetić, Hydrographic Institute of the Republic of Croatia, Split, Slobodna Dalmacija, Split, Croatia, 269 pp.

Hydrometeorologic Institute of Montenegro. 2010. Monitoring plaža u Crnoj Gori: Izvještaj za snimanje plaža Petrovac, Pržno, Mogren, Sutomore i Sv. Stefan (gradska plaža) u periodu 2009–2010. [*Beach monitoring in Montenegro: Report on measurements of Petrovac, Pržno, Mogren, Sutomore and Sv. Stefan (public beach) during 2009–2010*]. Hydrometeorologic Institute of Montenegro, Podgorica.

Juračić, M. and Prohić, E., 1991. Mineralogy, sources of particles and sedimentation in the Krka River estuary (Croatia). *Geološki Vjesnik*, 44, 195–200.

Juračić, M., Benac, Č., Pikelj, K. and Ilić, S., 2009. Comparison of the vulnerability of limestone (karst) and siliciclastic coasts (example from the Kvarner area, NE Adriatic, Croatia). *Geomorphology*, 107, 90–99.

Kolega, N., 2006. Slovenian coast sea floods risk. *Acta Geographica Slovenica*, 46/2, 143–169.

Lambeck, K., Antonioli, F., Purcell, A. and Silenzi, S., 2004. Sea-level change along the Italian coast for the past 10,000 years. *Quaternary Science Reviews*, 23, 1567–1598.

Magaš, D., 2002. Natural-Geographic characteristics of the Boka Kotorska area as the basis of development. *Geoadria*, 7/1, 51–81.

Marković, M., Pavlović, R. and Čupković, T., 2003. Geomorfologija [*Geomorphology*]. Faculty for mining and geology. Zavod za udžbenike i nastavna sredstva, Belgrade, Serbia.

Masselink, G. and Hughes, M. G., 2003. *Introduction to coastal processes and geomorphology*, Arnold, London, 354 pp.

Mavrič, B., Orlando-Bonaca, M., Bettoso, N. and Lipej, L., 2010. Soft-bottom macrozoobenthos of the southern part of the Gulf of Trieste: faunistic, biocoenotic and ecological survey. *Acta Adriatica*, 51/2, 203–216.

Menković, L., Marković, M., Čupković, T., Pavlović, R., Trivić, B. and Banjać, N., 2004. Glacial morphology of Serbia Yugoslavia, with comments on the Pleistocene glaciation of Montenegro, Macedonia and Albania, (In) *Quaternary glaciations: Extent and chronology. Part I: Europe* (eds.) Ehlers, J. and P. L. Gibbard, Elsevier, Amsterdam, pp. 379–384.

MONSTAT, 2011. Preliminarni rezultati Popisa stanovništva, domaćinstava i stanova u 2011 [First results of Population, Households and Dwellings Census in 2011]. Monstat, Podgorica.

Orlić, M., 2001. Croatian coastal waters, (In) *Physical oceanography of the Adriatric Sea: Past, present and future*, (eds.) B. Cushman-Roisin, M. Gačić, P.-M. Poulain and A. Artegiani, Kluwer Academic Publishers, Dordrecht, Netherlands, pp. 189–214.

PAP/RAC Synthesis Report, 2007. State of the art of coastal and maritime planning in the Adriatic region. INTERREG IIIB CADSES PlanCoast project, 58 pp.

Pasarić, M. and Orlić, M., 2001. Long-term meteorological preconditioning of the North Adriatic coastal floods. *Continental Shelf Research*, 21, 263–278.

Peterlin, M., Kontic, B. and Kross, B. C., 2005. Public perception of environmental pressures within the Slovene coastal zone. *Ocean and Coastal Management*, 48, 189–204.

Petković, S., 2004. Studija prihranjivanja plaža [*Beach nourishment study*]. Public Enterprise for coastal zone management of Montenegro.

Petković, S. and Jovanović, M., 2001. Projekat revitalizacije plaže u Petrovcu. [*Project for revitalization of Petrovac beach*]. Association for Waterways and Engineering, 'PIM Invest' Tivat, Montenegro

Petković, S. and Jovanović, M., 2002a. Projekat revitalizacije plaže u Sutomoru. [*Project for revitalization of Sutomore beach*]. Association for Waterways and Engineering, 'PIM Invest' Tivat, Montenegro.

Petković, S. and Jovanović, M., 2002b. Projekat revitalizacije plaže Mogren. [*Project for revitalization of Mogren beach*]. Association for Waterways and Engineering, 'PIM Invest' Tivat, Montenegro.

Petković, S. and Tatomir, U., 2006. Studija mogućnosti uređenja obale i plaže u Pržnom [*Study of the possibility for maintainance of Pržno beach and coast*]. Public Enterprise for Coastal Zone Management of Montenegro, Budva.

Petković, S., Petković S. and Tatomir, U., 2008. Monitoring plaža u Crnoj Gori: Izvještaj za 4 snimanja plaža u periodu 2007–2008 na plažama Petrovac, Pržno, Mogren, Sutomore i Sv Stefan (javna plaža). [*Beach monitoring in Montenegro: Report on 4 measurements during 2007–2008 at Petrovac, Pržno, Mogren, Sutomore and Sv. Stefan (public beach) beaches*]. Public Enterprise for Coastal Zone Management of Montenegro, Budva.

Picha, F. J., 2002. Late orogenic strike-slip faulting and escape tectonics in frontal Dinarides-Hellenides, Croatia, Yugoslavia, Albania and Greece. *American Association of Petroleum Geologists Bulletin*, 86/9, 1659–1671.

Poulain, P.-M. and Cushman-Roisin, B., 2001. Circulation, (In) *Physical oceanography of the Adriatric Sea: Past, present and future*, (eds.) B. Cushman-Roisin, M. Gačić, P.-M. Poulain and A. Artegiani, 67–109, Kluwer Academic Publishers, Dordrecht, Netherlands.

Radojičić, B., 1996. Geografija Crne Gore [*Geography of Montenegro*]. Unirex, Nikšić.

Rajčić, S. T., Faivre, S. and Buzjak, N., 2010. Promjene žala na području Medića i Mimica od kraja šezdesetih godina 20. stoljeća do danas. [*The changes in beaches surfaces in the Medići and Mimice area from 1960 till today; in Croatian*]. *Hrvatski geografski glasnik*, 72/2, 27–48.

Riđanović, J., 2002. *Geografija mora* [*Marine geography*; in Croatian], Hrvatski Zemljopis – naklada Dr. Feletar, Copygraf, Zagreb, Croatia, 214 pp.

Romić, D., Zovko, M., Romić, M., Ondrašek, G. and Salopek, Z., 2008. Quality aspects of the surface water used for irrigation in the Neretva Delta (Croatia). *Journal of Water and Land Development*, 12, 59–70.

Rupprecht Consult, 2006. *Evaluation of Integrated Coastal Zone Management (ICZM) in Europe: Final report*, Forschung & Beratung GmbH, Germany, 255 pp.

Skupština Crne Gore, 2007. Prostorni plan područja posebne namjene-morsko dobro [*Spatial Plan for the area of special purpose: The Coastal Zone*]. Službeni list Republike Crne Gore br. 24/08.

Sondi, I., Lojen, S., Juračić, M. and Prohić, E., 2008. Mechanisms of land-sea interactions: the distribution of metals and sedimentary organic matter in sediments of a river-dominated Mediterranean karstic estuary. *Estuarine, Coastal and Shelf Science*, 80, 12–20.

Sunamura, T., 2005. Cliffs, typology versus erosion rates. (In) *Encyclopedia of coastal science. Encyclopedia of Earth Sciences Series*, (ed.) M. Schwartz, Springer, Dordrecht, Netherlands.

Surić, M. and Juračić, M., 2010. Late Pleistocene-Holocene environmental changes: Records from submerged speleothems along the Eastern Adriatic coast (Croatia). *Geologia Croatica*, 63/2, 155–169.

Tatomir, U. and Petković, S., 2005. Monitoring plaža u Crnoj Gori: Izvještaj za 4 snimanja plaža u periodu 2004–2005 na plažama Petrovac, Pržno i Mogren. [*Beach monitoring in Montenegro: Report on 4 measurements during 2004–2005 at Petrovac, Pržno and Mogren beaches*]. Public Enterprise for Coastal Zone Management of Montenegro, Budva.

Turk, R., 1999. An assessment of the vulnerability of the Slovene coastal belt and its categorisation in view of (in)admissible human pressure, various activities and land-use [in Slovenian]. *Annales Series Historia Naturalis*, 15, 37–50.

Vahtar, M., 2002. Slovenian coast (Slovenia). Report of UAB Pilot Sites. EUROSION, Draft v. 3, 1–20.

Vlahović, I., Tišljar, J., Velić, I. and Matičec, D., 2005. Evolution of the Adriatic Carbonate Platform: Palaeogeography, main events and depositional dynamics. *Palaeogeography, Palaeoclimatology, Palaeoecology*, 220, 333–360.

Vranješ, M., Vidoš, D. and Glavaš, B., 2007. Stanje sedimenta u donjoj Neretvi. [*Conditions of sediment in lower Neretva;-in Croatian*]. *Proceedings of the 4th Croatian Water Conference*, Opatija, 338–344.

17 Albania

Gjovalin Gruda and Merita Dollma

Introduction

The Albanian coastline represents the western edge of the coastal lowland, which is the largest on the Balkan Peninsula's west coast (Figure 17.1). The coastline is 427 km long and 273 km of the Adriatic coastal zone is composed mainly of prograding beaches (between the Buna River mouth in the North and Vlorë in the South, including Sazan Island); whereas the remaining 154 km of coastline is almost completely erosional, extending from the Vlorë (on the bay of Vlorë) to the Cape of Stillo in Ksamil, near Sarandë. The width of the Adriatic coastal zone ranges from 8 to 35 km wide and is folded, basically because of the existence of the neogenic molasic structures of Rodon, Bishti i Palles Cape and Kryevidh, while to the North and South the limestone structures of Rrenc and Karaburun are more prevalent. The largest bays are Drini, Rodon, Lalezi and Durres as far as the entrance of Vlorë bay, where the Ionian coast begins. The coastline and beaches formed in these bays are characterized by rapid morphological development resulting in an ongoing westward growth of the coastal lowland.

Coastal morphology

Morphologic evolution of the lowland (comprising anticline and syncline structures striking northwest) and the Adriatic coast commenced during Pliocene and Quaternary times, when the structural basement formed on the tectonic zones of Ionic and Kruje. The lowland structures are traversed by several north-south and horizontal faults, such as those at Vlorë-Tepelena, Lushnje-Elbasan, Ulqin-Shkodra, and Mat, along which the Vjosa, Shkumbin, Drin and Mati rivers flow. Another tectonic fault with an almost north-south direction lies along the western border of the Narta and Karavasta lagoons, where there is a 5 to 10 m subsidence of the eastern side (Mathers, *et al.*, 1999). The morphological importance of this fault resulted in the formation of the two abovementioned largest lagoons (almost in the central Adriatic Lowland) and river mouth displacement southwards. In general, the lowland structural basement was characterized by tectonic uplifts, especially from the middle Quaternary, resulting in a westward coastline movement. Sectors with tectonic subsidence are limited in space and mainly situated in the river mouths of Seman, Mat and Porto Romano.

Evolution of the lowland areas and coastline advancement were made through the large amount of alluvial deposits obtained from river networks, which drain catchment areas of 43,305 km^2 (28,550 km^2 of which belongs to Albania). This morphological phenomenon was also driven by the prevalence of tectonic uplift and climate change in the Quaternary era. The Adriatic coastline advancement up to the present day has been made through formation of the coastal river delta cordons (the strip of sand separating lagoons from sea) which isolated coastal fragments, turning them into lagoons, marshes and swamps. The amount of riverine alluvial material transported into the Adriatic Sea reaches 59.8 $\times 10^6$ ton/yr (Pano, 2008), the Seman and Drin

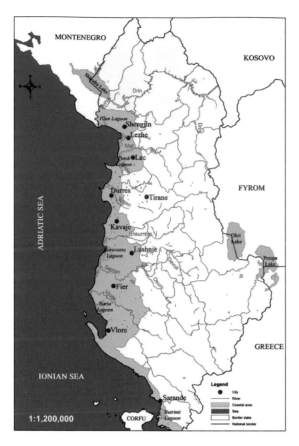

■ Figure 17.1 *Albanian coastal area*

rivers having the largest alluvial loads with respective values of 13.2×10^6 and 13.8×10^6 t/yr (Pano, 2008).

The Adriatic coastline's westward advancement accelerated, as evidenced by the wide extension of Pleistocene-Holocene deposits which built the major part of this lowland. Coastline evolution traces (evidenced by river deltas and continuous river mouth movement) are defined by aerial photos, satellite images, topographical and geomorphologic maps, as well as geological survey drillings. The conditions of the very active tectonic character of the Albanids (the Mediterranean alpine belt of Dinarid-Albanid-Helenid) have encouraged erosion and river transport. Based on current data, the major part of

this coastal lowland was created by river deposition from the sixteenth century to the present, with the growth mechanism being delta progression (Mathers *et al.*, 1999). Geological profiles based on heavy mineral analysis (rutile, zircon, chromium, etc.) show the coastline's progress and successive levels up to the present. The presence of these minerals in Quaternary lowland deposits indicates five successive steady growth stages, ranging from the Flandrian transgression to the present.

In general, coastal morphological features and coastline extension are similar for the river mouths of Vjosa, Seman and Shkumbin, all functions of the riverine solid load volume reflected in the number of deltas (7–8) and the size of the impact sectors through river mouth displacement. The maximum displacement distance of the Seman river mouth reaches about 33 km; at the Vjosa River it is 20 km while at all other rivers this value is 9–15 km. The evidence is the presence of old riverbeds in the form of abandoned meanders, the river Drin being the most characteristic. Typical also are the old riverbeds of the Seman, Vjosa and Shkumbin, which formed the major part of this lowland, especially after the fifteenth and sixteenth centuries. During this period the coastline constantly advanced westward to its present position, evidenced by the presence of lagoons, swamps and abandoned meanders in the form of a wide belt along its eastern side. Obviously, in this evolution, along with accumulation, there has also been erosion activity in the opposite sectors of the river mouth displacement, e.g. along the Capes of Rodon, Bishti Palles, Cape of Turra and Treporte (Berreti, 1998).

Human impact on the coastline evolution

Direct or indirect human impact on coastline morphological evolution mainly connected with river mouth displacements plays an important role (Drin and Ishem), as does construction of reservoirs for hydro power production (Drin and Mat), uncontrolled cutting of wood cover in the watersheds and extraction of inert materials from the river beds. The largest natural river mouth displacement is in the Drin River, about 36 km in length between Lezhe (southeast) and Ulcinj (Montenegro in the northwest). The last natural displacement of this river to the northwest occurred during 1859, leaving, until 1963, about 20 per cent of the river flow to Lezhe. After this date, construction of the Mjeda embankment completely stopped the river flow to Lezhe, the river bed becoming a drainage canal of the Zadrima Plain. The major part of the Drin River solid load was interrupted by reservoir construction at Vau i Dejes, Koman and Fierze after 1971; whereas on the Mat River, the Shkopeti and Ulza hydropower stations were built (1952–1956). The solid flow of the Mat River after 1956 is provided almost entirely by the Fan River which accounts for some 2 million tonnes per year of sediment. Under these conditions the rate of coastline accumulation and advancement increases unequally from north to south. The big reduction of solid flow of the Drin River after 1970 directly influenced development of coastal erosion along the coastal zone on both sides of the mouth, especially at Kuna/Vain. The Mat River mouth displacement to the south during this period also led to an increase of erosion at Tale beach; alluvial load interruption also caused an unstable coastline at Tale/Kune/Vain. Almost the same situation occurred in the Velipoja sector, where a big reduction in the Drin River solid load transport prompted beach erosion at Velipoja, especially in the southern part between Viluni and in Ranë e Hedhur.

Human intervention, basically carried out in order to acquire agricultural land, can also be found in the natural river mouth displacement of Ishem, from the southern edge of Rodon bay to Patok lagoon in the North. This intervention took place in 1967–1968 and was accompanied with the onset of erosion (2–3 m/yr) along the littoral cordon of Patok beach. Therefore, the 6 km coastline between the abandoned river mouth and the new one was subjected to erosion (150–300 m). The river flow movement back to its natural river mouth in 1985 was accompanied by the creation of a new strip some 150 to 200 m wide and about 500 m long.

Natural displacement of the Mat River after 1975–1978 to the present also has special morphological importance. In this period, the Mat River flowed some 2.5 km towards Patok lagoon in the South, forming a littoral cordon and a new lagoon almost parallel to that of Patok. The coastal strip is 3 km long and 50 to 100 m wide and represents a shield against beach erosion at Patok. During this period, to the north of the abandoned Mat River mouth, coastal erosion of 2–3 m/yr was initiated. However, in general the coastal lowland progressed significantly, especially in the southern sector between Mat and Ishmi.

In conclusion, the whole Adriatic coastline is advancing seawards, but in specific sectors erosion predominates with 0.3 to 20 m/yr, except at the Seman River impact area where this value is 30 m/yr (Meçaj, 2005). Erosion is concentrated mainly north of river mouths (except the Vjosa River), as well as all capes, whereas deposition is characteristic of southern river mouth sectors, indicating formation of the new lagoon of Patok, Kulari (to the south of the Shkumbin river mouth) and several other lagoons between the Vjosa and Seman rivers (Fouache, et al., 2001). Under these conditions the deposition/erosion ratio, or advancement and withdrawal of the Adriatic coastline, varies from 2:1 to 5:1, with an average value for the last 110 years being 2.5:1 Therefore, every year a total land area of about 105 ha is formed and nearly 42 ha eroded, resulting in an average addition of 63 ha/ year. Over the last 110 years some 2,200 ha in the Adriatic lowland has been eroded, producing a coastline withdrawal of 2.5 to 3 km (between the river mouths of Shkumbin and Seman and partly between the river mouths of Seman and Vjosa).

The high Ionian coastline or the Albanian Riviera (between Vlorë bay and Butrint bay, as far as the Cape of Stillo – 154 km) has quite different morphological features. Folding is found at the bays where some small beaches are formed where sediment accumulation has managed to partially compensate tectonic subsidence. The larger bays are Vlorë, Himara, Porto Palermo, Borshi, Sarandë, Butrint and Kakome. Beach dimensions are mainly related to the water regime, e.g. the field of Dukat is shaped by advancement of the Dukati stream mouth, which reaches as far as the Pashaliman lagoon and its littoral cordon. Even the mouths of the Kudhësi and Borsh Rivers, in the tectonic bay of Borsh, have specific morphological features. After Borshi beach are found the beaches of Palasa, Himara, Jala of Vuno and Livadhja. The bays of Portopalermo and Kakome again have an erosional coast. The major part of Butrinti bay (between the Capes of Shkalla and Stillos) is quite characteristic, where the alluvial load of the Pavlla River has created a depositional coastline of 2 km between these two Capes. This river flows along the limestone outcrop between Qafe Bote and the Cape of Stillo, which hosts an embankment to prevent the Mursi-Vrine field flooding. The Butrinti (Vivari) canal connecting the lake of Butrint with the sea is situated on the Southern edge of the Ksamil.

The beaches of Tale, Kuna, Vaini and Shëngjin, situated north of the Mat River mouth, are also subject to erosion. Southward river mouth displacement in 1978–1980 and northward displacement of the Drin River mouth to Velipoja has almost completely interrupted the beach alluvium supply of these rivers. After 1978–80, these beaches faced continuous erosion of 3–13 m/yr (Hyseni, 2009), particularly at Kuna and Vaini, situated within the impact area of the Drin of Lezhe River, which is currently converted to a drainage canal of the Zadrima Plain.

North of Tale beach, the 40–50-year-old sand dunes served as a natural beach protection after southward deviation of the Mat River mouth and now are also subject to erosion. Kune beach, lying between the two channels of the Merxhani lagoon to the sea, is severely endangered due to its low altitude (0.4–0.7 m). The littoral cordon of the Kune-Vain beach, separating the lagoon of Merxhani from that of Ceka, has narrowed to 40-50 m and there is a risk of these lagoons connecting to the sea, especially

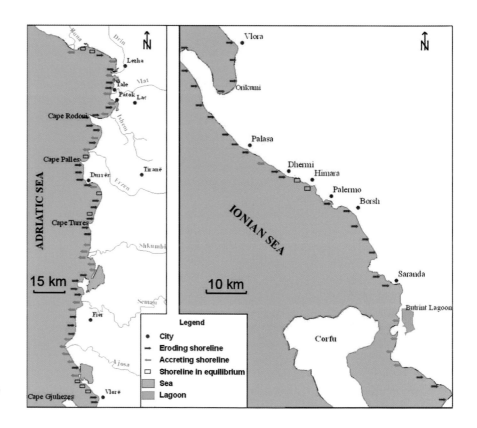

■ **Figure 17.2**
Erosion/accretion in Albania

near the Drin River mouth. This phenomenon is also prompted by beach erosion of Tale beach, south of Kuna, where the coastline withdraws some 2–3 m/yr over a length of 2 km.

Beach erosion/accretion

Although accumulation predominates at the Adriatic Lowland, beach erosion is present in some sectors (Figure 17.2). The most vulnerable beach is at Seman, where erosion rates of 7–30 m/yr over a length of 6 km occur south of the present river mouth and 10–30 m/yr over a length of 4 km north of this river mouth (Gruda *et al.*, 1999; Figure 17.3). Erosion of this beach started about 150 years ago, as a result of inland sea advancement up to 4 km leading to the destruction of 2,500 ha of land. This phenomenon will continue in the future, because construction of Banja reservoir has reduced the solid material load (about 40 per cent) of the Devoll River, which is the main branch of the Seman River.

Patok beach (between the Mat and Ishëm rivers) is another sector of the Adriatic coastline experiencing erosion, especially after 1978, when the Ishmi River mouth was displaced northward in order to drain Patok lagoon for agricultural land reclamation. An erosion rate of 2-3 m/yr eroded Patok beach of some 150–300 m over a length of 6 km. At the northern Patok beach sector, the erosion rate was 7–8 m/yr from 1980 to 2000. The present almost stable beach state (over the last decade) is related to the formation of a littoral cordon about 2 km long and 50 to 100 m wide (Hyseni, 2009) and the formation of a new lagoon from the Mat River, which is almost parallel to that of Patok.

Coastal erosion activity is also present at the Shkumbin River despite the fact that accumulation prevails in this area. A sector of the present Shkumbin river mouth is subject to erosion over a length of 4 km at a rate of 2–3 m/yr, especially the area from Spille beach, as far as an old meander located close to the Bashtova fortress.

The beach to the south of the Vjosa river mouth separates two small lagoons from the sea and is also subject to erosion over a 6 km length. The beach width is 100–220 m with an altitude of 0.5 to 3 m, which makes Kallënga lagoon extremely vulnerable over a 1.5 km length, as well as its sea connection. Breakwaters and groins and perched beaches may be seen in this area.

The Cold Water (Uji i Ftohtë) and other segments of Vlorë beach are in a very critical state and concrete blocks have been placed over a 500 m length to protect the coast road from further destruction, as well as from construction and demolition debris, which litter the beach in some sections (Figure 17.4)

■ **Figure 17.3** *The beach between the Seman and Vjosa river mouth, in Fier province (Photograph courtesy of Geocoste)*

■ **Figure 17.4** *Sand accumulation north of Vlorë fishing harbour, unusable for tourist activity due to scattered blocks from continuous construction and demolition (Photograph courtesy of Geocoste)*

After 1990, some hard defence structures (seawalls and revetments) have been built to counteract beach erosion (Figure 17.5). At Kavaja district beaches, revetments were built respectively at Mali i Robit (2), Golem (2), Karpen (4). At the southern edge of Lalzi Bay and Porto Romano beach (Durrës) seawalls and groins have been built (Figure 17.6). Also in the Bay of Vlorë, between the Old Port and the Plazhit të Ri (New beach), a seawall has been built to protect the road from erosion. However, despite these defence measures the erosion problem has not been completely solved and relevant coast protection management plans are needed besides hard structures (Figure 17.7).

■ **Figure 17.5** *Seawall and revetments south of Vlorë (Photograph courtesy of Geocoste)*

■ **Figure 17.6** *L groin south of Durrës (Photograph courtesy of Geocoste)*

■ **Figure 17.7** *Beach erosion and management problems at Durrës (Photograph courtesy of Geocoste)*

Evolution of settlements in the coastal area

Based on archaeological findings in Xara (the Sarandë District), Shkoder, Dajt, etc., it can be shown that humans have lived in the western regions of the Balkans from as early as the mid Palaeolithic period (100,000–40,000 BC). Early inhabitation of this region and continuation of residency until the present are shown by a number of archaeological findings, from the prehistoric settlements mentioned (Xara, Gajtan), to ancient cities (Albanopolis, Dimal, Bylis, Amantia, Apollonia, and Dyrrah), to Illyrian fortresses, to medieval castles, etc.

The territory and number of the settlements grew, with settlements covering almost the whole of the western part of the Balkans, but mainly the Albanian coast. The Illyrian tribes living on the coastal area were mainly Ardians, Taulants, Amants and Bulins. The early coastal population and continuation of life has been conditioned by several natural and historical factors, such as favourable geographical position, Mediterranean climate, low relief, fertile soils, water resources, long coastline, sand beaches, rich biodiversity, marine and land transportation.

The population of this zone has increased continuously throughout the centuries, but the highest increase of population became evident after the 1990s due to migration from the rural mountainous zones of the northern and northeastern regions. Under these conditions, the natural coastal area environment changed rapidly into an artificial environment and new settlements are still being created. Depending on relief, the population distribution in this region is uneven, with coastal areas having the highest in the country, but this varies from one district to another. According to the latest official census (INSTAT, 2001), cities with the greatest population densities are: Durrës, Vlorë and Shkoder (an average of 300–400 inhabitants/km^2). In the coastal area a younger age and urban population dominates. Coastal cities are numerous and most districts have a population of over 100,000 inhabitants (Fier, Vlorë, Shkoder, Durrës and Lushnja). The most important administrative and economic cities are respectively, Shkoder,

Durrës, Fier, Vlorë and Sarandë, but there are several rural settlements scattered throughout the western lowland region.

The coastal area is rich in biodiversity together with sand/gravel beaches, which makes it a priority zone in Albania, one targeted for tourism development. Today, this area is influenced by a combination of social, economic and environmental factors. The natural environment is facing the consequences of human impact on the environment from the past to the present. Due to fast urbanization of the coastal area, the natural Mediterranean bush forest and artificial alpine forest is being logged, dunes are disappearing, sand is being removed for construction industry purposes and the natural landscape is changing rapidly. The Bay of Shëngjin, Bay of Durrësi, Bay of Vlorë and Bay of Sarandë are facing extremely fast urbanization, with many buildings, hotels and restaurants located very close to the sea (10 m). Rapid urban development of these bays has resulted in the loss of several kilometres of sand dunes and hundreds of hectares of natural and artificial forests

Beach management

Before 1990, Albanian legislation concerning environmental protection was poor and had many gaps. Today, legislation has improved considerably and several laws and by-laws exist. In addition, numerous institutions and NGOs deal with management and protection. Since 1991, a number of laws have been approved representing a positive advancement in the legislative area, such as the: Law on Forestry and Forestry Service Police (no. 7623, 1992); Law on Environmental Protection (no. 7664, 1993); Law on Fishing and Aquaculture (no. 7908, 1995); Law on Water Resources (no. 8093, 1996); Law on Hunting and Wildlife Protection (no. 7875, 1994); Law on Protected Areas (no. 8906, 2002, 2006, 2008), etc.

The Ministry of Environment, Regional Environmental Agency, Public Health Institute, Interior Ministry, the Counties, Municipalities and Communes are the national and local authorities responsible for beach administration and protection.

According to law, the whole Albanian coast is state property and local authorities are the main administrators of the beaches. The Municipalities or Communes can rent them to private owners, who can use the beaches for tourism purposes for a certain period of time. In these so-called 'private' beaches, private businesses are responsible for beach management and maintenance. Administration of public beaches is the responsibility of the municipality or commune.

Beach management is limited to cleaning beaches, providing fresh water and electric power, improving water quality, reducing untreated waste water

flowing into the sea, increasing beach safety, etc. Unfortunately, little is contributed towards preventing beach erosion or beach nourishment. Although an Integrated Coastal Zone Management strategy and plan has existed for several years, nothing is done to stop or prevent beach erosion in the most affected coastal areas. Although there are many laws protecting the beaches and preventing damage, the negative human impact on the beaches is considerable, due to poor law enforcement. A regular monitoring system of beach erosion for the whole Albanian coast does not exist, only sporadic studies

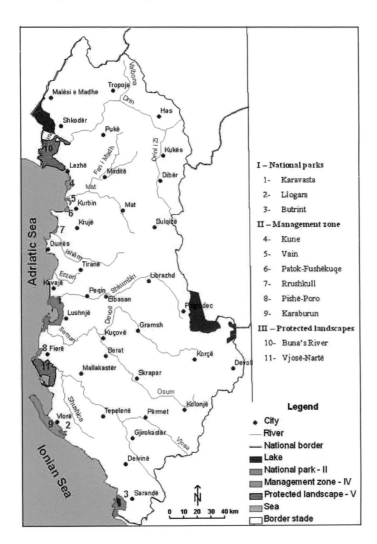

■ **Figure 17.8** Protected areas in the coastal zone of Albania

carried out by researchers and institutions within the framework of different projects. A monitoring system using high resolution satellite images could provide the authorities with accurate information and data of the erosion progress in the Albanian coast.

Along the Albanian coast, several areas are named as Protected Areas (Figure 17.8) of different categories (six categories of IUCN), such as the National Parks (Category II) of Butrint, Llogara, Divjaka; Managed Natural Reserves (Category IV) such as Karaburun, Pishe-Poro, Karavasta lagoon, Kune-Vain lagoon; Protected Landscapes (Category V) such as Vjose-Narte, Buna River; and several Natural Monuments (Category III) such as the Old Cliff of Borsh (Sarandë), Cliff of Sasan Island (Vlorë), General's Beach and Carina Beach (Kavajë), and Kallmi and Shën Pjetër Beaches (Durrës). However, very few of them have Protected Areas Management Plans. Even in those areas that already have management plans in place very little is done for their management. Poor management is due mainly to lack of financial resources for effective administration of the protected areas, shortage of personnel, lack of training and overlap of competencies.

Conclusions

Evolution of the present day Albanian coastline commenced during the Pliocene and Quaternary and formation of the current alluvial lowland relief was effected during the Holocene and Historic periods. The coastline and beaches are characterized by rapid morphological development, resulting in a westward coastal advancement. Coastline evolution is different over its length, being dependent on the alluvial load of each river. The presence of five coastline types ranging from that of the Flandrian transgression up to the present is evidence of this evolution. In general the coast is dynamic due to both natural and human factors, with impact of the latter on coastline evolution being very high. Due to these factors, every year a total land area of about 105 ha is formed and nearly 42 ha eroded. Beach erosion is present in some sectors of the Albanian coastline, with the most vulnerable being Seman, Patok, Tale, Kuna, Vaini and Shëngjin beaches. Although there is a relatively complete legislation for beach management and Integrated Coastal Zone Management, very little is done for protection and prevention of beaches from erosion.

References

Berreti, V., 1998. Le site antique de Treport, port des villes des Amantins in P.Caban (ed). L'Illyrie meridionale et l'Epir dans l'Antique. *Actes du colloque international de Chantilly* (16–19 octobre 1996), 181–185.

Ciavola, P. and Simeoni, U., 1995. A review of coastal geomorphology of Karavasta lagoon (Albania). Short term coastal change and implications for coastal conservation. *Directions in European Coastal Management*, 301–316.

Fouache, E., Gruda, Gj., Muçaj, S. and Nikolli, P., 2001. Recent geomorphological evolution of the deltas of the Seman and Vjosa (Albania), *Earth surface processes and landform* 26/7, 793–802.

Gruda, Gj. and Nikolli, P., 1995. Evolucioni morfotektonik dhe morfologjik i territorit bregdetar midis Semanit e Shkumbinit, Revista "*Studime gjeografike*", 6, 71–72. [In Albanian]

Hyseni, J. P., 2009. Evolucioni gjeotektonik dhe gjeomorfik i hapesires bregdetare Patok-Shengjin, unpub. PhD, Tirane, 107–115. Geotectonic and geomorphological evolution of the coastal area Pastok-Shen. [In Albanian]

INSTAT (Institute of Statistics, Albania), 2001. *Census report*, INSTAT, Tirana.

Mathers, S., Brew, D. S. and Russell, S. A., 1999. Rapid Holocene evolution and neotectonics of the Albanian Adriatic coastline, *Journal of Coastal Research* 15/2, 345–354.

Meçaj, N., 2005. River deltas, their morphology and the accompanying dynamic evolution of the Adriatic and Ionian coast of Albania. *Z. Geomorphology.N.F*, 141, 59-73, Berlin.

Pano, N., 2008. *Pasuritë ujore të Shqipërisë*, Akademia e Shkencave, Tiranë, 372–383. [In Albanian]

18 Greece

George Alexandrakis, George Ghionis, Serafim E. Poulos and Nikolaos A. Kampanis

Introduction

Greece is the country with the most extensive coastline among Mediterranean countries, the shoreline length exceeding 15,000 km. The coastal zone is almost equally divided between mainland and islands (60 per cent mainland, 40 per cent islands); 700 km of coastal zone corresponds to 3,053 islands, but only 227 are inhabited (NSSG, 2001). Every km² of the total area of Greece (131,957 km²) corresponds to 113 m of shoreline, whilst the global average does not exceed 5 m (Coccossis and Mexa, 2002).

Almost 38 per cent of the Greek population and 90 per cent of tourism and leisure activities are along the coastal zone of both mainland and islands, leading to increased urbanization of these areas (Coccossis and Mexa, 2002; Coccossis and Mexa, 1997), with a population density of 110 persons/km². Approximately 40 per cent of the coastal population work in economic activities, such as, fisheries, tourism, agriculture, services and industry and the population has changed during the past decades. The population decrease evidenced in the islands of the Northern Aegean Region, as well as the coastal areas of Eastern

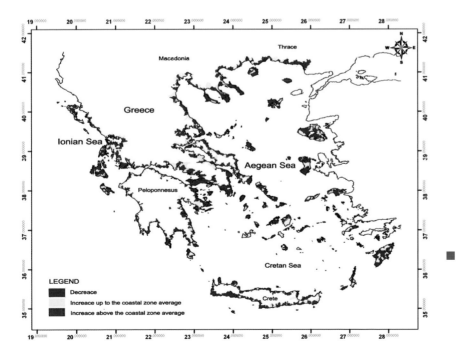

■ **Figure 18.1**
Changes in the coastal population for the period 1991–2001 (modified from Coccossis and Constantoglou, 2005)

Evia, Southern Crete and Sterea Hellas, has been attributed to the phenomenon of internal immigration from rural areas to urban centres or tourist districts. In contrast to the above, in the majority of the coastal tourist areas, such as the Cyclades and Dodecanese, Northern Crete, the Ionian Islands and the large urban centres (Athens, Thessaloniki and Patras), the population has remained constant or is increasing (Figure 18.1).

Tourism also plays an important role, as Greece is ranked among the 20 most important tourist destinations globally (UNWTO 2011), while a significant number of areas depend almost entirely (socially and economically) on tourism. Tourist growth is located mainly in coastal regions, making this zone a very important economic asset. This reveals the significance of the presence and sustainability of beach zones, where international visitors seek traditional 'sun, sea and sand' package vacations (Anastassopoulos *et al.*, 2009).

Hellenic coastal zone: Physiographic setting

Climate

Greece is located in the north temperate climate zone, characterized overall by a Mediterranean climate. Localized climate variation is due to different topographic and geographical factors. Generally, nearshore areas have a typical Mediterranean climate, which gradually becomes continental towards the central parts and in the mountains (Zabakas, 1981).

The annual climatic variability falls into two main periods: October to March, which is cool and rainy, and April to September, which is a hot and rather dry season. October and April can be characterized as intermediate months, between these two main periods. January and February, on the other hand, are colder and include more periods of snow. The warmest months are July and August. Generally, autumn is warmer than spring, by about 2–4°C (Zabakas, 1981). The mean annual air temperature varies between 14.5 and 19.5°C. The absolute

minimum temperatures occur in the mountains, where they have occasionally reached –25°C, whereas the absolute maximum temperatures of about +45°C occur on the mainland at low altitudes (Zabakas, 1981).

The mean annual rainfall over the country is 823 mm with areas receiving the most rain (>800 mm) being the high mountains and western Greece. The eastern part, including the Athens basin, receives the lowest amount of rainfall (400 mm). Snow occurs mainly on the mountains, generally commencing in September and lasting up to the middle of May. Most of the time, winds range between force 0 and 4 or 5 on the Beaufort scale; they rarely exceed force 8. Wind directions are affected greatly by local topography. During summer, the prevailing winds are from the north all over the country; in spring, there are westerly and northwesterly winds, whilst in autumn the emphasis shifts towards the north (Watts, 1975). Sea breezes occur during the summer along the coastlines.

Coastal geomorphology

The Hellenic coastal zone morphology, and in particular coastal landforms, are the natural outcome of marine processes operating during the last 7–8,000 years; this period corresponds to a phase of slow sea level rise, *i.e.* approximately 1–2 mm/yr (Lambeck and Purcell, 2005; Poulos *et al.*, 2009b). Four coastal types have been recognized, according to the geomorphological classification scheme of the EUROSION project (2004). These are (Figure 18.2):

- *Type I:* Rock coasts and/or cliffs made of hard rock (low level of erosion), sometimes with a rock platform.
- *Type II:* Cliffs consisting of conglomerates and/or soft rock (e.g. chalk), which are subject to low levels of erosion, with pocket beaches (<200 m long).
- *Type III:* Beach zones including small beaches (200 to 1,000 m long) usually separated by rocky capes (<200 m long); extensive beaches (>1 km long),

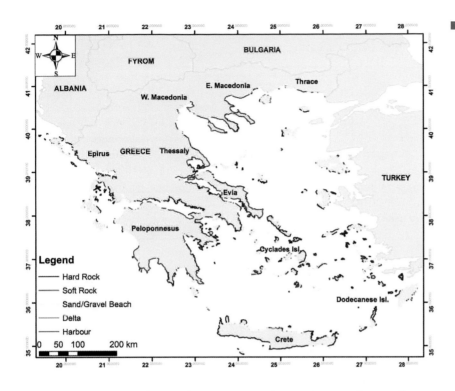

■ **Figure 18.2**
*Major coastal types
for the Greek
coastline (modified
from Alexandrakis
et al., 2010)*

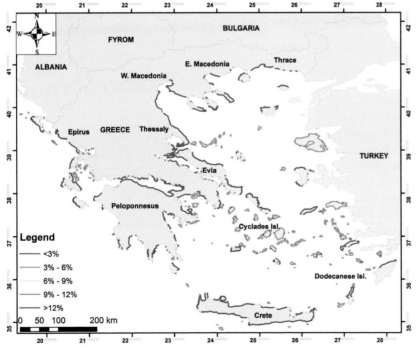

■ **Figure 18.3**
*Coastal slopes (%)
along the Greek
coastline based
on the distance
between the
shoreline and the 5
m elevation contour
(modified from
Alexandrakis
et al., 2010)*

■ **Table 18.1** *Spatial distribution of the Hellenic coastal slope categories*

AREA	COASTAL SLOPE				
	Very low (<3%)	*Low (3–6%)*	*Medium (6–9%)*	*High (9–12%)*	*Very high (>12%)*
Thrace / Eastern Macedonia	72%		13%	11%	4%
Central Macedonia	45%		2%	53%	
Thessaly	9%	39%	18%	17%	17%
Peloponnesus	31%		25%	44%	
Epirus	55%		15%	30%	
Ionian Islands	65%			35%	
Evia	15%		45%	31%	9%
North Aegean Islands	64%			36%	
Cyclades / Dodecanese Islands	51%		3%	46%	
Crete	50%			50%	

often with strands of coarse sediment (gravel or pebbles); and extensive beaches (>1 km long) with strands of fine to coarse sand. In addition, coastlines of soft non-cohesive sediments, e.g. barriers, spits and tombolos, are occasionally included together with artificial and nourished beaches.

• *Type IV:* Mud coasts, represented by strands of mud sediments, associated with deltaic deposits.

The above four coastal types represent 44 per cent, 14 per cent, 36 per cent and 6 per cent respectively of the total length of the Hellenic coast.

The coastal slopes presented in Figure 18.3 have been grouped into five categories: very high (>12 per cent), high (9–12 per cent), mean (6–9 per cent), low (3–6 per cent) and very low (<3 per cent; Alexandrakis *et al.*, 2010). Obviously, the high and very high slopes are associated either with rock coasts (Type I) or cliffs (Type II). Low and very low slopes are related to beach zones (Type III) that are the seaward limit of deltaic plains, coastal alluvial plains or marine terraces. The spatial distribution of the aforementioned five slope categories is given in Table 18.1.

Sea level variation

The astronomical tide is generally less than 10 cm, having as principal constituents the M2 and S2 (1–9 cm) with a phase angle between 30° and 330° (Tsimplis, 1994). However, the overall fluctuation of sea level exceeds 0.5 m, due to meteorological forcing (differences in barometric pressure, wind and wave setup). The mean sea level fluctuations (the sum of the meteorological and the astronomical tide) for selected locations in the Aegean, Cretan and Ionian Seas are given in Table 18.2. In general, the mean sea level variation of the northern Hellenic coast ranges from 0.68 m in the eastern part (Thrace) to 0.87 m in the western part (Thermaikos Gulf). These values are higher than those measured in the central (Cyclades, eastern coasts of Attica and the Peloponnesus) and south-eastern (Dodecanese islands) parts of the Aegean, which are of the order of 0.5–0.6 m. The northern coastline of Crete experiences the lowest sea level variations, *i.e.* <0.5 m (Table 18.2). This difference is attributed to wind forcing and the overall coastal configuration that enhances sea level rise in the Aegean Sea during persistent South winds, which obviously do not affect the northern coast of Crete.

■ **Table 18.2** *M2 and S2 amplitudes (cm) and phases (degrees) for selected locations in the Aegean, Cretan and Ionian Seas and the corresponding mean tidal levels (meteorological and astronomical)*

No.	Gauge station	M2 cm	degrees	S2 cm	degrees	Mean water level variation (m)
1	Alexandroupolis	7.1	73	5.0	92	0.66
2	Kavala	5.3	89	3.5	104	1.11
3	Thessaloniki	9.0	78	6.1	98	0.87
4	Chios	4.4	49	2.9	71	0.49
5	Rafina	2.0	74	1.0	86	0.63
6	Piraeus	1.0	63	0.5	63	0.61
7	Syros	2.0	42	1.0	57	0.52
8	Leros	2.1	305	1.3	316	0.40
9	Souda	1.0	348	1.8	368	0.44
10	Heraklion	1.5	304	1.1	69	0.45
11	Rhodes	4.4	257	2.7	268	0.59
12	Kalamata	2.2	69	1.1	69	0.60
13	Katakolo	3.3	62	1.7	35	0.50
14	Lefkas	4.0	79	2.2	85	0.46
15	Preveza	2.0	127	0.9	135	0.50

Wave regime

The offshore wave regime of the Aegean Sea has been measured by the POSEIDON environmental monitoring system of the Hellenic Centre for Marine Research and presented schematically in Figure 18.4.

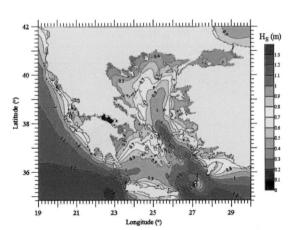

■ **Figure 18.4** *Mean annual significant wave height for the Aegean and Ionian Seas (modified from Soukissian et al., 2007)*

The average annual significant wave height is 0.4 m for Thrace and East Macedonia, 0.5 m for Central Macedonia, 0.6 m for Thessaly and 0.9 m for the North Aegean Islands. Along the coastline of the Cyclades and the Dodecanese islands, the average annual significant wave height is higher (from 0.9–1.1 m), becoming smaller (0.5 m) to the West, along the eastern coasts of Attica and the Peloponnesus. The average annual significant wave height in the Ionian Sea is 1.0 –1.1 m. Finally, the average annual significant wave height for the northern coastline of the island of Crete is 0.6 m in the central section, becoming 0.8 m towards its eastern and western sections.

Coastal erosion

Coastal erosion is defined as the long-term loss of coastal land relative to a fixed reference line. It is a very complex phenomenon, caused by a variety of natural processes and anthropogenic mechanisms. Furthermore, the time of occurrence and the rate of coastal erosion are not easy to predict with accuracy. Even in areas which are under constant erosion, the intensity of the phenomenon varies in time. Also, erosion rates are not the same over the whole coastline, as erosion depends on a variety of regional coastal factors, such as geomorphology, material (rock or sediment), slope, climatic and wave conditions, frequency of extreme events and anthropogenic factors (Pranzini and Rossi, 1995; Khalil, 1997; Berriolo *et al.*, 2001; Medina and Lopez, 1997).

Eustatic sea level change is related to natural factors, but also to human intervention, mainly as a consequence of the greenhouse effect (Wigley, 1999). However, anthropogenic factors often play an important role in coastal erosion. Human interventions in the natural processes may take the form of extraction of materials (sand and gravel) from beach zones or river banks, dam construction (e.g. Poulos *et al.*, 2002) and the construction of harbours and coastal defence structures, which usually interrupt the coastal zone sediment transport and alter the system's equilibrium.

Coastal erosion in Greece

The majority of the coastline appears to be erosional according to EUROSION (2004; Table 18.3) and the recent study by Alexandrakis *et al.* (2010; Figure 18.5).

Approximately 25 per cent of the Aegean coastline, consisting mainly of beach zones and low-lying coastal (including deltaic) plains, is undergoing erosion. The northern coastline of Crete, in particular, experiences extensive erosion, affecting 65.8 per cent of the coast; this large percentage of eroding coastline is due to the presence of extensive beach zones (from 1–10 km shoreline length) and the high frequency of occurrence of large waves generated by northern and northeastern winds in the Aegean Sea. In contrast, the South Aegean islands have the smallest percentage (~14.7 per cent) of coastal erosion.

■ **Table 18.3** *Erosion in the Greek coastal regions (from Eurosion, 2004)*

AREA	Coastline length (km)		
	Total	Eroding	%
E. Macedonia & Thrace	436.0	139	31.9
C. Macedonia	821.8	371	45.1
Thessaly	697.3	256	36.7
Epirus	313.5	106	33.8
Ionian Islands	1,065.9	260	24.4
W. Greece	859.3	198	23.0
Sterea Hellas	1,491.8	582	39.0
Peloponnesus	1,164.1	306	26.3
Attica	1,047.9	237	22.6
N. Aegean	1,311.3	231	17.6
S. Aegean	3,423.2	503	14.7
Crete	1148.3	756	65.8
Total	13,780.4	3,945	28.6

Coastal vulnerability to future sea level rise

A first step towards the protection of beach zones from erosion, induced by the anticipated rise of sea level, is the spatial identification and quantification of the problem. For this purpose, Alexandrakis *et al.* (2010) studied the vulnerability of the Aegean coastline to sea level rise, by using the Coastal Vulnerability Index (CVI) method developed by Thieler and Hammar-Klose (1999). Apart from the rate of sea level rise, which obviously plays an important role, the controlling factors for the CVI classification scheme are coastal geomorphology (*i.e.* coastal landforms and slopes) and offshore wave climate. Alexandrakis *et al.* (2010) found that, in the case of a sea level-rise up to 1.8 mm/yr, almost half of the Aegean coast is classified as moderately vulnerable with the remaining part being classified as highly vulnerable. Thrace and Eastern Macedonia, most of Central Macedonia, Thessaly and Evia are the less vulnerable coastal regions, while the northern coast of Crete and most of the remaining southern Aegean coasts (including Attica) present the highest

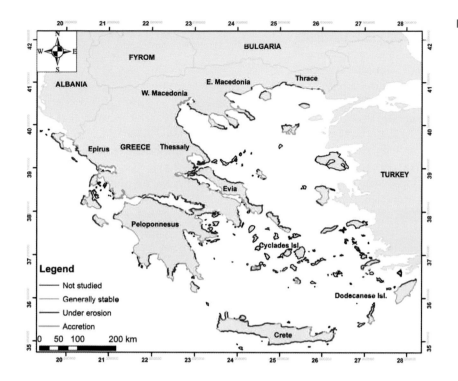

■ **Figure 18.5**
*Erosional trends
along the Hellenic
coastline (modified
from Alexandrakis
et al., 2010)*

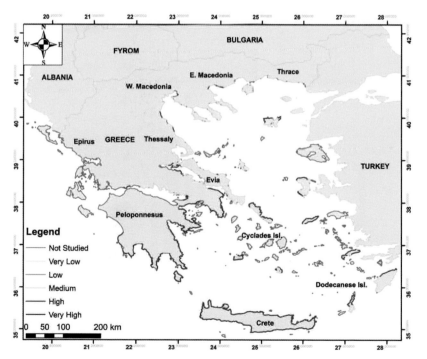

■ **Figure 18.6**
*Categorization of
the vulnerability of
the Aegean coastline
for sea level rise
(SLR) >3.5 mm/yr
(modified from
Alexandrakis et al.,
2010)*

vulnerability. This is attributed to the different coastal morphology types, mainly sand and gravel beaches (as in the case of Crete), and the higher incoming waves, compared to the northern part of the Aegean Sea. For the case of a sea level rise at a rate >3.5 mm/yr, as predicted by the IPCC (2007) report for the next 100 years, all the Hellenic Aegean coast, with the exception of some coastal segments of Thessaly and Evia, will have high to very high vulnerability (Figures 18.6 and 18.7). One-third of the total length of the Aegean coastline is very highly vulnerable and more than 50 per cent highly vulnerable, with the remaining being classified as medium vulnerable.

Even though the outcome of the above vulnerability assessment for the Hellenic Aegean coast provides a useful tool for coastal zone management, it has to be used with caution in the case of smaller-scale coastal studies, where local conditions may be the decisive factors.

Hellenic beach zones under erosion

Rationale

Beaches, characterized by their sediment material, wave and tidal regimes, range from narrow and steep (reflective) to wide and flat (dissipative), as grain size becomes finer and waves and tides larger; most beaches are intermediate between these extremes (Short, 1999; Finkl, 2004). Reflective beaches are coarse-grained and have no surf zones, whereas dissipative beaches have finer sediments and extensive surf zones. Filtration volumes are higher on permeable reflective beaches, mainly driven by wave action, and lower on dissipative beaches (McLachlan and Turner, 1994).

Beach systems erode only if the supply of sand or other sediment gained is less than the amount of sediment lost. Seasonal erosion, the reduction or complete loss of sand on the visible beach, during winter storms for example, does not usually imply a loss of sand from the system. Sediment can be supplied from rivers, cliff and shore erosion, offshore

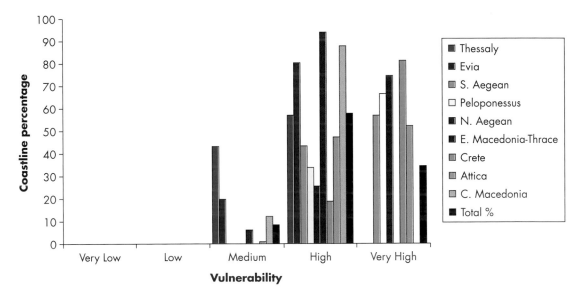

■ **Figure 18.7** *Classification of the coasts of the administrative prefectures of Greece into the five categories of vulnerability, for the case of SLR >3.5 mm/yr for the next 100 years (modified from Alexandrakis et al., 2010)*

or alongshore sources or from materials blown from the land. In addition, humans can supply beaches with sand via beach nourishment (Bird, 1995).

The most important factors that control beach erosion are waves generated by winds and storms, currents (longshore and cross-shore), relative sea level rise, beach grain size, beach slope and changes in riverine inputs (PAP/RAC, 2000). In addition to the above, climate change (e.g. SLR, storm frequency or intensity) and anthropogenic factors (including coastal defence structures, land use changes and dam construction in river drainage areas, dredging of the sea floor and removal of vegetation in dune fields) can intensify the beach erosion processes (e.g. Griggs et al., 1994; Sutherland et al., 2003; Hsu et al., 2007).

Modern erosion of the Greek (Hellenic) beaches

The beach zones of Greece are globally known for their scenic beauty and draw large numbers of tourists, both domestic and foreign. Beach zones represent 35.4 per cent of the total length of the Hellenic coasts, with 3.3 per cent being characterized as 'pocket' beaches (Table 18.4). In Epirus, beaches represent 19.6 per cent of the coast, which is the smallest percentage in Greece, whilst the highest percentage

is found in Macedonia (46.8 per cent). In the islands, due to irregular coastal morphology, beach zones are mainly pocket beaches (EUROSION, 2004).

One effect of large-scale socio-economic development in Greece has been a marked decrease in sediment supply from existing natural sources. Construction of dams, river channelization and intense coastal development (more than 1,000,000 houses built in 30 years) greatly reduced the riverine sediment supply by eliminating flows and trapping sediment behind dams, as well as reducing the ability of streams to further erode their channel beds. Extensive shoreline retreat has been associated with deltaic coast erosion, mainly following dam construction, as shown in the cases of the Gallikos River and the old mouth of the Axios river in the inner Thermaikos Gulf (Poulos et al., 2000; Kapsimalis et al., 2005) and the river mouth areas of the Alfios (Ghionis et al., 2004b), Sperchios (Psomiadis et al., 2005), Nestos (Xeidakis and Delimani, 2003), Kalamas and Arachthos rivers (Kapsimalis et al., 2005; for locations see Figure 18.8).

Beach erosion has also been profound for beaches open to (unprotected from) incoming offshore waves (see Table 18.5). Shoreline retreat since the 1990s in most of these beaches ranges from a few metres to more than 10 m and is partly attributed to human intervention in the nearshore hydrodynamics

■ **Table 18.4** Coastline lengths and types of beaches

Region	Coastline length (km)	Beaches	
		Pocket (%)	Open (%)
Epirus/Ionian	2158.3	2.69	19.60
Sterea Hellas	1491.8	4.42	34.40
Peloponnesus	1164.1	3.93	43.13
Thessaly	697.3	7.61	29.93
Macedonia	931.8	1.95	46.83
Thrace	326.0	0.25	45.43
Crete	1148.3	2.50	26.50
Cyclades	3423.2	4.60	45.08
Aegean Islands	1311.3	3.45	32.70
Total	13,700.0	3.33	35.40

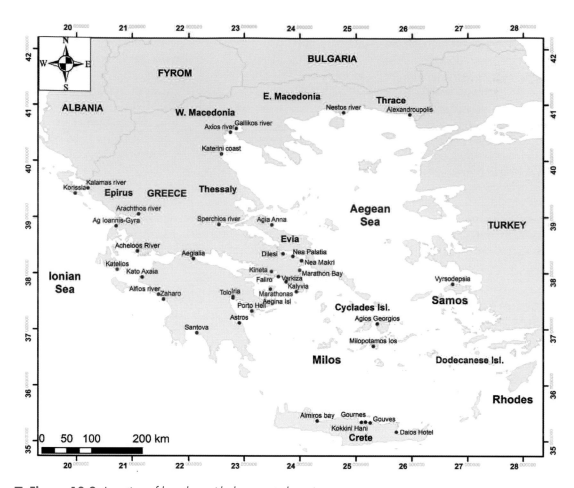

■ **Figure 18.8** *Location of beaches with documented erosion*

through artificial shoreline constructions. Beaches located in semi-enclosed embayments (*i.e.* protected from high waves) are also under erosion (see Table 18.5). In all cases, erosion has been induced mainly by human interference, the most characteristic being road construction close to the shoreline, establishment of fishing shelters and development of tourist marinas. Signs of beach erosion have been observed even in pocket beaches (see Table 18.5). Although shoreline retreat accounts for several metres (usually <10 m) over time periods of 20–30 years, erosion of pocket beaches is usually associated with direct and/or indirect human interference.

The geographical distribution of the eroding beach zones, excluding those related to deltaic retreat after damming and those with profound localized human interference, but including those located in tectonically stable areas, such as the Cyclades islands (Poulos *et al.*, 2009b), leads to the conclusion that their retreat may have been induced (at least partially) by a generalized cause, which could well be the recent increase of the sea-level rise rate from <1 mm/yr to >3 mm/yr since the 1990s (IPCC, 2007).

■ **Figure 18.9** *Examples of Hellenic beaches under erosion: (a) Iria (Argolis); (b, c) Skala Oropou area (Attica); (d) Zaharo beach; (e) Ag. Ioannis beach (Lefkada Island) and (f) Aegeira (Aegialia). For locations see Figure 18.8*

■ **Table 18.5** Examples of Hellenic beaches under erosion (for locations see Figure 18.8)

Open beaches	Semi-enclosed beaches	Pocket beaches
Alexandroupolis (N. Aegean) Xeidakis *et al.*, 2005	Dilesi (S. Evoikos) Poulos *et al.*, 2002	Katelios Bay (Kefalonia Isl., Ionian Sea) Tsoutsia *et al.*, 2010
Katerini (Thermaikos Gulf) Hatziopoulos, 2004	Nea Palatia (S. Evoikos) Poulos *et al.*, 2011	Varkiza beach (Saronikos Gulf) Karympalis *et al.*, 2002
Ag. Ioannis — Gyra (Lefkada Isl.) Verykiou *et al.*, 2008	Nea Makri (S. Evoikos) Chatzieleftheriou *et al.*, 2007	Santova (Messiniakos Gulf) Giannakopoulos *et al.*, 2005
Kokkini Hani - Gouves (N. Crete) Bouziotopoulou *et al.*, 2006	Marathon Bay (S. Evoikos) Poulos *et al.*, 2004a	Milopotamos (Ios Isl.) Karagkiozoglou *et al.*, 2010
Almiros Bay (N. Crete) Petrakis *et al.*, 2010	Iria (Argolikos Gulf) Giannouli *et al.*, 2010	Marathonas (Aegina Isl.) Adamopoulou and Dimopoulos, 2008
Korissia lagoon (Kerkyra Isl.) Milovanovic *et al.*, 2008	Tolo (Argolikos Gulf) Papadaki, 2001	Kalyvia (SE Attica) Koutsomichou *et al.*, 2009
Agia Anna - Krya Vrysi (NE Evia) Poulos *et al.*, 2004b	Kineta beach (Saronikos Gulf) Papadaki *et al.*, 2010	Vyrsodepsia (Samos Isl.) Bleta *et al.*, 2009
Aegialia (Korinthiakos gulf) Ghionis and Ferentinos, 2002	Astros Kynourias (Argolikos Gulf)	
	Kato Achaia (NW Peloponnessus) Poulos and Chronis, 2001	
	Porto Heli (Argolikos Gulf) Poulos *et al.*, 2009	

Assessment of beach zone retreat due to future SLR

Monioudi *et al.* (2009) have estimated the loss of coastal land of the Hellenic beach zones due to a sea level rise of 50 cm, for three different slopes and by the use of six different methodologies (Table 18.6).

On the basis of Table 18.6, a gross estimate of land loss of a substantial number of Hellenic beaches, for which morphometric characteristics are available, has been computed (Table 18.7).

Among open beaches, the one with the highest erosion potential is the Astros beach zone, located in the central part of the eastern coast of the Peloponnesus, with an estimated land loss of 50.9 per cent. This is due to the fact that the average beach width in the area where the data was taken is only 7 m. The minimum estimated value of land loss is found in the Falasarna beach zone (3.6 per cent), which has an average sub-aerial beach width of 120 m. The average percentage of land loss of the open beach zones is 29.1 per cent.

Among pocket beaches, the Afandou beach zone presents the least loss (10.2 per cent), when the average projected land loss for the pocket beaches is 23.4 per cent. The Firiplaka pocket beach, in Milos Island, is projected to lose 89.2 per cent of the beach zone, in the case of a 50 cm sea level rise.

■ **Table 18.6** *Estimated beach zone retreat (m) for different beach slopes due to a SLR of 50 cm (according to Monioudi et al., 2009)*

Method Slope	Bruun (1962)	Edelman (1972)	Kriebel and Dean (1993)	SBEACH (Larson and Kraus, 1989)	Energy Models (Leont'yev, 1997)	Boussinesq (Karambas and Koutitas, 1992)	Average
1:10	5.0	11.2	5.4	6.0	7.7	7.5	7.13
1:20	10.0	13.8	6.7	10.9	10.8	14.0	11.03
1:30	15.0	15.1	8.3	15.2	15.2	18.5	14.55

■ **Table 18.7** *A gross estimate of the land loss (expressed in percentages) of Hellenic beach zones of known morphometric characteristics and according to the average values of shoreline retreat given by Monioudi et al., (2009) (see Table 18.6). Beach zone locations are shown in Figure 18.10*

No	Beach zone Name	Type	Beach width (m) MIN	MAX	AVG	Beach slope (degrees) MIN	MAX	AVG	Land loss %
1	Kamiros [i]	Pocket	5	10	7.5	6	8	7	71.3
2	Ialysos [i]	Open	5	15	10	10	10	10	47.6
3	Elli [i]	Pocket	30	65	47.5	7	15	11	11.0
4	Afandou [i]	Open	50	70	60	5	7	6	10.2
5	Lardos [i]	Pocket	15	30	22.5	8	10	9	23.8
6	Falasarna [ii]	Open	40	200	120	9	14	11.5	3.6
7	Georgioupolis [iii]	Open	50	70	60	5	10	7.5	10.2
8	Rethymnon [ii]	Pocket	20	80	50	5	10	7.5	8.9
9	Ammoudara [iv]	Open	22	60	41	6	12	9	11.9
10	Gouves [v]	Open	13	35	24	7	8	7.5	20.4
11	Sitia [ii]	Open	5	15	10	5	6	5.5	47.6
12	Vai [ii]	Pocket	15	40	27.5	10	15	12.5	17.8
13	Ierapetra [ii]	Open	10	30	20	5	10	7.5	23.8
14	Agia Roumeli [ii]	Pocket	15	40	27.5	8	10	9	17.8
15	Elafonisi [ii]	Open	10	70	40	4	6	5	10.2
16	Marathonas-Aegina [vi]	Pocket	10	25	17.5	9	10	9.5	28.5
17	Alfios [vii]	Open	20	34	27	6	8	7	21.0
18	Ag. Petros Andros [viii]	Pocket	15	35	25	10	12	11	20.4
19	Astros [ix]	Open	4	15	7	6	9	7.5	50.9
20	Kineta [x]	Open	12	35	23.5	6	9	7.5	20.4
21	Marathon Bay [xi]	Open	6	41	23.5	7	11	9	17.4
22	Korissia Kerkyira [xii]	Open	15	30	22.5	8	11	9.5	23.8
23	Ag. Ioannis-Gyra [xiii]	Open	20	50	35	6	20	13	14.3
24	Paxaina [xiv]	Pocket	17	33	25	3	3	3	21.6
25	Firiplaka [xiv]	Pocket	6	8	7	2	4	3	89.2
26	Thiorixia [xiv]	Pocket	13	33	23	4	6	5	21.6
27	Achivadolimni [xiv]	Pocket	17	36	27	4	6	5	19.8
28	Vyrsodepsia [xv]	Pocket	11	21	16	7	9	8	34.0

■ **Table 18.7** *continued*

No	Beach zone		Beach width (m)			Beach slope (degrees)			Land loss %
	Name	Type	MIN	MAX	AVG	MIN	MAX	AVG	
29	Tsampou [xv]	Open	15.0	26	20.5	2	4	3.0	27.4
30	Potokaki [xv]	Open	19.4	28	23.7	5	7	6.0	25.5
31	Alexandroupolis [xvi]	Open	12.0	25	18.5	4	7	5.5	28.5
32	Agia Anna Evia [xvii]	Open	5.0	25	15.0	6	9	7.5	35.7
33	Pieria [xviii]	Open	19.0	40	29.5	5	9	7.0	17.8
34	Vatera [xix]	Pocket	22.0	50	36.0	6	9	7.5	14.3
	Average pocket beaches	17.6	45.7	31.6	5.9	9.4	7.6	23.4	
	Average open beaches		15.1	36.1	25.6	6.4	9.1	7.7	29.1
	Total Average		16.6	41.7	29.1	6.1	9.2	7.6	25.7

References. (i): Pyöokäari (1997); (ii): Pyöökaäri (1999); (iii): Petrakis *et al.*, 2010; (vi): Alexandrakis *et al.* (2007); (v): Bouziotopoulou *et al.* (2006); (vi): Adamopoulou and Dimopoulous (2008); (vii): Ghionis *et al.* (2004b); (viii): Karagiozoglou *et al.*, (2010); (ix) Markakis *et al.* (2010); (x): Papadaki *et al.*, (2010); (xi): Poulos *et al.* (2004a); (xii): Milovanovic *et al.* (2008); (xiii): Verykiou *et al.* (2008); (xiv): Andris and Poulos (2007); (xv): Bleta *et al.* (2009); (xvi): Karditsa (2006); (xvii): Chatziopoulos (2004); (xix) Vousdoukas (2007)

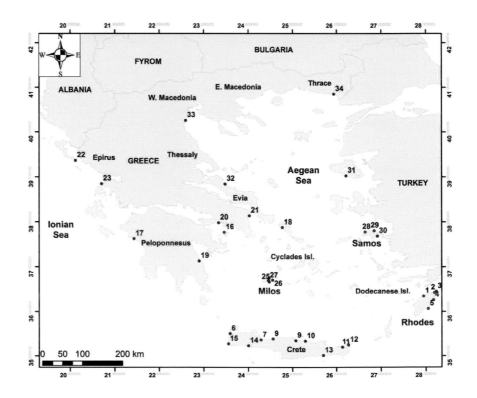

■ **Figure 18.10**
Locations of beach zones in Greece with estimated land loss, in relation to beach width and beach slope (numbers 1–34 correspond to the beach names given in Table 18.6)

Legislation and protection measures in Greece

The Greek state vision regarding protection and development of its coastal areas has been depicted in Law 2971/2001 (Greek Government, 2001). Among other things, the law regulates the inland limits of the seashore and beach zones, where private construction is strictly prohibited. The seashore limit is defined as a line connecting the points of the usual maximum wave run up on the beachface, and the beach zone extends up to 50 m landwards of the seashore limit. The beach zone is regarded as public property that has to be accessible by anyone. Permanent constructions in the seashore and beach zones are prohibited, with a few exceptions for public interest, safety and national security reasons. Protection of the beach zone from erosion is briefly addressed in article 12 (Law 2971/2001) that states: 'In the case of beach erosion, hard protection measures can be exercised and if a private owner is threatened by erosion, he can construct hard protection structures at his own expense. The structures belong to the state'. The constitutional mandate that the state guarantees the protection of life and private property of its citizens ensures permission for the construction of coastal protection structures, but only structures protecting public property are paid for by the state. In order to secure a sustainable coastal development and to protect life and property, a full environmental impact assessment study is required for any coastal zone construction.

Coastal defence covers two related but distinct issues: (i) coast protection, concerned with works designed to manage or prevent erosion of the coastline; and (ii) sea defence, which relates to schemes intended to manage or prevent flooding or inundation of the coastline. Today, coast protection works are the most common type of coastal defence, mainly consisting of structures designed to resist natural processes, such as wave action and sediment movement; commonly called 'hard engineering' options, they include various constructions such as vertical sea walls, groins, breakwaters, revetments, flood embankments, placement of gabions and rock armouring. Approximately 15 per cent of the eroding (4.2 per cent of the total) shoreline length of Greece is artificially protected and 4 per cent of the eroding (1.1 per cent of the total) shoreline length continues to erode despite the protection measures.

Along the Hellenic coast, vertical seawalls (built of concrete or rock) usually accompany and protect coastal roads (e.g. the coastal road at Nea Makri; Chatzieleftheriou et al., 2007). In many cases, especially where the beach zone is narrow, rock armouring is used to protect the seawall toe against undermining. The most common 'hard' defence currently employed to protect private property and infrastructure of low importance and economic value is designed walls and revetments constructed of boulders of a rather uniform size (rock armouring), typically many tonnes in weight.

Similar structures, formed from precast concrete blocks, are commonly used for protection against coastline retreat of reclaimed coastal lands, coastal installations of greater importance, or areas of higher aesthetic value, such as the 'Peace and Friendship' Stadium and the Olympic Beach Volley Stadium at Faliron Bay (Attica) (Figure 18.11g, h). In rare cases, gabions (wire baskets filled with stone and stacked vertically or damped horizontally) have been used temporarily for shore zone stabilization, as in the case of Xilokastro (Poulos, pers. com.). In other cases, the dumping of piles of rocks tipped over the coastal edge has been applied to prevent further erosion in areas of low conservation value.

The second most popular protection method during the past decades has been the construction of groins, generally perpendicular to the coastline, designed to intercept sand and gravel movement along the beach; these have usually been constructed of boulders with a concrete or asphalt-top pavement in areas of high wave activity, e.g. Gournes beach (Figure 18.11d), or of timber in beach zones that experience relatively low wave energy, such as Dilesi beach (Poulos et al., 2002).

Recently, concrete-filled geotextile tubes have been used for construction of less obtrusive groins in recreational beaches (Figure 18.11e), with limited success, especially in beaches with significant on–offshore transport.

■ **Figure 18.11**
*Common coastal
protection measures
applied along the
Hellenic coastal zone.
(a) groins at Kokkini
Hani, (b) Kato Achaia,
(c) Preveza and (d)
Gournes; (e) groins
made of concrete-filled
geotextile tubes at
Dilesi; (f) emerged
offshore breakwaters
at Skyros; (g) offshore
breakwater made
of precast concrete
blocks for the
protection of the
Kifissos River outfall
at Faliron Bay and
vertical seawall for
the protection of the
reclaimed land in the
area of the Olympic
Beach Volleyball
Stadium; (h) seawall
and breakwater in
front of the Olympic
Beach Volleyball
Stadium; (i) vertical
sea wall and beach
erosion at Gouves,
and (j) the same beach
after nourishment.
(Orthophotographs a,
b, c and g: Copyright
2012, Hellenic
Cadastre,
Ktimatologio S.A.
(modified); Image f
courtesy A. Ntouroupi)*

■ **Figure 18.12** *Replacement of groins by submerged offshore breakwaters: (a) Katerini beach zone with groins; (b) construction of submerged breakwaters; (c) Katerini beach zone with the groins removed; and (d) offshore breakwater crest (Photographs a, c and d. courtesy Assoc. Prof. Th. Karambas. Image b prepared by N. Danelon)*

Over the last one or two decades, groin construction has gradually been abandoned, mainly due to their limited success in areas with low longshore sediment transport rates, their obtrusiveness and problems with downdrift erosion. In open beaches with a significant onshore component in the nearshore wave energy field, construction of emerged shore-parallel offshore breakwaters has been quite successful (Figure 18.11f).

Submerged offshore breakwaters have becoming increasingly popular during the last two decades and in some cases they have successfully replaced existing groin systems, as in the coastal zone of Katerini (Figure 18.12).

In areas with a high conservation value, 'soft engineering' options are often preferred. These are designed to emulate, harness or manipulate natural processes. The most commonly used 'soft' protection methods are beach nourishment or recharge, sediment recycling and stabilization of coastal dunes with vegetation.

Beach nourishment (addition of sand or gravel to a beach to restore former levels or to improve current ones) has been used primarily for the maintenance or creation of beaches associated with large hotel units or coastal recreational facilities (Figure 18.13a, b), either as a standalone measure or in conjunction with a 'hard protection' measure, usually a detached

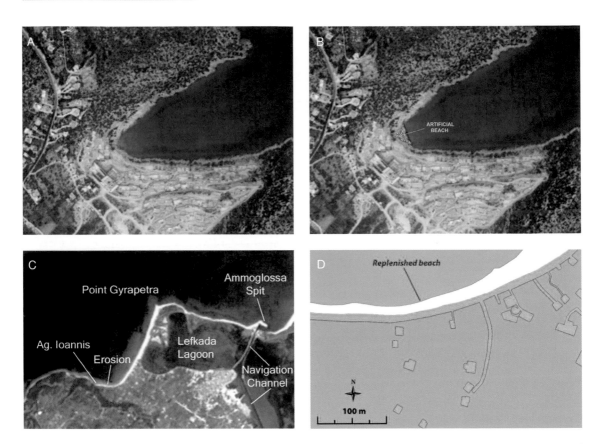

■ **Figure 18.13** *Beach nourishment for the creation of the artificial beach of Daios Hotel (NE Crete) (a, b); sand recycling in North Lefkada Island (c, d). Sediment is periodically transported from the Ammoglossa area (spit at the eastern end of the beach) to the Ag. Ioannis beach zone (at the western end of the beach). (Orthophotograph a: Copyright 2012, Hellenic Cadastre, Ktimatologio S.A. (modified); image b: same as image a with superimposed photograph of the 2010 artificial beach; image c courtesy of U.S. Geological Survey; image d prepared by N. Danelon)*

submerged breakwater. In many cases, such actions were undertaken during the low-season tourist periods either without the required permissions or by public authorities without proper studies (e.g. Gournes, Crete). Sediment recycling (transport of beach sediment from the down-drift end of a beach back to its up-drift end) has been used in beaches with significant longshore sediment transport, such as the Agios. Ioannis-Gyra beach zone on Lefkada Island (Figure 18.13c, d).

Additionally, other semi-technical semi-managerial schemes may be considered that incorporate coastal defences by combining elements of hard and soft engineering and a managed back retreat (removal of coastal defences inland to permit the natural evolution of a beach and, if the coastline is retreating farther, supply of beach sediment).

Conclusion

Nearly one-third (28.6 per cent) of the Hellenic coastline, approximately 36 per cent of which consists of beach zones, is under erosion. Approximately 15 per cent of the eroding (4.20 per cent of the total) shoreline length of Greece is artificially protected and 4 per cent of the eroding (1.1 per cent of the total) shoreline length continues to erode despite the protection measures. The geographical distribution of the eroding beaches (including tectonically stable coasts and taking into consideration the pronounced direct human intervention, such as river damming, coastal constructions, etc.) leads to the conclusion that their retreat is associated with a generalized cause, which could well be the recently increased rate of sea-level rise, attributed to climate change.

Mitigation of coastal erosion problems has to be associated with broader based beach management schemes that may have both a regional and a national scale of implementation. This is even more necessary in the case of the anticipated sea level rise (from 0.28 up to 0.60 m until the year 2100, according to the IPCC, 2007) that is expected to affect primarily low-lying coastal zones and in particular the beach zones. From a conservation perspective, proposals for coastal-defence should also assess matters related to preventing or minimizing loss of habitat, landforms or rock exposure, minimizing interference with natural processes, disturbance to wildlife and the impacts upon the landscape and amenity.

Even though the existing legislation for coastal zone protection is adequate, Greece has to develop a modern legal framework for sustainable management of its coastal zone; the latter may include a definition of coastal zones based on socio-economic grounds, spatial and terrestrial data collection, integration in a reference topographic frame, computation of natural (e.g. wave climate assessment and sediment balance estimation) and manmade pressures and integration of urban development data.

References

Adamopoulou, A. and Dimopoulou, N., 2008. *Hydrodynamic and sedimentological regime in sandy beaches of Egina Isl.* B.Sc. Thesis (unpublished), Department of Geology and Geoenvironment, National and Kapodistrian University of Athens, 86 pp.

Alexandrakis, G., Poulos, S., Ghionis, G. and Livaditis, G., 2007. A morphological study of a reef with beachrock characteristics in association with the recent evolution of the Ammoudara beach zone (Heraklion, Crete). *Bulletin of the Geological Society of Greece*, 39(3):146–155.

Alexandrakis, G., Karditsa, A., Poulos, S., Ghionis, G. and Kampanis, N. A., 2010. An assessment of the vulnerability to erosion of the coastal zone due to a potential rise of sea level: The case of the Hellenic Aegean coast. In Achim Sydow (ed.), *Encyclopedia of life support systems (EOLSS): Environmental systems*, developed under the auspices of UNESCO, EOLSS Publishers, Oxford, UK. [http://www. eolss.net]

Anastassopoulos, G., Filippaios, F. and Phillips, P., 2009. An eclectic investigation of tourism multinationals: Evidence from Greece. *International Journal of Hospitality Management*, 28(2):185–194.

Andris, P. D. and Poulos, S. E., 2007. Geoenvironmental study of the tourist beaches of the Milos Island. *Proceedings of the Eighth Panhellenic Geographical Congress, Athens, 5–8 October*, Geographical Society of Greece, 1:219–228. [in Greek with English abstract]

Berriolo, G., Fierro, G. and Gamboni, S., 2001. The evolution of the coast between Cape Farina and Cape Gamarth (Northern Tunisia). *Proc. of the Fifth International Conference on the Mediterranean Coastal Environment, MEDCOAST 01* (Ed. E. Özhan), MEDCOAST Secretariat, Middle East Technical University, Ankara, Turkey, 3:1497–1502.

Bird, E. C. F., 1995. Present and future sea level; the effects of predicted global changes. In: D. Eisma (ed.), *Climate change: Impact on coastal habitation*, Lewis Publishers, Boca Raton, CA, 29–56.

Bleta, A., Andris, P., Karditsa, K., Evelpidou, N., Poulos, S. E. and Leivaditis, G., 2009. A study of erosion of the beach zones of the N/NW coast of Samos Island. *Proceedings of the Ninth Symposium on Oceanography and Fisheries*, I:184–189. [in Greek]

Bouziotopoulou, N., Ghionis, G. and Poulos, S. E., 2006. Morphological changes in the shore zone of Gouves-Gournes (N. Crete) due to alterations of its sediment budget caused by human activities. *Bulletin of the Geological Society of Greece* 39/3: 135–145. [in Greek]

Bruun, P., 1962. Sea level rise as a cause of shore erosion. *Journal of Waterways and Harbors Division, ASCE*, 88:117–130.

Chatzieleftheriou, M., Alexandrakis, G., Gaki, K., Maroukian, H. and Poulos, S. E., 2007. Assessment of the vulnerability to a future sea level rise of the E and NE coast of Attica. *Proceedings of the Eighth Panhellenic Geographical Congress, Athens, 5—8 October,* Geographical Society of Greece, 1:298–305.

Chatziopoulos, E., 2004. *Sedimentological study and morphodynamic modeling of the Pieria coastline, with the use of a mathematical sediment transport model for non-cohesive sediments.* Unpublished MSc thesis. School of Science, National and Kapodistrian University of Athens, 146 pp.

Coccossis, H. and Constantoglou, M. E., 2005. The need of spatial typologies in tourism planning and policy making: The Greek case. *45th Congress of the European Regional Science Association: "Land Use and Water Management in a Sustainable Network Society",* Vrije Universiteit, Amsterdam. Available at: http://www-sre.wu-wien.ac.at/ersa/ersaconfs/ersa05/papers/693.pdf (accessed 25 August 2012).

Coccossis, H. and Mexa, A., 1997. *Coastal management in Greece,* Hellenic Ministry for the Environment, Physical Planning and Public Works, Athens.

Coccossis, H. and Mexa, A., 2002. The coastal zone. In: H. Coccossis (ed.), *Man and the environment in Greece,* Hellenic Ministry for the Environment, Physical Planning and Public Works, Athens, 74–81. [in Greek]

Edelman, T., 1972. Dune erosion during storm condition. *Proceedings of the 13th International Conference on Coastal Engineering,* ASCE, 1305–1312.

EUROSION, 2004. *Living with coastal erosion in Europe: Sediment and space for sustainability. Part II – Maps and statistics.* Directorate General Environment, European Commission, 25 pp.

Finkl, C. W., 2004. Coastal classification: Systematic approaches to consider in the development of a comprehensive scheme. *Journal of Coastal Research,* 20:166–213

Ghionis, G. and Ferentinos, G., 2002. Wave-induced shoreline changes along the coast of Aigialeia (Greece). *Proceedings of the 6th Panhellenic Geographical Congress, Thessaloniki, 3–6 October,* Geographical Society of Greece, 1:46–53. [in Greek with extended English abstract]

Ghionis, G., Poulos, S. E., Bouziotopoulou, N. and Dounas, K., 2004a. The role of a natural submerged breakwater in controlling the hydrodynamic and sedimentological conditions of Ammoudara beach (Herakion, Crete). *Proceedings of the 7th Panhellenic Geographical Congress, Mytilini, 14–17 October,* Geographical Society of Greece, A:436–443. [in Greek with an English abstract]

Ghionis, G., Poulos, S. E., Gialouris, P. and Giannopoulos, Th., 2004b. Recent morphological evolution of the deltaic coast of R. Alfios, due to natural processes and human impact. *Proceedings of the 7th Panhellenic Geographical Congress, Mytilini, 14–17 October,* Geographical Society of Greece, A:302–308. [in Greek with an English abstract]

Gianakopoulos, E., Barakos, N. and Tsakalos, D., 2005. *Coastal geomorphological, sedimentological study of the Santova beach (NE Messinian Gulf).* BSc Thesis (unpublished), National and Kapodistrian University of Athens, 138 pp.

Giannouli, D.-I., Poulos, S. E., Andris, P. and Petrakis, S., 2010. A study of the erosion of the Iria beach zone (Argolic Gulf), with emphasis on human interference. *Proceedings of the 9th Panhellenic Geographical Congress, Athens, 4–6 October,* Geographical Society of Greece, 68–75.

Greek Government, 2001. Law 2971: Seashore, beach and related issues. *Bulletin of the Government of the Hellenic Republic,* A(285):3985–3998. [in Greek]

Griggs, G. B., Tait, J. F. and Corona, W., 1994. The interaction of seawalls and beaches: seven years of monitoring, Monterey Bay, California. *Shore and Beach,* 63(4):32–38.

Hatziopoulos, E., 2004. *A sedimentological study and simulation of morphodynamic characteristics of Pieris coast with the application of mathematical simulation of non-cohesive coastal sediment transport.* M.Phil. Thesis (unpublished), School of Science, National and Kapodistrian University of Athens, 178 pp.

Hsu, T., Lin, T. and Seng, I., 2007. Human impact on coastal erosion in Taiwan. *Journal of Coastal Research*, 23:961–973.

IPCC, 2007. Climate change: Impacts, adaptation and vulnerability. In M. L. Parry, O. F. Canziani, J. P. Palutikof, P. J. van der Linden and C. E. Hanson (eds), *Contribution of Working Group II to the Fourth Assessment Report of the Intergovernmental Panel on Climate Change, 2007*, Cambridge University Press, Cambridge, UK.

Kapsimalis, V., Poulos, S. E., Karageorgis, A. P., Pavlakis, P. and Collins, M., 2005. Recent evolution of a Mediterranean deltaic coastal zone: Human impacts on the Inner Thermaikos Gulf, NW Aegean Sea. *Journal of the Geological Society*, 162: 897–908.

Karagiozoglou, K., Poulos, S. E., Verykiou, E. and Alexandrakis, G., 2010. Physico-geographical study of the beach zones Gialou and Mylopotamou of the Ios island. *Proceedings of the 9th Panhellenic Geographical Congress, Athens, 4–6 November*, Geographical Society of Greece, 84–91.

Karambas, T. V. and Koutitas, C., 1992. A breaking wave propagation model based on the Boussinesq equations. *Coastal Engineering*, 18:1–19.

Karditsa, A., 2006. *Environmental study of the coastal zone of the Alexandroupolis Gulf in relation to the successive development steps of the local harbor*. M.Phil. Thesis (unpublished), School of Science, National and Kapodistrian University of Athens, 124 pp.

Karympalis, Th., Kiousis, I. and Ntoutoulaki, I., 2002. *Seasonal oceanographic and sedimentological observations in Varkiza beach zone (Vari Bay)*. BSc Thesis (unpublished), Department of Geology, National and Kapodistrian University of Athens, 94 pp.

Khalil, S., 1997. Critical problems of the Egyptian Mediterranean coastal zones. In E. Özhan (ed.), *Proc. of the Third International Conference on the Mediterranean Coastal Environment, MEDCOAST 97*, MEDCOAST Secretariat, Middle East Technical University, Ankara, Turkey, 1:513–521.

Koutsomichou, I., Poulos, S. E., Evelpidou, N., Anagnostou, C., Ghionis, G. and Vasilopoulos, A., 2009. The role of beach rock formations in the evolution of embayed coastal zones of Attica

(Greece) in relation to sea level rise: The case of Kalyvia beach zone. *Geografica Fisica Dinamica Quaternaria*, 32:49–56.

Kriebel, D. L. and Dean, R. G., 1993. Convolution method for time-dependent beach-profile response. *Waterway, Port, Coastal and Ocean Engineering*, 119(2):204–227.

Lambeck, K. and Purcell, A., 2005. Sea-level change in the Mediterranean Sea since the LGM: Model predictions for tectonically stable areas. *Quaternary Science Reviews*, 24:1969–1988.

Larson, M. and Kraus, N. C., 1989. SBEACH: Numerical model for simulating storm-induced beach change. Report 1: Empirical foundation and model development, *Technical Report CERC- 89-9*, US Army Coastal Engineering Research Center, Vicksburg, MS.

Leont'yev, I. O., 1997. Short-term shoreline changes due to cross-shore structures: A one-line numerical model. *Coastal Engineering*, 31(1–4):59–75.

Markakis, E., Bleta, A. and Poulos, S. E., 2010. A near–shore sedimentological and oceanographic study of the beach zone of Paralio Astros Kinourias. *Proceedings of the 8th Panhellenic Geographical Congress, Athens, 5–8 October*, Geographical Society of Greece, 1:257–265. [in Greek with an English abstract]

McLachlan, A. and Turner, I., 1994. The interstitial environment of sandy beaches. *PZNI Marine Ecology*, 15:177–211.

Medina, J. M. and Lopez, J. S., 1997. Strong erosion scenario due to disequilibrium of solid transport rate: The case of Torrox Beach (Malaga). In E. Özhan (ed.), *Proc. of the Third International Conference on the Mediterranean Coastal Environment, MEDCOAST 97*, MEDCOAST Secretariat, Middle East Technical University, Ankara, Turkey, 2:1234–1239.

Milovanovic, M., Parharidis, I., Vasilakis, E. and Poulos, S. E., 2008. Study of the horizontal changes of the strip of land separating Korissia lagoon from the open sea using remote sensing methods. *Bulletin of the Geological Society of Greece*, 42(1):114–118.

Monioudi, I., Velegrakis, A. F., Karambas, Th. and

Koutsouvela, D., 2009. Coastal retreat due to the sea level rise: Morphodynamic models predictions. *Bulletin of the Geological Society of Greece,* 42(1):153–170. [in Greek with English abstract]

National Statistical Service of Greece, 2001. *Population census 1991 and 2001.* National Statistical Service of Greece, Athens.

PAP/RAC, 2000. *Report on the Albanian coastal erosion: PAP/RAC Mission to Albania to assess the problem of coastal erosion.* Consultants' Reports, UNEP/ MAP Priority Actions Program Regional Activity Centre, Split, Croatia, October.

Papadaki, P., Poulos, S. E., Kaleantopoulou, O., Koutelidaki, K. and Verykiou, E., 2010. A physico-geographical study of the coastal zone of Kineta (Saronikos Gulf). *Proceedings of the 8th Panhellenic Geographical Congress, Athens,* 5–8 October, Geographical Society of Greece, 1:275–283. [in Greek with an English abstract]

Papadaki, P., 2001. *Erosion of the Tolo coast (Nafplio).* M.Sc. Thesis (unpublished), School of Science, National and Kapodistrian University of Athens, Athens, 64 pp.

Petrakis, S., Alexandrakis, G. and Poulos, S. E., 2010. Investigation of coastline retreat due to anticipated sea-level rise in the case of beach zones exposed to high wave energy: Case study of Almiros Bay, N. Crete, Aegean Sea. *2nd EFMS Conference,* 15–16 October 2010, Athens, Greece, European Federation for Marine Science (poster).

Poulos, S. E., Alexopoulos, J. D., Karditsa, A., Giannia, P., Gournelos, T. and Livaditis, G., 2009a. Formation and evolution of Ververonda Lagoon (Porto Heli Region, SE Argolic Gulf) during historical times, on the basis of geophysical data and archeological information. *Zeitschrift für Geomorphologie,* 53(1):151–168.

Poulos, S. E. and Chronis, G., 2001. Coastline changes in relation to longshore sediment transport and human impact, along the coastline of Kato Achaia (NE Peloponnese, Greece). *Mediterranean Marine Science,* 2(1):5–13.

Poulos, S. E., Chronis, G., Collins, M. and Lykousis, V., 2000. Thermaikos Gulf Coastal-System, NW Aegean Sea: An overview of water/sediment fluxes in relation to air–land–ocean interactions and human activities. *Journal of Marine Systems,* 25:47–76.

Poulos, S. E., Ghionis, G. and Maroukian, H., 2009b. Sea-level rise trends in the Attico–Cycladic region (Aegean Sea) during the last 5000 years. *Geomorphology,* 107:10–17.

Poulos, S. E., Iordanis, K., Gourdoubas, I. and Pavlopoulos, K., 2004a. The sedimentary environment of the shore zone of the Schinias Bay (Marathonas Gulf). *Proceedings of the 7th Panhellenic Geographical Congress,* Mytilini, 14–17 October, Geographical Society of Greece, ∞:238–245. [in Greek with an English abstract]

Poulos, S., Maroukian, H., Leontaris, S. N., Neofotistos, D. and Rouskas, G., 2002. Natural processes and human interference in the evolution of the shore zone of Dilesi (south Evoikos Gulf). *Proceedings of the 6th Panhellenic Geographical Conference,* Thessaloniki, 3–6 October, Geographical Society of Greece, π:296–303. [in Greek with an English abstract]

Poulos, S., Petrakis, S., Kalyva, D. and Pourharidou, M., 2011. Estimation of the loss of coastal land, due to the anticipated sea-level rise, in the case of the coastal zone from Chalkoutsi to Nea Palatia (S. Evoikos Gulf). *Bulletin of the Geological Society of Greece,* 44:47–54.

Poulos, S., Skoubri, M., Kiamou, V. and Myronidou, P., 2004b. Study of the sedimentological and oceanographical conditions of the beach zone of Agia Anna – Krya Vrisi (NE Evia). *Proceedings of the 7th Panhellenic Geographical Congress,* Mytilini, 14–17 October, Geographical Society of Greece, 197–204. [in Greek with an English abstract]

Pranzini, E. and Rossi, L. 1995. A new Bruun-rule-based model: An application to the Tuscany Coast, Italy. In: Ozhan, E. (ed.). *Proceedings of the Second International Conference on the Mediterranean Coastal Environment MEDCOAST 95,* October, 1995, Tarragona, Spain, 1145-1159.

Psomiadis, E., Parcharidis, I., Poulos, S., Stamatis, G., Migiros, G. and Pavlopoulos, A., 2005. Earth observation data in seasonal and long term coastline changes: Monitoring the case of Sperchios river delta (central Greece). *Z. Geomorphologie N.F.*, 137:159–175.

Pyökäri, M., 1997. The provenance of beach sediments on Rhodes, southeastern Greece, indicated by sediment texture, composition and roundness. *Geomorphology*, 18(3–4):315–332.

Pyökäri, M., 1999. Beach sediments of Crete: Texture, composition, roundness, source and transport. *Journal of Coastal Research*, 15(2):537–553.

Short, A. D. (Editor), 1999. *Handbook of beach and shoreface morphodynamics*, John Wiley, London, 379 pp.

Soukissian, T., Hatsinaki, M., Korres, G., Papadopoulou, A., Kallos, G. and Anadranistakis, E., 2007. *Wind and wave atlas of the Hellenic Seas*. Hellenic Centre for Marine Research Publ., 300 pp.

Sutherland, J., Brampton, A., Motyka, G., Blanco, B. and Whitehouse, R., 2003. Beach lowering in front of coastal structures: Research scoping study. Defra/EA R&D Report FD1916/TR1. Available via http://sciencesearch.defra.gov.uk.

Thieler, E. R. and Hammar-Klose, E. S., 1999. *National assessment of coastal vulnerability to sea-level rise: U.S. Atlantic coast*. U.S. Geological Survey, Open-File Report, 99–593.

Tsimplis, M. N., 1994. Tidal oscillations in the Aegean and the Ionian Seas. *Estuarine, Coastal Shelf Science*, 39:201–208.

Tsoutsia, A., Karditsa, A., Alexandrakis, G., Poulos, S. and Petrakis, S., 2010. Coastal and underwater geomorphological and sedimentological survey of Katelios bay, Kefallonia Island. *Proceedings of the 9th Panhellenic Geographical Congress*, Athens, 4–6 November, Geographical Society of Greece, 280–287.

United Nations World Tourism Organization (UNWTO), 2011. *Tourism Highlights, 2011 Edition*. United Nations World Tourism Organization, 12 pp. Available at: http://www.e-unwto.org/content/u27062/fulltext.pdf (accessed 25 August 2012).

Verykiou-Papaspyridakou, E., Andris, P. D., Karditsa, A., Alexandrakis, G., Poulos, S. E. and Ghionis, G., 2008. Study of the beachrock formations in the Gyrapetra region of the Lefkada island. *Bulletin of the Geological Society of Greece*, 42/I:105–113.

Vousdoukas, M., 2007. *Coastal dynamics of beachrock-infected beaches*. PhD Thesis (unpublished), Department of Marine Sciences, School of Environment, University of the Aegean, 236 pp.

Watts, A., 1975. *Wind Pilot: Eastern Mediterranean Coasts*. Nautical Pub. Co. Ltd., Lymington, UK, 496–590.

Wigley, T., 1999. *The science of climate change: Global and U.S. perspectives*. Pew Center on Global Climate Change, June. Available at: www.c2es.org/docUploads/env_science.pdf (accessed 25 August 2012).http://www.pewclimate.org/projects/env_science.cfm

Xeidakis, G. S. and Delimani, P., 2003. Coastal erosion problems along the northern Aegean Coastline: The case of the Nestos River Delta and the adjacent coastlines. In: C. Goudas, G. Katsianis, V. May and Th. Karambas (eds.), *Soft Shore protection: An environmental innovation in coastal engineering*, Kluwer Academic Publishers, Dordrecht, 337–348.

Xeidakis, G. S., Delimani, P. and Skias, S. 2005. Erosion problems in Alexandroupolis coastline, North-Eastern Greece. *Environmental Geology*, 53(4):835–848.

Xenidis, Y., Angelides, D. C., Tziavos, I., Koutitas, C. and Savvaidis, P., 2005. An integrated framework for sustainable Coastal Zones Management in Greece, *IASME Transactions*, 2(5):813–822.

Zabakas, J. D., 1981. *General climatology*, National and Kapodistrian University of Athens, 493 pp. [in Greek]

19 Bulgaria

Margarita Stancheva

Introduction

In common with many European coasts, the Bulgarian Black Sea coast is exposed to the combined effects of natural and anthropogenic hazards, the most crucial being persistent coastal erosion and human occupation. Erosion, both natural and human induced, has posed the most critical coastal threat and this process has become further aggravated in the last decade. In an attempt to manage erosion, many hard defence structures have been widely used since the 1980s, including dikes, seawalls and solid groins. Despite numerous defence measures that have been applied for over a century, erosion and landslide problems have not been completely solved and there is still a lack of relevant coast-protection management plans, which invariably involve hard engineering structures. Current cliff and beach erosion is associated with these structures, which have reduced sediment inputs and interrupted sand movement. The mushrooming growth of population and tourism are other hazard factors contributing to coastal zone instability resulting in large areas of sand dunes/beaches being illegally overbuilt putting major recreational resources at risk of degradation.

Coastal morphology

Bulgaria is situated in the Balkan Peninsula, South-East Europe, between 41°14'–44°13' N and 022°21'–028°36' E, covering an area of 110,842 km². The coast is located in the Black Sea between Cape Sivriburun to the North, bordering Romania and southwards, the Rezovska River mouth, which borders Turkey (Figure 19.1)

The coastline is 412 km long (Stanchev, 2009) with a general eastward aspect. It is composed of *circa* 60 per cent rock cliffs, *circa* 30 per cent sand beaches, some with backing dunes, and the remaining 10 per cent is formed of low-lying parts of firths and lagoons together with two large bays (Varna and

■ **Figure 19.1** *Bulgarian Black Sea coast*

Burgas). The 26 firths and five lagoons are typical of the coastal zone, especially along the southern coast. Firths are former river valleys drowned by rising Holocene sea levels and at present their configuration almost repeats old river valley contours (Dachev, 2000). Hard stabilization structures and harbour development cover some 10 per cent of the coast (Figure 19.1). The coefficient of coastline crenulation varies from 1.02 in the north to 3.79 in the south (Popov and Mishev, 1974; Peychev, 2004; Keremedchiev and Stancheva, 2006) and the coast is characterized by its geology.

An erosion coast is prevalent between Capes Sivriburun and Shabla (Figure 19.1), the geology including loess sediments and limestones (Figure 19.2). Between Cape Shabla and Kavarna town, dense shell and oolitic limestone outcrops. The cliffs are high (70 m at certain sites) and steep. Between Kavarna town and the resort of Albena, the coast consists of limestone or clay, clay-sands and marls, as far as Cape St. George (Figure 19.1). Aleurolites, sandstones and clays are predominant in the geology between Albena and Cape St. George (Peychev and Stancheva, 2009). Different classes of landslides also occur here; the largest landslide complex is located at Frangenski (Popov and Mishev, 1974). Southwards is Varna Bay, situated between the Capes of St. George and Galata, where sandstones and sands predominate. Between Capes Galata and Emine (Figure 19.1), the geology is one of sandstone, limestone, marl, clay, aleurolites and argillaceous rocks. Large sand stretches are found here, the longest being a 12 km beach named Kamchia-Shkorpilovtzi (Figure 19.1).

Between Cape Emine and Nessebar town, flysch sediments outcrop and on Nessebar Peninsula, Crimean-Caucasian limestone is found. Between Pomorie and Cape Foros is Burgas, the largest bay in Bulgaria. Volcanic rocks (potassium-alkaline trachytes, latites, and psammitic and psephitic tuffs, and pyroclastic flysch) and volcanites, andesite-basalts, and basalts outcrop at the southernmost part between Cape Foros and the Rezovska River (Figure 19.1). These rocks are resistant against wave erosion and the

■ **Figure 19.2** *Loess erosion on the northern coast (between Capes Sivriburun and Shabla)*

■ **Figure 19.3** *Volcanic rocks, the south coast (Cape Maslen nos)*

average cliff retreat rate is low (around 0.01 m/yr; Peychev and Stancheva, 2009; Figure 19.3).

Beaches and dunes are widely distributed along the Black Sea coast (Figure 19.4 and 19.5), with more than 70 sand beaches having a total length of 121 km (Stancheva, 2009). In the northernmost area they consist of organic medium-sized sands with high (93 per cent) carbonate contents (Dachev *et al.*, 2005), because of large mussel beds found on the nearshore. Beaches here have a low heavy mineral content.

In the central coast, sand beach sediment input is basically from landslides and small rivers. Generally, beach sands are coarse to medium-sized, with a low carbonate content consisting predominantly of quartz. At the southernmost section, beaches are composed of medium- and fine-sized magnetite-titanite sands, with a high content of heavy minerals (up to 75 per cent) due to the volcanic rocks (Sotirov, 2003; Peychev, 2004).

Sand dune systems occur behind the larger beaches or in small inlets, where their development has been favoured by the combined factors of coastline orientation, wind direction and sufficient sediment supply. Commonly, such beach-dune systems or separate dunes consist of beaches, foredunes, parallel mobile and more stabilized or fixed dunes located landwards. Extensive dune fields are developed at the northern and middle sections and a number of dune complexes are located along the southern coastline (Figure 19.5).

However, due to increased human impact, the dune landscape is continually diminishing and today the dunes run along only 39 km of shoreline, covering an area of 9 km^2 (Stancheva, 2010a). Although sand dunes are protected environmental areas, they have been exposed to expanded anthropogenic pressure, leading to degradation.

One hundred and two rivers disgorge into the Black Sea sector. Of these, 19 enter directly, while 14 end in nearshore lakes and the other 69 rivers are inflows. The largest Black Sea river with the most important sediment input is the Kamchia (Figure 19.1). After construction of three large artificial lakes during the 1970s, its solid discharge was reduced

■ **Figure 19.4** *Sand beach, the north coast (south of Cape Galata)*

■ **Figure 19.5** *Large dune complex, the south coast (south of Cape Maslen nos)*

from 2,106,000 to 462,000 t/yr (Peychev and Stancheva, 2009).

Wave climate

The Black Sea is almost tideless (10–15 cm) and wind wave energy is the primary factor affecting the coast. Due to the general eastward exposure of the coastline, waves are mainly generated by winds from the N, NE, E, SE and S directions. Waves approaching from the NE prevail 72.9–76.3 per cent of the time (Cherneva *et al.*, 2003). This dominant wave direction produces

a general southward direction for longshore sediment transport (Dachev and Cherneva, 1979).

There are six wave observation stations along the Bulgarian Black Sea coast: Shabla, Kaliakra, Varna, Emine, Burgas and Ahtopol (Figure 19.6) and annual values of significant wave heights ($H_{1/3}$) vary between 0.12 m in summer to 1.7 m during winter (Grozdev, 2006). The lowest wind wave values are found at Varna Bay station (0.12 m), while the highest are at Ahtopol coastal station (1.7 m). Extreme waves are lowest in Varna and Burgas bays and the highest are predicted for the southernmost station of Ahtopol.

■ **Figure 19.6** *Significant wave height H_s (m) with one and ten-year return periods (After Grozdev, 2005)*

Extreme wind waves are the main cause of flooding in low-lying coastal areas and these also cause erosion and landslides with consequent damage to buildings and developments, as well as people (Dachev, 2000).

Sea level changes along the Bulgarian Black Sea coast have been recorded for over 100 years. Based on mareograph records located at Varna and Burgas, a continuous sea level increase has been found, particularly over the last few decades (Peychev, 2004). The mean values of average sea level rise for the western Black Sea vary between 1.5 and 3 mm/yr (Pashova and Jovev, 2007). Although such rates alone are not dramatic for the Bulgarian coast, there would be a case for a sudden sea level rise under certain meteorological conditions, e.g. during severe coastal storms, which can have devastating effects on the environment and coastal infrastructures, both on and offshore (Stanchev *et al.*, 2009). When combined with additional events, such as surge waves, tsunamis or heavy rainfall, they could be disastrous (Boyd *et al.*, 2002; Grozdev, 2008). Over the past few decades, there has been an increased frequency in occurrence and intensity of extreme meteorological events, such as a sudden sea level rise (Stanchev *et al.*, 2009). For example, the February 1979 storm had a sea level increase (maximum rise of 1.37 m) and the June 2006 storm combined with heavy rain caused much infrastructural damage (Grozdev, 2008). Predictions and assessments of potential occurrence of extreme sea level rise have significant importance and could serve as base information in coastal decision-making and prevention/mitigation measures in hazard-prone coastal zones (Table 19.1).

■ **Table 19.1** *Probability of occurrence of extreme sea level rise (cm) along the Bulgarian coast (After Grozdev, 2008)*

Stations/ Period	1	10	25	50	100
Shabla	38	104	139	168	202
Kaliakra	23	50	63	73	83
Varna	49	121	159	192	229
Emine	43	92	116	136	158
Burgas	62	92	121	142	162
Ahtopol	44	63	71	77	84

About 20 per cent (83 km) of the Bulgarian coast has been identified as vulnerable to inundation at given scenarios of extreme sea level rise (0–5 m). These are mostly low-lying territories – firths, lagoons, river mouths and wetlands. For example, 14 towns, 17 villages, 13 sea resorts and seven small campsites would be potentially flooded by an extreme sea level rise of 5 m and the total number of local coastal residents at these sites constitutes almost 7 per cent of the entire country's population (Palazov *et al.*, 2007; Stanchev *et al.*, 2009).

Evolution of coastal settlements

Over the past century, the Bulgarian coastal zone has shown a trend of increasing population compared to inland areas, due mainly to the economic and recreational opportunities that exist. This process has also been driven by migration of citizens to larger coastal towns after political and economic reforms during the early 1990s. Since then the coast has been highly developed and urbanized, e.g. from 1934 to 2001 the number of local residents 10 km from the coast increased by 215 per cent and now its population density is 211 people/km^2 (NSI, 2002). By contrast, for 30 km and 60 km zones away from the coast, the population density is 30 and 28 people/km^2 respectively (Palazov and Stanchev, 2006).

There are 262 municipalities (the smallest administrative-territorial units) in Bulgaria and 14 are classified as coastal municipalities being located entirely or partially within the Black Sea coast. The total area of all 14 is 5,784 km^2 comprising about 5 per cent of the entire country's territory, accommodating almost 9 per cent of the national population. By comparison, in 2001 the average population density for the entire country was 72 people/km^2, whereas in all coastal municipalities the average density value was 122 people/km^2 (Palazov and Stanchev, 2006). Coastal tourism development has posed a number of threats to shoreline areas. Over the last two decades, there has been no significant residential growth along the shore, but recent tourist arrivals have dramatically increased (more than 20 per cent) the total number of summer coastal inhabitants. However, in some municipalities with large seaside resorts this number could exceed 200 per cent (Palazov and Stanchev, 2006; Figure 19.7).

■ **Figure 19.7** *The largest Bulgarian Black Sea resort, Sunny beach*

Population concentration is lowest in the coastal municipalities of Shabla and Biala, while the large seaside towns of Varna and Burgas are the most populous sites (Figure 19.8). These two cities, equipped with the largest harbours in the country, have continually expanded due to diverse economic, service and labour opportunities, which attract more people to the shore. This population growth has had a significant impact on the coastal geomorphology of Varna and Burgas bays, since the higher population concentration has demanded a larger number of shore protection measures to mitigate coastal hazards. In some municipalities, existing facilities are not able to cope with the additional population and this has adversely affected the marine ecosystems. Furthermore, the increased coastal infrastructure required by the additional population is also threatened by acceleration of erosion-landslide processes.

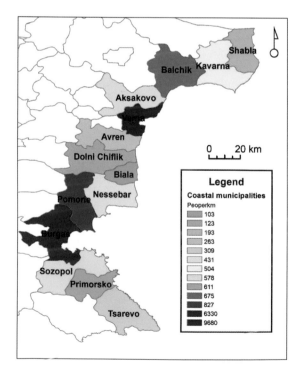

■ **Figure 19.8** *Sensitivity map at 14 Bulgarian Black Sea municipalities (from Stancheva, 2010b; with permission from Comptes rendus de l'Academic bulgare des Sciences)*

The erosion process: Natural vs. anthropogenic factors

Various research institutes have carried out investigations of the scale and rate of cliff erosion since the 1960s: the Institute of Geology, the Institute of Geography, and the Institute of Water Problems, all affiliated to the Bulgarian Academy of Sciences (BAS) (Marinski, 1998), and later by the Institute of Oceanology, BAS (Peychev, 2004). Large-scale erosion studies started in the 1980s by the Geozashtita organization, which became an executive agency responsible for Bulgarian shore protection. Field measurements of cliff erosion rates have been carried out since 1983, by the Institute of Oceanology (BAS) applying a specific methodology suggested by Parlichev (1986) which includes regular measurements of the distance to the cliff edge using a preliminary established network of constant benchmark stations. Based on these studies, average cliff erosion rates have been identified and the controlling influences of anthropogenic, geological and other natural factors have been identified (Peychev, 2004). Beach changes in terms of shoreline dynamics and erosion/accretion processes have also been studied for over 30 years by the Institute of Oceanology, BAS (Dachev *et al.*, 2005; Peychev, 2004; Stancheva *et al.*, 2008).

Natural and human factors causing coastal erosion

Managing erosion through various hard engineering methods often has adverse effects on the natural landscape and coastal dynamics. The choice of coastal protection options requires detailed knowledge both on physical processes leading to erosion and on the prevention options (Stancheva and Marinski, 2007). In Bulgaria, different factors are responsible for the high erosion rate along the coast:

- environmental factors, such as sea level rise, extreme waves, geological setting, changes in sediment supply, etc.;
- factors related to human activities, such as rapid

coastal urbanization, port developments and extended coastline armouring by hard engineering structures.

Erosion rates have increased since the 1960s, mainly because of uncontrolled urbanization and various human activities in the coastal zone, the major causes including:

- building artificial lakes;
- river-sand mining;
- installing port/coast-protection structures; and,
- dredging works.

Previous studies noted that accretion was the dominant process with erosion on beaches or cliffs occurring only during severe storms (Rozhdestvensky, 1969). In 1974, Popov and Mishev calculated that 60 per cent of coastal cliffs were undergoing erosion, whilst the proportion of sand beaches constituted at least 28 per cent of the shoreline. Recent studies have also identified that erosion is a progressive process dominant along 70.8 per cent of the coast (Keremedchiev and Stancheva, 2006). A number of defence measures have been applied to solve the erosion/landslide problems. Most include hard stabilization options such as solid groins, dikes and seawalls. As a result, from 1960 to 2008, the amount of sediments from cliff erosion, river solid discharge and aeolian drift has decreased from 4,979,700 t/yr to 1,221,300 t/yr. Sporadic natural erosion in the near past has now become critical at many coastal sites (Peychev and Stancheva, 2009).

Current state: Rates of beach and cliff erosion

Recently, a number of large beach areas have experienced continuous reduction of levels (Dachev et al., 2005; Peychev, 2004; Stancheva, 2009). Based on data from more than 30 years of investigations on beach dynamics, it was found that *circa* 48 per cent of sand beaches have been eroding (Figure 19.9a) and varying amounts of cliff (Figure 19.9b). There are a number of contributory factors. At some beaches, an active cliff at the beach rear erodes (Dachev et al., 2005). In other cases because of human activities, there is a deficit of new beach material (Peychev and Stancheva, 2009). Large parts of the coast have been armoured with dikes thus causing sediment loss by cliff erosion to adjacent beaches. Furthermore, construction of concrete groin systems or harbour breakwaters, with lengths of 100–130 m offshore, has obstructed sand longshore movement. In addition, large sand beaches and dunes, especially in the most famous seaside resorts, have been illegally reclaimed, as hotels and second homes have built directly on active beach areas.

Today, the above changes have resulted in severe beach erosion downdrift from groins and harbours, e.g. Varna central beach in Varna Bay, at the re-entry angle of the 1 km long harbour breakwater built in the 1900s. This beach had constantly increased in volume until the 1980s, when the large groin system was placed on its northern part. At present, sand accretion is observed only in close vicinity to the harbour breakwater area (which acts as a trap for sand); while at the northern part, the beach has been disappearing (Stancheva et al., 2008).

With regard to the need for beach monitoring and proper management, Dachev et al. (2005) separated beaches into three groups with different priorities:

- Priority 1: Beaches where erosion processes prevail and/or beaches where the largest tourist resorts and infrastructures are developed, and/or beaches for which long-term data sets have been collected;
- Priority 2: Almost stable beaches, where small tourist complexes are built;
- Priority 3: Stable beaches (with a prevailing accretion process).

Environmental factors responsible for increased erosion at the coast are highly variable, both temporally and spatially. Cliff retreat extent depends on local wave climate and storm surges, dominant coastline orientation or exposure and coastal geological settings. Generally, rock properties that are relevant

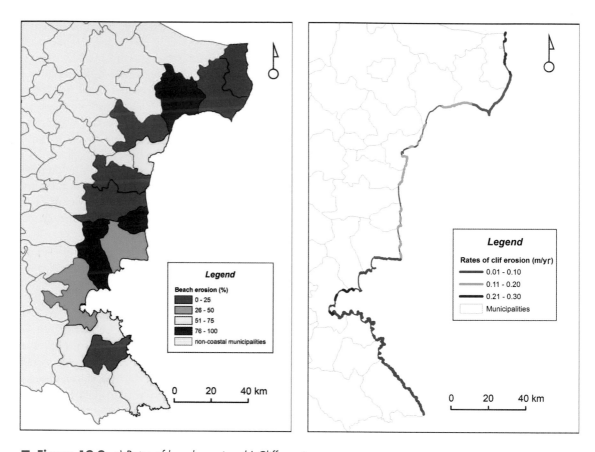

■ **Figure 19.9** *a) Rates of beach erosion; b) Cliff erosion*

in controlling the erosion process include hardness of deformation or pressure for igneous/metamorphic rocks and shear strength for unconsolidated sedimentary rocks (Zenkovich, 1967; Pilkey and Thieler, 1992). The average rate of cliff retreat at the coast is 0.08 m/yr. Highest erosion rates occur on those areas of the north coast formed of loess, between Capes Sivriburun and Shabla (0.30 m/yr), and those areas formed of clay on the south coast between Pomorie town and Cape Lahna – 0.22 m/yr (Figure 19.1 and Figure 19.9b). In the most southerly area, rates are much lower, being around 0.01 m/yr on volcanic rocks resistant to erosion (Peychev and Stancheva, 2009).

Previous cliff erosion measurements have indicated a constant decrease in erosion area/volume, as well

as length of eroding coast, *i.e.* there is not only a decrease in the length of coast that is eroding, but actual volumes of material being released from these areas are also decreasing. There are a number of natural factors contributing to this trend; however, the predominant cause is coastal armouring through hard protection structures, mostly dikes. From 1960 to 2008, the volume of eroded material decreased from 1,344,000 m³/yr to 221,000 m³/yr and the length of eroded cliff coast reduced from 271 km to 219 km resulting in insufficient sediment supply from cliffs to adjacent beaches and nearshore region (Peychev and Stancheva, 2009). Differential cliff erosion rates are initially predetermined by the complex coastal geology. Similar to beach erosion since the 1980s, activation of cliff retreat has been mainly induced by accelerated anthropogenic impact on the coastal zone.

Shore protection

From the beginning of the twentieth century to the present day, control over coastal erosion/landslide processes has been implemented via traditional hard defence options, *i.e.* groins, dikes and seawalls. The need to protect the coast at most erosion-prone areas has been driven by the growing coastal zone population and installation of protection structures has particularly increased over the last few decades. Data from 1:25,000 scale topographic maps (1994) indicates some 217 port and coast-protection structures comprising a length of 71 km for the entire Bulgarian coast (Stancheva, 2009).

Administrative competencies on shore protection

Coastal protection against waves and erosion/landslide processes is implemented by the institutional organization 'Geozashtita', established in the early 1980s, that created a long-term programme for shoreline defence and management. Under this programme, a range of multipurpose engineering structures have been designed, related mostly to industrial activities, rather than coastal defence. Because of such complex approaches to shore protection, a large number of hard structures with mutually exclusive functions have been constructed (Marinski, 1998). The Black Sea Coast Development Act (enforced January 1, 2008), is intended to encourage development of improved strategies for shoreline management. This is based on accepted new standards and guidelines for the design of maritime constructions. However, there are a number of ambiguous texts in the law, related to precise definitions of coastal zone, shoreline, different types of port and protection structures and their standardization, including precise definition and delineation of beaches. All these issues need to be further elaborated in secondary legislation instruments in order

that the Act delivers the intended benefits (Marinski *et al.*, 2008).

Groins

Groin construction is a widely used method for protection and commenced at the end of the nineteenth century as a general defence option (Marinski, 1998). Since the 1980s, a large number of solid T- and Γ-shaped groins have been built in an attempt to protect the coast and retain beaches. A typical design is a system of three or more structures in combination with dikes or seawalls. The groins are impermeable, are of different lengths between 130 m and 220 m, and are usually built of rubble, while either the front or all the sides are protected with manufactured elements, such as tetrapods and tetrahedrons (Parlichev, 1994). These unusual constructions are still termed 'groins' (Marinski, 2006).

Many large structures with mutually exclusive functions (port and coast protection) have been built (Figure 19.10); mostly creating protected water-areas for small vessels. In the past, the coastal protection impact of these schemes has not been central to their design. Such structures should be classified as jetties, due to the fact that they do not function as classic groins. Thus, instead of traditional types of groins along the coast, breakwaters with T-, Γ- and Y-shapes have been designed (Marinski, 1998). These solid groin systems have obstructed long-shore sediment transport and caused a sediment reduction in adjacent downdrift beaches, as is the case at the abovementioned beach in Varna Bay (Stancheva *et al.*, 2008).

Coastal dikes

Both erosion and landslide processes have also been controlled by construction of dikes (or embankments), as another very common defence option. Using the Netherlands experience and expertise in protective dikes, coastal engineers have built a number of huge dikes in order to manage cliff erosion. A common construction is a combination of dike and

■ **Figure 19.10** *'T-Groin' with a mutually exclusive function (north from Cape St. George)*

■ **Figure 19.11**
3 km long dike on the coast at Varna town

groin systems, as well as dikes functioning as road connections (Figure 19.11).

These structures are built between groin compartments on the narrow beach, as their external part is based in up to a 1.5–2 m water depth (Parlichev, 1994). Application of projects for artificial beaches at these depths requires large amounts of sand, which consequently makes such protection methods expensive. The biggest dike so far, with a length of 8 km, has been recently constructed between Balchik and Albena, on the northern coast (Stancheva et al., 2010). In general, dikes provide effective wave breaks and sufficiently protect low-lying coasts against flooding. However, such structures stop the exchange between land and sea, and *vice versa*; disrupt sediment input from the cliff, provoking active or passive erosion; reduce beach access; and have negative visual effects (Griggs, 2005; Stancheva et al., 2010).

Seawalls

Seawalls are another common protection method; designed seawalls are mostly vertical and smooth, which promotes scouring of the structure's fundament and leads to their destruction. Bottom erosion immediately in front of the structure can be enhanced due to increased wave reflection caused by the seawall, resulting in a steeper bottom profile, which in turn allows larger waves to reach the structure and beach sand to be carried away (Burcharth and Hughes, 2005). It is surmised that seawalls with wave-breaking walls, which have been elaborated over the recent years, would impact less adversely on the adjacent beach. Such a structure was placed at one of the most erosion-prone sections at the north coast (Figure 19.12), where the average rate of erosion reaches 1.2–1.6 m/yr (Peychev and Stancheva, 2009). The seawall is built on the cliff at the rear of a 10 m wide beach. The structure has effectively protected the coast, but has caused beach reduction, especially during storms (Parlichev, 1994; Marinski, 1998).

Massive structures, such as groins, dikes and seawalls, are not only expensive, but in most cases are inefficient and require significant ongoing maintenance (Stancheva and Marinski, 2007). In fact,

■ **Figure 19.12** *Seawall with wave-breaking cells (north coast between Capes of Sivriburun and Shabla)*

there are currently many structures along the coast which are partially/completely broken (Figures 19.13 and 19.14). The main reason for this inadequate and poorly planned practice is the lack of specialized technical standards and guidelines for designing coastal defence structures (Marinski, 2006). Adverse effects are especially found when extreme storms impact on areas where beach widths have been reduced by uncontrolled reclamation and where there are no well-planned shoreline protection measures (Stancheva and Marinski, 2007).

An evaluation report showed that there is no Integrated Coastal Zone Management (ICZM) or beach management (Williams and Micallef, 2009) equivalent policy. At present, there are only a limited number of fragmented tools in place to address coastal issues (EICME, 2006). In general, there is a lack of strategy implementation and lack of awareness by both public and state sectors on the importance of ICZM. Involvement of scientific expertise, public discussions and education programmes on critical problems have not been conducted and

■ **Figure 19.13**
Collapsed parts of a massive coastal dike

■ **Figure 19.14**
Broken seawall without beach in front of the structure

research results have rarely been incorporated in the strategies for coastal zone planning and shoreline management.

Experimental and innovative solutions adopted for coastal protection

There have been isolated cases of applying soft defence measures along the coast over the years, which have been more associated with recreational rather than protection purposes. These mostly include creation of artificial beaches or beach nourishment. For example, a transverse sand bypass was successfully applied to fill the artificial beach in groin compartments at Varna Bay (Figure 19.15; Dachev and Leont`yev 2005). One of the main problems in applying nourishment projects is the shortage of appropriate sizes and amounts of sand, whether from terrestrial or offshore sources. Beach fill or nourishment is also a very expensive method, one that must be periodically repeated.

■ **Figure 19.15** *Artificial beach between groin fields (Varna Bay, sand bypassing)*

BOX 19.1   COASTAL DEGRADATION INDUCED BY ANTHROPOGENIC IMPACTS ALONG THE NORTH BULGARIAN BLACK SEA SHORE

The study site is a 10 km long section between the resort of Albena and the town of Balchik, on the north Bulgarian Black Sea coast, where an 8 km dike road connection has recently been constructed. There are also other structures along the study area: a port, solid groins, supporting wall and seawall, and the renewed coastal dike. Before dike implementation, there were a few narrow sand beaches in front of the cliff, which were then completely destroyed.

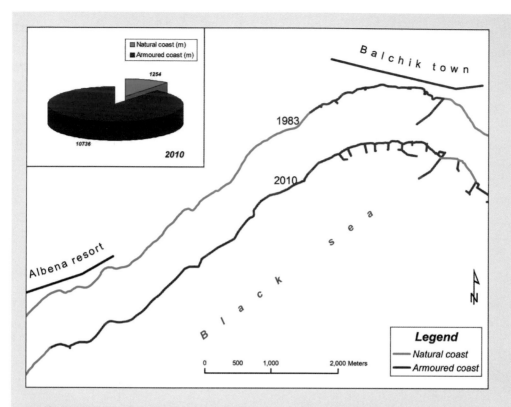

■ **Figure 19.16** *Coastline changes 1983–2010*

To identify modifications of the coastline two available shoreline positions from 1983 and 2010 were compared and analyzed by GIS.

As a result of dike construction, the coastline advanced seaward by 28 m and new land created in the place of the former water-area. This shoreline advance seaward could not be considered as accretion as the coast was asphalted and armoured.

In 2010, the entire coastal section was 9,150 m long and the dike structure comprised 7,550 m. The length of natural coastal segments (cliffs and partially beaches) is only 1,254 m and the total length of armoured coast reaches 10,736 m (including all structures such as dike, seawalls, groins and port moles) and exceeds the coastline length of 9,150 m. The coast has been completely armoured.

Although dikes have been constructed under a project aimed to stop erosion/ landslide processes and to fortify the coast, the negative impacts caused may reach far beyond the expected protection effects for which they have been designed. The narrow sand beaches that existed previously were destroyed and sand from beaches and shallow water areas lost. Another important aspect of adverse dike construction effects could be associated with irreversible coastline modifications; interrupted sediment input from the cliff, aggravation of erosion downdrift, negative visual impacts and lost public access to the water-area.

(Reference: Stancheva *et al.*, 2010b)

■ **Figure 19.17**
An 8 km long dike completed in 2010

Conclusions

Over the last few decades, a major cause for eroding beaches and associated cliff retreat is increased human impact, e.g. the large number of port and hard defence structures, tourist development boom and existent coastal sediment deficit. In the future, as coastal population grows, the demand for measures to protect the coast and infrastructures is likely to increase. Instead of soft protection alternatives, construction of hard stabilization structures, such as groins, dikes and seawalls, has been used as a common erosion solution. Although numerous defence measures have been implemented over the last 100 years, the increased rates of erosion that has occurred suggests that these measures have been ineffective. It is important for coastal authorities to develop sustainable management plans. In particular, attention should be given to soft defence measures, such as preservation of beaches and dunes, as these are one of the best and most effective way of protecting coastal areas.

References

Black Sea Coast Development Act, 2008. *Promulgated, State Gazette No* 48/15.06.2007, effective 1.01.2008.

Boyd, K., Hervey, R. and Stradtner, J., 2002. Assessing the vulnerability of the Mississippi Gulf coast to coastal storms using an on-line GIS-based coastal risk atlas, *Oceans '02 MTS/IEEE*, 2, 1234–1240.

Burcharth, H. F. and Hughes, S. A., 2005. *Coastal Engineering Manual: Types and functions of coastal structures*, Chapter 2, Part VI, EM 1110-2-1100 Washington, DC, US Army Corps of Engineers.

Cherneva, Z., Valchev, N., Petrova, P., Andreeva N. and Valcheva, N., 2003. Offshore wind wave distribution in Bulgarian part of the Black Sea, *Proc. of Institute of Oceanology*, 4, 10–18.

Dachev, V., 2000 Implications of accelerated sea-level rise (ASLR) for Bulgaria, *Proc. of SURVAS Expert Workshop on 'European Vulnerability and Adaptation to Impacts of Accelerated Sea-Level Rise (ASLR)'*. Hamburg, Germany, 19–21 June, 25–28.

Dachev, V. and Cherneva, Z., 1979. Longitudinal-coastal transfer of the deposits in the coastal region of the Bulgarian Black Sea coast between the cape of Sivriburun and the Bourgas bay, *Oceanology*, 4, 30–42. [In Bulgarian]

Dachev, V. and Leont`yev, I., 2005. Mechanism and theoretical background of the transverse bypass of sea sands, *Proc. of Institute of Oceanology*, 5, 161–169. [In Russian]

Dachev, V. Z., Trifonova, E. V. and Stancheva, M. K., 2005. Monitoring of the Bulgarian Black Sea beaches, in, C. Guedes Soares, Y. Garbatov and N. Fonseca (eds.) *Maritime Transportation and Exploitation of Ocean and Coastal Resources*, London, Taylor & Francis Group/ Balkema, 1411–1416.

EICME, 2006. *Evaluation of Integrated Coastal Zone Management (ICZM) in Europe: Final report*, 43 pp., available at: http://ec.europa.eu/environment/ iczm/pdf/evaluation_iczm_report.pdf (last accessed on 20.05.11).

Griggs, G., 2005. The impacts of coastal armouring, *Shore and Beach*, 73, 1, 13–22.

Grozdev, D., 2005. Wind wave climate of the nearshore part of the Bulgarian Black Sea sector, PhD Thesis, Sofia, 180 pp.

Grozdev, D., 2006. Wind wave climate along the Bulgarian Black Sea coast, Proc. of Annual Scientific Conference of the Bulgarian Geological Society. *"Geosciences 2006"* (Sofia, Bulgaria), 296–299. [In Bulgarian]

Grozdev, D., 2008. Extreme characteristics of the Black Sea level along the Bulgarian coastline, *Proc. of Annual Scientific Conference of the Bulgarian Geological Society: Geosciences 2008*, Sofia, Bulgaria, 125–126. [In Bulgarian]

Keremedchiev, S. and Stancheva, M., 2006. Assessment of geo-morphodynamical coastal activity of the Bulgarian Black Sea part, *Compt. Rend. Acad. Bulg. Sci.*, 59, 2, 181–190.

Marinski, J., 1998. Coastal erosion: Causes for activation and the struggle against it. Protection and long-term stabilization of the slopes of the Black Sea coast, Marin Drinov Academic Publishing House, 138–142. [In Bulgarian]

Marinski, J., 2006. On the need for improving technical standards and guidelines in coastal engineering, *International Conference on Civil Engineering Design and Construction*, 14–16 September, Varna, Bulgaria, 587–591. [In Bulgarian]

Marinski, J., Droumeva, G. and Stancheva, M., 2008. Legislation and Integrated Coastal Zone Management, *1st PoCoast Seminar on Coastal Research*, FEUP, Porto, Portugal, May, 26–28.

NSI (National Statistical Institute) Staff, 2002. *Population*, Volume 1, Part 3: *Populations according to the censuses by districts, municipalities and settlements.* Publication of National Statistical Institute, Sofia, 523pp. [In Bulgarian]

Palazov, A. and Stanchev, H., 2006. Evolution of human population pressure along the Bulgarian Black Sea coast, *Proc. of the 1st Biannual Scientific Conference: Black Sea Ecosystem 2005 and Beyond*, Istanbul, Turkey, 158–160.

Palazov, A., Stanchev, H. and Stancheva, M., 2007. Coastal population hazards due to extremal sea level rise: Sunny beach resort case study, *Proc. of the 4th International Conference on Global Changes and Problems: Theory and Practice*, 20–22 April, Sofia, Bulgaria, 93–97.

Parlichev, D., 1986. Method for investigation of coast and shallow bottom area dynamic, *Oceanology*, 15, 58–65 [In Bulgarian]

Parlichev, D., 1994. Methods applied to eco-protection of the Bulgarian Black Sea zone, *Geoecology`94: Proc. of National Scientific-Practical Conference of Geography*, 9–10 April, Sofia, 119–121. [In Bulgarian]

Pashova, L. and Jovev, I., 2007. Geoid modelling for the Black Sea and future prospects, in C. Guedes Soares and P. Kolev (eds.), *Maritime Industry, Ocean Engineering and Coastal Resources.* Taylor & Francis Group, Abingdon, 761–768.

Peychev, V., 2004. Morphodynamical and litho-dynamical processes in coastal zone, Slavena Publishing House, Varna, 231 pp. [In Bulgarian]

Peychev, V. and Stancheva, M., 2009. Changes of sediment balance at the Bulgarian Black Sea coastal zone influenced by anthropogenic impacts, *Compt. Rend. Acad. Bulg. Sci*, 62, 2, 277–285.

Pilkey, O. H. and Thieler, E. R., 1992. Erosion of the United States shoreline, *Quaternary coasts of the*

United States: Marine and lacustrine systems, SEPM Special Publication, 48, 4–7.

Popov, V. and Mishev, K., 1974. *Geomorphology of the Bulgarian Black Sea coast and shelf*, Bulgarian Academy of Sciences, Sofia, 267pp. [In Bulgarian]

Rozhdestvensky, A., 1969. Dynamics of the Bulgarian Black Sea coast, *Fish Economy*, 1, 5–7. [In Bulgarian]

Sotirov, A., 2003. Division of the Bulgarian Black Sea coast according the type of the beach sands and their supplying provinces, *Review of the Bulgarian Geological Society*, 64, 1–3, 39–43.

Stanchev, H., 2009. Studying coastline length through GIS techniques approach: A case of the Bulgarian Black Sea coast, *Compt. Rend. Acad. Bulg. Sci.*, 62, 4, 507–514.

Stanchev, H., Palazov, A. and Stancheva, M., 2009. 3D GIS model for flood risk assessment of Varna Bay due to extreme sea level rise, *Journal of Coastal Research*, Special Issue 56, 1597–1601.

Stancheva, M., 2009. Indicative GIS-based segmentation of the Bulgarian Black Sea coastline for risk assessment, *Compt. Rend. Acad. Bulg. Sci.*, 62, 10, 1311–1318.

Stancheva, M., 2010a. Sand dunes along the Bulgarian Black Sea Coast, *Compt. Rend. Acad. Bulg. Sci.*, 63, 7, 1037–1048.

Stancheva, M. 2010b. Human-Induced Impacts along the Coastal Zone of Bulgaria. A Pressure Boom versus Environment. *Compt. Rend. Acad. Bulg. Sci*, Volume 63, 1, 137–146.

Stancheva, M. and Marinski, J., 2007. Coastal defence activities along the Bulgarian Black Sea coast methods for protection or degradation? in L. Franco, G. Tomasicchio, and A. Lamberti, (eds), *Coastal structures 2007: Proc. of the 5th International Conference*, Venice, Italy, 2–4 July, 480–489.

Stancheva, M., Marinski, J. and Stanchev, H., 2008. Evaluation of maritime construction impact on the beaches in Varna bay (Bulgarian Coast), *Proc. of 9th International Conference on Marine Sciences and Technologies: Black Sea 2008*, October 23–25, Varna, 190–195.

Stancheva, M., Stanchev, H. and Palazov, A., 2010. Implications of increased coastal armouring for the Bulgarian Black Sea shoreline, *Proc. of 10th International Conference on Marine Sciences and Technologies: Black Sea 2010*, 7–9 October, Varna, Bulgaria, 224–229.

Williams, A. T. and Micallef, A., 2009. *Beach management: Principles and practices*, Earthscan, London, 445 pp.

Zenkovich, V. P., 1967. *Processes of coastal development*, Interscience (Wiley), New York, 738 pp.

20 Romania

Adrian Stănică, Nicolae Panin and Glicherie Caraivan

Introduction

The Romanian coast, situated on the northwestern part of the Black Sea, has an overall length of *circa* 243 km, stretching from the Ukraine border in the north (Musura Bay, Danube Delta) to Bulgaria in the south (Vama Veche, north of Cape Sivriburun; Figure 20.1). The coast is divided into two major physiographic units. The northern unit is part of the Danube Delta Biosphere Reserve, Europe's largest nature reserve, rich in wildlife and with the lowest population density in Romania. The southern part comprises Constanta, one of the biggest cities of Romania with the biggest Black Sea harbour, plus an almost uninterrupted chain of tourist resorts alternating with towns and harbours. This massive development has generated a 'vicious circle' in that coastal assets have required protection, which in turn has generated negative side effects, especially in the pattern of sediment dynamics.

Coastal morphology and sedimentology

The northern unit has a coastal length of some 160 km and consists of low-lying wild beaches, generally composed of fine sands brought by the river Danube and redistributed by waves and currents. The morphology and sedimentology, as well as genesis and evolution, have been extensively studied and described in a series of works, e.g. Panin (1996, 1998, 1999), Giosan *et al.* (1999), Ungureanu and Stănică

(2000), Bhattacharya and Giosan (2003), Stănică *et al.* (2007) Vespremeanu-Stroie *et al.* (2007) and Dan *et al.* (2009). There are three main coastal sedimentary cells fronting the Danube delta (Figure 20.1) and from north to south they are:

1. The Kilia sedimentary cell: located in front of the Kilia secondary delta. Most of this sedimentary cell is on Ukrainian territory, the state boundary with Romania being in Musura Bay, an area marked by active sedimentation that is currently being transformed into a lagoon. The southern boundary of this cell is represented by the jetties of the Sulina Canal, built to protect the navigation waterway along the middle arm of the Danube; Figure 20.2). The jetties extend 8 km offshore; block sediments transported by longshore currents from the north; and represent an impermeable boundary between the Kilia and Sulina–Sf. Gheorghe sedimentary cells.

2. The Sulina–Sf. Gheorghe sedimentary cell: has a length of almost 60 km, some 34 km existing between the Danube mouths of Sulina and Sf. Gheorghe, while Sahalin Spit, at the southern part of the Sf. Gheorghe Danube mouth, represents the rest. The southern tip of Sahalin Spit represents the southern cell boundary. The general longshore circulation pattern is from north to south, except for the northern section. Here an eddy-like current generated by the Sulina jetties (Figure 20.2) has reversed the longshore drift from south to north. Most sediment from this cell is Danube-borne alluvium, less than 10 per cent being

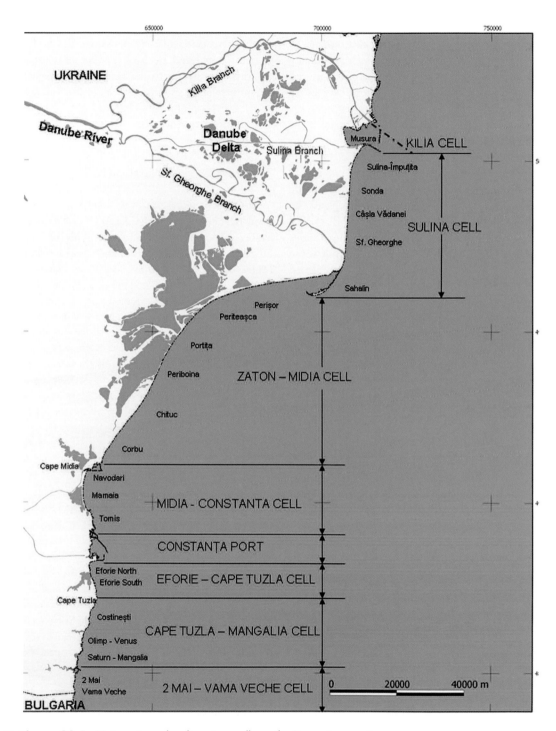

■ **Figure 20.1** *Main units and sedimentary cells on the Romanian coast*

represented by the calcareous fraction from shell fragments. The southern cell boundary is permeable, some sediments being occasionally transported to the neighbouring sedimentary cell under specific conditions. The shoreline may be divided into the following coastal sectors: Sulina, Canalul cu Sonda, Casla Vadanei (Figure 20.3) and Sf. Gheorghe (between the Danube mouths) and Sahalin Spit.

3. The Zatoane–Midia Harbour sedimentary cell: has a length of some 100 kilometres, mainly represented by barrier beaches, which separate the Razelm–Sinoe Lagoon system from the Black Sea. There were three inlets of the lagoon system, Portita, Periboina and Edighiol, all built some four decades ago; Portita inlet has been closed and developed for tourism (Figure 20.4), while Periboina and Edighiol have lock systems that controls water circulation between the sea and the lagoon. Most sediments are Danube alluvia, many being remobilized by erosion of previous littoral bars. There is a gradual southwards increase (up to about 50 per cent) in the calcareous fraction (sands from mollusc shell fragments). The last few kilometres of this coastal cell are seen by the first cliffs that outcrop at Cape Midia, the headlands being separated by long pocket beaches. The northern jetty of Midia Harbour represents the southern limit of this cell. This jetty, built around 1980 has a length of 5 km and blocks sediment longshore transfer towards the south. Coastal dynamics here have been assessed along the beach sectors of Zatoane, Perisor, Periteasca, Portita, Periboina, Chituc and Cape Midia (Corbu).

The southern unit, with a total length of *circa* 80 km, is located between Cape Midia and Vama Veche, and has several coastal sedimentary cells. This unit consists of cliffs, separated by low sand shores (Mamaia, Eforie, Costinesti and Olimp-Mangalia). Most of the information presented here has been synthesized from Constantinescu (2005), Kuroki *et al.* (2006) and Caraivan (2010), as well as in the JICA/ECOH (2006) report.

• The Midia–Constanta (northern jetty of Constanta harbour) sedimentary cell contains Mamaia Bay, which was initially a barrier beach

■ **Figure 20.2** *Sulina Canal jetties, close to the river mouth. The jetties represent an artificial barrier for the natural longshore transfer of sediments. Picture taken from lighthouse, located 7 km offshore*

■ **Figure 20.3** *Danube Delta coast: a natural environment. Casla Vadanei sector*

■ **Figure 20.4** *Portita beach: the only place developed for tourism along Danube Delta beaches, also the only place with breakwaters*

supplied by Danube-borne sediments transported by longshore currents. The situation changed after the building of Midia harbour (northern part of the bay), as this blocked all southbound longshore transfer of sediments. Over the past three decades, there has been a continuous increase in the role played by mollusc shells fragmented and ground down by waves as a sediment source. Long impermeable jetties built for harbour protection (Midia and Constanta) represent the northern and southern boundaries of this cell.

- In the area south of Constanta Harbour up to the southern limit of the Romanian coast, beaches consist of medium to coarse-grained sands, most of them (almost 100 per cent) of calcareous origin. The sediment sources have been mollusc shells, ground down by waves, as well as fragments derived from the sea bottom, which in some parts consists of Upper Neogene limestone. Most beaches are under the direct influence of human interventions, due to the presence of extensive hard coastal defence structures (described later).

- The South Constanta state border with the Bulgarian coastal stretch is divided into three coastal cells: the southern jetty of Constanta harbour–Cape Tuzla Sedimentary Cell, which is separated from the Cape Tuzla–Mangalia sedimentary cell by a significant headland, Cape Tuzla (Figure 20.1), with a submerged shallow limestone platform, which makes longshore sediment transfer very difficult under average hydro-meteorological con-

ditions. The harbour of Mangalia represents the anthropogenic boundary with the 2 Mai–Vama Veche coastal cell, which is also the southernmost one, extending into Bulgaria.

In the southern unit, shores with active eroding cliffs, are predominant. These types occur in front of Eforie North town, as well as almost from the town of Eforie South to Vama Veche (Brătescu, 1933). In this latter section, the cliff is discontinuous due to the presence of several sand barrier beaches in front of Costinesti, Tatlageacul Mare and Mangalia lakes, as well as in front of the Mangalia and Comorova coastal wetlands. The headlands are preferentially developed in hard rock areas and are under the direct influence of marine factors. From north to south, the most prominent Capes are Midia, Ivan, Turcului, Tuzla and Aurora. The uppermost Quaternary cliff consists of friable deposits, being composed of alternate layers of loess and palaeosoil layers on a level of Lower Quaternary (Villafranchian) impermeable red clay (Caraivan, 2010). The clay in turn lies on the Upper Neogene (Sarmatian) limestone. The predominant silt and clay grains are taken offshore by currents and out of the coastal system, whilst the few sand and gravel sediments are transported along the shore.

More detailed statistics of the coastal sectors, morphological types and lengths for the entire Romanian coast are presented in Table 20.1.

■ *Table 20.1* Lengths and types of the Romanian beach sectors

	Sector name	length (km)	type
Northern Unit	Musura	15	spit
	Sulina	10	deltaic beach
	Canalul cu Sonda	8	deltaic beach
	Casla Vadanei	10	deltaic beach
	Sf. Gheorghe	6	deltaic beach
	Sahalin	19	spit
	Ciotic-Zatoane	12	deltaic beach
	Perisor	12	deltaic beach
	Periteasca	13	deltaic beach

	Sector name	length (km)	type
	Portita North	6	barrier beach
	Gura Portitei	1	coastal structures
	Portita South	4	barrier beach
	Periboina	10	barrier beach
	Edighiol	11	barrier beach
	Chituc	17	deltaic beach
	Corbu	3	pocket beach
	Cape Midia	0.3	cliff
	Midia harbour	7	coastal structures
Southern Unit	Mamaia Bay	12	barrier beach
	Constanta	7	coastal structures
	Tomis marina, Casino	2	coastal structures
	Constanta harbour	15	coastal structures
	Agigea	1	coastal structures
	Eforie North	2	coastal structures
	Eforie North Belona	1	barrier beach
	Techirghiol centre	0.3	coastal structures
	Techirghiol	1	barrier beach
	Eforie South	4	coastal structures
	Eforie South-Tuzla	3	cliff
	Cape Tuzla	0.2	coastal structures
	Tuzla-Costinesti	5	cliff
	Costinesti North	0.5	barrier beach
	Costinesti	1	coastal structures
	Costinesti South	0.7	cliff
	Frenchman's Bay	2	pocket beach
	23 August	3	cliff
	Tatlageac	0.4	barrier beach
	North Olimp	1	cliff
	Olimp	2	coastal structures
	Neptun North	1	coastal structures
	Neptun South	1	barrier beach
	Jupiter-Venus	5	coastal structures
	Venus - North Saturn	1	barrier beach
	Saturn	1.5	coastal structures
	Mangalia	2	coastal structures
	Mangalia harbour	4	coastal structures
	2 Mai	1	pocket beach
	2 Mai - Vama Veche	3	cliff
	Vama Veche	1	pocket beach
	South Vama Veche	1	cliff

Hydro-meteorological factors

Wind regime

The Romanian Black Sea coast is a region with relatively high wind speed values, due to the complex inter-connections between the Atlantic storm track, the Mediterranean area and atmospheric circulation over the Black Sea. The multi-annual (1961–2000) mean wind speed along the coast is: Constanta: 5.15 m/s, Mangalia: 4.2 m/s, Sf. Gheorghe: 5.2 m/s and Sulina: 6.95 m/s. At Constanta, the predominant wind direction is from the west (16.4 per cent), followed by the north (13.1 per cent) and north-east (12.0 per cent); for the Danube Delta coast, the predominant wind directions are north and north-east. Winter (December–February) is the windiest period of the year and the annual frequency of the number of days with wind speeds >16 m/s is between two and three weeks on the Danube Delta coast and about ten days on the southern coast. The highest values are registered in December–February (with a maximum in January) and the smallest values in April–September. The maximum annual wind speed values, registered from 1961 to 2000 exceeded 40 m/s and the directions associated with these maximum values are, generally, north and north-east (National Administration of Meteorology, 2008).

Wave climate

Under the action of these winds, wind-generated waves and swell waves occur along the coast. The calm state of the Black Sea along this coastal stretch occurs on average about 1.9 per cent per year, the remainder, being wind waves, of *circa* 50.7 per cent, swell waves of about 20.1 per cent and combined waves (wind and swell), of about 27.3 per cent (Bondar *et al.*, 1973). Wind directions can be divided into three main groups:

1. Northerly directions: from N to ENE. These conditions cover some 102 days/yr; mostly in wintertime and host the largest wind speeds (34–40 m/s from north) and, consequently, the largest average wave heights.
2. Easterly directions: from E to SE. These conditions occur on average for 27.8 days/yr. These produce the smallest average wave heights.
3. Southerly directions: from SSE to WSW. These conditions cover on average some 90.8 days/yr.

According to Kuroki *et al.* (2007), the mean significant wave height for the winter season (November to March) is 1.16 m; while for the summer season (April to October) it is 0.79 m. The 100-year significant wave height is 7.8 m with a period of 11s (Kuroki *et al.*, 2007). Storm wave energy reached values of 12,242 kWh/m recorded on February 17, 1979), but generally the energy value is about 2,000 kWh/m (Spătaru, 1984). Storm surges from N, NE, E and SE directions induce a water level rise of 1.2–1.5 m. Black Sea tides have an average period of 12h 25min with a maximum amplitudes of only 11 cm (Bondar *et al.*, 1973). The general relative sea-level rise in the delta-front area (at Sulina gauge) is estimated at 3.7 mm/yr, of which subsidence accounts for 1.5–1.8 mm/yr, while for the southern Romanian unit the general sea-level rise is about 2.5 mm/yr (Bondar, 1989).

Human settlements

The first human settlements appeared on the north-western Black Sea coast more than 26 centuries ago, in the era of the Ancient Greeks. Greek colonists from Milet established a series of ports/coastal towns, most of which developed over time, surviving empires and changing cultures, from the Ancient Greeks, Romans, Byzantines, Barbarians, and Ottomans, to present-day Romania. This history can be found in the JICA/ECOH Report (2006).

After the mid-fifteenth century, when Genovese colonies and Moldovan and Wallachian towns were conquered by Ottoman Turks and most of the Black Sea became a Turkish internal lake, international maritime trade dramatically decreased in intensity. A true revival occurred towards the late eighteenth to early nineteenth centuries, when the decay of the Ottoman Empire allowed development of maritime trade and activities (including military) by other empires, such as the Russian, the Habsburg and the British. This competition generated tensions, which peaked during the Crimean War (1853–1856). At the end of this war, the Peace Treaty, signed in Paris in 1856, shaped a new framework for development of trade around the Western Black Sea and along the Danube River. This commenced the drive for rapid development of coastal settlements along what in 1878 became the Romanian coast. The jetties protecting navigation in Sulina go back to this period, while the town of Sulina was about ten times its current population (30,000 inhabitants compared to about 3,000 today). Sulina's development reverted (and its population rapidly dropped) after Soviet occupation and the beginning of the Communist regime, when trade with other countries was blocked or strictly controlled. The city of Constanta had a happier fate. From the ancient Greek colony of Tomis and the Byzantine town of Constantiniana, it

remained a small fishing village until after the Crimean War. Since then, and especially after World War II, the population has boomed, and it has become one of the largest cities in Romania, with the metropolitan area having around 400,000 inhabitants (Figure 20.5). The overall population of the southern unit of the Romanian coast approaches 650,000 permanent inhabitants. Besides the harbours of Midia, Constanta and Mangalia, there are also two marinas, developed industries, trade and tourism.

In contrast, a very different situation exists in the northern unit, as no human settlement was built on these 160 km of wild, remote Danube delta beaches. Only two settlements – Sulina and Sf. Gheorghe – exist in the vicinity of the coast, both located about 2 km inland from the sea. The former inlet of Razelm Lake (the former lagoon) closed about four decades ago and, protected by breakwaters, Gura Portitei is the only tourist resort on northern unit beaches, but it only has a small tourist capacity. Therefore, population density in this part is the lowest in the country.

■ **Figure 20.5**
The city of Constanta has had rapid development over the past century. Dikes from the obsolete coastal protection system are also visible

Coastal erosion

Romanian coastal geology and geomorphology, as well as its dynamics, has been a target of research since the 1960s, when the study of beach morphology and sediment sampling and analysis started. Beach sectors in front of the Danube delta were measured by the Laboratory of Marine Geology and Sedimentology (transformed in 1993 into the National Institute of Marine Geology and Geoecology, or GeoEcoMar) of the Institute of Geology and Geophysics in co-operation with the marine geology group of the Faculty of Geology and Geophysics, University of Bucharest. Therefore, measured information covers a time span of about half a century. Erosion, accretion and stable beach sectors were identified in front of the Danube delta, as well as in the southern part of the coast in several sea resort areas: Capes Midia, Navodari, Mamaia, Agigea (until the late 1980s), Eforie, Costinesti, 2 Mai and Vama Veche.

Simultaneously, measurements of coastal morphology and beach profile variations started to be undertaken by other research organizations and groups. The Romanian Institute for Marine Research (nowadays the Grigore Antipa National Institute for Marine Research and Development) has established its own network of measurements/beach profiles along the entire Romanian coastline. The coastal geomorphology group of the Faculty of Geography, University of Bucharest, commenced work on the Danube delta coastline in the early 1980s. Unfortunately there are only a small number of studies, mainly developed by the Faculty of Geography, University of Bucharest, that target the unstable Romanian loess cliffs. As is the case for most European coasts, a mixture of natural and human factors controls the evolution of the Romanian littoral.

The most significant natural factors are:

- *Climate and hydrology*: Besides the data presented in the previous section on the hydro-meteorological factors, the coastal dynamics of the coast in front of the Danube delta are closely related to the seasonal and multiannual variation of the Danube River liquid and solid discharges. Short-term

erosion occurs frequently, as the shoreline retreats swiftly – as much as tens of metres in a matter of hours (max. a few days), mainly induced by storms, with related waves and sea level setup (Stănică *et al.*, 2011). Long periods with extremely low temperatures result in freezing phenomena, with related changes in the dynamics of coastal sediments.

- *Geomorphology and sedimentology*: Of significant importance is the overall N–S orientation of the coast, exposed to the action of major storm winds and related waves, mostly from the NE (which has the strongest impact of the mobilization and transport of coastal sediments during storms). Another significant factor is cliff stratigraphy from the southern unit of the Romanian coast. Generally, eroding cliffs are sources of sediments for neighbouring beaches, but this does not apply to the Romanian cliffs, where sediments from cliff erosion sections are mainly silt and clay washed offshore.

The most significant human factors have developed since the second half of the nineteenth century and have significantly shaped the natural evolution and distribution patterns and volumes of sediments discharged into the sea by the Danube. These have been studied in many papers, by workers in the field such as Panin (1996), Giosan *et al.* (1999), Ungureanu and Stănică (2000), and Stănică *et al.* (2007). In summary:

- Embankments along the entire Danube basin and damming of the Danube in its entire watershed, including dams at Iron Gates I and II, generated a reduction in sediment supply of over 50 per cent to the coastal waters.
- Hydro-technical interventions in the Danube delta, e.g. cutting meanders to shorten distances (Sulina in the second half of the nineteenth century; Sf. Gheorghe in the 1980s), embankments of the Sulina Canal, hydro-technical works at Ceatal Izmail, which generated changes in distributing liquid and solid discharges between the three main arms.

- Coastal engineering works represent another major category of direct interventions. These include navigation and harbour jetties, locks, piers, seawalls, groins, etc. These works have changed the natural coastal sediment circulation, blocked longshore sediment transfer, etc; the overall impacts on coastal dynamics range from the large scale, such as Sulina Jetties, to the local, e.g. coastal defence structures in the southern part. A series of significant human interventions is presented later. Besides these hard defence works, a significant role in altering natural conditions is represented also by dredging activities at the Sulina mouth bar in order to maintain safe navigation depths. Dredged materials are discharged offshore, taking sediments out of the coastal system.

- Human impacts vary from direct interventions on the beach, to introduction of alien species of molluscs, and pollution. Others, less obvious, but which when summed, produce significant effects, are:

  - abstraction of beach sand, even though prohibited by national laws;
  - 'manufactured' long sand mounds in front of beach bars/restaurants, which help mobilize sand due to wind transfer and storm waves;
  - a positive effect in the local beach sediment budget is represented by constant winter interventions carried out by coastal managers, e.g. placement of reed curtains on southern tourist beaches in order to keep sand from being transported by wind out of the beach system;
  - introduction of alien species to the Black Sea coast by cargo ships/oil tankers has also affected the supply of coastal sediments (shell fragments ground by waves on the sea bottom);
  - pollution, impacting the mollusc population, has an effect on shell fragment volumes.

The northern unit

The southernmost stretch of the Kilia sedimentary cell (Musura Bay in Romanian territory) has been an active accumulation zone. The appearance and rapid evolution of a spit in Musura Bay started to transform the bay into a lagoon, whilst aggradation occurred in the spit-protected area. The new spit development (elongation and inland shifting) was extremely active (about 100 m/yr) during the 1980s and 1990s (Stănică et al., 2007), but the process has recently decreased in intensity (Vespremeanu-Stroe et al., 2007).

In the northern part of the Sulina sector (the first 3 km), a rapid progradation of about 10–12 m/yr has taken place but the next 2.5 km of coast remains in equilibrium. In contrast, the natural erosion trends in the Imputita–Casla Vadanei sector has dramatically accelerated, producing the highest rates of shoreline retreat along the entire Romanian coast, i.e. some 28 m in 1997, between Imputita and Canalul cu Sonda, and 30 m to the north of Casla Vadanei (Stănică and Panin, 2009). Recent annual erosion rates (from the 1980s) vary between 5 and 25 m (Panin, 1999). Finally, most of the Sf. Gheorghe Sector is in equilibrium, as the shoreline generally maintains its position on an inter-annual scale, with the exception of the northern part and the extreme south, which are eroding.

Sahalin Spit moves westwards with an average speed of more than 15 m per year and elongates southwards. Its dynamics have been the subject of a long series of scientific studies (e.g. Panin, 1996; Giosan et al., 1999; Bhattacharya and Giosan, 2003; Vespremeanu Stroe et al., 2007; Dan et al., 2009; 2011).

The Zatoane–Midia sedimentary cell has also been subject to mixed dynamics. The northern sectors, in Zatoane and Perisor, have been eroding at rates between 2 and 5 m per year, but this is reversed towards Periteasca, where the average trend is one of accumulation of some 2 metres per year. From the southern part of Periteasca, all beach barriers in front of the Razelm–Sinoe Lagoon System are in an erosion phase of about 3 m per year on average.

The final stretch of the coast, from the southern part of Chituc Sector to Midia Harbour, is accumulating at a rate of about a metre per year (Vespremeanu-Stroe et al., 2007).

The southern unit

Most of these beaches have been artificially stabilized by a long series of coastal defence works; however, the chronic lack of sediments from the coastal system has generated erosion. The coastal cell delimited by the harbour jetties of Midia (north) and Constanta (south) has suffered from the building of these coastal structures. In Mamaia Bay (Figure 20.6) the beach suffered from accelerated erosion of up to 2 m/yr after development of the Midia Harbour jetties at the northern cell limit, except for the northernmost part of the bay, which is sheltered by these jetties (JICA/ECOH, 2006).

In the coastal cell between Eforie and Cape Tuzla, beaches have suffered from erosion of up to 2.5 m/yr, with local sheltering effects from the Belona Marina in Eforie North (Constantinescu, 2005). Constantinescu (2005) also showed the mild erosion phenomenon in Costinesti (lower than 1 m/yr) as well as shoreline retreat between 1 and 2 m/yr on average in the area north of Mangalia. The southern pocket beach of 2 Mai suffered from erosion at a rate of about 2 m/yr after the building of the Mangalia Harbour protection jetties. Vama Veche, the southernmost beach, is stable.

Cliff erosion has been an active process during the past century, with rates varying from 0.3 to 0.7 m/yr on average. Cliff erosion rates were obtained through comparison of topographic maps in 1924 and Ikonos satellite images taken in 2002 (Constantinescu, 2005). In the late 1950s, a widespread system of stabilizing cliff erosion was developed from Constanta South, Eforie North, Eforie South, Costinesti, Mangalia North (Olimp, Cape Aurora, Venus, Saturn) and Mangalia. Most of these coastal protection works (systems of drains combined with seawalls) have until recently been successful. The past decade started to witness failures of the cliff coastal defence systems, as no maintenance investment has been made for more than five decades. The rest of the cliffs have been left in a natural state, some (like the sections of Cape Tuzla–Costinesti and 2 Mai–Vama Veche) suffering from very active erosion, with retreat rates reaching even 1 m/yr.

■ **Figure 20.6**
The southern unit of the Romanian coast is almost entirely anthropogenic. Mamaia resort has the highest tourist capacity among all Romanian holiday resorts

Shore protection

All Romanian beaches have been for the past 15 years under the unitary management and coordination of the Romanian Waters National Administration and Dobrogea Littoral Water Basin Administration (ABADL). The northern unit, as part of the Danube Delta Biosphere Reserve, has been under the jurisdiction of the Danube Delta Biosphere Reserve Administration. Both administrations are under the coordination of the Romanian Ministry for Environment and Forests. The use and administration of tourist beaches from the southern unit are then periodically rented during a public bidding processes. Nevertheless, the idea of a centralized coastal management has been to avoid fragmented coastal development and protection strategies and plans. After some two decades, with no serious investment in coastal protection against erosion, ABADL is now managing a significant project aiming at creating an overall master plan for coastal protection against erosion, and selection of priority areas for coastal protection and organization of the necessary documentation for the design of coastal protection plans. This process, which is now in development, has put into practice the condition of applying modern concepts of coastal protection by adopting soft methods as much as possible, to avoid the negative effects of hard defence works. This has also come as a lesson learned from past interventions.

Nevertheless, except for two minor experiments in artificial nourishment (Mamaia and Venus, in 1989 and 2005), there has not been any significant work that embraces such a philosophy. The only large-scale soft interventions have been made by coastal managers during windy (winter) seasons, *i.e.* the placement of reed fences along tourist beaches.

Hard coastal defence works

From north to south, the following categories of coastal defence works are present:

The northern unit

- The Sulina navigation canal jetties, extending 8 km offshore, block southerly longshore sediment drift. The jetties are made of massive stone blocks, whose construction started in 1858 and continued in several stages until the 1960s.
- Groins closing the Portita Inlet (which had been the communication between the former Razelm Lagoon and the Black Sea until the 1970s), impacting nearby coastal dynamics and totally altering the salinity and water cycles in the lagoon system.
- Locks have controlled the water and sediment exchanges at Periboina and Edighiol inlets in Sinoe Lagoon since the 1970s.
- Midia Harbour jetties block longshore sediment transfer southwards and practically remove southerly coastal stretches out of the influence of the Danube's jetties (built in stages between the 1940s and 1981).

The above has been extensively commented upon by Panin, 1996; Giosan, *et al.*, 1999; Ungureanu and Stănică, 2000.

The southern unit

- Constanta (Figure 20.7) and Mangalia Harbour jetties have impacts on water and sediment circulation. The jetties, developed to the present stage in the period 1970s–1980s are made of concrete tetrapods (locally named *stabilopods*).

In total, hard coastal defence works represent 21.56 per cent of the entire length of the Romanian coast from Mamaia, Tomis (Constanta City), Eforie North, Centre (barrier beach separating Tekirghiol Lake from the Black Sea), South Costinesti, Olimp, Neptun, Jupiter, Aurora, Venus, Saturn, Mangalia, 2 Mai, with typical local impacts on coastal dynamics (their lengths are presented in Table 20.1).

The above has been extensively commented upon by Constantinescu, 2005; Caraivan, 2010; and JICA/ECOH, 2006)

■ **Figure 20.7**
*Harbour jetties
dramatically altered
coastal circulation.
Detail of Constanta
Harbour*

■ **Figure 20.8**
*Southern unit cliffs
and pocket beaches
almost entirely built
over, together with
extensive protection
systems developed
about 50 years ago*

Except for a permeable groin at Mamaia bay, built in the mid 1920s, the Romanian coast does not have such 'traditional' structures. Groins nevertheless exist in complex coastal defence systems, with various shapes (J, L and G), also combined with shoreline connected breakwaters and with seawalls in some parts of the coast. The most significant coastal defence systems, consisting of mixed J-shaped and L-shaped groins, breakwaters and seawalls are present in front of the city of Constanta (built in the 1980s), Eforie North (1950s–1960s), Eforie Middle on the barrier beach in front of Tekirghiol Lake, in combination with submerged breakwaters (1970s–1980s), Eforie South (1950s–1960s). Olimp, Neptun, Jupiter, Aurora, Venus, Saturn and Mangalia North coastal defence systems were put in place in the 1960s to the early 1970s, while the small L-shaped groin in 2 Mai also dates back more than 30 years (Figures 20.9 to 20.12).

■ **Figure 20.9** *Shore protection: systems of G-shaped groins*

■ **Figure 20.10** *Coastal protection system in Mangalia North, Aurora: small bays formed between groins and other defence works suffered from water quality issues during summers due to the lack of good circulation*

■ **Figure 20.11** *Coastal defence system at Mangalia: groins*

■ **Figure 20.12**
Coastal defence system at Mangalia: breakwaters and seawall (behind)

Detached breakwaters are found in the southern part of Mamaia Bay, as well as in the Saturn and Mangalia protection systems and as submerged structures in Constanta and Eforie, most dating from the mid-1980s.

Seawalls are extensively developed in some parts of Constanta, Eforie North, Eforie South and Mangalia and go back more than 50 years. The earliest date to the 1920s (Eforie) and were rebuilt in the late 1950s and 1960s. Other seawalls were built in the late 1960s in the resorts of Olimp and Venus and Mangalia North.

This extensive system of hard defence works was put in place several decades ago and due to changes in Romania's political regime and society, with the fall of a centralized economy, there have been extremely little or no investment made for maintenance of these coastal protection works; most are now at the end of their serviceable life. Besides their role in stabilizing the shoreline, accelerated erosion has appeared downdrift at each of these coastal protection works. Significant problems of water quality because of very low water velocity appear during summer (in areas such as at Aurora and Saturn – Mangalia North resorts) and in some of the small bays created between breakwaters.

Conclusions

The Romanian coast is divided into two different littoral units. The northern one, in front of the Danube delta, consists of low-lying beaches made of terrigenous fine sands and is part of a Nature Reserve, having the lowest density of inhabitants in Romania. In contrast, the southern Romanian unit, consisting of cliffs separated by barrier beaches and pocket beaches, has been almost entirely under anthropogenic intervention. Sources of sands here are mainly mollusc shells, the area is starved of siliclastic sediment and the entire littoral has suffered from human intervention. The Danube delta coast has been affected by the dramatic decrease in sediment supply caused by damming of the Danube and its tributaries and by other hydro-technical interventions along riverbanks, as well as by development of long jetties at the middle (Sulina) Danube mouth. These interventions have changed both the volume of sediments arriving to the coastal zone, as well as water and sediment circulation patterns. The limit between the two main units is at Midia Harbour. Jetties made for protection of this harbour three decades ago practically blocked any longshore drift that brought Danube alluvia towards Mamaia Bay. The protective jetties at Constanta and Mangalia Harbours also changed the natural circulation of water and sediments on the southern Romanian coast. Extensive coastal protection systems, consisting of breakwaters, groins and seawalls were built generally in the 1960s to the 1970s, and also induced adverse effects on the neighbouring coastal dynamics. Most of these works have reached the limits of their serviceable life and should be replaced, especially as, because of the political and economic changes during the past two decades, there have been no major investments in maintenance. New plans are under development to put a new coastal protection system into place by applying more up-to-date philosophies regarding technical solutions.

Acknowledgements

All photographs were taken by GeoEcoMar and University of Bucharest staff during field campaigns for R&D projects funded by the National Authority for Scientific Research, mainly Project CLASS – 32130/2008. Many thanks also to Dr. Stefan Constantinescu from the Faculty of Geography, University of Bucharest, for help in drawing Figure 20.1.

References

Bhattacharya, J. P., Giosan, L., 2003. Wave-influenced deltas: Geomorphological implications for facies reconstructions, *Sedimentology*, 50, 187–210.

Bondar, C., 1989. Trends in the evolution of the mean Black Sea level, *Meteorology and Hydrology*, 19, 2, 23–28

Bondar, C., Roventa V., State I., 1973. *La Mer Noire dans la zone du littoral roumain Monographie hydrologique*, Editura Technica, Bucharest, Romania, 516 pp. [in French]

Bratescu, C., 1933. Quaternary profiles in the Black Sea cliffs, *Bul. Soc. Rom. Geogr.*, 52, 24–58. [in Romanian]

Caraivan, Gl., 2010. *Sedimentological study of the beach and inner Romanian shelf deposits between Portita and Tuzla*, Ex. Ponto Printing house, Constanta, Romania, 171 pp. [in Romanian]

Constantinescu, St., 2005. Analiza geomorfologică a țărmului cu faleză intre Capul Midia și Vama Veche pe baza modelelor numerice altitudinale, Unpublished Ph.D. thesis, Faculty of Geography, University of Bucharest, 220 pp. [in Romanian]

Dan, S., Stive, M., van der Westhuysen, A., 2007. Alongshore sediment transport capacity computation on the coastal zone in front of the Danube Delta using a simulated wave climate, *Geo-Eco-Marina*, 13, 21–30.

Dan S., Stive M. J. F., Walstra D. J., Panin N., 2009. Wave climate, coastal sediment budget and shoreline changes for the Danube Delta, *Marine Geology*, 262, 39–49.

Dan S., Walstra D. J., Stive M. J. F., Panin N., 2011. Processes controlling the development of a river mouth spit, *Marine Geology*, 280, 116–129.

Giosan, L., Bokuniewicz, H., Panin, N., Postolache, I., 1999. Longshore sediment transport pattern along the Romanian Danube Delta coast, *Journal of Coastal Research* 15, 4, 859–871.

JICA/ECOH report (Japan International Cooperation Agency and Ministry of Environment and Water Management, Romania), 2006. *The study of protection and rehabilitation of the southern Romanian Black Sea Shore in Romania*, 129 pp. Available at: http://www.mmediu.ro/gospodarirea_apelor/zona_costiera/proiect_jica.htm (last accessed 26 August 2012). [in Romanian]

Kuroki, K., Goda Y., Panin, N., Stănică, A., Diaconeasa, D., Babu, Gh., 2006. Beach erosion and its countermeasures along the Southern Romanian Black Sea shore, *Coastal Engineering 2006* (Proceedings of the 30th International

Conference, San Diego, California, USA, 3–8 September), 4, 3788–3799.

National Administration of Meteorology, 2008. *Clima Romaniei*, Ed. Academiei Romane, Bucharest, 365 pp. [in Romanian]

Panin, N., 1996. Impact of global changes on geo-environmental and coastal zone state of the Black Sea, *Geo–Eco–Marina*, 1, 7–23.

Panin, N., 1998. *Danube Delta: Geology, Sedimentology, Evolution*, Association des Sedimentologistes Francais, Paris, 65 pp.

Panin, N., 1999. Global changes, sea level rise and the Danube Delta: Risks and responses, *Geo-Eco-Marina*, 4, 19–29.

Spătaru, A., 1984. Research programme for coastal protection works, *Hydraulics Research*, 30, 159–214.

Stănică, A., Dan, S., Ungureanu, G. V., 2007. Recent sedimentological processes and coastal evolution as effects of human activities in the area of the of Sulina mouth of the Danube–Black Sea, Romania, *Marine Pollution Bulletin*, 55, 10–12, 555–563.

Stănică, A., Dan, S., Jimenez, J., Ungureanu, Gh., 2011. Dealing with erosion along the Danube Delta coast: The CONSCIENCE experience towards a sustainable coastline management, *Ocean and Coastal Management*, 54, 898–906.

Stănică, A., Panin, N., 2009. Present evolution and future predictions for the deltaic coastal zone between the Sulina and Sf. Gheorghe Danube river mouths (Romania), *Geomorphology*, 7, 41–46.

Ungureanu, Gh., Stănică, A., 2000. Impact of human activities on the evolution of the Romanian Black Sea beaches, *Lakes and Reservoirs: Research and Management*, 5, 111–115.

Vespremeanu-Stroe, A., Constantinescu, fi., Tatui, F., Giosan, L., 2007. Multi-decadal evolution and North Atlantic oscillation influences on the dynamics of the Danube Delta shoreline, *Journal of Coastal Research*, SI 50 (Proceedings of the 9th International Coastal Symposium), 157–162.

21 Ukraine

Yuri N. Goryachkin

Introduction

Ukraine is the largest European country (603,700 km^2) and has a population of 45.6 million people. The 2,700 km coastline includes the northern and western shores of the Black Sea together with the Sea of Azov. With *circa* 40 per cent being beaches, the traditional places for summer recreation, the most popular area is the Crimean peninsula, whose coastline length comprises about one-third of the total coastline. Crimea has some five million tourists per year in addition to its two million inhabitants. Significant parts of the coast are used not for recreational purposes, but for economic activities (ports, agriculture and others), so in the holiday season the real figure reaches some 20 people per metre of coastline. One significant problem is coastal erosion, which in some areas has been accelerating over recent decades.

Coastal morphology

The current shape of the Black Sea coastal zone and the Sea of Azov was formed during the Holocene period, as a result of interaction between land and sea. Traditionally, the Black Sea coastal zone and Sea of Azov is divided into eight areas (Figure 21.1, a to h), based on coastal morphology and processes (Zenkovich, 1960; Shuyskiy, 2000).

(a) *The northern area of the Danube delta.* The total length is 75 km, formed of typical late Quaternary sediments. In the nearshore zone a sand and silt longshore bar occurs, which moves southward with

a volume of more than 1 million tonnes per year. The region has a complex structure of several parallel bars. At the edge, an increase in horizontal extent of 3–40 m/yr occurs, which in some areas reaches 130–180 m/yr as accumulation processes predominate (Lihosha *et al.*, 2004; Cheroy and Lihosha, 2007).

(b) *The north-western firth area.* With a length of 355 km stretching from the Danube estuary to the Bug firth, it is located within the boundaries of the coastal base-level plain, the southern part of which was flooded during the Holocene transgression. The lagoons are separated from the sea by sandbars, which alternate with erosion areas (up to 3 m/yr) resulting in frequent landslides, but in spite of this little sediment is produced for beach formation, because the rock composition is basically clay. Sandbars are retreating normally in a landward direction, therefore keeping their original volume rather than being washed away (Vykhovanets, 1993). If the level of the Black Sea keeps rising, the coast will be menaced by small and large scale flooding.

(c) *The Dnieper-Karkinita area.* This has a total length of 660 km, stretching from the Bug firth to Bakalskaya spit (Karkinita bay, North-Western Crimea). The alluvial-marine depositional plain region is characterized by relatively small actual elevations (up to 10 m). The coast represents the largest accumulative entities (the spits) of the North-West Black Sea coast, as well as a predominance of abrasion shores on the head of Karkinita bay. A meandering coastline and shallow underwater slope is typical of this area, with several depositional forms, mainly sand bars. Erosion (0.7–1.8 m/yr) takes place

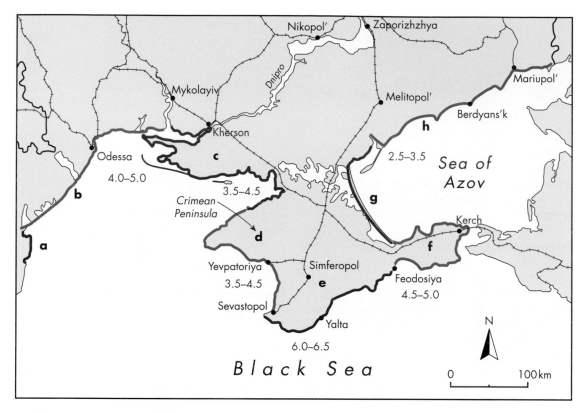

■ **Figure 21.1** *Geomorphological zone division of the Ukrainian coast (described in text). (a) Northern area of the Danube delta; (b) North-Western firth area; (c) Dnieper-Karkinita area; (d) West-Crimean area; (e) South Crimean area; (f) Kerch area; (g) Arabat-Sivash area; (h) North Azov area. Red figures in sea: significant wave heights [m] for 10– 50 years return storms (Efimov and Komarovskaya, 2010). Red (deposition) and blue (erosion) lines separate the different dominant processes*

mainly during storm surges (height, 0.7–3.0 m). Where the shoreline comprises bluffs, wave erosion triggers landslides with dissected blocks.

(d) *The West Crimean area.* This has a total length of 275 km and stretches from Bakalskaya spit to Sevastopol city. The erosional coast in the Tarkhankut peninsula region and in the south-western spurs of the Crimean Mountains is composed mainly of limestone. The northern segment of this coast is formed by highly erodible rocks, very susceptible to landslides. To the south, the coast is mainly erosional with caves cut into clay rocks, whilst the centre is occupied by sandbars. The coastal zone is significantly sinuous, due to a well-developed

network of gulfs, bays and lakes separated from the sea by narrow sand bars. In recent decades, significant anthropogenic pressure has been the main cause of erosion, as this is the most promising area for resort development and the area has recently acquired great importance.

(e) *The South Crimean area* extends from Sevastopol to Feodosiya, covering a total length of 350 km and is the main Ukrainian coastal resort region. It is adjacent to the Crimean Mountains, an Alpine mountain system, 700–1,300 m high. The coast is characterized by small bays, with separate smooth concave areas having weak sinuosity. A series of capes in the Western and Eastern parts are composed of massive limestone and volcanic rocks, the bays being mainly developed in argillite and sandstone areas. As the coast consists mainly of solid rocks it is hardly subject to any change. Only a relatively small area of the coast from Alushta to Sudak is made of complex clay rocks. Landslides are widespread with up to 50 per cent of the coastline being affected. A ria coast occurs in the West, in contrast to an eastern mountainous erosional coast with small bays cut into the hard rocks. One of the features of the central coastline is the number of coastal protection structures, a significant part of the beaches being artificial (Goryachkin, 2010).

(f) *The Kerch area* covers the Kerch peninsula from the Feodosia to Arabatskaya spit and has a total length of 320 km. In the south, it is washed by the Black Sea, in the east by the waters of the Kerch strait, and in the north by the Sea of Azov. The area lies at the periphery of the Crimean Mountains, and rocks of the upper Paleogene-Neogene and Quaternary deposits are predominant. The area is characterized by large coastline variations and the presence of numerous curved bays and capes, made up, mainly, by outliers of reef limestones. The Holocene later created bays separated by sand/shell and pebble/sand bars (the main depositional shoreline forms – about 20 per cent) to form numerous lagoons. A significant part of the coast is again subject to landslides in clays and a variety of erosional and depositional forms is typical. The coastal region is characterized by a high development of mud volcanoes and a significant area is used as military firing ranges.

(g) *The Arabat-Sivash area* is situated in the Western part of the Sea of Azov, with a total length of approximately 180 km (excluding Sivash). The area received its name from the Arabatskaya spit, which separates the cove of Sivash from the Sea of Azov and a narrow strait connects it northwards with the sea. The Arabatskaya spit is an accumulative surface bar more than 120 km long, 2–5 m high and 0.5–5 km wide, composed mainly of sand and shells. Its underwater coastal slope is gentle, with prevailing depths of up to 10 m but certain parts suffer from erosion. The Sivash shoreline is low and flat, being mainly made of Quaternary loams and is highly dissected, with a network of bays, peninsulas and lakes and longshore drift prevails. The main source of sediment is bottom erosion and wave emission of the shells. Wind drying of areas up to 6–7 km wide is typical.

(h) *The North-Azov area* is located on the northern coast of the Sea of Azov, on the Ukraine border, and has a length of some 480 km. It stretches along the southern slopes of the Ukrainian shield and Azov massif. The indigenous seashore, up to 15 m high, is made of Neocene-Quaternary deposits. The shores have accumulation bodies, *i.e.* spits which stretch their free distal 30–45 km into the sea. The basement spits' width is about 5–7 km, at the distal, 0.5–1.5 km, and comprises sand and shells, usually 2–3 m above sea level. The main sediment source is biogenous and erosion; however, this area suffers from a deficiency in these matters (Shuyskiy, 1987). The coastline between spits is characterized by active erosional landslide processes.

Wave climate and sea level rise

North and North-East winds prevail during the year but due to the coastal configuration, they do not cause storms which would have a significant coastal impact, but drift sand sediments to the sea, which is a significant factor in the coastal dynamics (Vykhovanets, 2003). Storm waves from the southern, south-west and south-east have the greatest impact. Despite its relatively small size, the Black Sea has large waves and when there are hurricane force winds; waves can reach heights of 8–10 m in the open sea, and 4–7 m near the coastline (red numbers in Figure 21.1). In the Sea of Azov, wave maximum heights are 3.5–4 m. Autumn and winter cyclones, which have especially devastating consequences for the coast and coastal facilities, take place every seven to ten years and differ from ordinary cyclones. An example of such a severe storm was the Balaklava storm on 14–15 November 1854, which triggered the creation of a network of meteorological stations in Europe. Analysis of the long-term variability of repetitive strong winds and storm waves from hazardous wave directions proved that the 1950–1960s period was characterized by high intensities of winds

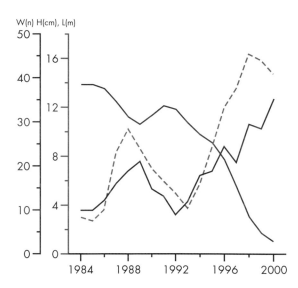

W(n) H(cm), L(m)

■ **Figure 21.2** *Changes of the Black Sea level (B – blue), Kalamita bay coastline (L – red) and number of storm wave cases (W – black)*

and waves, replaced by a significant recession in the 1970–1980s. In the 1990s, it increased again, especially for South-West and West storm winds, which are the most dangerous for the major part of the coast (Figure 21.2; Goryachkin and Ivanov, 2010).

The natural variability of the present coast is by sea transgression. During the last 60 years, the Black Sea water level has increased by *circa* 15 cm and reached a historic high in 2010 (for 160 years of observation). The average rate of the Black Sea level rise can be estimated as 0.25 cm/yr, 0.1 cm/yr being due to tectonic movements (land subsidence) of the coast (Goryachkin and Ivanov, 2006), especially in the Odessa region (0.4 cm/yr). However, the main reason for the Black Sea level rise is the current increase in fresh water input which exceeds the sea water evaporation rate (Goryachkin and Ivanov, 2006). Coastal retreat under the influence of sea level rise can be estimated as 0.2–0.3 m per 1 cm sea level rise (Goryachkin, 2011).

Evolution of coastal settlements

States and settlements did not emerge on the present coast until about the fifth century BC; these were colonies of the Mediterranean civilizations, the ancient Greek city-states. During this colonization, a number of city-colonies were founded, e.g. Olbia and Chersonesus. Later, in the second century BC, they came under the control of the Roman Empire. The location of ancient settlements often coincides with that of the modern ones, especially in the Crimea, e.g. Chersonesus (the modern name is the city of Sevastopol), Kerkinitida (Yevpatoria), Theodosia (Feodosiya). In ancient times, economic activity and trade were very intense, with trade routes linking Europe with Asia. The fall of the Eastern Roman Empire (Byzantium) in the fifteenth century resulted in the present-day Ukrainian coastline coming under the Ottoman Empire and its vassals (the Crimean Khanate) and coastal trade and economic activities was greatly reduced. With the accession of the Black Sea Coast to the Russian Empire in 1783, economic activity began to grow again. New cities were established – e.g. Odessa, Nikolayev and Sevastopol – and industry and trade became highly developed. In the late nineteenth to early twentieth centuries the coast began to be used for resort purposes – the summer residence of the Russian Czars was in the Crimea and the coastal population was constantly expanding; in 1897, only 0.5 million people lived in the Crimea, but in 1937 more than 1 million.

USSR industrial development, followed by reconstruction of cities destroyed during World War II, took place from 1930 to 1950, resulting in a great amount of shoreline sand and gravel being excavated from beaches. In 1963, a USSR Government Resolution prohibited beach sand extraction. Significant development of coastal resorts and business activities in general is associated with the 1960–1980s, in tandem with erosion problems. By 1989, the seaside population had reached a high of 6.8 million people (2.5 million in the Crimea). With the formation of the Ukraine, the population reduced and on the coast it decreased by 0.6 million. After

1991, there was a reduction in the number of industries and simultaneously an increase in construction on the coast, of many private houses, hotels, sanatoriums and other recreational facilities. In many cases, this construction was carried out without necessary engineering and scientific support, creating significant problems, for private owners and society as a whole.

Beach erosion

Historical times

In ancient times, the level of the Black Sea was lower than it is now (Berenbeym, 1959) and the coastline was different, especially on the Kerch strait, the north-western Black Sea and the Sea of Azov. Some of the rivers flowing into the Black Sea and the Sea of Azov exited far from their present-day mouths. Evidence can be found from:

- detailed studies of the ancient Chersonesus harbour, which showed that 2,500 years ago sea level was much lower than the modern one and the old coastline is now *circa* 4 m under water (Zolotarev, 2004);
- Roman roads, which follow the shallow bottom isobaths around Cape Chersonesus;
- Danube river valley excavations, where artefacts of the Dac and Greek cultures were found 4 m below the modern river channel (Banu, 1969);
- numerous finds in cities, e.g. moorings, are lower than present-day levels, clearly indicating sea level rise (Gorlov *et al.*, 2004).

Judging the modern rate of coastal submersion under tectonic factors indicates a rise of 2.5 to 4.0 m during the past 2,500 years (9.0 m in the Odessa region; Goryachkin and Ivanov, 2006). The slower periods of sea level rise date back to 400–600 and 1200–1300 AD. Evolution of erosional-depositional coastal systems took place when sea level was rising due to small amplitude fluctuations, which are apparently linked with climate-induced fluctuations

of the Black Sea water balance, which can be observed at present (Goryachkin and Ivanov, 2006).

The various relations in the rates of level change and the quantity of sediment arriving into the coastal zones define the dynamics of these systems. In conditions of a considerable sediment budget imbalance sea level rise may:

- determine the rate of coastal retreat (in the case of a sediment lack);
- not have a great influence on shore behaviour, or may lead to its increase (in conditions of sediment abundance).

The nineteenth and twentieth centuries

In the nineteenth to the beginning of the twentieth century, coastal economic activity was relatively small and, therefore, evolution was under the influence of natural factors. Analysis of old maps (since 1817) have been made via systematic topographic mapping. Although co-ordinate accuracy compared with modern technologies is relatively low, they can indicate qualitative shoreline changes (*Oceanographic Atlas*: Eremeev and Symonenko, 2009). They showed that the greatest changes took place on the depositional shores of the North-Western segment of the Black Sea and on the shores of the Sea of Azov (maximum erosion of the shore 12 to 16 m/yr) and different trends were observed. Some spits divided into several islands; some, on the contrary, actively grew; small bays turned into closed lagoons, etc. The Crimean North-Western coast actively retreated and several coastal settlements were flooded. In 1925, a severe storm washed out a spit in the Kerch strait; its distal part formed the island of Tuzla, and later (in 2003) caused a territorial dispute and political crisis between Russia and Ukraine.

Since the middle of the twentieth century, coastal erosion has greatly increased, due to intensive coastal zone development. In most cases, anthropogenic impact has led to violation of natural coastal zone processes and in a number of cases, has determined coastal dynamics. This caused some reaction by the former USSR authorities and coastal research became active, resulting in the launch of a coastal observation network. In 1974, an erosion rate map for the whole coast was published, and for the first time causal analyses were made (Shuyskiy, 1974), which showed the prevalence of erosion processes. Erosion rates varied greatly, but was usually 0.5–1.0 m/yr, sometimes reaching 15–20 m/yr. This map was later updated and complemented (*Oceanographic Atlas*, 2009).

The highest erosion rates (on average being 5–8 m/yr) are on landslide eroding shores in cliffs made of clay rocks, typical for the north-western Black Sea coast and the Sea of Azov. The lowest erosion rate was a few centimetres per year, observed on strong limestone conglomerates, sandstones or other rock coasts, such shores being located on the southern Crimean coast on the Tarkhankut and the Kerch peninsulas. Abnormally high rates of erosion of up to 12 m/yr were registered in Kalamita bay, possibly the result of aggregate extraction from the nearshore. Here approximately 15 million m³ of materials were quarried, triggering a 200 m shoreline retreat. Along with other actions, this has caused a serious problem for health resorts on Kalamita bay, which has yet to be solved.

In the twentieth century to the present, the following anthropogenic impact factors have accelerated beach erosion (Goryachkin, 2010):

- Construction of coastal hydrotechnical facilities. In most cases, their purpose is protection from erosion, which in some cases has led to negative consequences.
- Operation of coastal zone construction material quarries has caused beach/bottom changes and lack of sediment. As a result, beaches are becoming much narrower and currently such quarries have been closed almost everywhere. However, there are some cases of 'pirate' production from sand mining ships.
- Dredging carried out in ports which leads to changes in the natural direction of sediment movement, which can cause a lack of sediment leading to beach volume reduction in certain areas, e.g. the commercial port of Yevpatoria.

- Removal of sand and pebbles away from beaches, causing a sediment shortage. It was banned in 1963, but has been carried on everywhere until now.
- Construction on beaches and coastal dunes leads to beach changes and increases the wave influence, finally causing beach reduction. Such construction can be seen everywhere.
- Regulation of river flow by reservoirs leads to reduction of solid runoff, lack of sediment, and to reduction of the natural longshore sediment flow. By the 1970s, almost all rivers had been regulated.
- Artificial cutting of sandbars, which leads to a reduction of beach length and change in sediment dynamics of the adjacent areas. It is typical of the western Crimea.
- Pollution of water and bottom sediments by communal and industrial wastewater. This leads to reduction of benthic vegetation, as well as a reduction of the number of shellfish, the shells of which are a source for sand formation.
- Biological invasive species. In the mid twentieth century, a mollusc called *Rapana venosa* appeared in the Black Sea from its natural habitat in the Sea of Japan and due to a lack of natural enemies, its population has grown rapidly. It has destroyed colonies of local shellfish, which has reduced the potential for beach material formation.
- Unauthorized activity in the coastal zone. Such activity is observed everywhere on the coastline and includes construction of various kinds of structures on beaches and coastal cliffs without permission by the authorities.

Present status

In the first decade of the twenty-first century, significant coastline changes have occurred associated with the intensification of economic activities, construction in particular. These resulted in exhaustion of the natural and artificial beach material, in washing out depositional forms and in active development of landslides along the greater part of the coast. There are plans, not sufficiently implemented, to expand resorts in both the Black Sea and the Sea of Azov, with special attention being paid to developing the high recreational potential of the Western Crimea and Sea of Azov.

Scientific research on coastal processes has recently expanded. After the collapse of the USSR, some scientific and practical organizations dealing with coastal studies were closed and financing for others reduced dramatically. However, current society and governmental structures understand the importance for scientific research into coastal processes. In recent years, new research tools have appeared, such as satellite imagery, GPS, etc. Information from space allows scientists to obtain exact values of shoreline changes over large coastal areas and an example of the use of such information is presented in Figure 21.3. This shows the perilous state of beaches on Yevpatoria, one of the most popular resorts. Analysis of satellite images has indicated that up to 70 per cent of the Crimean west coast is suffering from significant erosion The condition of other coasts is little better.

Shore protection

Administrative competencies on shore protection

The legal and actual chaos, typical of the coastal situation since Ukraine obtained independence, posed significant problems for society as a whole and caused much criticism and debate. It led to adoption

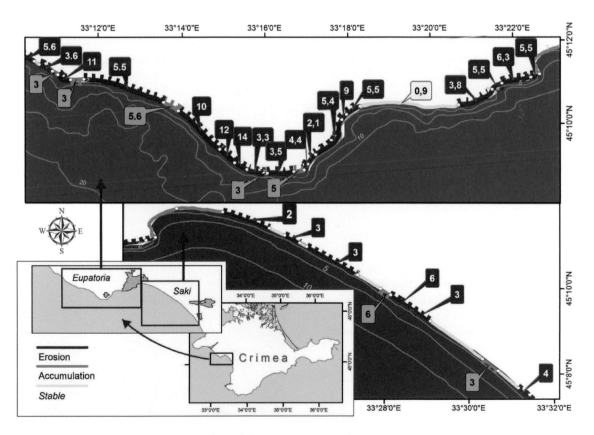

■ **Figure 21.3** *Changes in the coastline of the Western Crimea for 2006–2010 (metres)*

by Parliament of a new law (amendments to the Water Code), which entered into force on 1 January 2011, stating the concept of coastal protection belts. These strips are to be on all land plots (except maritime transport land), belong to the state and are communal property. The law introduced the concept of the beach area – a part of the coastal protection belt, not less than 100 m in width from the coastline. Free access to this area should be provided (exceptions are sites with hydrometric structures, communications, sanatoria, medical establishments, health centres and children's camps). Within the limits of the beach area construction is forbidden (except for hydraulic engineering, hydrometric structures and communications). Local authorities must develop the beach zone within the borders of cities and villages. The term 'beach area' is not applicable to land belonging to maritime transport and military facilities. Financing is a significant problem, as state budget financing is not sufficient and the possibility of help from municipal budgets is low.

History and modern state of shore protection in Ukraine

Most coastal structures are situated on the Crimean peninsula coast. They were first built at the beginning of the twentieth century and were primitive retaining sea walls made from unstable materials, which were quickly destroyed. Later, people started to protect the coast not only from the sea, but also from the land, *i.e.* landslides and mudslides. A need for protection arose when the coast began undergoing active development and expansion of recreational activities and during the 1960s–1980s large amounts of money were allocated for these purposes. About 700 ferroconcrete groins were built in the Crimea, and 80.6 km of the coast fixed. These were mainly built around public beaches and resort facilities, as well as for the protection of bays with a high landslide risk. Simultaneously, experimental coastal protection facilities were built (breakwaters, transverse dikes). The most preferred kinds of coastal protection for the Crimean southern coast were groins (25–30 m long);

■ **Figure 21.4** *A typical view of coast protection on the southern coast of the Crimea*

nourished beaches (up to 20 m wide); and complex seawalls (first level 2 m high, embankment 10 m wide, and a sloping buttress 2 m high).

Many of these structures were built with various modifications and important principles for their successful operations have been established. After the collapse of the USSR and formation of an independent Ukraine, financing of all coastal protection works stopped. Since then, during the past 20 years, there has been a progressive deterioration in structures: e.g. about 25 per cent of groins have been destroyed or require urgent repair. A significant part of the artificial beaches, a major element of coastal protection created in the 1970s and 1980s, are now in a critical condition, as their area has considerably decreased. In some inter-groin sections, the artificial beaches have disappeared completely, which often

■ **Figure 21.5** *Embankment in Yevpatoria*

■ **Figure 21.6** *Total disappearance of the beach*

leads to structural destruction, even with small storms. In order to replenish and restore beaches along the whole coastal protection area, 800,000 m³ of crushed stone, costing about US$7 million is needed. Another US$16 million is needed to prevent further destruction of coast protection structures, but for many years even such relatively small sums of money have not been granted by the government.

Assessment of the effectiveness of coastal protection structures on the Crimean southern coast taking into account operational experience was carried out according to the following criteria: engineering, geological, technical, economic and ecological status. As a result, it was found that beach nourishment with groins and complex seawalls were most effective; artificial beaches with a system of breakwaters proved to be the least effective.

While on the Crimean southern coast protective constructions have played a positive role in general, on the Crimean west coast, where their number is the second largest in the area, they turned out to be less effective. For example, construction of a concrete embankment with a sea wall and breakwater in the port of Yevpatoria resulted in the total disappearance of a city centre sand beach and erosion of adjacent coastal areas. Today, the embankment is being destroyed (Figure 21.5) and the city is gradually losing its significance as the main children's health resort in Ukraine. Formerly magnificent sand beaches are being replaced by a 'stone chaos', because of unauthorized coastal protection works (Figure 21.6). Recently, nourished beaches were stabilized by wave-absorbing structures, *i.e.* concrete elements with openings through which waves enter and are absorbed through multiple reflections, thus reducing offshore reflection that induces structure scouring (Figure 21.7). Since installation is quite recent, it is impossible to evaluate their efficiency.

One of the peculiarities of the Crimean western coast is the large cliffs made up of easily eroded clay rocks and each year landslides injure or kill people. On a vast part of the coastline, concrete step slopes were used as a means of coastal protection, but proved completely ineffective. At present, they either are destroyed or are near to being destroyed (Figure 21.8). Their further operation may be a risk hazard for people.

■ Figure 21.7 *Coast protection with wave-absorbing structures (West Crimea)*

■ Figure 21.8 *Concrete step embankment (West Crimea)*

Wave-absorbing structures were also used on the eastern coast of the Crimea (Figure 21.9). Experience has shown that a retaining sea wall eventually forms a water area with very shallow depths and a sand beach. Currently, coastal protection structures here are in ruins.

Development of beaches in the unique bay of Koktebel (eastern Crimea) has been instructive. It is a closed system, which has favourable conditions for wide natural beaches. Several watercourses flowing into the bay served as a source of material and only the capes were subject to erosion. In 1966, because of ten years of uncontrolled export of beach material for construction purposes, the bay was under pressure. Sand was extracted mainly in the bay centre where the beach was wide, it being a convergence area, and the beach quickly recovered due to sediment transfer from adjacent bay segments. Although quarrying was carried on only at the bay centre, all the system was exploited and the beach was not large enough to absorb storm waves, which induced severe erosion. In 1968 a beach nourishment project was carried out with 150,000 m³ of sediment deposited on the west side of the bay, and longshore transport was expected to redistribute this material. As a result, where the beach width was 2–4 m in 1966, it increased to 30–35 m by 1969. In 1981, the artificial beach was slowly moved by longshore currents back

■ Figure 21.9 *Coastal protection structures of different types (East Crimea)*

to the bay centre (the convergence zone) and erosion again commenced on the flanks. A beach fill for more than 360,000 m³ of sand has been designed, but this project has not been completed.

The main part of the city of Odessa and its adjacent suburbs is located on a landslide coast. The 'struggle' with the sea started almost in the first decades of the city's existence. However, active measures against coastal erosion began only at the end of the nineteenth century, after landslides had seriously damaged and even destroyed many luxurious villas and in the mid-1960s the problem became even more serious. The government made a decision to allocate money

■ **Figure 21.10** *Types of coast protection in Odessa bay: sloping wall with riprap and reef-ball*

for the construction of coast protection structures on the greater part of the Odessa coast. The cost of actions was astronomical, even for the USSR. About a dozen beaches were created and protective dikes, breakwaters and sea walls surrounded the shores. As a result, Odessa became *de facto* a resort apart from being the largest industrial centre. The project construction had a safety factor of 25 years and in general, structures have withstood this term. Today the expensive coastal protection and anti-landslide systems are in a state of disrepair and some 20 per cent of coastal protection structures have been destroyed, mainly groins, transverse dykes and piled wharfs, but local authorities have no money to keep the structures operating. Consequently, beaches in Odessa and its suburbs are completely or partially disappearing again; coastal slopes and the shoreline are being destroyed. Currently, one can get permission for construction close to the sea only if relevant coastal protection facilities are constructed. However, in most cases such a policy does not work, as very often companies break their obligations, or utilize unauthorized construction. Various methods are being used: for example, cliffs and slopes are cut back, and on sand beaches, artificial reefs are constructed (Figure 21.10), all too often without permits, etc. In general, there is no thought-out strategy of coastal protection.

As an example of bad management, in 2007 there was an attempt to restore the Odessa region's artificial beach. The beach was completely washed away within two years causing great damage to the marine biota (Figure 21.11), as construction had violated the coastal zone's natural processes. Due to destruction of natural habitats, species diversity, population density and biomass for almost all groups of aquatic life has decreased. Many species of crabs, fish and other sea creatures have either disappeared from the area of the hydrotechnical engineering structures, or their number has decreased astronomically (Adobovskiy, 2001). Continuous degradation of the water areas and coastal protection construction system could lead to a significant loss of value and attractiveness for resort areas on the coast near to Odessa.

■ **Figure 21.11** *Artificial beach creation in the Odessa region*

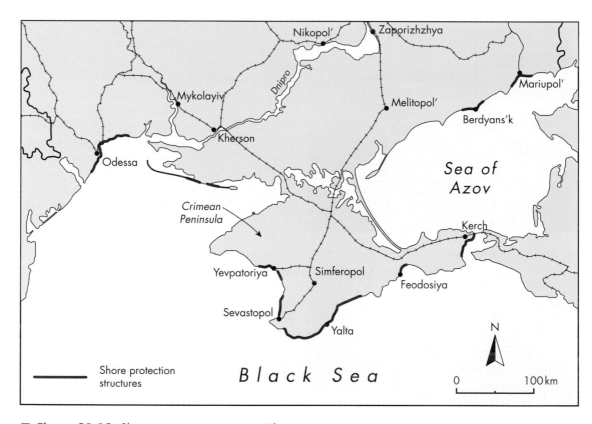

■ **Figure 21.12** *Shore protection structures in Ukraine*

In the Berdyansk and Mariupol regions, retaining sea walls and various types of breakwaters, including segmented breakwaters are used. On landslide cliffs, slope trimming and subsequent consolidation is used. At the same time, a lack of funds allocated for these purposes result in measures that do not provide timely protection and preservation for coastal areas. Location of the main areas of Ukraine, where there are coastal protection structures, is shown in Figure 21.12. In the Sea of Azov, coastal protection constructions are confined mainly to port structures.

Conclusions

In recent decades, due to natural and human factors, acceleration of erosional rates has occurred. When assessing the tasks of coastal protection, it is necessary to take into account not only local processes, but also phenomena occurring at the global level (sea level changes, storm activity, a fall in productivity of sediment sources, etc.) which directly influences the coast. It is time to re-examine the current viewpoint of protection strategy, taking into account sea level change forecasts for the twenty-first century. Coastal protection in Ukraine has been given to technical organizations for a long time and the coastal zone has actually turned into a testing ground for different types of constructions. Positive examples of coastal protection in Ukraine are rare and do not affect the bleak overall picture. The main reasons for such failures seem to be the following:

- lack of knowledge or ignorance of natural processes in the given area of shore protection;
- a technical approach to coastal protection, allowing replacement of the natural landscape with a mass of reinforced concrete constructions;
- ignoring world experience and advanced technologies;
- inactivity and inability of administrations of all levels, who are obliged to carry out monitoring of economic activities in the coastal zone, to provide positive action; the difficult economic condition of the country.

It is impossible to protect the entire Ukrainian coast. In the current conditions, it is necessary to define the most valuable sites from an economic and social viewpoint, as well as the damaged parts of the coast. There is a need to develop rational projects for protection and stabilization, and sacrifice less valuable areas. In other words, there is a need to advance and implement an optimal strategy of coastal protection. Scientists are ready for such actions; it is for administrations to decide.

Acknowledgement

Thanks to the Marine Hydrophysical Institute at the National Academy of Sciences of Ukraine for providing all the figures.

References

Adobovskiy, V. V., 2001. Sovremennoe sostoyanie i nekotoryie ekologicheskie problemyi akvatoriy sistemyi beregozaschitnyih sooruzheniy Odessyi [Current state and some environmental problems of Odessa coast protection structures (in Russian)], in *Ustoychivoe razvitie turizma na Chernomorskom poberezhe*, OTsNTEI, Odessa, 344–352.

Banu, A. C., 1969. Preuves historiques sur lesoscillations du niveau de la mer noire pendant les ages derniers millenaires. *Rapp. Pr.-Verb. Reun. C. I. E. S. M. M*, 19 (4), 617–618.

Berenbeym, D.Ya., 1959. Kerchenskiy proliv vo vremena Strabona v svete noveyshih dannyih ob izmenenii urovnya Chernogo morya [Kerch Strait at the time of Strabo in the light of new data on the variation of the Black Sea (in Russian)]. *Sov. Arheologiya*, 4, 42–52.

Cheroy, A. I. and Lihosha, L. V., 2007. Protsessyi deltoobrazovaniya v uste Dunaya [Processes of delta formation in the Danube estuary (in Russian)]. *Ekologiya morya*, 74, 91–94.

Efimov, V. V. and Komarovskaya, O. I., 2010. *Atlas ekstremalnogo vetrovogo volneniya Chernogo moray [Atlas of extreme wind-generated waves of the Black Sea* (in Russian)], EKOSI-Gidrofizika, Sevastopol, 59 pp.

Eremeev, V. M. and Symonenko, S. V. (eds), 2009. *Oceanographic Atlas of the Black Sea and the Sea of Azov*, State Hydrographic Service of Ukraine, Kiev, 356 pp.

Gorlov, Yu.V., Porotov, A.V. and Stolyarova, E.V., 2004. Kotsenke izmeneniy urovnya Chernogo morya v antichnyiy period po arheologo-paleogeograficheskim dannyim [Evaluating the changes of the Black Sea in ancient times on archaeological and paleogeographic evidence (in Russian)]. *Drevnosti Bospora*, 7, 197–207.

Goryachkin, Yu. N., 2010. Antropogennoe vozdeystvie na chernomorskie berega Kryima [Anthropogenic impact on the Black Sea coast of Crimea (in Russian)]. *Ekologicheskaya bezopasnost pribrezhnoy i shelfovoy zon i kompleksnoe ispolzovanie resursov shelfa*, 23, 193–197.

Goryachkin, Yu. N., 2011. Otklik beregovoy linii Kalamitskogo zaliva na izmeneniya urovnya moray. [The response of the shoreline to the Gulf Kalamitsky sea level change (in Russian)] *Ekologicheskaya bezopasnost pribrezhnoy i shelfovoy zon i kompleksnoe ispolzovanie resursov shelfa*, 23, 56–69.

Goryachkin, Yu. N. and Ivanov, V. A., 2006. *Uroven Chernogo morya: proshloe, nastoyaschee i buduschee [The level of the Black Sea: Past, present and future* (in Russian)], EKOSI-Gidrofizika, Sevastopol, 210 pp.

Goryachkin, Yu. N. and Ivanov, V. A., 2010.. Sovremennoe sostoyanie chernomorskih beregov Kryima [Current status of the Black Sea coast of Crimea (in Russian)]. *Dopovidi Natsionalnoyi akademiyi nauk Ukrayini*, 10, 78–82.

Lihosha, L.V. 2004. Kiliyska chastina delti Dunayu ta ii suchasna dinamika [Chilia part of the Danube Delta, and its contemporary dynamics (in Ukranian)]. *Vestnik Odesskogo universiteta*, vol. 9, 4, 56–67.

Shuyskiy, Yu. D., 1974. Protsessyi i skorosti abrazii na ukrainskih beregah Chernogo i Azovskogo morey [Processes and the speed of abrasion on the Ukrainian coast of the Black and Azov Seas (in Russian)]. *Izv. AN SSSR*, ser. geograf., 6, 107–117.

Shuyskiy, Yu. D., 1987. Sovremennaya dinamika akkumulyativnyih beregovyih form relefa [Modern dynamics of coastal accumulative forms of relief (in Russian)]. *Prirodnyie osnovyi beregozaschityi*, 116–131.

Shuyskiy, Yu. D., 2000. *Tipi beregiv svitovogo okeanu [World ocean types of coasts* (in Ukrainian)], Astroprint, Odessa, 479 pp.

Vykhovanets, G. V., 1993. Sandy accumulative forms within the Black Sea coastal zone, in *Coastline of the Black Sea*, (eds.) R. D. Kosyan and O. T. Mogoon, 452–466, American Soc. Civil Engs., New York.

Vykhovanets, G. V. 2003. *Eolovyiy protsess na morskom beregu [Wind processes on the sea coast* (in Russian)], Astroprint, Odessa, 368 pp.

Zenkovich, V. P. 1960. *Morfologiya i dinamika sovetskih beregov Chernogo morya [Morphology and dynamics of the Soviet Black Sea coast* (in Russian)], vol. 2, 216 pp, Izd-vo AN SSSR, Moskva.

Zolotarev, M. I. 2004. Portovyie sooruzheniya Hersonesa Tavricheskogo v Karantinnoy buhte [Chersonese Taurian Port facilities in Karantinnaya Bay (in Russian)]. *Hersonesskiy sbornik*, XIII, 55–66.

22 Conclusions

Enzo Pranzini, Lilian Wetzel and Allan Williams

Introduction

Coastal processes include natural phenomena, e.g. storms, tides, currents, waves and subsidence, as well as human-induced ones, e.g. damming of rivers, insertion of hard coastal protection structures (sea-walls, groins, etc.), land reclamation, reforestation, etc. and both categories are factors that can result in erosion/deposition. Responses associated with erosion processes are many and varied and represent functions of the coastal typology, resilience and sediment amount available/not available. To eliminate or slow down erosion rates, many forms of protection have been involved, the costs usually being high and coastal mitigation measures that enable protection to be carried out are numerous. 'Hard' solid structures (seawalls, revetments, etc.) have been traditionally built to prevent wave action from eroding material, in essence, to manage storms and tides for the protection of developments within the coastal zone (Figure 22.1).

Historically, anthropogenic coastal development has utilized this traditional type of protection. For example, Roman rip-rap defended the Via Severiana; murazzi were built by the Republic of Venice to stabilize barrier islands closing the lagoon (Franco, 1996); and the Netherlands and Belgium use century-old defences against polder flooding (this book). Pre-existing coastal activities, such as fishing, agriculture and maritime trading, affect coastal evolution, in many cases favouring settlement growth. Seaside resorts with ubiquitous seawalls and piers sprang up in elegant Victorian times in the UK (Figure 22.2),

■ **Figure 22.1** *top: Texel, Netherlands; bottom: Brean revetment, Somerset, UK*

but also developed in the nineteenth century in Russia, Poland and the Ukraine, to name but a few countries. Some winter resorts had already been created for the French bourgeoisie by the nineteenth century, development being made easier by railway

■ **Figure 22.2**
*Llandudno, Wales,
a Victorian town
complete with
cobble beach and
pier*

access. This continued until the recent expansion of international airports and resorts, for example, in the Mediterranean Riviera and more recently of resorts in countries formerly belonging to the Eastern bloc. These 'new' coasts were no longer viewed as an insalubrious area populated by mosquitoes, subject to flooding, storms and piracy and under military threat. Intense coastal human migration followed and the service sector started to dominate. Beaches were in favour and hence protection strategies were introduced with a vengeance.

Hard structures on a local basis may reduce erosion rates (Figures 22.1 and 22.3), but invariably interfere with the natural sediment budget and cause erosion further downstream (van der Weide *et al.*, 2001). Cipriani *et al.* (2004) have indicated that 2.2 km of hard structures protect every kilometre of coastline at Marina di Pisa, Tuscany, Italy. Europa (2007) indicated that of the 875 km of European coast that has eroded in the past two decades, 63 per cent can be linked to being less than 30 km from coastal engineering works, which is quite an extent! However Basco (1999; Basco *et al.*, 1997) was of the opinion that policy decisions to locate infrastructures on the coast are the cause of erosion, rather than seawalls themselves. Inappropriate construction policies and

execution to protect areas from storm damage has resulted in seawalls being blamed for consequent beach erosion, but these structures are essentially retroactive rather than pro-active. In 1988, Kraus summarized the viewpoint relating to seawalls, and little has changed. In this context, some 100 years ago, it was noted that sea walls, unless properly constructed, are,

'agents of their own destruction' inasmuch as, if a wall be not carefully designed and erected, the waves breaking against it, when recoiling from its face, tend to scoop out the beach material at its foot; removal of this material, if not held in place by groins or other adequate means, causes an undermining of the structure, which not infrequently leads to destruction of the wall.

(RCCEA, 1911, p. 83, point 20)

One century later, this quote is still valid. Other structures can slow down the longshore transport of littoral drift, e.g. the ubiquitous groins and jetties, but these do not solve the cause of erosion; they simply pass the problem downstream.

'Soft' techniques (e.g. nourishments of sand, gravel or other materials – china clay in England), makes

■ **Figure 22.3** *Top: Bettystown, Ireland (Photograph by A. Cooper); Bottom: south of Marina di Pisa, Italy (Photograph by Provincia di Livorno)*

up for a deficiency in beach sediment supplies. This improves a beach's appearance and is very attractive for tourists, but the technique has a less than respectable origin, the beach frequently being used as a dump site for materials excavated during building operations on the French Riviera (Anthony, this book). At Liguria, Italy, Fierro *et al.* (2010) showed that several pocket beaches were formed because of longshore sediment transport and the sorting process operated by waves from material discharged for this purpose from a coastal road. In addition, several beaches have been built because of mining waste discharge, as happened at the island of Elba, Italy (Nordstrom *et al.*, 2009) and Buckhaven, Scotland (Duck, this book) where mine closures triggered beach erosion. Since the 1970s, soft techniques have tended to be the protection measure of choice by many coastal engineers and authorities and are the preferred US coastal erosion response (Harris *et al.*, 2009). Miami Beach is the oft-quoted case of successful nourishment and infrastructure improvement, where every $1 invested brings a return of *circa* $500 annually (Houston, 2002, 2008). Dunes may also be enhanced and form 'natural' protection barriers, as they are part of the beach/dune symbiosis (Nordstrom *et al.*, 2009). Beach drainage schemes tried out in France, Spain, Italy and Denmark are also in this category, but have had a mixed reception. Per Sørensen (this book) has written about the initial success then halting of such schemes in Denmark, but at Les Sables (France), a macrotidal area, it seems to be successful.

As may be seen from the many examples given in this book, rates of coastal erosion vary enormously in Europe and consequently protection measures associated with high and low erosion rates vary as well. Erosion is currently affecting many EU member states with coastlines retreating on average between 0.5 and 2 m/yr and by 15 m/yr in a few extreme cases (Europa, 2007), with large extents of stable coastlines, such as northern Spain being rare. Hard rock coasts, such as in Scotland, tend to be very resistant to erosion and have negligible erosion rates. Soft coasts, for example made of boulder clay bluffs, erode at a very fast rate, e.g. the Holderness bluff coastline of Eastern England, which recedes at some 10 m/yr. Eurosion

(2004a, b, c, d) and the EEA (2006) have shown that some 20 per cent of European coasts being affected by erosion. These issues are not well addressed within the European Community with regard to any environmental assessment procedures. Other European countries, not covered by EU analyses, suffer similar problems, as shown in this book.

Specifically, the European coastal population has increased enormously, resulting in a 'coastal squeeze', where new developments, such as housing, harbours, marinas, industries and recreational facilities, e.g. 'link' golf courses, have all created demands on the land/water interface; several have affected the physics of water and sediment movement in the nearshore environment. The tendency towards erosion, which is global in effect, has in many cases been aggravated by the very protection strategies deemed necessary to eliminate it. Gillie (1997) cited costly seawall construction in the Pacific islands, which exemplifies this point.

Erosion

Goldberg (1994) pointed out that coastal areas have always attracted humans for the economic, social and recreational opportunities, which are larger than inland areas (Figure 22.4). However, high risks of erosion, subsidence, tsunami and flooding from sea and river are evident, e.g. tropical cyclone hazards killed 250,000 people between 1980 and 2000 (Nicholls *et al.*, 2007). The same authors postulate an accelerated rise in sea level of up to 0.6 m or more by 2100. Church and White (2006), suggested that the rise in sea level from 1990 to 2100 could range from 280 to 340 mm. The consensus amongst scientists seems to be that there will be an intensification of tropical and extra-tropical cyclones, larger extreme waves and storm surges, all impacts being overwhelmingly negative and enhancing erosion tendencies. This is even truer for low coasts, developed near the base level and now protected by sand/gravel barriers and dune belts. For many European eroding beaches, the causes of this phenomenon are well away from the sea. Many examples

■ **Figure 22.4** *Monte Carlo: extreme coastal zone anthropogenic pressure*

can be found within this book, e.g. where agriculture was abandoned in Italy; dams built which closed sediment access to the coast – Portugal, Albania, Bulgaria, Ukraine and Poland; and rivers used as aggregate quarries – Greece, Italy and Romania. Harbours have also been excavated and these invariably intercept longshore sediment transport, e.g. Latvia, Portugal, Croatia, Greece and Bulgaria.

Bruun (1962) long ago proposed that long-term erosion of sand beaches was a consequence of sea level rise – the Bruun Rule. His model predicts that beaches will erode by 50 to 200 times the rate of increase of sea level (Douglas, 2001). It does not suggest that sea level rise actually causes erosion; rather, increased sea level enables high-energy, short-period storm waves to attack further up the beach and transport sand offshore, although a part can be excluded from the beach budget if sea level rise puts sand deeper than the closure depth. Leatherman (2001, p. 183) would argue that 'while many factors contribute to shoreline recession, sea level rise is considered the underlying factor accounting for the ubiquitous coastal retreat'. However, when subsidence is added to this process (Sestini, 1996), relative sea level rise can be several centimetres per year, becoming the main agency responsible for shoreline retreat, from the North Sea coast, e.g. the Netherlands, to the Mediterranean, e.g. Rimini. When storm surges and high tides are coupled with sea level rise, the result is an even more significant rate of erosion. Ciscar *et al.* (2009) estimated that without public adaptation and if the climate of the 2080s occurred today, the impact of EU climate change, in GDP terms, would range from 0.2 to 0.5 per cent depending upon the climate scenario (2.5°C and 5.4°C) and land value losses for the latter would be US$1.548.21 million. Damage to the economy would be €20 billion for a 2.5°C temperature rise, up to €65 billion for a scenario of a 5.4°C rise with an 88 cm sea level rise. Coastal protection with adaptation to a high sea level rise by 2085 would cost 0.012 per cent. However, absolute sea level rise, although being a cause of beach erosion all over Europe, is a minor item, explaining approximately 10–15 per cent of the shoreline retreat in Portugal or Italy (Pranzini, 1994).

Coastal protection

For the period 1998–2015, the EC (2009) estimates for European total coastal protection expenditure is €15.8 billion, with a normal (maintenance, new construction, etc.) expenditure of €10,465 billion, the remainder being spent on 'hot spots', especially in the case of Italy. The cost of coastal protection escalates. For instance, EU public expenditure for protection in 2007 was €3.2 billion against €2.5 billion in 1986 and it is estimated that for 1990–2020 it will be €5.4 billion (CEC, 2007). The cost breakdown for European protection is 63 per cent by national authorities, 32 per cent by sub-national ones, 4 per cent by the EU and 1 per cent privately (CEC, 2007). The Netherlands Delta Commission estimates that it will cost €1.0 to 1.5 billion/yr to protect against flooding from 2010 to 2100. In the past, private owners in almost all European countries intervened to defend their properties or economic activities, generally with poorly designed projects, having limited effect on the site and negative response from downcoast segments. For example, revetments built in the mid 1970s by private owners at Loderups Strandbad, Sweden, caused downdrift erosion resulting in a call for extension of coastal defences (Larson and Hanson, this book), whereas in the Gulf of Follonica (Italy), a few detached breakwaters built in front of two tourist villages expanded the beach considerably, thereby triggering extensive erosion along the bay (Pranzini, this book). Cliff stabilization is another practice that on a European scale can be responsible for beach erosion; in Jastrzebia Gòra (Poland), beach erosion was induced because of seawall gabion strengthening of a 30 m high cliff. Similarly, Cayeaux gravel spit (France) had to be stabilized with 85 groins due to sediment input reduction. The cliffs at Llantwit Major (Wales, UK) provide an interesting example of a failed protection stabilization scheme. The project blasted away some 800 m of almost vertical cliff face in order to produce a talus cone (150,000 tonnes) which would act as a revetment and protect against further cliff erosion (Williams and Davies, 1980). The talus cone lasted some four years, as a result of longshore drift efficiency and erosion rates in the blast area are some three times that of adjacent cliffs (Figure 22.5).

Protection problems and solutions exist that are specific for each country, frequently for opposite reasons, e.g. geography, politics, local issues, etc. The Netherlands, blessed with an abundance of sand and non-availability of shoreline rock, commenced sediment dredging (or bypass) very early on, as their first and cheapest solution was moving sand, either across or longshore. In Italy, where rock sources were available (and free) on the coast, hard solutions were preferred. Ex Soviet Union countries, where the coastal area was closed for military reasons, are experiencing coastal defence failure because of maintenance interruption, which has given rise to economic difficulties. In many cases, these military-driven protection structures – huge concrete interlocking blocks – are now proving extremely hard to remove and are impossible to repair, e.g. Yevpatovia (Ukraine), Liepāia (Latvia) and Pionerskiy (Russian Baltic coast). Are technical solutions based on wave energy? From the chapters given in this book, it appears not to be so. Site specificity does play a part, e.g. permeable groins in the North Sea and Baltic; and submerged groins in Italy, but in the Netherlands pure nourishment rules in complete contrast to Belgium, their next door neighbour, where hard coastal protection

■ **Figure 22.5** *Top: Llantwit Major, Wales. The talus cone formed as a result of the blast; bottom: the beach area four years later*

(seawalls, groins, etc) are in vogue, even though both countries have almost the same wave energy and sand resources.

Many protection strategies have been shown within this book and it can be argued on socio-economic grounds that hard engineering protection will always be necessary, as stakeholders would be unwilling to allow areas of high-value real estate to be abandoned to the sea unless cost benefit analysis and more importantly, political will, prove otherwise. These structures include: construction of submerged breakwaters that reduce effective depth offshore and consequently reduce wave power and erosion of the beach (Aminti *et al.*, 2002); and groin field

techniques, which promote sediment deposition and evolution of the shoreline (Kunz, 1999), e.g. submerged groins to intercept sediments without shifting the longshore transport offshore, as occurs with traditional groins (Jackson *et al.*, 2002). However, because of problems associated with hard engineering in the nearshore zone, not least cost and maintenance, alternative soft engineering techniques that work in conjunction with natural coastal processes have increasingly been used since the 1970s.

The beach nourishment method for shoreline protection is a soft engineering solution, extensively used primarily for the benefit of tourism. Some 40 per cent of the EU population live on the coast and 60 per cent of EU holidaymakers go to the coast bringing in €75 billion per annum, the Mediterranean alone receiving 170 million visitors (EC, 2009). Although not a permanent solution, beach nourishment can be a sustainable way to manage coastal erosion, if coastal processes are appropriately considered. Following nourishment, the new wider beach serves as shore protection from the impacts of storms and increases recreational benefits with new tourism related opportunities (Benassai *et al.*, 2001). Another soft engineering solution, which has been tested on eroding coastlines, is beach drainage (Turner and Leatherman, 1997). Bowman *et al.* (2007) showed moderate beach accretion and shoreline advance following installation of beach drainage at Alassio Beach, northern Italy. However, effectiveness of this method is governed by beach thickness and sediment type. All coastal protection decisions should be based on sound data.

Management problems/issues

Coastal Protection measures were usually introduced in the past as *ad hoc* measures to counteract specific issues associated with coastal erosion. The late twentieth century saw the rise of Integrated Coastal Management, which involved strategies such as Strategic Environmental Assessments, which tend to incorporate protection issues into more general management approaches to the coast – the options being essentially to hold the line, to advance/retreat the line (managed realignment), or make no active intervention. The latter is the philosophy of the National Trust in UK.

Intermediate solutions are applied as well: in the Maremma Regional Park (Tuscany, Italy) shoreline retreat is approximately 20 m/yr at the River Ombrone mouth, and 1 km of coast has been lost since 1881. 'Managed' erosion has been actuated here: buried groins are under construction in a trench some 150 m inland from the present shoreline which will be gradually uncovered by waves and can help in reducing beach retreat, but not halting it, since eroded sediments are feeding more than 30 km of tourist beaches. Within 30 years, the shoreline will reach the groin root, where a dyke has been built, and by then the Ombrone River basin authority should have taken measures to increase the river sediment input (Figure 22.6).

The type of strategy adopted is a function of the prevalent issue, *i.e.* if a recreation beach needs protection, nourishment is usually the first option considered; if an urban area is at risk, seawalls tend to be favoured, running the risk of losing the beach, which could have a social value for the less privileged part of the population. As stated previously, the fundament for any coastal management protection scheme has to be a reliable data base; the building blocks for a European data bank were the CORINE (2004) and Eurosion (2004a–d) projects and many other EC directives.

The financing of effective protection measures will probably be a future significant factor as to which type of protection measure to inaugurate (Cipriani *et al.*, 1999; van der Weide *et al.*, 2001). The main aim will probably be to work with natural forces and it is heartening to see many European coastal forums working on the sediment cell concept. The essential element to be considered is the cause of erosion and then mitigating strategies can be discussed (Phillips, 2011). Since one of the main causes of beach erosion is reduction in river sediment input, (see e.g. Portugal, Italy and Romania, in this book), integrated coastal zone management should work with drainage basin management. However, a

■ **Figure 22.6** *Beach erosion management at the Ombrone river mouth (Rendering: Nevio Danelon, DST-UNIFI. Top: seawall at the back of the beach with groins built inland; middle: partial groin exposure; bottom: beach retreat down to the seawall, which is not likely to occur*

conflict arises between populations living inland *vs.* those living on the coast, as any measure to reduce landslides and flooding limits sediment amounts delivered to the coast. Expansion tanks, the most frequently used device to reduce peak river flow, cuts most of the bedload transported to only a few days per year.

Designing coastal management interventions requires a sound knowledge of coastal processes in the area. The legacy of coastal defences around the world's coast invariably arises due to a lack of understanding of flood and coastal erosion risk in the past, and of the impacts of defences, *i.e.* there are now extensive urban areas in which no choice exists but that of coastal defence at ever-increasing cost. These interventions, as mentioned frequently in this book, can include hard structures, such as seawalls, revetments and groins; and other measures, such as beach recharge. Coastal process understanding is required to ensure that interventions are capable of withstanding the forces acting on them, as well as minimizing the intervention impact on coastal processes. In this context, relevant coastal processes include waves, water level variations (e.g. tides, atmospheric surges), the movement of sediment and resulting change in coastal form (morphology), ideally all through a long time span. Additionally it may be important to predict future impacts due to climate changes and extreme conditions and these can be taken into account when designing interventions.

Coastal science is presently developing a range of high-resolution hydrodynamic models, which target sites to assess coastal processes. Coastal engineers typically make use of a range of physical (Figure 22.7) and numerical (Figure 22.8) models to investigate these processes. A model needs data availability in order to be verified, and a rigorous means of simulating these processes and predicting conditions for the present as well as for any future proposed changes. These models can be classified to include one-line models, coastal profile models and coastal area models.

- One-line models represent sections of coastline with the assumption that the contours are straight and parallel (examples include LITPROF by DHI). These models can be used to simulate the impact of various interventions on the longshore transport and resulting beach position over time.
- Coastal profile models represent the shoreline as individual cross-shore transects and predict the behaviour of these profiles to different driving forces such as waves and tides (examples include UNIBEST by Deltares, LITPROF by DHI).
- Coastal area models can be either two or three dimensional and cover both alongshore and off-shore regions (examples include MIKE21, MIKE 3D, and Delft 3D by Deltares). Both the Delft 3D and MIKE by DHI 2D and 3D coastal modelling suites are routinely applied to assessment of engineering and environmental studies within the coastal community, as well as in a wide range of other engineering works. These models are typically used to investigate the interactions

■ **Figure 22.7** *Left: Force measurement on a piled type coastal protection structure at the Middle East Technical University, Civil Engineering Department, Ocean Engineering Research Centre Laboratory, Ankara, Turkey; right: 3D mobile-bed physical model at Politecnico di Bari (Italy)*

of management interventions with waves, tides, sediment transport, and ultimately morphology (Figure 22.8). Two-dimensional models represent depth-averaged currents; three-dimensional layered currents through the vertical. Depending upon the important processes, coastal area models can include interactions between tides and waves to simulate wave-driven currents and associated sediment transport.

As regards coastal protection, two methods have been widely applied in European countries for a reliability based risk assessment of coastal structures. The first method is the partial coefficient system (PIANC, 1989), in which the First Order Mean Value Approach (FMA) is utilized (Burcharth, 1992).

The second method comprises the Hasofer-Lind second order reliability (HL) index, which has been employed to compare risk levels of rubble mound and vertical wall structures (Christiani *et al.*, 1996; Oumeraci and Kortenhaus, 1998; Burcharth and Sorensen, 1999). As an example, a reliability structural safety study at a Colhuw beach revetment, Wales, was evaluated at the limit-state, by modelling random design variables with common probability distributions (Williams *et al.*, 1998). Resistance variables were generally modelled by beta distributions, since with determination of their location and scale parameters, the density function can be used to fit a wide variety of frequency shapes.

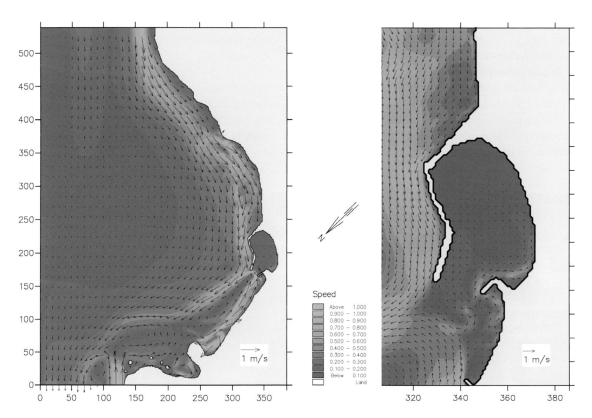

■ **Figure 22.8** *Longshore current velocities in a pocket beach (Cavo, Elba, Italy) hosting a marina, retrieved with MIKE21 (DICEA-UNIFI)*

The European coastal scene

In essence, development and urbanization at the coast has immobilized shorelines and reduced sediment input to beaches, requiring implementation of technical engineering solutions to maintain the increased demand for recreational areas by the sea, under the sun, and on the sand. In the post 1950s, the economic boom triggered a mushrooming growth in coastal usage and pressure especially for leisure development, coastal mass/high quality and eco tourism, activities in which eastern bloc countries have only recently participated. For example, Italian industry competed with tourism for use and occupation of the coastal zone; contrariwise, during Soviet occupation of its adjacent countries, many ports declined due to restrictions on maritime trade, reducing population and activities by the sea (e.g. Romania). Construction of hard structures, along with other maritime works necessary to maintain this market (such as marinas) created a cascade effect and required continuous monitoring and maintenance to keep those areas out of risk. A 'stone chaos', as mentioned in Chapter 21 on Ukraine in this book, was created in many cases.

Many countries started to realize that their 'golden egg' would be at risk if they did not change direction, and went from hard to soft solutions, choosing strategies that would not impact the landscape, and, better still, would cause less impact on habitats and biodiversity. Environmental aspects were then progressively taken into consideration, through creation of protected areas along the coast. In fact, countries such as Wales rely on heritage sites of historical, cultural or environmental value for leisure by the coast. The Curonian spit in Lithuania is a UNESCO heritage site, whereas the Danube Delta (Romania) is a Biosphere reserve. Many other Regional and National parks, Natura 2000 territories and other types of protected areas exist along European coastal areas. Curiously, some areas remained in a protected status due to their closure for military reasons during Soviet occupation (e.g. Lithuania, Latvia). Due to their more pristine coastal environments, these countries are now experiencing a move to higher quality tourism, which might have outstanding economic potential but will require even greater planning.

Summary of European protection:

- Seawalls: Only Lithuania has an absence of these, but there is a good spread from large to small with respect to concrete seawalls; brick ones being mainly absent with a few countries having them on a medium scale and only the Netherlands and UK on a small scale. Stone walls fare better, with only the Baltic countries having none, or only on a small scale. Wooden ones were unpopular, with the majority of countries not constructing them. Russia has an experimental site where fibreglass modules are tested. The gabion alternative was similar but at present, in the Ukraine, Russia and Poland, three experimental works are being carried out on this protection measure.
- Revetments were popular as long as they were constructed of natural stone or wire cage gabions. The Eastern Adriatic countries appeared not to favour this type or that of rubble mound rip-rap structures. Large/medium scales seemed to be the most favoured option in countries adopting this method.
- Island platforms were very uncommon, Italy, Romania, Spain, France and Great Britain being the few countries to have such structures. Surfing reefs were also unpopular except for France and Great Britain, but Montenegro and Germany are experimenting with such structures.
- Few countries construct detached breakwaters, both in their emerged and submerged form. Italy, Romania and the Ukraine were the only counties that have built detached emergent ones on a frequent scale. In Italy along with groins, they are the most frequently utilised coastal defence structure, but most breakwaters are now being lowered to -0.5 m. In Russia and the Ukraine concrete blocks are used for these structures. The dominant feature for detached submerged breakwaters was also their general absence in most countries; they have been built infrequently and only on a small scale. Montenegro has some structures built on an experimental site.

- Emerged groins appear to be ubiquitous, with Estonia and Latvia being the only countries not to have them in any quantity. Wood and stone appear to be the most popular material used. For the submerged variety, Germany is the only country that has them on a very frequent scale, France and Italy on a medium scale, with many countries having either none or infrequent. Permeable groins are not popular, Denmark, Germany and Poland being the only countries to have them. Linear, T and fishtail shapes were utilized in all countries except for Greece (linear and T only) and the Ukraine (L only).
- Sediment bypassing is mainly absent or exists on a very small scale, with Montenegro and Germany carrying out experimental work. Sand beach nourishment with marine aggregates is being carried out experimentally in several countries, e.g. Russia, and is either a very frequent occurrence or carried out on a small scale, with few absentees, e.g. the Eastern Adriatic countries. Gravel nourishment is essentially small scale. For gravel-sourced terrestrial nourishment, the only medium-scale work is to be found in France, Italy and Slovenia, but absenteeism is the norm; Italy and Bulgaria utilize sand on a medium scale. Large-scale sand nearshore nourishment is the prerogative of the Netherlands but only a handful of countries practise this method.
- With respect to dune reconstruction, there is a marked absence of such structures in the Eastern Adriatic countries, as dunes have a very limited extension, and the ex-Soviet bloc ones, except for Lithuania, but experimental stabilization is being trialled in Estonia, although the method is popular in many countries. Apart from the Netherlands, dune construction appears to be only occurring in Germany and in Italy (within the Venice Lagoon protection scheme); Denmark is carrying out trials, but it appears to be an unpopular choice.
- Beach dewatering is very uncommon but several experimental prototypes are in place (Denmark, Sweden, Netherlands, Spain, Italy and Bulgaria).
- Wave attenuators are even more uncommon, with experimental floating types in Spain and Denmark and fixed ones in France and Italy.

- Configurational dredging has a similar history and apart from an experimental site in Germany, there is a marked tendency not to follow this course.
- *Posidonia oceanica* planting follows a similar course with France, Portugal and Italy having field sites; Slovenia and Croatia are also experimenting with the technique (artificial not natural), but very few countries seem to have undertaken large scale trials.
- Other very small-scale techniques include wire mesh protection (Italy and Eastern Adriatic), tyres (Netherlands and Greece) and bitumen coatings (GB), but these are all on a very minor local scale.

The 25 countries under consideration in this book vary greatly in terms of geomorphology, environmental dynamics, habitats, human occupation, history, economic activities, social values, government systems and regimes, political activity, degree of technical expertise, research funds, etc. This implies varied forms of perceiving and dealing with the advance of the sea on the shore – from technical experimentations, strategic choices and related political and legal frameworks. In spite of that, a few trends can be perceived, which may be of importance, as countries tend to move towards higher integration under a political and economic European veil, which emphasises technical and political cooperation and standardization for better management practices.

Integrated Coastal Zone Management (ICZM) is a recommendation for European countries (Barcelona Convention). Some countries have already passed laws or created programmes to some extent related to ICZM, e.g. Portugal, Lithuania, Montenegro, Spain and Denmark; some have a clear demand for a coastal programme, e.g. Greece; whereas in other cases, a shoreline protection management plan may exist, although lacking in coordination with other sectors, e.g. Bulgaria, Poland, Albania and Slovenia.

Institutional, intergovernmental, legal and financial arrangements are necessary to overcome inadequate or conflicting competencies. Disarticulation between government levels and institutions, and lack of coordination resulting in piecemeal approaches (case-by-case) carried out by non-experts may be observed in countries such as Portugal, Scotland, Greece, Spain, Albania, Sweden and France (this book). Governmental inertia, coupled with an inadequate legal framework, has led to private owners defending the shoreline facing their properties, independently and 'extra-locally', lacking coordination with the 'rest of the shore' and often with no input from experts, creating or highlighting deleterious effects on neighbouring areas. Funds for shore protection usually come from government or private owners, whereas lack of public funding may contribute to the 'DIY' approach (Sweden). Lack of funds, e.g. due to the decline of the Soviet empire, may increase the 'stone chaos' in a state of abandonment on previously protected coasts.

With respect to management scales and levels, there is decentralization regarding coastal and beach management in some countries (France, except for port issues; Italy, Albania, Denmark and Spain: this book). History also shows that regional or local peculiarities must also be accounted for when designing a 'master plan', as seen from the uniform USSR coastal protection scheme. On the other hand, a trans-boundary approach may be necessary for countries which recently gained independency and which must develop cross-border cooperation mechanisms and strategies to deal with similar development trends and problems. International agreements are also being developed, such as the Black Sea Coastal Development Act, whereas European countries need to conform to EU regulations.

Initiatives for innovative projects dealing with risk management, especially in face of climate change, are taking place in many countries, as shown for example, in the chapters in this book on Great Britain, Italy, France, Netherlands, Greece, Bulgaria, Denmark and Germany. Although uncertainty may be involved in decision-making, systematic monitoring (Pranzini and Wetzel, 2008) appears to be another factor important for development of a new view in many countries, as longer time-series data may play an important role in modelling coastal processes, and in evaluating success or failure of chosen strategies. Public participation is also characteristic of ICZM processes, and helps achieve new solutions for coastal

conflicts providing more solid support to the decision making process. Some multi-party forums are being created to discuss and point out key aspects for public policy related to coastal use and protection, e.g. in France, Italy and Great Britain.

Research and technology have been traditionally economically oriented and coastal protection is historically associated to an increase in the economic use of the coast, so that erosion became a problem associated with money, welfare, properties and investments. This is often backed by more or less articulated/planned political choices. Apart from other economic sectors associated to shipping and commerce (port construction), or to agriculture (empolderment/embankment/reclamation), in many countries it is only when the coastal tourism boom appeared as an economic/most promising alternative that coastal retreat began to be dealt with. Beach width had to be maintained/increased, thereby securing a comfortable leisure surface for increasing human densities on a beach.

After many years of hard – and soft – engineering, countries are recognizing that the whole coast, and not only the beach, needs to be managed in an articulated form. The economic value of the shore – be it gravel, sand, rock or concrete beach – is triggering investments on maintenance or even creation of beach areas. Table 22.1 gives an idea of the costs involved for the Dutch coast, but it should be realized that great variability can occur, induced by rock availability (not only sand), wave energy, pre-construction erosion rate and local labour costs. For the Dutch coast there appears to be little cost difference between hard and soft engineering solutions and secondary considerations, e.g. amenities, play significant roles.

This is true for countries with a clear beach tourism vocation, but also for countries wishing to develop this potential on shores previously used for other purposes. The economic value of these areas will be dependent though on its environmental quality. The need for considering other aspects to coastal defence (such as environmental values, e.g. disturbance to wildlife, loss of habitat, impacts on landscape, interference with natural processes, and recuperation of

■ **Table 22.1** Indication of shoreline protection measure investment costs (Marchand, 2010)

Type of structure	Construction + maintenance costs over 50 years (in Euros per m coastline per year)
Straight rock groins	50–100
Rock revetments	100–200
Shoreface nourishments (every five years)	100–200 (if sand is readily available)
Seawall	50–300
Beach fills (every three years)	200–300 (if sand is easily available)
Submerged breakwaters	200–400
Emerged breakwaters	250–500

degraded areas) has been pointed out in many of the chapters in this book, e.g. Greece, France, Lithuania, Portugal, Netherlands, Latvia, Belgium and Sweden. Under stronger environmental values, old practices tend to be abandoned. Granted access to the beach, such as in Spain, Denmark, Greece and Portugal, and limits imposed to construction ('technical belts', in Poland, Ireland, France, Ukraine, Denmark, Spain and Greece), all may act as important measures against coastal immobilization and decline, in spite of removing homes and other buildings.

Many examples of environmental protection conflict can be given. Dunes are seen as a major protection structure, and have been long protected in many countries, as exemplified in Denmark, Portugal, Poland, Spain and the UK (Figure 22.9). Nevertheless, conflict may arise between the need to conserve protected habitats (such as grey dune habitats, according to EU habitat and species directive) and the need to keep sand mobile for erosion mitigation working 'with' natural processes, e.g. in Wales and the Netherlands. Another example is disposal of construction waste on Croatian beaches, which has an immense negative visual impact, but also provides material to cover sediment deficits at certain beaches.

However, what is the sustainable perspective? Erosion is usually seen as a problem, a threat to human activities and properties. There is a generalized lack of understanding of erosion as a function of the ecosystem, and a natural preference for permanent, hard, fixed and visible strategies – named

'structures'. Today, a large part of coast is indeed still 'technogenous' (artificial beaches, marinas, ports, reclamation sites, hard structures, etc.). A long-held preference for engineering approaches is evident in most countries. In spite of moving towards softer strategies, it must be said that even these must be carefully considered as to the environmental impacts they may cause. In that respect, it is also necessary to clarify choices to society and obtain their support to those normally not perceived as 'adequate' by the public, which seems to prefer harder, visible and fixed structures. Environmental education, including a 'natural aesthetic education', might be able to shift preference from 'artificial' coastal environments – as with groins – for natural ones.

Legal frameworks need to be deeply reviewed, as the law is often seen as inadequate, insufficient and not rigorously implemented as has been shown in Greece, Croatia, France, Denmark and Portugal.

Major challenges include:

- the need for new urbanism, economic and social planning, and new defence technology;
- the need to incorporate stronger environmental (sustainability) requirements as a limit to defence permits and as a value for coastal planning: 'softer' is not necessarily equal to 'sustainable';
- the need for broad cooperation among European countries in the face of failure of emerging countries linked to limited experience in defence protection;
- the need for a new approach that moves from beach to coastal management, and considers the

■ **Figure 22.9** *top: dune planting and fencing, Paredes da Vitória, Portugal; bottom: walkway at La Barrosa, Andalusian Atlantic coast, Spain*

whole watershed, at a more appropriate timescale (considering climate change scenarios) – ecological functions (impacts from works) should be considered along with the choice and implementation of protection structure (which areas to sacrifice and which to protect and how);

- the creation of comprehensive coastal management strategies under ICZM principles in conformity to international guidelines and regulations;
- the change from shoreline management to coastal zone management, integrating ecology, morphodynamics, civil engineering, and with a wide public participation. It is important to take into account the various functions of the coast, which often conflict, from which coastal protection is but one of many.

However coastal erosion and protection is viewed, the shoreline is an area where sometimes the 'rocky shore beats back the envious siege/ Of watery Neptune' (Shakespeare, *Richard II*), but sometimes anthropogenic help is needed!

References

Aminti, P., Cipriani, L. E., Iannotta, P. and Pranzini, E., 2002. Beach erosion control along the Golfo di Follonica (Southern Tuscany): Actual hard protection vs. potential soft solutions. In F. Veloso-Gomes, F. Taveira-Pinto and L. das Neves (eds), *Littoral 2002: The Changing Coast*, Vol. 2, 355–363.

Basco, D. R., 1999. Misconceptions about seawall and beach interactions. In E. Ozhan (ed.). *Land –Ocean interactions: Managing coastal ecosystems* (Proc. of the Medcoast–EMECS Joint Conf.), Medcoast, Ankara, Turkey, Vol. 3, 1565–1578.

Basco, D. R., Bellomo, D. A., Hazelton, J. M. and Jones, B. N., 1997. The influence of seawalls on subaerial beach volumes with receding shorelines. *Coastal Engineering*, 30, 203–233.

Benassai, E., Calabrese, M. and Uberti, G. S. D., 2001. A probabilistic prediction of beach nourishment evolution. In E. Ozhan (ed.), *Medcoast 01: Proceedings of the Fifth International Conference on the Mediterranean Coastal Environment*, Medcoast, Ankara. Vol. 1, 1323–1332.

Bowman, D., Ferri, S. and Pranzini, E., 2007. Efficacy of beach dewatering: Alassio, Italy. *Coastal Engineering*, 54, 791–800.

Bruun, P., 1962. Sea-level rise as a cause of shore erosion. *Journal of the Waterways and Harbors Division*, 88, 117–130.

Burcharth, H. F., 1992. Reliability evaluation of a structure at sea: Design and reliability of coastal structures. In A. Lamberti (ed.), *Proceedings of the short course on design and reliability of coastal structures, and structural integrity*, 23rd International Conference on Coastal Engineering (ICCE), ASCE, Venice, Italy, 511–545.

Burcharth, H. F. and Sorensen, J. D., 1999. The PIANC safety factor system for breakwaters. In I. J. Losada (ed.), *Coastal structures'99*, Balkemare, Spain; Lisbon, Portugal, 1125–1144.

Commission of the European Communities (CEC), 2007. Communication from the Commission to the European Parliament, the Council, the European Economic and Social Committee and the Committee of the Regions. *An Integrated Maritime Policy for the European Union*. COM (2007) 575 final.

Christiani, E., Burcharth, H. F. and Sorensen, J. D., 1996. Reliability-based optimal design of vertical breakwaters modeled as a series system of failure, *Proceedings of International Conference on Coastal Engineering*, ASCE, Orlando, Florida, 2, 124, 1589–1602.

Church, J. A. and White, N. J., 2006. A 20th-century acceleration in global sea-level rise. *Geophysical Research Letters*, 33: L01602, doi:10.1029/2005 GL024826.

Cipriani, L. E., Pelliccia, F. and Pranzini, E., 1999. Beach nourishment with nearshore sediments in a highly protected coast, in E. Ozhan (ed.), *Land–ocean interactions: Managing coastal ecosystems* (Proceedings of the Medcoast–EMECS Joint Conference), Vol. 3, 1579–1590, METU, Medcoast, Ankara, Turkey.

Cipriani, L. E., Wetzel, L., Aminti, D. L. and Pranzini, E., 2004. Converting seawalls into gravel beaches, in A. Micallef and A. Vassallo (eds.), *Management of coastal recreational resources*, 3–12, ICoD, Malta.

Ciscar, J. C., Iglesias, A., Feyen, L., Goodess, C. M., Szabó, L., Christensen, O. B., Nicholls, R., Amelung, B., Watkiss, P., Bosello, F., Dankers, R., Garrote, L., Hunt, A., Horrocks, L., Moneo, M., Moreno, A., Pye, S., Quiroga, S., van Regemorter, D., Richards, J., Roson, R., Soria, A., 2009. *Climate change impacts in Europe*. Final report of the PESETA research project, EUR Number: 24093 EN, EU Publishing Office, Luxembourg, 115 pp.

CORINE, 2004. Coastal erosion database of the European Environment Agency. Information and related publications available at: http://www.eea.europa.eu/ (accessed 26 August 2012).

Douglas, B. C., 2001. An introduction to sea level, in B. C. Douglas, M. S. Kearney, and S. P. Leatherman (eds), *Sea level rise: History and consequences*, 1–11, Academic Press, San Diego, USA.

EC, 2009. *The economics of climate change in EU coastal areas*, Policy Research Corporation (in association with MRAG), Antwerp, Belgium, for the Directorate-General for Maritime Affairs and Fisheries, EC Brussels, 153 pp.

EEA, 2006. *The changing faces of Europe's coastal areas*. EEA Report No 6/2006, European Environment Agency, Copenhagen.

Europa, 2007. *Coastal Zone Policy*. Available at: http://ec.europa.eu/environment/iczm/home.htm (accessed 26 August 2012).

EUrosion, 2004a. *Living with coastal erosion in Europe: Sediment and space for sustainability: Guidelines for incorporating coastal erosion issues into Environmental Assessment (EA) procedures*, prepared for European Commission Directorate General Environment, Service contract B4-3301/2001/329175/MAR/B3.

EUrosion, 2004b. *Living with coastal erosion in Europe: Sediment and space for sustainability: A guide to coastal erosion management practices in Europe*, prepared for European Commission Directorate General Environment, Service contract B4-3301/2001/329175/MAR/B3.

EUrosion, 2004c. *Living with coastal erosion in Europe: Sediment and space for sustainability: Part IV. A guide to coastal erosion management practices in Europe: Lessons learned*, prepared for European Commission Directorate General Environment, Service contract B4-3301/2001/329175/MAR/B3. Available at: http://www.EUrosion.org/reports-online/part4.pdf (accessed 26 August 2012).

EUrosion, 2004d. *Living with Coastal Erosion in Europe: Sediment and Space for Sustainability: Part V. Guidelines for implementing local information systems dedicated to coastal erosion management: Executive Summary*, prepared for European Commission Directorate General Environment, Service contract B4-3301/2001/329175/MAR/B3.

Fierro, G., Berriolo, G. and Ferrari, M., 2010. *Le spiagge della Liguria occidentale*, Genova University Press, Genoa, Regione Ligura, 174 pp.

Franco, L., 1996. History of coastal engineering in Italy, in *History and heritage of coastal engineering*, Amer. Soc. Civ. Eng., New York, 275–335.

Gillie, R. D., 1997. Causes of coastal erosion in Pacific Island nations. *Journal of Coastal Research*, SI24, 173–204.

Goldberg, E. D., 1994. *Coastal zone space: Prelude to conflict*, UNESCO, Paris, 138 pp.

Harris, M. S., Wright, E. E., Fuqua, L. and Tinker, T. P., 2009. Comparison of shoreline erosion rate derived from multiple data types: Data compilation for legislated setback lines in South Carolina, USA, *Journal of Coastal Research*, SI56, 1222–1228.

Houston, J. R., 2002. The economic value of beaches: A 2002 update. *Shore and Beach*, 70, 1, 3–10.

Houston, J. R., 2008. The economic value of beaches: A 2008 update. *Shore and Beach*, 76, 3, 22–26.

Jackson, L. A., Tomlinson, R. B. and d'Agata, M., 2002. The challenge of combining coastal processes and improved surfing amenity, in F. Veloso-Gomez, F. Taveira-Pinto, and L. das Neves (eds), *Littoral, 2002: The changing coast*, Vol 1. Eurocoast, Portugal and EUCC, 257–263.

Kraus, N. C., 1988. The effects of seawalls on the

beach: An extended literature review. *Journal of Coastal Research*, SI 4, 1–28.

Kunz, H., 1999. Groin field technique against the erosion of salt marshes, renaissance of a soft engineering approach, in E. Ozhan (ed.), *Land–ocean interactions: Managing coastal ecosystems* (Proc. of the Medcoast–EMECS Joint Conf.), Vol. 3, 1477–1490, METU, Medcoast, Ankara, Turkey.

Leatherman, S. P., 2001. Social and economic costs of sea level rise, in B. C. Douglas, M. S. Kearney and S. P. Leatherman (eds.), *Sea level rise: History and consequences*, Academic Press, San Diego, USA, 181–223.

Marchand, M. (ed.), 2010. *Concepts and science for coastal erosion management: Concise report for policy management,* Deltares, Delft, 31 pp.

Nicholls, R. J., Wong, P. P., Burkett, V., Cogignottoe, J., Hay, J., McLean, R., Ragoonaden, S. and Woodroffe, C. D. 2007. Coastal systems and low lying areas, in M. L. Parry, O. F. Canziani, J. P. Palutikof, P. J. van der Linden and C. E. Hanson (eds), *Climate change 2007: Impacts, adaptation and vulnerability.* Contribution of Working Group II to the fourth assessment report of the Intergovernmental Panel on Climate Change, Cambridge, UK, Cambridge University Press, 315–356.

Nordstrom, K. F., Jackson, N. and de Butts, H. A., 2009. A proactive programme for managing beaches and dunes on a developed coast: A case study of Avalon, New Jersey, USA, in A. T. Williams and A. Micallef (eds), *Beach management: Principles and practice*, Earthscan, London, 307–316.

Oumeraci, H. and Kortenhaus, A., 1998. *Probabilistic Design Tools for Vertical Breakwaters (PROVERBS)*, Commission of the European Communities: MAST Days and EUROMAR Project Reports, 1.

Phillips, M. R., 2011. Managing the coastal zone: Learning from experience, in Andrew Jones and Michael Phillips (eds), *Disappearing destinations*, CABI, Wallingford, UK, 30–46.

PIANC (Permanent International Association of Navigation Congresses), 1989. *Construction deviations and reliability of construction,* WG-12D Working Group Report 45, Belgium.

Pranzini, E., 1994. The erosion of the Ombrone River delta, in G. Soares de Caevalho and F. Veloso-Gomez (eds), *EUROCOAST, Littoral 94, 2nd Int. Symposium*, University of Porto, Portugal, Vol. 1, 133–147.

Pranzini, E. and Wetzel, L., Eds. 2008. *Beach erosion monitoring.* BeachMed-e/OpTIMAL Project, Nuova Grafica Fiorentina, Florence, 230 pp.

RCCEA, 1911. *Royal Commission Coastal Erosion, UK: Third (and Final) report of the Royal commission appointed to inquire into and to report on certain questions affecting coastal erosion, the reclamation of tidal lands, and afforestation in the United Kingdom.* HMSO, London, 177 pp.

Sestini, G., 1996. The case of the Po Delta Region Italy. In John D. Milliman and Bilal U. Haq (eds), *Sea-level rise and coastal subsidence: Causes, consequences, and strategies*, Kluwer Academic Publishers, Dordrecht, 235–242.

Turner, I. L. and Leatherman, S. P., 1997. Beach dewatering as a 'soft' engineering solution to coastal erosion: A history and critical review. *Journal of Coastal Research* 13, 4, 1050–1063.

Van der Weide, J., de Vroeg, H. and Sanyang, F., 2001. Guidelines for coastal erosion management, in E. Ozhan (ed.), *Proc of the 5th Int. Conf for the Med Envir.*, Vol 3, 1399–1414, Medcoast, Ankara, Turkey.

Williams, A. T. and Davies, P., 1980. Man as a geological agent: The sea cliffs of Llantwit Major, Wales, UK. *Zeitschrift fur Geomorphologie*, Suppl Band, 34, 129–141.

Williams A. T., Davies P., Ergin A. and Balas, C. E., 1998. Coastal recession and the reliability of planned responses: Colhuw Beach, the Glamorgan Heritage Coast, Wales, UK. *Journal of Coastal Research*, SI26, 72–79.

Index